Advances in Experimental Medicine and Biology

Volume 1175

Advances in Experimental Medicine and Biology presents multidisciplinary and dynamic findings in the broad fields of experimental medicine and biology. The wide variety in topics it presents offers readers multiple perspectives on a variety of disciplines including neuroscience, microbiology, immunology, biochemistry, biomedical engineering and cancer research.

Advances in Experimental Medicine and Biology has been publishing exceptional works in the field for over 30 years and is indexed in Medline, Scopus, EMBASE, BIOSIS, Biological Abstracts, CSA, Biological Sciences and Living Resources (ASFA-1), and Biological Sciences. The series also provides scientists with up to date information on emerging topics and techniques. 2018 Impact Factor: 2.126.

Content of this series is reviewed.

More information about this series at http://www.springer.com/series/5584

Alexei Verkhratsky · Margaret S. Ho · Robert Zorec · Vladimir Parpura
Editors

Neuroglia in Neurodegenerative Diseases

Springer

Editors
Alexei Verkhratsky
Faculty of Biology, Medicine and Health
The University of Manchester
Manchester, UK

Margaret S. Ho
School of Life Science and Technology
ShanghaiTech University
Shanghai, China

Robert Zorec
Institute of Pathophysiology
University of Ljubljana
Ljubljana, Slovenia

Vladimir Parpura
Department of Neurobiology
The University of Alabama at Birmingham
Birmingham, AL, USA

ISSN 0065-2598 ISSN 2214-8019 (electronic)
Advances in Experimental Medicine and Biology
ISBN 978-981-13-9912-1 ISBN 978-981-13-9913-8 (eBook)
https://doi.org/10.1007/978-981-13-9913-8

This Springer imprint is published by the registered company Springer Nature Singapore Pte Ltd.
The registered company address is: 152 Beach Road, #21-01/04 Gateway East, Singapore 189721, Singapore

Preface

Over the last century the mankind suddenly became old, as the average life span increased from ~45 years in 1900 to approximately 75 years in the year 2000 and to almost 79 years in 2019; in Japan, Spain, Singapore and Switzerland the life expectancy in 2019 reached in excess of 85 years. This abrupt ageing of the population changed the epidemiological landscape and brought forward something entirely unknown to the humanity over the history—the growing epidemic of neurodegenerative diseases, which end up with senile dementia. The numbers are almost impossible to accurately define; rough estimates account for ~ 50–70 million of senile people worldwide, with each year adding extra 10 million cases. The neurodegenerative diseases are several, of which the most known are Alzheimer's disease, Parkinson disease, Huntington disease, frontotemporal dementia and tauopathies; the common denominator is senility as the outcome and absence of therapeutic hopes; all available pharmacological options remain merely symptomatic.

The ultimate result of neurodegenerative process is the atrophy of the brain and loss of brain function. The direct link between a decrease in the size (i.e. atrophy) of brain tissue and a decrease in cognitive capabilities (i.e. dementia) was suggested by Thomas de Willis at the end of the seventeenth century. Despite a remarkable progress in understanding the biochemistry and genetics of neurodegenerative processes, the genesis and pathobiology of the majority of sporadic cases remain obscure. Neurodegeneration, as well as the majority of other neurological diseases, are commonly believed to reflect pathological evolution that is conceived and developed primarily, if not solely, in neurones. This neurone-centric view emerged and expanded with the triumphal spread of the neurone doctrine born at the end of the nineteenth century. This development was, however, at odds with neuropathological thoughts of earlier neuroanatomists who recognized the remarkable pathological potential of neuroglia that represent the homeostatic arm of the nervous tissue. Rudolf Virchow, who invented the concept of neuroglia was convinced that 'This very interstitial tissue of the brain and spinal marrow is one of the most frequent seats of morbid change' (R. Virchow, (1858). *Die Cellularpathologie in ihrer Begründung auf physiologische and pathologische Gewebelehre.* August

Hirschwald, Berlin; p. 317 of the 1st English Edition, 1860). The fundamental contribution of neuroglia to neuropathology was also considered by such towering figures as Santiago Ramón y Cajal, Alois Alzheimer, Franz Nissl, Pió del Rió-Hortega and William Lloyd Andriezen, to name only a few.

The term Neuroglia defines the class of neural cells of ectodermal (neuroepithelial) and mesodermal (myeloid) origin, which are responsible for support, maintenance and defence of the nervous tissue. In the brain and in the spinal cord (which together form the central nervous system, CNS) neuroglia are represented by astrocytes (main homeostatic cells of the CNS), oligodendrocytes (cells myelinating and supporting axons) and microglia (cell of mesodermal origin, which, in the guise of foetal macrophages, migrate into the neural tube in the early embryonic development and disseminate throughout the CNS); microglial cells represent innate immune and defensive system of the nervous tissue. The neurone-centric neuropathological ideas have been challenged in the recent decade, when contribution of neuroglia to neurological disorders begun to be widely considered. It emerged that complex neuroglial reactions and metamorphoses contribute to most if not all neurological diseases, either as primary factors, or as factors instigated by exogenous pathology and neuronal abnormalities or (most frequently) as combination of both. Neurogliopathology is manifested in many forms from degeneration, loss of function, glial asthenia and paralysis to glial reactivity in which neuroglial cells undergo context-specific biochemical, morphological and functional remodelling producing neuroprotective or neurotoxic phenotypes. Understanding neuropathology ultimately requires in-depth analysis of glial pathological changes, which in turn may stimulate the development of novel, glia-specific therapies. In this volume we collected essays, written by leading gliologists. These essays highlight glial contribution to neurodegenerative diseases and we hope that this book (being first of its kind on the market) may be appreciated by academics, professionals and students of medicine, neuroscience, biology, biochemistry and pharmacology.

Manchester, UK — Alexei Verkhratsky
Shanghai, China — Margaret S. Ho
Ljubljana, Slovenia — Robert Zorec
Birmingham, USA — Vladimir Parpura

Contents

Contributors

Rebecca Atkinson-Dell StrataStem Ltd., Warrington, UK

Liliana Brambilla Laboratory for Research on Neurodegenerative Disorders, IRCCS Istituti Clinici Scientifici Maugeri, Pavia, Italy

Marianna Bugiani Department of Pathology, Free University Medical Centre, Amsterdam, The Netherlands

Arthur M. Butt School of Pharmacy and Biomedical Science, University of Portsmouth, St. Michael's Building, Portsmouth, UK

Micaël Carrier Axe Neurosciences, Centre de Recherche du CHU de Québec, Québec, QC, Canada

Irene Chacon De La Rocha School of Pharmacy and Biomedical Science, University of Portsmouth, St. Michael's Building, Portsmouth, UK

Michelle Gray Department of Neurology and Center for Neurodegeneration and Experimental Therapeutics, The University of Alabama at Birmingham, Birmingham, USA

Giulia Guidotti Laboratory for Research on Neurodegenerative Disorders, IRCCS Istituti Clinici Scientifici Maugeri, Pavia, Italy

Margaret S. Ho School of Life Science and Technology, ShanghaiTech University, Shanghai, China

M. S. Jorge Department of Pathology, Free University Medical Centre, Amsterdam, The Netherlands

Lisa Mohamet StrataStem Ltd., Warrington, UK

Maria Papanikolaou School of Pharmacy and Biomedical Science, University of Portsmouth, Portsmouth, UK

Vladimir Parpura Department of Neurobiology, The University of Alabama at Birmingham, Birmingham, AL, USA

Andrea Rivera School of Pharmacy and Biomedical Science, University of Portsmouth, St. Michael's Building, Portsmouth, UK

Jose Julio Rodriguez-Arellano BioCruces Health Research Institute, Achucarro Center for Neuroscience, IKERBASQUE, Basque Foundation for Science, Bilbao, Spain;
Department of Neuroscience, The University of the Basque Country UPV/EHU, Barakaldo, Bizkaia, Spain

Daniela Rossi Laboratory for Research on Neurodegenerative Disorders, IRCCS Istituti Clinici Scientifici Maugeri, Pavia, Italy

Tuan Leng Tay Cluster of Excellence BrainLinks-BrainTools, Institute of Biology I, Institute of Biology III, University of Freiburg, Freiburg, Germany

Marie-Ève Tremblay Axe Neurosciences, Centre de Recherche du CHU de Québec, Québec, QC, Canada

Chiara F. Valori Department of Neuropathology, German Centre for Neurodegenerative Diseases (DZNE), Tübingen, Germany

Nina Vardjan Laboratory of Neuroendocrinology-Molecular Cell Physiology, Faculty of Medicine, Institute of Pathophysiology, University of Ljubljana, Ljubljana, Slovenia;
Celica Biomedical, Ljubljana, Slovenia

Alexei Verkhratsky Faculty of Biology, Medicine and Health, The University of Manchester, Manchester, UK;
Faculty of Health and Medical Sciences, Center for Basic and Translational Neuroscience, University of Copenhagen, Copenhagen, Denmark;
Achucarro Center for Neuroscience, IKERBASQUE, Basque Foundation for Science, Bilbao, Spain

Robert Zorec Laboratory of Neuroendocrinology-Molecular Cell Physiology, Faculty of Medicine, Institute of Pathophysiology, University of Ljubljana, Ljubljana, Slovenia;
Celica BIOMEDICAL, Ljubljana, Slovenia

Chapter 1
The Concept of Neuroglia

Alexei Verkhratsky, Margaret S. Ho, Robert Zorec and Vladimir Parpura

Abstract Neuroglia represent a diverse population of non-neuronal cells in the nervous systems, be that peripheral, central, enteric or autonomic nervous system. Arguably, these cells represent about half of the volume of the human brain. This volumetric ratio, and by extension glia to neurone ratio, not only widely differ depending on the size of the animal species brain and its positioning on the phylogenetic tree, but also vary between the regions of an individual brain. Neuroglia derived from a dual origin (ectoderm and mesodermal) and in an assorted morphology, yet their functional traits can be mainly classified into being keepers of homeostasis (water, ions, neurotransmitters, metabolites, fuels, etc.) and defenders (e.g., against microbial organisms, etc.) of the nervous system. As these capabilities go awry, neuroglia ultimately define their fundamental role in most, if not, all neuropathologies. This concept presented in this chapter serves as a general introduction into the world of neuroglia and subsequent topics covered by this book.

A. Verkhratsky (✉)
Faculty of Biology, Medicine and Health, The University of Manchester, Manchester M13 9PT, UK
e-mail: Alexej.Verkhratsky@manchester.ac.uk

Faculty of Health and Medical Sciences, Center for Basic and Translational Neuroscience, University of Copenhagen, 2200 Copenhagen, Denmark

Achucarro Center for Neuroscience, IKERBASQUE, Basque Foundation for Science, 48011 Bilbao, Spain

M. S. Ho
School of Life Science and Technology, ShanghaiTech University, Shanghai 201210, China

R. Zorec
Laboratory of Neuroendocrinology-Molecular Cell Physiology, Faculty of Medicine, Institute of Pathophysiology, University of Ljubljana, Ljubljana, Slovenia

Celica BIOMEDICAL, Ljubljana, Slovenia

V. Parpura
Department of Neurobiology, The University of Alabama at Birmingham, Birmingham, AL, USA

A. Verkhratsky et al. (eds.), *Neuroglia in Neurodegenerative Diseases*,
Advances in Experimental Medicine and Biology 1175,
https://doi.org/10.1007/978-981-13-9913-8_1

Keywords Neuroglia · Origin · Morphology · Function · Homeostasis · Defence · Physiology · Pathophysiology

1.1 The Birth of the Concept of Homoeostatic Neuroglia

The complexity of the human brain is remarkable: a population of more than 200 billion (i.e. 2×10^{11}) neural cells (neurones and neuroglia) is packed within a limited volume (average human brain occupies 1200–1400 cm^3). These neural cells form complex networks, connected through 15–20 trillions of chemical and electrical synapses that provide for computing power of this organ. The logistical support underlying this highly complex organ is provided by a specific class of cells known as neuroglia.

The concept of connective tissue of the nervous system emerged in the nineteenth century [16, 47]; this concept was initially formalised by Rudolf Virchow who introduced the term neuroglia in the 1850s [100, 101]. According to Virchow the neuroglia was '…connective substance that forms in the brain, in the spinal cord and in the higher sensory nerves a type of putty (neuroglia), in which the nervous elements are embedded…' [100]. Prominent neuroanatomists of the second half of the nineteenth century characterised the cellular nature of glia in great detail, and described many types of glial cells [16]. At the same time numerous theories have considered the functional role of neuroglia in the brain homeostasis, nutritional support, regulation of blood flow, sleep and conscience, as well as in neuropathology [6, 29, 30, 72, 79]. The first major type of glia, the astroglia, has been defined in 1895, when Michal von Lenhossék suggested to name a sub-population of parenchymal glia astrocytes, star-like cells (from Greek αστρον κυτος). At the same time the parenchymal glia was also sub-classified into protoplasmic (grey matter) and fibrous (white matter) cells [6]. The myelinating cells of the central nervous system (CNS) were first drawn by the Scottish pathologist William Ford Robertson [74, 75], and subsequently Pío del Río Hortega named them oligodendrocytes and recognised their myelinating function [24]. It was also Pío Del Río Hortega who identified and named microglia as the defensive cellular elements of the CNS, by demonstrating that these cells undergo remarkable metamorphosis in pathology and suggesting their role as 'garbage collectors' [21–23]. Finally, in the 1980s the fourth type of neuroglia, the NG2 glia (also known as oligodendrocyte progenitor cells or polydendrocytes), was discovered by William Stallcup and colleagues, after they developed an antibody to a chondroitin sulphate proteoglycan, dubbed NG2 [88]. Based on their developmental origin (neuroepithelial or mesodermal), neuroglia of the CNS have been classified as macroglia (astrocytes, oligodendrocytes, NG2) and microglia, respectively.

1.2 The Definition of Neuroglia

The definition of neuroglia is based on the unifying fundamental function of these cells, which, regardless of their origin, is homeostasis of the nervous system. This function is fundamental for both physiological context, when glial cells perform their routine housekeeping duties, as well as for pathological context, when glial cells can undergo reactive remodelling in order to preserve, repair and restore brain homeostasis. Failure in this function results in the development of the neurological disease and damage to the nervous tissue. Therefore, neuroglia can be defined as **homeostatic and defensive cells of the nervous system**, represented by highly heterogeneous cellular populations of different origin, structure and function [94].

In this sense neuroglia are the **ultimate supportive cells** of the nervous system, keeping it in a functional state. This reflects upon evolution of the nervous system, which resulted in the division of labour: the information processing and electrical excitability became confined to the neuronal networks, whereas homeostatic support and defence became the sole prerogative of the neuroglia [95]. This homeostatic support occurs at all levels of organisation of the nervous system: at molecular level (control over homoeostasis of ions, neurotransmitters, protons, reactive oxygen species, metabolites, etc.), at cellular level (astrocytes involvement in neurogenesis), at network level (both astroglia and microglia regulate synaptogenesis, synaptic maturation and extinction), connectome level (which is maintained by myelinating oligodendroglia and Schwann cells), organ level (astrocytes control blood-brain barrier and glymphatic flow and regulate functional hyperaemia) and systemic level (glial cells emerge as central chemoceptors and contribute to systemic control over ventilation, ion homeostasis and energy metabolism); for comprehensive coverage of neuroglial homeostatic capabilities see [1, 2, 7, 19, 31, 34, 37, 41, 45, 46, 49, 50, 59, 68, 70, 93, 96, 97, 103, 105].

This ultimate homeostatic capability of neuroglia underlies their fundamental role in neuropathology, the latter being broadly defined as a homeostatic failure of the nervous system. Environmental stress and/or pathological insults trigger glial homeostatic response (when glial cells attempt to keep homeostatic equilibrium or steady state) and glial reactivity, which represents an evolutionary conserved programme of glial cells remodelling aimed at mounting defence of the nervous tissue. Neuroglial reactivity is manifested in reactive astrogliosis, microglial activation and Wallerian degeneration (for oligodendrocytes). At the same time glial asthenia or atrophy, which is observed in numerous neurological conditions, facilitates evolution of the disease because of compromised homeostatic and defensive capabilities. Although the fundamental role of neuroglia in neuropathology has been predicted by prominent neurologists of the nineteenth and the beginning of the twentieth centuries (such as Rudolf Virchow, Carl Frommann, Alois Alzheimer, Nicolas Achucarro and Franz Nissl), the pathophysiological role of glia begun to be universally recognised only in the recent decade; for references and concepts see [11, 12, 53, 66, 67, 76, 81, 84–86, 99]. The concept of astrotauopathology, recently introduced by Kovacz [51], supports the notion that the neuroglia emerges in the limelight when considering the evolution of neurological diseases.

1.3 Glial Numbers

The numerical preponderance of glial cells in the brains and spinal cords of different species and glial to neurone ratio (GNR) have been a matter of the most common fallacy over recent decades. The notion of glial cells outnumbering neurones in the human brain by a factor of 10 or even 50 [10, 18, 44] represented an undisputed general knowledge that has been repeatedly proclaimed in glial literature (for critical analysis see for example [39, 98, 102]). The story of exceedingly high number of glial cells in the human brain goes back to Franz Nissl [58]; this idea became rather popular and reached the climax in writings of Robert Galambos who considered that neuroglia represent the primary seat of intelligence, consciousness, emotions and are overall responsible for our 'humanity'. 'Glia is … conceived as genetically charged to organize and program neuron activity so that the best interests of the organism will be served; the essential product of glia action is visualized to be what we call innate and acquired behavioural responses. In this scheme, neurons in large part merely execute the instructions glia give them' [28]. The notion was further promoted by the finding that the Einstein's brain had a rather higher glia to neurone ratio in his associated cortex than that found in the control human population [25], leading to speculations that this could be the reason for his remarkable intellectual abilities (https://www.theguardian.com/science/2007/feb/21/neuroscience.highereducation) (https://www.npr.org/templates/story/story.php? storyId=126229305). The public myth of glia has extended into that of an untapped part of the brain that we may not use, perhaps gloriously captured in Starbuck's The Way I see it? (http://www.stevekmccoy.com/blog/2005/08/starbucks_the_w) #236 quote 'Scientists tell us we only use 5% of our brains. But if they only used 5% of their brains to reach that conclusion, then why should we believe them?' Of course, based on any functional imaging, this myth has been debunked and the authors would like to assure the readers that we had used the vast majority if not all of our brains to write this chapter.

None of these concepts had experimental confirmation. Exceptionally high glia to neurone ratio of the human brain was not related to actual cell counts; to the contrary most of stereological investigations produced the GNR values in the neocortex somewhere around 1.5 (see Table 1 in [42, 102]), with the number of neuronal counts in the range of 20–30 billion and glial cells in the range of 27–38 billion. In the cerebellum, which contains the largest number of brain neurones (around 70 billion) the number of glial cells is much smaller, with GNR not exceeding 0.1 [3, 4]. These stereological data obviously made the total GNR estimate of 10:1 unrealistic. Further advances in defining the glial numbers are associated with the application of isotopic fractionation technique, which counts nuclei of neurones and non-neuronal cells in the homogenates of the nervous tissue [8, 40, 54]. This technique demonstrated that the total numbers of neuronal and non-neuronal cells in the human brain are more or less on par, both being in the range of ~80–100 billion. After subtracting the population of endothelial cells which may account for about 20% of all non-neuronal cells, the true number of glial can be estimated at ~60 billion and total GNR for the whole brain is therefore less than 1:1. The density of glia is quite different in various brain areas. For example, the GNR varies between 1.2 in the grey matter of the occipital cortex and 3.6 in the grey matter of frontal cortical regions [73, 82], it is technically an infinity in the white matter that does not contain neuronal cell bodies, and hence inclusion of white matter counts increases the total GNR in the cortex to ~3. As already alluded above the GNR in cerebellum is very low probably not exceeding 0.1. Much higher GNR values were reported for the striatum (3.7:1), for the superior colliculus (10:1), for the ventral pallidum (12.2:1), for the lateral vestibular nucleus (30–50:1), while for the globus pallidus a very high GNR of 160:1 has been calculated from stereological counts [5, 8, 64, 71, 80, 91, 102]. Similarly, the GNR for the spinal cord was determined at 5:1 in cynomolgus monkey and almost 7:1 in humans [9].

Evolution of the nervous system paralleled with an increase in GNR, which however was not entirely linear and was not directly related to the intelligence. The nervous system of invertebrates has, as a rule, relatively smaller numbers of glial cells, with a GNR between 0.01:1 and 0.2:1 (50 glial cells derived from neuronal/epithelial progenitors and six glial cells that are mesodermally derived per 302 neurones in *Caenorhabditis elegans* [63, 89]; 10 glial cells per 400–700 neurones in every ganglion of the leech [20]; ~9000 glial cells per 90,000 neurones in the CNS of Drosophila [26, 52]). At the same time, the buccal ganglion of the great ramshorn snail *Planorbis corneus* contains 298 neurones and 391 glial cells [69], thus having a GNR of 1.3, very similar to that of humans. Similarly, in vertebrates there is no hard-and-fast increase in the GNR with an increasing brain size; for example, in the cortex, the GNR is about 0.3–0.4 in rodents, ~1.1 in cat, ~1.2 in horse, 0.5–1.0 in rhesus monkey, 2.2 in Göttingen minipig, ~1.5 in humans and as high as 4–8 in elephants and the fin whale [15, 27, 38, 43, 55, 65, 92]. The largest absolute number of glial cells has been counted in the neocortex of whales [27, 56]; stereological cell counts in the neocortex of the long-finned pilot whale (*Globicephala melas*) brain determined there are approximately 37.2 billion neurones and 127 billion glial cells and this gives a GNR of 3.4 [56]. The largest GNR was found in the neocortex of the common Minke whale (*Balaenoptera acutorostrata*), which contained ~12.8 billion neurones and 98 billion of glia giving therefore a GNR of ~7.6 [27].

1.4 Classification and Main Functions

Neuroglia (Fig. 1.1, see also [94]) are classified into glia of the peripheral nervous system (PNS) and of the CNS. The glial cells of the PNS originate (similarly to peripheral neurones) from the neural crest and are classified into:

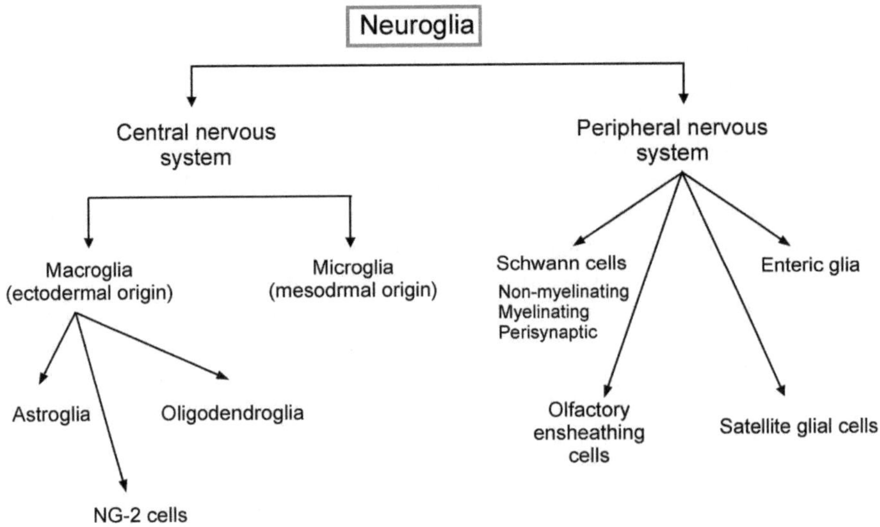

Fig. 1.1 Classification of neuroglia

(1) Schwann cells [48] associated with sensory, motor, sympathetic and parasympathetic axons; Schwann cells are further subdivided into (i) myelinating Schwann cells that myelinate peripheral axons; (ii) non-myelinating Schwann cells that surround multiple non-myelinating axons and (iii) perisynaptic Schwann cells, which enwrap peripheral synapses and maintain homoeostasis in the perisynaptic milieu.
(2) Satellite glial cells [35, 36], which are surrounding neurones in sensory, sympathetic and parasympathetic ganglia. These satellite glial cells control local homeostasis and are capable of reactive remodelling in pathology.
(3) Olfactory ensheathing cells [77], which are a part of the olfactory system. These cells extend very fine processes that enclose large numbers of unmyelinated olfactory axons
(4) Enteric glia [32, 33], represented by homeostatic glial cells of the enteric nervous system.

Neuroglia of the CNS are subdivided into macroglia (cells of ectodermal, neuroepithelial origin) and microglia (cells of mesodermal, myeloid origin). The macroglia is further classified into:

(1) Astroglia or astrocytes. Astrocytes are heterogeneous population of primary homeostatic glia residing throughout the brain and the spinal cord, in both grey and white matter. Astroglia include [94, 96]:
   (i) protoplasmic astrocytes of grey matter;
   (ii) fibrous astrocytes of white matter
   (iii) surface-associated astrocytes associated with the cortical surface in the posterior prefrontal and amygdaloid cortex;

(iv) Velate astrocytes, which are localised in the parts of the brain that are densely packed with small neurones, for example in the olfactory bulb or in the granular layer of the cerebellar cortex;
(v) Radial glia, which are the pluripotent neural cells precursors that generally disappear at birth in mammals
(vi) Radial astrocytes, which include Bergmann glia in the cerebellum, Müller glia of the retina, radial glia-like neural stem cells of the neurogenic niches and tanycytes of the hypothalamus, hypophysis and the raphe part of the spinal cord;
(vii) Pituicytes, which are the glial cells of the neurohypophysis;
(viii) Gomori astrocytes rich in iron and positive for Gomori's chrome alum hematoxylin staining identified in the hypothalamus and in the hippocampus;
(ix) Perivascular and marginal astrocytes, which are placed near the pia mater, where they form endfeet with blood vessels. These astrocytes do not establish contacts with neurones and their main function is in establishing the pial and perivascular glia limitans barriers.
(x) Ependymocytes, choroid plexus cells and retinal pigment epithelial cells. These cells line up the ventricles and the subretinal space; the choroid plexus cells produce the cerebrospinal fluid. Ependymocytes possess small movable processes (microvilli and kinocilia), which by rhythmic movements produce a stream of cerebrospinal fluid.

In addition, the brain of higher primates (including humans) contains several types of specialised astrocytes [17, 98], which include:

(xi) Interlaminar astrocytes;
(xii) Polarised astrocytes;
(xiii) Varicose projection astrocytes.

Function of these hominoid astroglia remain unknown.

Parenchymal astrocytes of the human brain are substantially larger and more complex compared with astroglial cells of rodents, and have distinct gene expression pattern [60–62, 87, 104]. Human protoplasmic astrocytes have about 10 times more primary processes and a more complex secondary process arborisation, with an average volume about 16.5 times larger than that of the corresponding astrocytes in a rat brain [61]. The larger human protoplasmic astrocytes also have extended outreach onto neuronal structures, on average contacting and encompassing up to 2 million synapses residing within astrocytic territorial domains, significantly more than the integrating capacity of rodent protoplasmic astrocytes, which covers ~20,000–120,000 synaptic contacts [13, 61]. Similarly, human fibrous astrocytes have a 2.14-fold larger domain compared to that in rodents [61].

(2) Oligodendroglia or oligodendrocytes, the myelinating cells of the CNS are subdivided into 4 classes [94]:

(i) *Type I oligodendrocytes* are most numerous in the cortex and grey matter; they have small rounded somata and fine branching processes that myelinate 30 or more small diameter axons;
(ii) *Type II oligodendrocytes* are similar to type I, but have parallel arrays of intermediate length internodes (100–250 μm), and are most common in white matter, such as the corpus callosum, optic nerve, cerebellum and spinal cord;
(iii) *Type III oligodendrocytes* have larger (than type I and II) irregular cell bodies, with one or more thick primary processes that myelinate a small number of large diameter axons with long internodes (250–500 μm). These cells are localised in the cerebral and cerebellar peduncles, the medulla oblongata, and the spinal cord funiculi;
(iv) *Type IV oligodendrocytes*, are somewhat similar to Schwann cells, being directly associated with a large diameter axon to form a single long internodal myelin sheath (as long as 1000 μm), and are restricted to tracts containing the largest diameter axons near the entrance/exit of nerve roots into the CNS.

(3) NG-2 glia also known as oligodendrocyte progenitor cells or OPCs, or synantocytes, or polydendrocytes [14, 57]. The NG2 glia can have homeostatic role and contribute to adulthood myelination, albeit their functions are yet to be better characterised.

Microglia originate from the foetal macrophages that migrate into the neural tube very early in the embryonic development; arguably, microglia represent the first parenchymal glia to populate neural tissue in development. Microglial cells carry numerous physiological functions, including shaping neuronal synaptic connectivity, removing of redundant or apoptotic neurones in the development and regulating synaptic transmission [45, 46, 90]. Microglia form the main defence system of the CNS through evolutionary conserved programme of activation (microgliosis) which can produce numerous neuroprotective and neurotoxic phenotypes [78, 83].

In terms of numbers, the most numerous glia are oligodendrocytes and NG2 cells combined (40–60%), with astrocytes accounting for 20–40% and microglia for ~10% of neuroglia population, although there is, of course, a considerable variability between the brain regions, developmental stage and species.

1.5 Envoi and Outlook

One of the two goals of this chapter is to serve as a general introduction into the world of Neuroglia. The other goal is to pique an interest of the reader into subsequent chapters in this book. As we tersely reviewed Neuroglia we establish the origin of these cells, their classification and their general functions in homeostasis and defence of the brain. In the following chapters, we explore the role of these cells in the progression of neuropathologies, especially neurodegenerative disorders. For a long time, the neurone-centric view dominated neuropathological thinking, and only recently the role of glia has been reassessed and the perception is mounting of specific importance of neuroglia that to a very large extent defines the progression and outcome of most (if not all) neurological diseases.

Acknowledgments VP's work is supported by a grant from the National Institute of General Medical Sciences of the National Institutes of Health (R01GM123971). VP is an Honorary Professor at University of Rijeka, Croatia.

References

1. Abbott NJ, Patabendige AA, Dolman DE, Yusof SR, Begley DJ (2010) Structure and function of the blood-brain barrier. Neurobiol Dis 37:13–25
2. Abbott NJ, Ronnback L, Hansson E (2006) Astrocyte-endothelial interactions at the blood-brain barrier. Nat Rev Neurosci 7:41–53
3. Andersen BB, Korbo L, Pakkenberg B (1992) A quantitative study of the human cerebellum with unbiased stereological techniques. J Comp Neurol 326:549–560
4. Andersen K, Andersen BB, Pakkenberg B (2012) Stereological quantification of the cerebellum in patients with Alzheimer's disease. Neurobiol Aging 33(197):e111–e120
5. Andrade-Moraes CH, Oliveira-Pinto AV, Castro-Fonseca E, da Silva CG, Guimaraes DM, Szczupak D, Parente-Bruno DR, Carvalho LR, Polichiso L, Gomes BV, Oliveira LM, Rodriguez RD, Leite RE, Ferretti-Rebustini RE, Jacob-Filho W, Pasqualucci CA, Grinberg LT, Lent R (2013) Cell number changes in Alzheimer's disease relate to dementia, not to plaques and tangles. Brain 136:3738–3752
6. Andriezen WL (1893) The neuroglia elements of the brain. Br Med J 2:227–230
7. Attwell D, Buchan AM, Charpak S, Lauritzen M, Macvicar BA, Newman EA (2010) Glial and neuronal control of brain blood flow. Nature 468:232–243
8. Azevedo FA, Carvalho LR, Grinberg LT, Farfel JM, Ferretti RE, Leite RE, Jacob Filho W, Lent R, Herculano-Houzel S (2009) Equal numbers of neuronal and nonneuronal cells make the human brain an isometrically scaled-up primate brain. J Comp Neurol 513:532–541
9. Bahney J, von Bartheld CS (2018) The cellular composition and Glia-neuron ratio in the spinal cord of a human and a nonhuman primate: comparison with other species and brain regions. Anat Rec (Hoboken) 301:697–710
10. Bear MF, Connors BW, Paradiso MA (2007) Exploring the brain. Lippincott Williams & Wilkins, Philadelphia
11. Burda JE, Bernstein AM, Sofroniew MV (2016) Astrocyte roles in traumatic brain injury. Exp Neurol 275(Pt 3):305–315
12. Burda JE, Sofroniew MV (2014) Reactive gliosis and the multicellular response to CNS damage and disease. Neuron 81:229–248
13. Bushong EA, Martone ME, Jones YZ, Ellisman MH (2002) Protoplasmic astrocytes in CA1 stratum radiatum occupy separate anatomical domains. J Neurosci 22:183–192
14. Butt AM, Kiff J, Hubbard P, Berry M (2002) Synantocytes: new functions for novel NG2 expressing glia. J Neurocytol 31:551–565
15. Christensen JR, Larsen KB, Lisanby SH, Scalia J, Arango V, Dwork AJ, Pakkenberg B (2007) Neocortical and hippocampal neuron and glial cell numbers in the rhesus monkey. Anat Rec (Hoboken) 290:330–340
16. Chvatal A, Verkhratsky A (2018) Early history of neuroglial research: personalities. Neuroglia 1:245–257
17. Colombo JA (2018) Interlaminar glia and other glial themes revisited: pending answers following three decades of glial research. Neuroglia 1:7–20

18. Darlington CL (2009) The female brain. CRC Press, Boca Raton
19. Deitmer JW, Rose CR (1996) pH regulation and proton signalling by glial cells. Prog Neurobiol 48:73–103
20. Deitmer JW, Rose CR, Munsch T, Schmidt J, Nett W, Schneider HP, Lohr C (1999) Leech giant glial cell: functional role in a simple nervous system. Glia 28:175–182
21. Del Rio-Hortega P (1919) El tercer elemento de los centros nerviosos. I. La microglia en estado normal. II. Intervencíon de la microglia en los procesos patológicos. III. Naturaleza probable de la microglia. Bol de la Soc esp de biol 9:69–120
22. Del Rio-Hortega P (1920) La microglia y su transformacíon en células en bastoncito y cuerpos gránulo-adiposos. Trab del Lab de invest biol 18:37
23. Del Rio-Hortega P (1932) Microglia. In: Penfield W (ed) Cytology and cellular pathology of the nervous system, vol 2. Hoeber, New York, pp 482–534
24. Del Río-Hortega P (1921) Estudios sobre la neuroglia. La glia de escasas radiaciones oligodendroglia. Biol Soc Esp Biol 21:64–92
25. Diamond MC, Scheibel AB, Murphy GM Jr, Harvey T (1985) On the brain of a scientist: Albert Einstein. Exp Neurol 88:198–204
26. Edwards TN, Meinertzhagen IA (2010) The functional organisation of glia in the adult brain of Drosophila and other insects. Prog Neurobiol 90:471–497
27. Eriksen N, Pakkenberg B (2007) Total neocortical cell number in the mysticete brain. Anat Rec (Hoboken) 290:83–95
28. Galambos R (1961) A glia-neural theory of brain function. Proc Natl Acad Sci USA 47:129–136
29. Golgi C (1870) Sulla sostanza connettiva del cervello (nevroglia). Rendiconti del R. Instituto Lombardo di Scienze e Lettere. serie 2, 3:275–277
30. Golgi C (1903) Opera Omnia. Hoepli, Milano
31. Gourine AV, Kasymov V, Marina N, Tang F, Figueiredo MF, Lane S, Teschemacher AG, Spyer KM, Deisseroth K, Kasparov S (2010) Astrocytes control breathing through pH-dependent release of ATP. Science 329:571–575
32. Grubisic V, Gulbransen BD (2017) Enteric glia: the most alimentary of all glia. J Physiol 595:557–570
33. Grubisic V, Verkhratsky A, Zorec R, Parpura V (2018) Enteric glia regulate gut motility in health and disease. Brain Res Bull 136:109–117
34. Halassa MM, Florian C, Fellin T, Munoz JR, Lee SY, Abel T, Haydon PG, Frank MG (2009) Astrocytic modulation of sleep homeostasis and cognitive consequences of sleep loss. Neuron 61:213–219
35. Hanani M (2005) Satellite glial cells in sensory ganglia: from form to function. Brain Res Rev 48:457–476
36. Hanani M (2010) Satellite glial cells in sympathetic and parasympathetic ganglia: in search of function. Brain Res Rev 64:304–327
37. Hansen DB, Garrido-Comas N, Salter M, Fern R (2015) HCO3–independent pH regulation in astrocytes in situ is dominated by V-ATPase. J Biol Chem 290:8039–8047
38. Hawkins A, Olszewski J (1957) Glia/nerve cell index for cortex of the whale. Science 126:76–77
39. Herculano-Houzel S, Dos Santos SE (2018) You do not mess with the Glia. Neuroglia 1:193–219
40. Herculano-Houzel S, Lent R (2005) Isotropic fractionator: a simple, rapid method for the quantification of total cell and neuron numbers in the brain. J Neurosci 25:2518–2521
41. Hertz L, Dringen R, Schousboe A, Robinson SR (1999) Astrocytes: glutamate producers for neurons. J Neurosci Res 57:417–428
42. Hilgetag CC, Barbas H (2009) Are there ten times more glia than neurons in the brain? Brain Struct Funct 213:365–366
43. Jelsing J, Nielsen R, Olsen AK, Grand N, Hemmingsen R, Pakkenberg B (2006) The postnatal development of neocortical neurons and glial cells in the Gottingen minipig and the domestic pig brain. J Exp Biol 209:1454–1462

44. Kandel ER, Schwartz JH, Jessell TM (2000) Principles of neural science. McGrawhill, New York
45. Kettenmann H, Hanisch UK, Noda M, Verkhratsky A (2011) Physiology of microglia. Physiol Rev 91:461–553
46. Kettenmann H, Kirchhoff F, Verkhratsky A (2013) Microglia: new roles for the synaptic stripper. Neuron 77:10–18
47. Kettenmann H, Verkhratsky A (2008) Neuroglia: the 150 years after. Trends Neurosci 31:653–659
48. Kidd GJ, Ohno N, Trapp BD (2013) Biology of Schwann cells. Handb Clin Neurol 115:55–79
49. Kirischuk S, Kettenmann H, Verkhratsky A (2007) Membrane currents and cytoplasmic sodium transients generated by glutamate transport in Bergmann glial cells. Pflugers Arch 454:245–252
50. Kofuji P, Newman EA (2004) Potassium buffering in the central nervous system. Neuroscience 129:1045–1056
51. Kovacs GG, Ferrer I, Grinberg LT, Alafuzoff I, Attems J, Budka H, Cairns NJ, Crary JF, Duyckaerts C, Ghetti B, Halliday GM, Ironside JW, Love S, Mackenzie IR, Munoz DG, Murray ME, Nelson PT, Takahashi H, Trojanowski JQ, Ansorge O, Arzberger T, Baborie A, Beach TG, Bieniek KF, Bigio EH, Bodi I, Dugger BN, Feany M, Gelpi E, Gentleman SM, Giaccone G, Hatanpaa KJ, Heale R, Hof PR, Hofer M, Hortobagyi T, Jellinger K, Jicha GA, Ince P, Kofler J, Kovari E, Kril JJ, Mann DM, Matej R, McKee AC, McLean C, Milenkovic I, Montine TJ, Murayama S, Lee EB, Rahimi J, Rodriguez RD, Rozemuller A, Schneider JA, Schultz C, Seeley W, Seilhean D, Smith C, Tagliavini F, Takao M, Thal DR, Toledo JB, Tolnay M, Troncoso JC, Vinters HV, Weis S, Wharton SB, White CL 3rd, Wisniewski T, Woulfe JM, Yamada M, Dickson DW (2016) Aging-related tau astrogliopathy (ARTAG): harmonized evaluation strategy. Acta Neuropathol 131:87–102
52. Kremer MC, Jung C, Batelli S, Rubin GM, Gaul U (2017) The glia of the adult Drosophila nervous system. Glia 65:606–638
53. Lanciotti A, Brignone MS, Bertini E, Petrucci TC, Aloisi F, Ambrosini E (2013) Astrocytes: emerging stars in leukodystrophy pathogenesis. Transl Neurosci 4
54. Lent R, Azevedo FA, Andrade-Moraes CH, Pinto AV (2012) How many neurons do you have? Some dogmas of quantitative neuroscience under revision. Eur J Neurosci 35:1–9
55. Lidow MS, Song ZM (2001) Primates exposed to cocaine in utero display reduced density and number of cerebral cortical neurons. J Comp Neurol 435:263–275
56. Mortensen HS, Pakkenberg B, Dam M, Dietz R, Sonne C, Mikkelsen B, Eriksen N (2014) Quantitative relationships in delphinid neocortex. Front Neuroanat 8:132
57. Nishiyama A, Komitova M, Suzuki R, Zhu X (2009) Polydendrocytes (NG2 cells): multifunctional cells with lineage plasticity. Nat Rev Neurosci 10:9–22
58. Nissl F (1898) Nervenzellen und graue Substanz. Munch Med Wochenschr 45:988–992; 1023–1029; 1060–1062
59. Noda M, Hiyama TY (2015) The Na_x channel: what it is and what it does. Neuroscientist 21:399–412
60. Oberheim NA, Goldman SA, Nedergaard M (2012) Heterogeneity of astrocytic form and function. Methods Mol Biol 814:23–45
61. Oberheim NA, Takano T, Han X, He W, Lin JH, Wang F, Xu Q, Wyatt JD, Pilcher W, Ojemann JG, Ransom BR, Goldman SA, Nedergaard M (2009) Uniquely hominid features of adult human astrocytes. J Neurosci 29:3276–3287
62. Oberheim NA, Wang X, Goldman S, Nedergaard M (2006) Astrocytic complexity distinguishes the human brain. Trends Neurosci 29:547–553
63. Oikonomou G, Shaham S (2011) The glia of Caenorhabditis elegans. Glia. 59:1253–1263
64. Pakkenberg B, Gundersen HJ (1988) Total number of neurons and glial cells in human brain nuclei estimated by the disector and the fractionator. J Microsc 150:1–20
65. Pakkenberg B, Gundersen HJ (1997) Neocortical neuron number in humans: effect of sex and age. J Comp Neurol 384:312–320

66. Parpura V, Heneka MT, Montana V, Oliet SH, Schousboe A, Haydon PG, Stout RF Jr, Spray DC, Reichenbach A, Pannicke T, Pekny M, Pekna M, Zorec R, Verkhratsky A (2012) Glial cells in (patho)physiology. J Neurochem 121:4–27
67. Pekny M, Pekna M, Messing A, Steinhauser C, Lee JM, Parpura V, Hol EM, Sofroniew MV, Verkhratsky A (2016) Astrocytes: a central element in neurological diseases. Acta Neuropathol 131:323–345
68. Pellerin L, Magistretti PJ (2012) Sweet sixteen for ANLS. J Cereb Blood Flow Metab 32:1152–1166
69. Pentreath VW, Radojcic T, Seal LH, Winstanley EK (1985) The glial cells and glia-neuron relations in the buccal ganglia of Planorbis corneus (L.): cytological, qualitative and quantitative changes during growth and ageing. Philos Trans R Soc Lond B Biol Sci 307:399–455
70. Pfrieger FW (2010) Role of glial cells in the formation and maintenance of synapses. Brain Res Rev 63:39–46
71. Ponomarev VS (1966) Glial index in vestibular nuclei of humans, macacos, and dogs. Arkh Anat Gistol Embriol 51:100–104
72. Ramón y Cajal S (1895) Algunas conjeturas sobre el mechanismoanatomico de la ideacion, asociacion y atencion. Imprenta y Libreria de Nicolas Moya
73. Ribeiro PF, Ventura-Antunes L, Gabi M, Mota B, Grinberg LT, Farfel JM, Ferretti-Rebustini RE, Leite RE, Filho WJ, Herculano-Houzel S (2013) The human cerebral cortex is neither one nor many: neuronal distribution reveals two quantitatively different zones in the gray matter, three in the white matter, and explains local variations in cortical folding. Front Neuroanat 7:28
74. Robertson W (1899) On a new method of obtaining a black reaction in certain tissue-elements of the central nervous system (platinum method). Scott Med Surg J 4:23
75. Robertson W (1900) A microscopic demonstration of the normal and pathological histology of mesoglia cells. J Ment Sci 46:733–752
76. Rodriguez-Arellano JJ, Parpura V, Zorec R, Verkhratsky A (2016) Astrocytes in physiological aging and Alzheimer's disease. Neuroscience 323:170–182
77. Ruitenberg MJ, Vukovic J, Sarich J, Busfield SJ, Plant GW (2006) Olfactory ensheathing cells: characteristics, genetic engineering, and therapeutic potential. J Neurotrauma 23:468–478
78. Savage JC, Picard K, Gonzalez-Ibanez F, Tremblay ME (2018) A brief history of microglial ultrastructure: distinctive features, phenotypes, and functions discovered over the past 60 years by electron microscopy. Front Immunol 9:803
79. Schleich CL (1894) Schmerzlose Operationen: Örtliche Betäubung mit indiffrenten Flüssigkeiten. Psychophysik des natürlichen und künstlichen Schlafes, Julius Springer, Berlin, p 256
80. Schroder KF, Hopf A, Lange H, Thorner G (1975) Morphometrical-statistical structure analysis of human striatum, pallidum and subthalamic nucleus. J Hirnforsch 16:333–350
81. Seifert G, Steinhauser C (2013) Neuron-astrocyte signaling and epilepsy. Exp Neurol 244:4–10
82. Sherwood CC, Stimpson CD, Raghanti MA, Wildman DE, Uddin M, Grossman LI, Goodman M, Redmond JC, Bonar CJ, Erwin JM, Hof PR (2006) Evolution of increased glia-neuron ratios in the human frontal cortex. Proc Natl Acad Sci USA 103:13606–13611
83. Sierra A, Beccari S, Diaz-Aparicio I, Encinas JM, Comeau S, Tremblay ME (2014) Surveillance, phagocytosis, and inflammation: how never-resting microglia influence adult hippocampal neurogenesis. Neural Plast 2014:610343
84. Sofroniew MV (2014) Astrogliosis. Cold Spring Harb Perspect Biol 7:a020420
85. Sofroniew MV (2014) Multiple roles for astrocytes as effectors of cytokines and inflammatory mediators. Neuroscientist 20:160–172
86. Sofroniew MV (2015) Astrocyte barriers to neurotoxic inflammation. Nat Rev Neurosci 16:249–263
87. Sosunov AA, Wu X, Tsankova NM, Guilfoyle E, McKhann GM 2nd, Goldman JE (2014) Phenotypic heterogeneity and plasticity of isocortical and hippocampal astrocytes in the human brain. J Neurosci 34:2285–2298

88. Stallcup WB (1981) The NG2 antigen, a putative lineage marker: immunofluorescent localization in primary cultures of rat brain. Dev Biol 83:154–165
89. Stout RF Jr, Verkhratsky A, Parpura V (2014) Caenorhabditis elegans glia modulate neuronal activity and behavior. Front Cell Neurosci 8:67
90. Tay TL, Savage JC, Hui CW, Bisht K, Tremblay ME (2017) Microglia across the lifespan: from origin to function in brain development, plasticity and cognition. J Physiol 595:1929–1945
91. Thorner G, Lange H, Hopf A (1975) Morphometrical-statistical structure analysis of human striatum, pallidus and subthalamic nucleus. II. Globus pallidus. J Hirnforsch 16:401–413
92. Tower DB (1954) Structural and functional organization of mammalian cerebral cortex; the correlation of neurone density with brain size; cortical neurone density in the fin whale (Balaenoptera physalus L.) with a note on the cortical neurone density in the Indian elephant. J Comp Neurol 101:19–51
93. Tremblay ME, Stevens B, Sierra A, Wake H, Bessis A, Nimmerjahn A (2011) The role of microglia in the healthy brain. J Neurosci 31:16064–16069
94. Verkhratsky A, Butt AM (2013) Glial physiology and pathophysiology. Wiley-Blackwell, Chichester, p 560
95. Verkhratsky A, Nedergaard M (2016) The homeostatic astroglia emerges from evolutionary specialization of neural cells. Philos Trans R Soc Lond B Biol Sci 371
96. Verkhratsky A, Nedergaard M (2018) Physiology of Astroglia. Physiol Rev 98:239–389
97. Verkhratsky A, Nedergaard M, Hertz L (2015) Why are astrocytes important? Neurochem Res 40:389–401
98. Verkhratsky A, Oberheim Bush NA, Nedergaard M, Butt AM (2018) The special case of human astrocytes. Neuroglia 1:21–29
99. Verkhratsky A, Parpura V, Pekna M, Pekny M, Sofroniew M (2014) Glia in the pathogenesis of neurodegenerative diseases. Biochem Soc Trans 42:1291–1301
100. Virchow R (1856) Ueber das granulirte Ansehen der Wandungen der Gehirnventrikel. In: Virchow R (ed) Gesammelte Abhandlungen zur wissenschaftlichen Medicin. Meidinger Sohn & Comp., Frankfurt A.M., pp 885–891
101. Virchow R (1858) Die Cellularpathologie in ihrer Begründung auf physiologische und pathologische Gewebelehre 20 Vorlesungen, gehalten während d. Monate Febr., März u. April 1858 im Patholog. Inst. zu Berlin. August Hirschwald, Berlin, 440 pp
102. von Bartheld CS, Bahney J, Herculano-Houzel S (2016) The search for true numbers of neurons and glial cells in the human brain: a review of 150 years of cell counting. J Comp Neurol 524:3865–3895
103. Wenker IC, Kreneisz O, Nishiyama A, Mulkey DK (2010) Astrocytes in the retrotrapezoid nucleus sense H^+ by inhibition of a $K_{ir}4.1$-$K_{ir}5.1$-like current and may contribute to chemoreception by a purinergic mechanism. J Neurophysiol 104:3042–3052
104. Zhang Y, Sloan SA, Clarke LE, Caneda C, Plaza CA, Blumenthal PD, Vogel H, Steinberg GK, Edwards MS, Li G, Duncan JA 3rd, Cheshier SH, Shuer LM, Chang EF, Grant GA, Gephart MG, Barres BA (2016) Purification and characterization of progenitor and mature human astrocytes reveals transcriptional and functional differences with mouse. Neuron 89:37–53
105. Zorec R, Horvat A, Vardjan N, Verkhratsky A (2015) Memory formation shaped by astroglia. Front Integr Neurosci 9:56

Chapter 2
Evolution of Neuroglia

Alexei Verkhratsky, Margaret S. Ho and Vladimir Parpura

Is evolution a theory, a system or a hypothesis? It is much more - it is a general postulate to which all theories, all hypotheses, all systems must henceforward bow and which they must satisfy in order to be thinkable and true. Evolution is a light which illuminates all facts, a trajectory which all lines of thought must follow-this is what evolution is.
Pierre Teilhard de Chardin.
quoted from Theodosius Dobzhansky, *Nothing in Biology Makes Sense Except in the Light of Evolution*, The American Biology Teacher, Vol. 35, No. 3 (Mar., 1973), pp. 125–129.

Abstract As the nervous system evolved from the diffused to centralised form, the neurones were joined by the appearance of the supportive cells, the neuroglia. Arguably, these non-neuronal cells evolve into a more diversified cell family than the neurones are. The first ancestral neuroglia appeared in flatworms being mesenchymal in origin. In the nematode *C. elegans* proto-astrocytes/supportive glia of ectodermal origin emerged, albeit the ensheathment of axons by glial cells occurred later in prawns. The multilayered myelin occurred by convergent evolution of oligodendro-

A. Verkhratsky (✉)
Faculty of Biology, Medicine and Health,
The University of Manchester, Manchester M13 9PT, UK
e-mail: Alexej.Verkhratsky@manchester.ac.uk

Faculty of Health and Medical Sciences, Center for Basic and Translational Neuroscience,
University of Copenhagen, 2200 Copenhagen, Denmark

Achucarro Center for Neuroscience, IKERBASQUE, Basque Foundation for Science,
48011 Bilbao, Spain

M. S. Ho
School of Life Science and Technology,
Shanghai Tech University, #230 Haike Road, Shanghai 201210, China

V. Parpura
Department of Neurobiology, The University of Alabama at Birmingham, Birmingham, AL, USA

University of Rijeka, Rijeka, Croatia

A. Verkhratsky et al. (eds.), *Neuroglia in Neurodegenerative Diseases*,
Advances in Experimental Medicine and Biology 1175,
https://doi.org/10.1007/978-981-13-9913-8_2

cytes and Schwann cells in vertebrates above the jawless fishes. Nutritive partitioning of the brain from the rest of the body appeared in insects when the hemolymph-brain barrier, a predecessor of the blood-brain barrier was formed. The defensive cellular mechanism required specialisation of bona fide immune cells, microglia, a process that occurred in the nervous system of leeches, bivalves, snails, insects and above. In ascending phylogeny, new type of glial cells, such as scaffolding radial glia, appeared and as the bran sizes enlarged, the glia to neurone ratio increased. Humans possess some unique glial cells not seen in other animals.

Keywords Astrocytes · Blood/haemolymph-brain barrier · Brain size · Complexity of glia · Glia to neuron ratio · Microglia · Myelination · Oligodendrocytes · Radial glia

2.1 Evolution of the Nervous System

Several taxonomy charts are currently in use. Here we use the system that classifies all living forms into Superkingdoms or Empires of Prokaryota and Eukaryota (Fig. 2.1). The Empire of Prokaryota comprises a single Kingdom of Bacteria, whereas the Empire of Eukaryota includes the Kingdoms of Protozoa, Animalia,

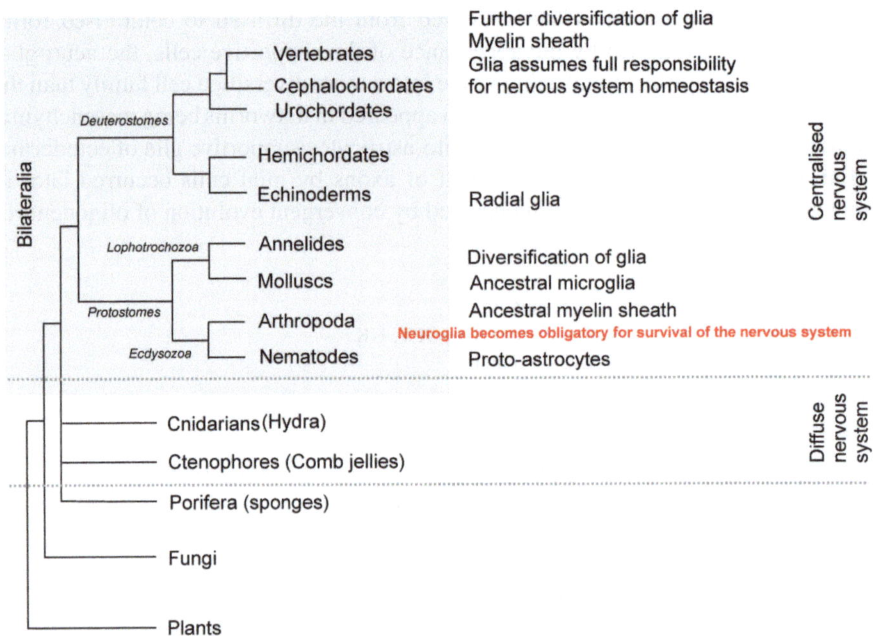

Fig. 2.1 Tree of life and evolution of the nervous system and of neuroglia. Adapted from Verkhratsky and Butt [143]

Fungi, Plantae and Chromista [17, 18]. The nervous system is the sole prerogative of the Kingdom of Animalia, which is represented by radially symmetrical Cnidaria and Ctenophora, and Bilateralia. The bilateralia are further subclassified into Protostomia and Deuterostomia, the latter including Echinodermata, Hemichordata and, finally, Chordata to which vertebrates belong.

Arguably, the most ancient nervous systems appeared in Ctenophora (comb jellies) and Cnidarians (hydras and sea jellies). This was represented by the so-called diffuse nervous system, formed by homogeneously distributed network of neurones connected with their processes [45]. The principal cells of the diffuse nervous system are multipolar and unipolar neurones, organised in several semi-independent networks connected through chemical synapses that mainly used peptides as neurotransmitters. The neurones remain the sole elements in the diffuse nervous system with no evidence of (and obviously no need for) any kind of specialised supportive cells. Of note, the primordial neurones evolved from epithelial cells [58], which expressed ion channels and transporters (that formed basis for electrical excitability) and the secretory vesicular system that was the morphological substrate for chemical neurotransmission [123].

Further evolution of the nervous system saw the change from diffuse nervous system to the centralised nervous system. At that stage the first conglomerates of neuronal somatas forming ganglia had emerged. The tendency to concentrate nervous elements was observed already in some Cnidarian polyps where neural networks appear denser around the oral opening. The first true centralised nervous systems, however, appear in early Bilateralia, in flatworms, earthworms and roundworms. For example, in roundworms the centralised nervous system is composed of several ganglia surrounding the oral orifice. In more advanced protostomes (e.g., insects and crustacea), the centralisation advanced further, with the polyganglionic brain. In vertebrates the new brain organisations in layers have appeared and progressed. The centralisation of the nervous system also signalled the appearance of supportive neural cells, the neuroglia.

Characterisation and morphological analysis of glial cells in early invertebrates and in phylogenetically lower taxa have been very much hampered by the absence of specific markers similar to those found for higher species. The main criteria for identification of glia is their close association with and coverage of neuronal elements, these being fundamental features of supportive cells. It is most probable that neuroglia has evolved several times in different species. Despite very similar functions, the appearance of glial cells is rather different between non-vertebrates and vertebrates, for example. Even genes responsible for glial differentiation can be distinct. In the insects, for example, development of glia is controlled by the gene glial cells missing (*gsm*—[60, 63]). In contrast, the *gsm* homologue gene is not even expressed in the CNS of mammals [69].

The very first glial cells provided for support of neuronal cell bodies within the ganglia (primeval astrocytes) as well as support of axons (primeval oligodendrocytes/Schwann cells). Another important function of primordial glia was formation of peripheral sensory organs, or sensilla. The glia-like cells are found in Acoelomorpha, the primitive flatworms, which are generally considered to be the earliest (or one of the earliest) Bilateralia. Electron microscopy of the brains of *Symsagittifera roscoffensis*, *Convoluta psammophila*, *Amphiscolops sp.* and *Otocelis rubropunctata* (free-living Acoela worms) characterised non-neuronal cells with electron-dense cell bodies in which nuclei occupy most of the cytosol, and lamellar processes extend into neuropil and surround groups of neurites [10, 11]. More advanced neuroglia is present in the nematode *Caenorhabditis elegans*. For example, the sheath glia of the cephalic sensilla in *C. elegans* possess some anatomical and functional characteristics that parallel those of astrocyte and oligodendrocyte lineages in the mammalian nervous system [133]. In Platyzoa (another member of early Bilateralia), glial cells have been found in polyclad flatworms and in some (but not in all) triclad planaria, whereas neuroglia seem to be absent in Rotifera (wheel animals) and in many platyhelminthes (for example, in tubellarian flatworms, Catenulida or Macrostomida). Neuroglia are generally present throughout Ecdysozoa and Lophotrochozoa, being well developed in molluscs, in Annelida, and even more developed and rather diverse in Arthropoda. In Deuterostomes the organisation of the brain as well as newly emerged spinal cord has changed form polyganglia to the layers, as a result of the appearance of the radial glia, a new type of neuroglia, which provided both for neurogenesis and migration of neuronal precursors to their appropriate layers. In early Chordata, the radial glial cells predominate and are present throughout life, while the parenchyma is quite underdeveloped. An increase in brain thickness instigated the emergence of parenchymal glia that diversified and became responsible for major homeostatic tasks in the CNS of mammals. The radial glial, however, remain active only in the prenatal period and largely disappear after birth. Below, we shall provide an account of the evolution of the main types of neuroglia.

2.2 Neuroglia in Invertebrates

2.2.1 Primitive Glia of Flatworm

The flatworms are the most primitive bilateralia with clearly formed centralised nervous system with the 'brain' represented by cerebral ganglia. The cerebral ganglia of at least two flatworms, *Fasciola hepatica* and *Notoplana acticola*, apparently contain some type of supportive cells that might be considered the ancestral neuroglia [135, 136]. These primitive glia, defined as mesenchymal cells, have long processes emanating from the cell body; some of these processes encircle the cerebral ganglion, some invaginate into giant nerve cells processes, whereas some other send processes into the ganglion and enclose clusters of neuronal processes. The 'glia-like' mesenchymal cells most likely originate from parenchymal cells of the worm and undergo morphological specialisation after contacting the nervous elements. Some of these cells contain glycogen and may act as a source of energy substrates [139].

2.2.2 Complex Neuroglia of the Earthworm

In the earthworm *Eisenia fetida* several types of glia have been characterised according to the morphology and localisation of the supportive cells. These neuroglia types were classified as neurilemmal-, subneurilemmal-, supporting-nutrifying- and periaxonal sheath-forming glial cells [26]. The neurilemmal glia are elongated with long processes. Subneurilemmal glia are small spindle-like cells with few processes. Supporting-nutrifying glial cells are positive for glial fibrillary acidic protein (GFAP, a well-accepted marker for mammalian astrocytes) and appear as brushes on the surface of the neuronal perikarya. Sheath-forming glial cells are found around giant axons [26]. The membranes of glial cells form close contacts with neuronal membranes, these possibly being involved in glia- to-neurone transport. Glial cells in the earthworm contain multiple intracellular vesicular structures with diameters between 200 and 400 nm, some with dense cores, perhaps indicating glial secretion.

2.2.3 Proto-astrocytes in Caenorhabditis Elegans

The nervous system of C. *elegans* has been precisely mapped with a wealth of details and meticulously categorised with structural information allowing identification of each neural cell [147]. The neuroglia of *C. elegans* were described and characterised based on light and electron microscopy (none of these glial cells express markers of mammalian glia), with these cells being defined as neural in origin lacking morphological characteristics of neurones, for example pre-synaptic structures [138, 147]. The nervous system of *C. elegans* contains 302 neurones, 50 supportive (glial) cells derived from the ectoderm and six supportive cells of the mesodermal origin [101, 134, 147]. The central nervous system of *C. elegans* comprises the nerve ring located in the frontal part of the body. The nerve ring receives processes of peripheral sensory neurones and also contains cephalic and motor neurones, axons of which convey efferent signals through the ventral and dorsal nerve cords.

Most of the neuroglial cells (that is 46 cells out of 56) of *C. elegans* are associated with the sensory system (Fig. 2.2). These cells are classified into 26 socket cells and 20 sheath cells that (together with neuronal processes) form sensory organs known as sensilla [112]. Four glial cells of the ectodermal origin known as cephalic sheath (CEPsh) cells are localised in the nerve ring (Figs. 2.2 and 2.3). The anterior processes of CEPsh cells cover cephalic neuronal dendrites and form sensilla in the lips of the animal. Posterior processes of CEPsh cell with lamellar morphology ensheath the nerve ring and send processes to the neuropil, where they contact and possibly enwrap synapses [101, 134]. The CEPsh cells control ion homeostasis in perisynaptic regions and are involved in neuronal development and morphogenesis. Complete ablation of glia in of *C. elegans* results in complex morphological, developmental, sensory and behavioural deficits, although it does not affect survival of the worm [4].

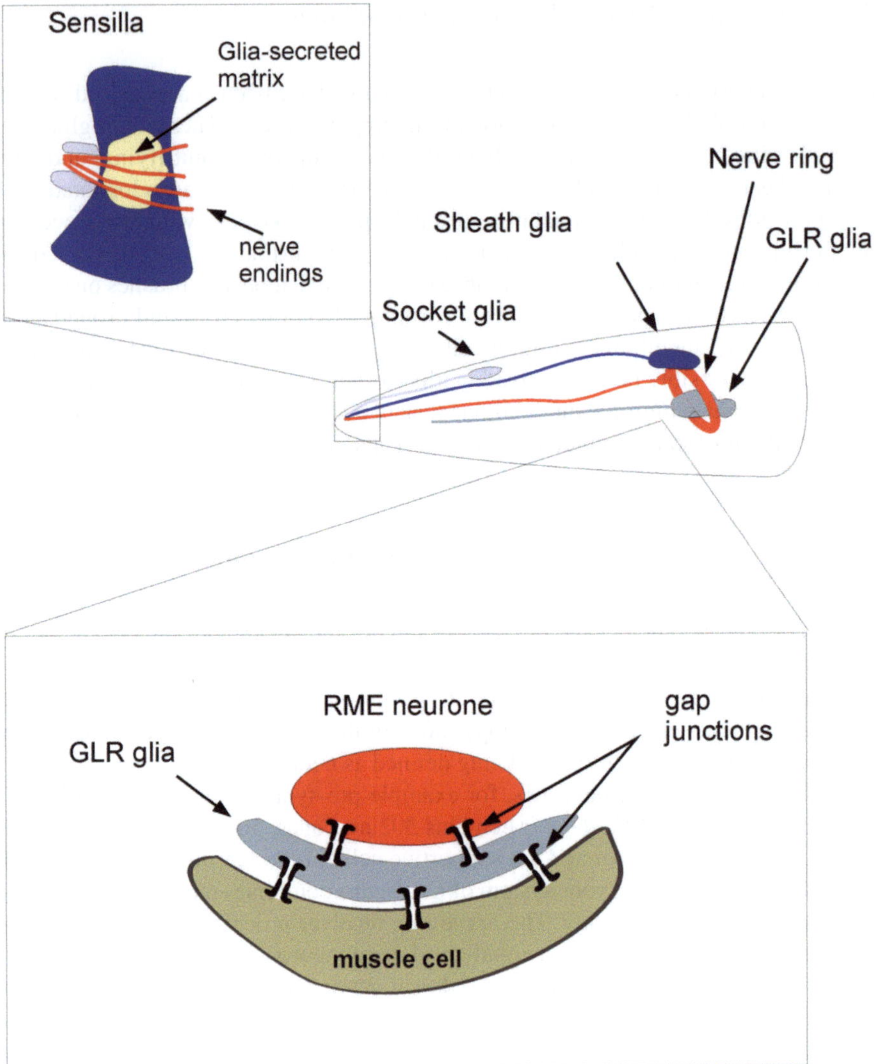

Fig. 2.2 Glial cells in *Caenorhabditis elegans*. The "brain" of *C. elegance* is represented by the nerve ring. Most of the glial cells are part of sensory organs known as sensilla. Each sensilla has two glial cells: the sheath cell and socket cell. In the anterior part there are 4 CEP (cephalic) glial cells that ensheath nerve ring. The nerve ring also has 6 GLR glial cells which establish gap junctional contacts between ring motor neurones (RME) and muscle cells

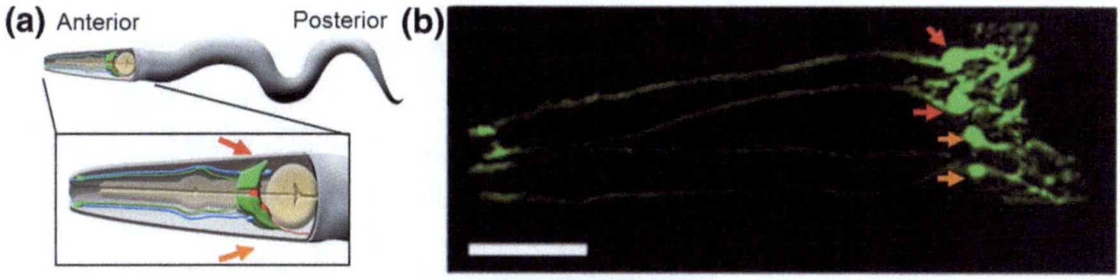

Fig. 2.3 The CEPsh glia. **a** A cartoon of an adult worm showing the four CEPsh glial cells (green) positioned in the anterior of the worm (inset). The CEPsh cell bodies with their velate processes are positioned around the central nerve ring (red) which they enwrap along with the proximal section of the ventral nerve cord. Additionally, each CEPsh glial cell possesses a long anterior process, emanating to the anterior sensory tip, which closely interacts with the dendritic extension of a nearby CEP neuron (blue). Arrows indicate the dorsal (red arrow) and ventral (orange arrow) side of the worm. **b** A confocal image showing green fluorescent protein expression driven by the *hlh-17* promoter to visualize the four CEPsh glial cells (worm strain VPR839). The anterior (head) of a juvenile (larval stage 4) worm is shown; the worm is turned ~45° from "upright" such that all four CEP sheath cells are visible. The sheath portion of the cells that form a tube around the dendritic endings of the CEP neurones are seen at the left of the image. The dorsal (red arrow) and ventral (orange arrow) CEPsh cell bodies are seen. The thin sheet-like extensions that surround and invade the nerve ring are seen in the rightmost part of the image. Scale bar, 20 μm. Image adapted from [134]

The glia of *C. elegans* have numerous differences from neuroglial cells of higher animals. Porto-astrocytes of the worm do not express GFAP; their physiology has both neuronal and non-neuronal properties—for example, Ca^{2+} signals in *C. elegans* glial are generated by Ca^{2+} entry through voltage-gated channels, while functional intracellular Ca^{2+} stores appear to be rudimentary or absent (Fig. 2.4, [133]). Glial cells of the worms, of course, do not form the glia limitans barrier because of the absence of the circulatory system. Main homeostatic functions of *C. elegans* glial cells similarly remain unknown; arguably, they may include K^+ buffering and neurotransmitter catabolism [134]. At the same time several genetic pathways controlling development and differentiation of glia are shared between the worm and mammals. For example, the transcription factor LIN-26 contributes to glial cell development and ablation of the *lin-26* gene may turn glial cells into neurones [75]. Another gene expressed in worms, the *hlh-17* gene (the promotor of which was used to generate markers for CEPsh glia [86]) has homology to the mammalian regulator of glial development *Olig2* [101, 154].

Neuroglia in *C. elegans* is responsible for the following functions: (i) establishment of the location of neuronal structures; (ii) regulation of size and morphology of sensory endings; (iii) creation of a barrier that bundles and separates neuronal elements from other cells; and (iv) modulation of neuronal activity. Conceptually, these functions are similar to the functions of glia in higher invertebrates and in vertebrates. Glial cells in *C. elegans* support development of the nervous system. In particular, glial cells modulate sensory activity by controlling the development of cellular compartments surrounding sensory cilia [114]. The sensory organs of the worm are singular aspects of the nematode nervous system, and their correct

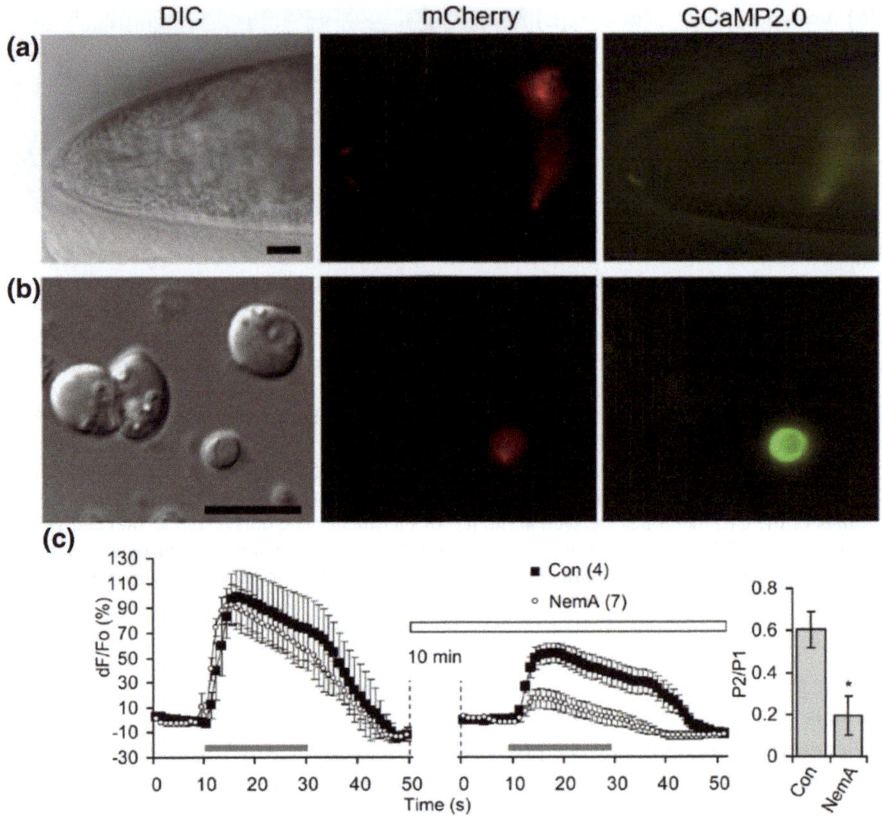

Fig. 2.4 L-type voltage-gated Ca^{2+}channels (VGCCs) play a role in depolarization-induced intracellular Ca^{2+} elevations in CEPsh glial cells. **a** The hlh-17 promoter can be used to drive expression of a red fluorescent protein marker (red, mCherry) in the CEPsh glia along with a fluorescent-protein based Ca^{2+} sensor (green, GCaMP2.0). DIC, differential interference contrast. An anterior portion of an L4 stage worm (VPR108 strain) is shown. **b** CEPsh glial cells in mixed culture prepared from embryos can be identified based on their mCherry/GCaMP2.0 expression. **c** Time-lapse of GCaMP2.0 fluorescence emission from CEPsh glial cells. Paired-pulse application of a depolarization stimulus, high extracellular potassium (HiK^+, 100 mM; horizontal bar), to CEPsh glial cells results in an elevation of intracellular Ca^{2+} levels (black squares). Nemadipine-A (NemA), a pharmacological L-type VGCC blocker, can be used to test the channels present in glial cells in culture; Con, sham stimulated control. (right, bar graph). Ratio of the peak Ca^{2+} level in response to the second HiK^+ application (P2) over the first application (P1). *Indicates a significant difference. Adapted from [133]

development is impaired in the absence of glia; moreover, development of sensory structures requires the expression of gene sets both in neurones and in glia [102]. Factors released by glia control sensory dendrites attachment during migration of neurones in development [53].

The four CEPsh glial cells (see above and Figs. 2.2, 2.3) differentially express netrin (UNC-6). Two ventral CEPsh cells express netrin, which regulates axon guidance, while the dorsal pair of CEPsh glia lacks the expression of this protein [52, 146, 154, 52, 146, 154].

There are some hints indicating that the CEPsh cells modulate dopamine-dependent behaviours in the worm, including feeding pattern and certain forms of learning [41]. Disruption of the *hlh-17* gene, expressed in CEPsh cells affected the egg-laying behaviour, instigated deficits in feeding behaviour and plasticity, and disrupted gustatory associative learning. The CEPsh glia are closely associated with four CEP neurones, which mediate the aforementioned behaviours through release and up-take of dopamine, and hence CEPsh glia seem to modulate neuronal function [41, 134].

The *C. elegans* is also in possession of a rather unique class of glial cells, which connect neurones with myocytes. These so-called GLR cells are of the mesodermal origin [1]. The GLR cells are integral part of the nervous system of the *C.elegans*; these cells are involved into the development of the nerve ring and the muscle-based feeding organ of the worm known as pharynx. The most idiosyncratic feature of GLR cells is that the cells are being sandwiched between and connected to both neurones and muscle cells in the head by gap junctions. This arrangement arguably represents a signalling circuit for producing specialised fine motor movements of the frontal part of the worm during foraging [120, 148]. The GLR-neuronal circuit contains both coincidence detection and shunting activity that is based on gap junction intercellular communication [115]. Gap junctional connectivity between glial cells and neurones are not limited to invertebrates only; there are some indications of this form of neuronal-glial communications in higher animals [113] as well as in developing nervous system of the mammals [95, 105]. The special set of cells connecting neurones and muscle cells is present in vertebrates including mammals; these are the telocytes initially discovered by Santiago Ramón y Cajal in the gut as interstitial cells of Cajal at the beginning of the twentieth century; the telocytes are found now in various tissues [113].

2.3 Homeostatic Proto-astrocytes in Annelida

The nervous system of the medicinal leech *Hirudo medicinalis* includes the anterior and posterior brains and the chain of 21 ganglia that are positioned in between (Fig. 2.5a). The anterior brain is made of six ganglia fused into two neuronal masses, while seven fused ganglia in the tail form the posterior brain. Somatic ganglia innervate corresponding segments of the leech body [22, 30]. Every ganglion contains ~400 neurones (with exception of the 5th and 6th ganglia innervating the reproduc-

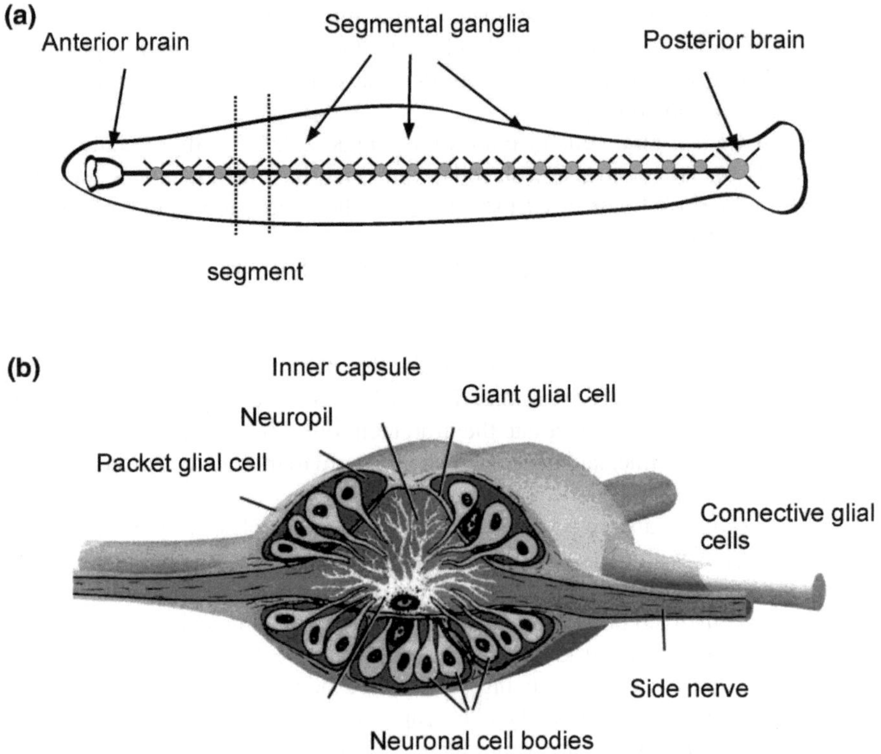

Fig. 2.5 Neuroglia in medicinal leech *Hirudo medicinalis*. **a** General structure of the nervous system. **b** Structure of a segmental ganglion, which contains three types of glial cells: the giant glial cell; packet glial cells and connective glial cells. Adapted from Verkhratsky and Butt [143]

tive system, which have about 700 neurones) and ten glial cells. These glial cells are (Fig. 2.5b): (i) two connective glial cells, which ensheath axons; (ii) six packet cells covering neuronal cell bodies and (iii) two giant glial cells [30]. Glial cells of the leech nervous system are interconnected by gap junctions assembled from innexins [37, 65, 79] thus creating a panglial syncytium. The nervous system of the leech additionally contains some amount of microglial cells that can be activated in response to lesions [76]. Within the ganglia the homeostatic support is maintained by packet glia and giant glial cells which in this function resemble mammalian astrocytes. The packet glial cells buffer extracellular K^+, especially at high extracellular K^+ concentrations [97, 124]. The giant glial cells have somata of 80–100 μm in diameter; these somata are localised in the centre of the ganglion. The processes of giant glial cells are 300–350 μm long; they extend through the entire neuropil and contact neuronal dendrites [93]. The neuropil is partitioned by these processes into several functional domains. In addition, glial processes sometimes invaginate into neuronal somata creating a structure described as 'trophospongium' [59]. The membrane of giant glial cells is highly permeable to K^+, which underlies hyperpolarised resting potential of about−75 mV. Giant glial cells express multiple types of neurotransmitter receptors including ionotropic glutamate, acetylcholine and serotonin receptors as well as metabotropic receptors to glutamate, serotonin, myomodulin and possibly P2Y-like purinoceptors and A_1-like adenosine receptors [30, 92]. Giant glial cells regulate many homeostatic responses, including regulation of pH involving plasmalemmal Na^+–HCO_3^- co-transporter, Na^+–H^+ and Cl^-–HCO_3^- exchangers [28, 31, 32]. They also regulate neurotransmitters uptake and catabolism through plasmalemmal Na^+-dependent glutamate and Na^+-dependent choline transporters [33, 56, 149]. Giant glial cells respond to neuronal activity and to evoked behaviours by changes in membrane potential [29] and by generation of cytosolic Ca^{2+} signals that occur in both somata and processes [34]. In contrast to mammalian glia, and rather similar to *C. elegans* CEPsh glial cells, the main source of Ca^{2+} signal generation in leech glia is associated with the opening of plasmalemmal Ca^{2+} channels. Termination of Ca^{2+} signal is mediated by plasmalemmal Ca^{2+} pump and Na^+/Ca^{2+} exchanger. The intracellular Ca^{2+} stores seem to play a minor role in Ca^{2+} dynamics of leech glia [34, 80].

2.4 Proto-astrocytes in Insects

The brain of *Drosophila* contains ~90,000 cells of which 10% belong to neuroglia; some glial cells also exist in the peripheral nervous system. Neuroglia of *Drosophila* are classified (Fig. 2.6) into the following classes [39, 43, 49, 108]: (i) Wrapping glia of the peripheral nervous system; (ii) Surface glia, which make the brain-hemolymph barrier, is further subclassified into perineural glia (small cells lying on the ganglionic

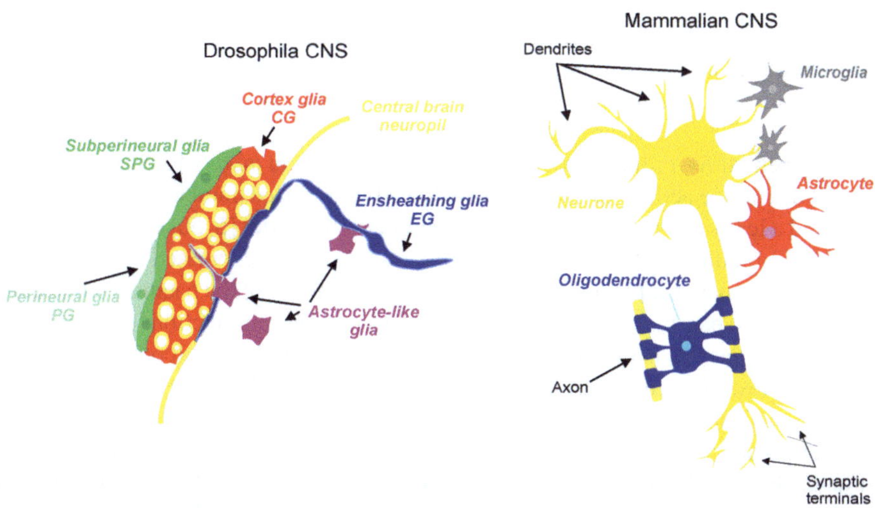

Fig. 2.6 Neuroglia in Drosophila and mammals

surface) and subperineural or basal glia (which are large sheet-like cells connected with septate junctions that form the barrier); (iii) Cortex glia that contact neuronal cells somata in the CNS, with each glial cell establishing contacts with many neurones; (iv) Neuropil glia, which are located in the neuropil and cover axons and synapses. The neuropil glia are further subdivided into ensheathing or fibrous and astrocyte-like glia, which forms a perisynaptic glial cover. Finally, (v) Tract glial cells cover axonal tracts connecting different neuropils. The cell mapping using the glia-specific GAL4-UAS system derived numerical calculation of various glial cell types in the Drosophila CNS. The perineural glia accounted for ~17%, subperineural—2%, cortex glia—20%, astrocyte-like glia—34% and ensheathing glia for 27% of total glial cells [70].

The cells of the above classes ii, iii and iv (i.e. surface glia, cortex glia and neuropil glia) are homeostatic cells being thus similar to astrocytes in mammals; the degree of morphological specialisation is, however, greater with specialised cell types responsible for distinct set of functions. A layer of perineural glia covers and delineates the brain as an organ. This coverage is provided by ~100 perineural glial cells, which create a physical barrier of the CNS. Perineural glia and primary surface glia of the ventral nerve cord are derived from embryonic neuro-glioblasts. Both types of primary glial cells increase in size, but not in number, during the larval stage. These cells are retained throughout metamorphosis and become the functional adult glia [35]. Primary perineural glia are mitotically active in the larva, undergo a late phase of proliferation during late larval stage, and differentiate into the optic lobe wrapping glia and then into optic lobe distal satellite glia [104].

Immediately beneath the perineural glial layer, the subperineural glia forms the hemolymph-brain barrier that controls exchange of substances between the CNS and the rest of the body [3, 131]. The subperineural glial cells have a flat morphology and are connected with innexin-based septate junctions, which seal the barrier. Molecular components of septate junctions include Neurexin IV, Gliotactin, Neuroglian and both of α and β subunits of Na^+/K^+ ATPase [8, 44, 132]. Loss of function of subperineural glia results in aberrant physiological properties such as the alternation in the permeability of extracellular dextrans [132] and reversed polarity of the electroretinogram [150, 151]. The G-protein coupled receptors moody expressed in the surface glia is essential for proper barrier formation. Glial cells in *moody* locus mutant flies exhibit disruptive actin cytoskeleton, which leads to reduced number of septate junctions and a leaky and malfunctional brain-hemolymph barrier [6, 50, 126]. Mutations in the ATP-binding cassette (ABC) transporter gene *mdr65* expressed in surface glia alters the passage of substrates and increases the sensitivity of the haemolymph-brain barrier to toxic pharmaceuticals, thus playing a role in neuroprotection [85]. The glial barrier in the insects is functionally analogous to the endothelial blood-brain barrier in vertebrates. This glial barrier in Drosophila is particularly important in guarding the CNS against substantial fluctuations in systemic K^+ that occur after feeding [39].

The specialised type of parenchymal glia, known as cortex glia covers neuronal somata; these cells are very much different from mammalian CNS glia in that a single cell can provide coverage for many neuronal cell bodies. The cortex glial cells also make contact with the hemolymph-brain barrier and provide trophic support to neurones. The cortical glial cells generate activity-dependent intracellular Ca^{2+} oscillations and regulate seizure susceptibility [88]. Similarly to the perineural glia, primary cortex glial cells enter a phase of proliferation during the late larval period, likely forming the secondary population of adult cortical glia. Glial cells in the insect CNS also plaster tracheoles, which deliver oxygen to the nerve tissue [111].

The neuropil glia in *Drosophila* are represented by the ensheathing glia and the astrocyte-like glia, both of which are critical elements for the neuropil homeostasis, synaptogenesis and synaptic transmission. Physical contacts between neuropil glia and axons as well as with axon fascicles allow these glial cells to function in a variety of synaptic contexts to regulate neuronal activity and survival. The flat ensheathing glial cells line the borders of the neuropil thus segregating it from the cortex. In addition, ensheathing glia enwrap glomeruli in the antennae lobe. The astroglia-like cells are morphologically similar to parenchymal astrocytes in vertebrates with elaborated arborisation; these cells extend processes into the neuropil and provide for synaptic coverage [36]. Astrocyte-like glia express plasmalemmal amino acid transporters to regulate uptake and release of neurotransmitters to modulate synaptic activity. The ensheathing glia express the engulfment receptor Draper and dCed-6, which control clearing axonal debris due to injury in adult brains. The astrocyte-like glia, also express Draper, which regulates pruning of axons during the development [137] Both types of neuropil glia, the ensheathing glia and astrocyte-like glia share the same origin being the progenies of the embryonic lateral glioblast. The lateral glioblast-generated primary glia undergo several rounds of mitotic divisions to produce a cluster of cells that differentiate into the ensheathing glia and astrocyte-like glia. The primary glial cells are subject to programmed cell death and are not retained into adulthood [9]. Instead, a secondary wave of proliferation from larval neuro-glioblasts is responsible for generating adult ensheathing and astrocyte-like glia [104].

Neuroglial cells in *Drosophila* exhibit classical intracellular ionic excitability, which contributes to neuronal-glial interactions. Different types of insect glia demonstrate spontaneous and evoked Ca^{2+} signalling mediated by both intracellular Ca^{2+} release (mostly in soma) and plasmalemmal Ca^{2+} entry [88]. Mutation in glial Na^{+}/Ca^{2+}-K^{+} exchanger (NCKX), which arguably mediates plasmalemmal Ca^{2+} flux in cortical glia, results in seizures [88]. The astrocyte-like glial cells in *Drosophila* demonstrate prominent Ca^{2+} oscillations, seen as fast fluctuations of intracellular Ca^{2+} in processes. Acute induction of Ca^{2+} influx into these astroglia-like cells triggered rapid behavioural paralysis and suppressed neuronal activity [156].

Homeostatic functions of glial cells in *Drosophila* CNS include regulation of ionic balance and control over clearance, recycling and metabolism of neurotransmitters. In the retina, for example, glial cells are providing for recycling of the principal neurotransmitter histamine. Histamine, released from photoreceptors, is accumulated by glial cells, processes of which enwrap relevant synapses [87]. After entering the glial cytoplasm histamine is converted into β-alanyl-histamine (also known as carcinine) by Ebony (N-β-alanyl-biogenic amine synthetase)-catalysed reaction [13]. This carcinine is then shuttled back to photoreceptors, which is mediated by a plasmalemmal carT (in humans—organic anion transporter family *Slc22a*) transporter [130]. In the photoreceptor cell, carcinine is hydrolysed to histamine by Tan (acyltransferase) protein [140]. Mutations in the components of this histamine/carcinine shuttle impair vision of the fly *Drosophila* [19, 157]. Genetic alterations of glial cells in *Drosophila* glia which interfere with vesicle trafficking (by specific expression of temperature-sensitive dynamin) and ionic transport (by glia-specific expression of bacterial Na^+ channel) alter circadian rhythm [96].

Another important function of homeostatic neuroglia lies in regulation of neurotransmitters turnover and catabolism. For example, *Drosophila* neuropil glia express plasmalemmal transporters for excitatory amino acid transporters dEAAT1and dEAAT2 [66, 127], as well as glutamine synthetase [27], all these being key components of glutamate-glutamine shuttle. The plasmalemmal glutamate transporters are preferentially localised at glial perisynaptic processes [121]; decreased expression of these transporters triggers neurotoxicity, degeneration of the neuropil and premature death [78]. In addition, *Drosophila* glia regulate the homeostasis of inhibitory neurotransmitter γ-aminobutyric acid (GABA) through activity of relevant transporters [103]. The glia-specific cascades regulating glutamate transport are involved in control of sexual behaviour and courtship of the flies. These cascades are represented by glial cystine-glutamate transporter, which controls ambient glutamate concentration and therefore affects the strength of glutamatergic transmission. Loss of function mutation of these transporters (observed in a mutant known as a genderblind, *gb*) results in 'homosexual' courting [47].

The neuroglia in insects is fundamental for metabolic support of neurones and for neuroprotection. In the retina of the honeybee, glial cells supply photoreceptors with alanine, which subsequently is converted to pyruvate for use in the Krebs cycle and production of energy [141]. Targeted ablation of glial cells in the fly instigates extensive neuronal death [12]. Similarly, neuronal loss and progressive neurodegeneration are observed in mutants with aberrant or non-functional glia (mutants designated as *drop dead*, *swiss cheese* and *repo* [14, 71, 150]). Insect neuroglia provides for nervous tissue defence through reactive gliosis and phagocytosis activated in response to lesions [72, 81].

Neuroglia in *Drosophila* form the neurogenic niche, a specialised anatomical location where stem cells reside, proliferate and differentiate [20, 90]. The cortex glial cells in particular regulate neuroblast proliferation; the neuroblasts establish a specific adhesive contact with cortex glia, this process involving phosphoinositide 3-kinase (PI3-kinase) and *D*E-cadherin. The cortical glia also regulate neuroblast proliferation through secretion of nutritional signal molecules, such as insulin-like peptides (ILPs) or the glycoprotein encoded by *anachronism* locus [38, 128, 129]. Cortex glia also produce and secrete ILPs in response to nutrition; these peptides activate the PI3 K/AKT signalling in neuroblasts, thus stimulating neuroblast growth and proliferation. The insulin/insulin-like growth factor (IGF) receptor pathway is necessary for neuroblasts to exit quiescence [20]. Likewise, a fat-body-derived signal required for neuroblast activation is linked to rapamycin (TOR) signalling cascade [128]. Another type of glia, the optic lobe-associated cortex glia, promote neuroepithelial proliferation and neuroblast formation by activating epidermal growth factor receptor. *Drosophila* microRNA mir-8 (the homolog of vertebrate miR-200 family) is expressed in a subpopulation of optic-lobe-associated cortex glia processes, which ensheath the neuroepithelium. In the absence of glial mir-8, excess proliferation and ectopic neuroblast transition were detected, suggesting that optic-lobe-associated cortex glia use signalling via mir-8 to communicate with the neuroepithelium. The optic-lobe-associated miR-8-positive cortex glia thus acts as a niche component that contributes signals for the growth and morphogenesis of the neuroepithelium [89]. Taken together, these findings suggest that glia are indispensable components of the neurogenic niche in insects; glial cells regulate formation, proliferation and differentiation of neuroblasts.

Drosophila glia also positively regulates and promotes synaptogenesis as well as synaptic maturation through activation of a homologue of the Gabapentin receptor $\alpha2\delta1$ [73].

The glia-regulated signalling cascade involving the peripheral glia expressing Eiger, the first invertebrate tumour necrosis factor (TNF) superfamily ligand [61], and the motorneurone-specific *Drosophila* TNF-α receptor (TNFR) Wengen [64], regulates neuromuscular junction synaptogenesis. This glia-initiated TNF signalling depends on caspase and mitochondria to regulate neuromuscular junction degeneration, further demonstrating the importance of glia in regulating neuromuscular junction synaptogenesis [67].

2.5 Astrocytes in Chordata and Low Vertebrates

In the CNS of Chordata, Hemichordata and Echinodermata, the main (and often the only) type of neuroglia is represented by radial glial cells. The radial glia, although occurring at some developmental stages in the insects and being identified in some protostomes (for example, in Annelida and Scalidophora [54]), is mainly associated with vertebrates [116]. In the Echinodermata (sea urchin, star fishes or sea cucumber), radial glial cells are the only glia in the CNS. These radial glial cells produce and secrete the Reissner's substance [84, 145], which mainly contains the glycoprotein known as SCO-spondin, that acts as cell adhesion modulator [46]. This Reissner's substance has been identified in radial glia throughout Chordata from cephalochordates to *Homo sapiens*. Glia of Echinodermata have a characteristic radial morphology with elongated shape; these cells have long processes penetrating the whole thickness of the neural parenchyma, and orienting perpendicularly to the surface of the neuroepithelium and high level of expression of intermediate filaments in the cytoplasm [83, 84].

Radial glia represent the main type of neuroglia in the CNS of many early vertebrates, which are almost completely devoid of other types of parenchymal glia. This in particular is characteristic for the brains with thin parenchyma. In Elasmobranchii (chondrichthyan/cartilaginous fish, such as sharks and rays), the brains are sub-classified into the type I, or 'laminar' brain (with thin brain wall and large ventricles) and the type II or 'elaborate' brain (with thicker parenchyma and smaller ventricles) [2, 16]. In type I brains radial glia predominate, whereas type II brains contain numerous well-developed parenchymal glia resembling astrocytes [2]. Emergence of parenchymal astrocytes in elaborate brains probably reflects an increased homeostatic challenge of the enlarged nervous tissue that cannot be met by radial glia. This constrains homeostatic capabilities of the radial glia and hence prompts an increase in numbers and complexity of parenchymal astrocytes [118].

Similarly radial glia are well developed in bony fish, with teleosts (e.g., zebra fish) being a particular example. The radial glia of the zebrafish, extend their processes through the entire thickness of the brain from the ependymal cells of the ventricles to the pial surface. In these radial glial cells high expression of GFAP has been detected; in addition, these cells possess glutamine synthetase contributing to glutamate homeostasis, and express aquaporin-4 contributing to water homeostasis, [48]). Zebrafish does not possess parenchymal glia (i.e., astrocyte-like cells) and hence radial glial cells are specially important for the responses of zebrafish nervous tissue to injury. Brain lesions in the teleost do not instigate astrogliosis. The stab wounds are closed in several days without formation of the scar; because of rapid increase in the neurogenesis, that generates new cells to fill up the wound [7]. Of note the blood-brain barrier in teleosts is formed by endothelial cells lining brain capillaries, which is similar to all higher vertebrates.

2.6 Astrocytes in Higher Vertebrates and Hominids

An increased complexity of the brain, which developed in parallel with increased intellectual power and increased homeostatic and energy demands stipulated high diversification of neuroglia in mammals [99, 100, 117, 119]. Glial cells became heterogeneous in their form and function. This evolutionary advance in astroglial complexity is specifically prominent in the brains of higher primates and humans [99, 144].

The number of glial cells varies substantially between different species and the GNR does not simply increase with increasing brain size (Fig. 2.7). Albeit already discussed in Chap. 1, we here revisit the glial-to-neurone ratio (GNR) in the nervous system. In invertebrates, it varies between 0.001 and 0.1 (56 glia per 302 neurones in *C. elegans* [101]; 10 glial cells per 400–700 neurones in every ganglion of the leech [30]; ~9000 glial cells per 90,000 neurones in the central nervous system (CNS) of Drosophila [39, 70]). There are exceptions though: the buccal ganglia of the great ramshorn snail *Planorbis corneus* contains 298 neurones and 391 glial cells with GNR of 1.3 [110]. In vertebrates the GNR is about 0.3–0.4 in rodents, ~1.1 in cat, ~1.2 in horse, 0.5–1.0 in rhesus monkey, 2.2 in Göttingen minipig, ~1.5 in humans, and as high as 4–8 in elephants and the fin whale [21, 40, 51, 62, 77, 107]. The largest absolute number of glial cells has been identified in the neocortex of whales [40, 31, 35]; for example the neocortex of the long-finned pilot whale (*Globicephala melas*) contains ~ 37.2×10^9 neurones and 127×10^9 glial cells with GNR of 3.4 [91]. In the human brain, the total number of glia is more or less the same as number of neurones (about ~80 billion of neurones and ~60 billion of glia) with remarkable

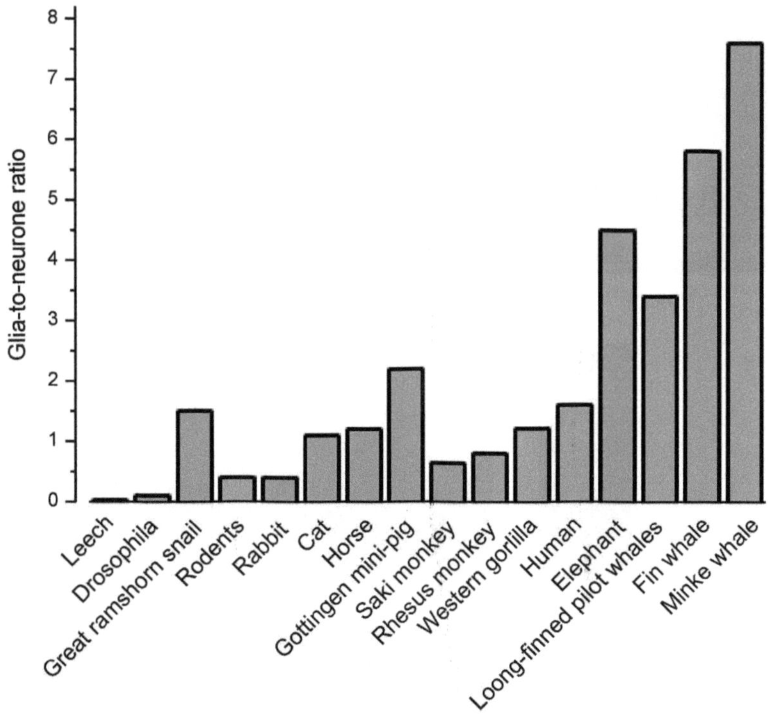

Fig. 2.7 Phylogenetical advance of neuroglia. Glia-to-neurone ratio in the nervous system of invertebrates and in the cortex of vertebrates. Glia-to-neurone ratio is generally increased in phylogeny; more or less this ratio linearly follows an increase in the size of the brain

regional differences. Of note, the more primitive parts of the human brain have a higher GNR of 7–10 in the brainstem, or even more according to some studies [106]. The GNR of ~5 was determined for the spinal cord of cynomolgus monkey and GNR of almost 7 for spinal cord in humans [5]. These trends argue against the concept that a high GNR reflects evolutionary advance and increased intelligence. Nonetheless, it is important to be aware that evolution brought with it substantial changes in the morphology and complexity of astroglia in the human cortex, which also contains several highly specialised types of glial cell which are absent in the brains of lesser vertebrates.

Human astrocytes are much larger and far more complex than astroglial cells in, for example, rodent brain (Fig. 2.8). In the human brain the average diameter of the domain belonging to a human protoplasmic astrocyte is ~2.5 times larger than the domain formed by a rat astrocyte (142 vs. 56 μm). The volume of the human protoplasmic astrocyte domain was ~16.5 times larger than that of the corresponding domain in a rat brain. Human protoplasmic astrocytes have ~10 times more primary processes, and correspondingly much more complex processes arborisation than rodent astroglia. Similarly, fibrous astrocytes, localised in the white matter are ~2.2

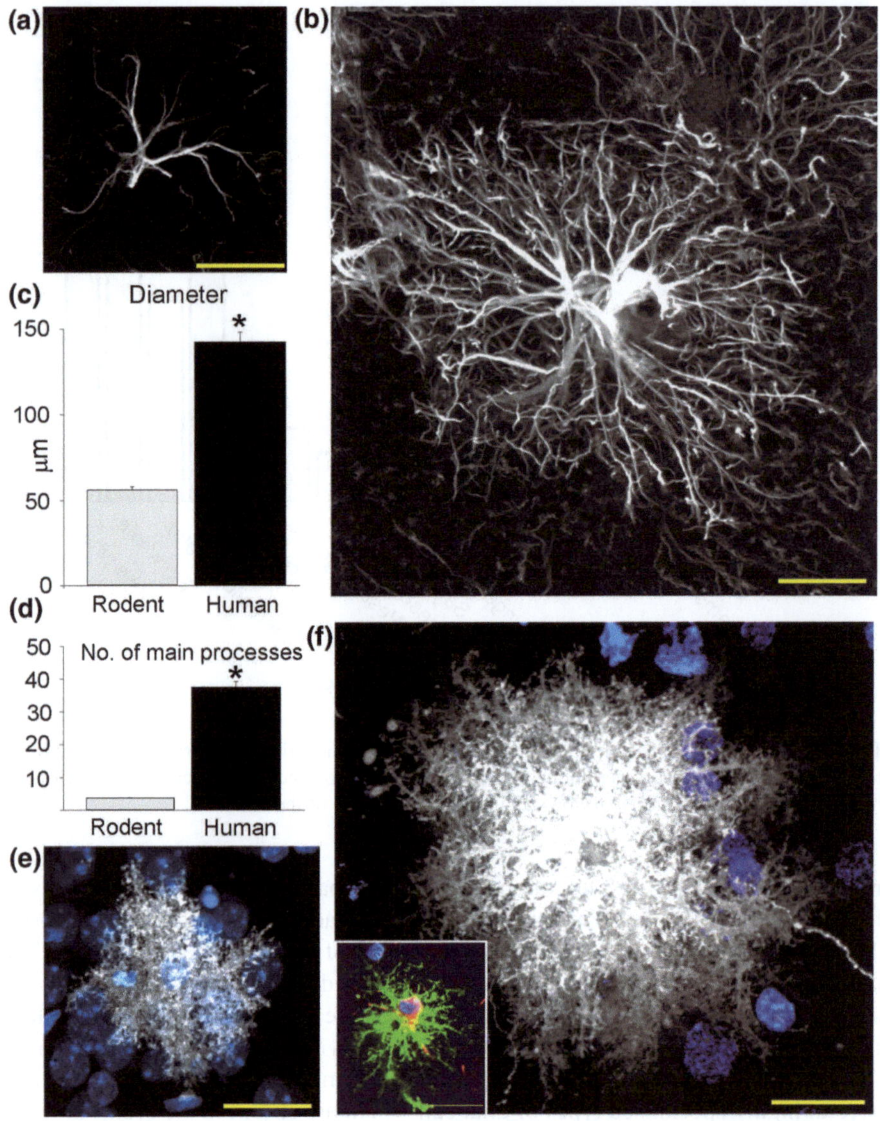

Fig. 2.8 Comparison of rodent and human protoplasmic astrocytes. **a** Typical mouse protoplasmic astrocyte. GFAP, White. Scale bar, 20 μm. **b** Typical human protoplasmic astrocyte at the same scale. Scale bar, 20 μm. **c**, **d** Human protoplasmic astrocytes are 2.55-fold larger and have 10-fold more main GFAP processes than mouse astrocytes (human, n = 50 cells from 7 patients; mouse, n = 65 cells from 6 mice; mean ± SEM; *$p < 0.005$, t test). **e** Mouse protoplasmic astrocyte diolistically labelled with DiI (white) and sytox (blue) revealing the full structure of the astrocyte including its numerous fine processes. Scale bar, 20 μm. **f** Human astrocyte demonstrates the highly complicated network of fine process that defines the human protoplasmic astrocyte. Scale bar, 20 μm. Inset, Human protoplasmic astrocyte diolistically labelled as well as immunolabelled for GFAP (green) demonstrating colocalisation. Scale bar, 20 μm. Reproduced, with permission from [99, 144]

times larger in humans when compared to rodents. Due to this increased complexity human protoplasmic astrocytes enwrap ~2 millions of synapses localised in their territorial domains, whereas in rodents single astrocyte covers only 20,000–120,000 synaptic contacts [15, 99].

Special subpopulations of astrocytes are found in the brains of higher primates and humans (Fig. 2.9). The first type of these specialised astroglial cells is known as interlaminar astrocytes [23–25]. The somata of these interlaminar astrocytes are localised in the layer I of the cortex, which has low density of neuronal cell bodies and high density of synaptic connections. Somata of interlaminar astrocytes are rather small not exceeding 10 μm, from these somata emanate several short and one or two very long processes. These long processes, which can be as long as 1 mm pen-

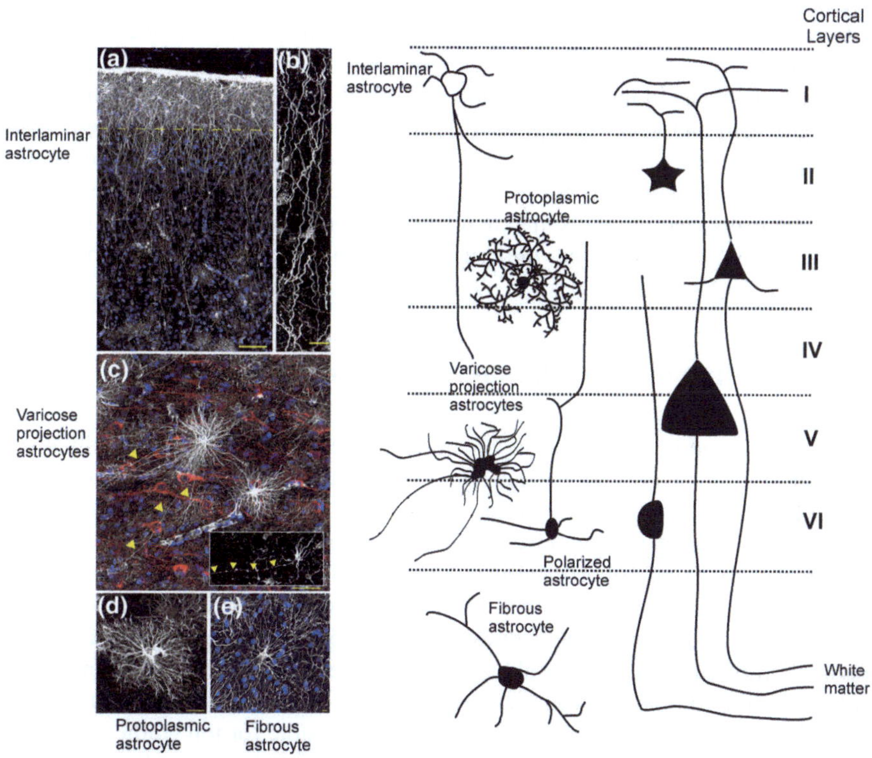

Fig. 2.9 Morphological heterogeneity and subtypes of astrocytes in the cortex of higher primates. **a** Pial surface and layers 1–2 of human cortex. GFAP staining in white; DAPI, in blue. Scale bar, 100 μm. Yellow line indicates border between layer I and II. **b** Interlaminar astrocyte processes. Scale bar, 10 μm. **c** Varicose projection astrocytes reside in layers V and VI and extend long processes characterized by evenly spaced varicosities. Inset: Varicose projection astrocyte from chimpanzee cortex. Yellow arrowheads indicate varicose projections. Scale bar, 50 μm. **d** Typical human protoplasmic astrocyte. Scale bar, 20 μm. **e** Human fibrous astrocytes in white matter. Scale bar, 10 μm. (modified with permission from [98]. Left panel schematically shows different astrocytes and their relatuons to cortical layers. Adapted from [143]

etrate through the cortex, and terminate in layers III and IV. At the tips of these long processes special structures known as the 'terminal mass' or 'end bulb' have been detected. These terminal masses often contain mitochondria. Physiological properties of interlaminar astrocytes remain completely unknown. It has been suggested that they may form astroglial component of neuronal columns that are regarded as cortical functional units. It has been further speculated that interlaminar astrocytes may contribute to long distance signalling and integration within cortical columns.

The second type of specialised human astrocytes is represented by polarised astrocytes, represented by uni- or bipolar cells, somata of which are in cortical layers V and VI close to the white matter. These cells possess one or two very long (up to 1 mm) and very thin (2–3 μm in diameter) processes that end in the neuropil. In addition these processes have numerous varicosities with yet unknown functions [99, 144].

In contrast to neuroglia neurones did not increase their size that much. Similarly, the density of synaptic contacts in rodents and primates is similar (the mean density of synaptic contacts in the rodent brain is ~1397 millions/mm^3, whereas in humans synaptic density in the cortex is around 1100 millions/mm^3). Likewise, the number of synapses per neurone is not much different between primates and rodents. The average size of human neurones is ~1.5 times larger than in rodents. Thus, at least morphologically, evolution resulted in much more prominent changes in glia than in neurones, which most likely has important, although yet undetermined, significance.

2.7 Evolution of Myelination and Oligodendroglia

The emergence and evolution of myelination is linked to an increase in animal size that requires faster nerve conductance; increase in action potential propagation velocity can be achieved either through an increase in axon diameter, or through introduction of saltatory nerve impulse propagation. Increase in axon diameter reduces resistance of the axon proportionally to the square of diameter, with the conductance velocity being directly proportional to the square root of the axon diameter [57]. In the *Loligo* squid, for instance, thick axons of 0.5 mm in diameter sustain action potential propagation velocity of about 30 m/s. Large axons, however, present several problems to the complex nervous system. Firstly, the conduction through large axons is energetically costly, because Na^+/K^+ pump responsible for ion gradients maintenance consumes substantial amounts of ATP [82]. Secondly, large diameter axons are associated with prominent space constrains incompatible with a rather compact design of advanced CNS.

The saltatory propagation of action potential, which allows high speed nerve impulse conductance, solves both problems due to restriction of ion fluxes to small portions of axonal membrane. Namely, axons are covered with multiple layers of lipid membranes, interrupted by gaps known as nodes of Ranvier. The nodal membranes have high densities of voltage-gated Na^+ and K^+ channels responsible for action potential generation [94]. The lipid-rich membranous lamellae insulate parts of the axons in between the nodes, thus increasing axonal transverse resistance and reducing transverse capacitance.

The very first structures unsheathing the axons and allowing saltatory propagation of action potentials emerged early in evolution. Arguably, the first organisms in possession of this mechanism are prawns, which appeared in the Cambrian period (500–540 million years ago). At that time, the giant prawn, *Anomalocaridids*, was the sole and the most dangerous predator of the sea [142]. These predatory prawns had about 1 m in length, and they were endowed with large compound eyes. These eyes were exceptionally big (according to the fossil studies the visual surface was 22 mm long and 12 mm wide) being composed of tens of thousands of hexagonal ommatidial lenses ~70–110 μm in diameter [109]. Feeding thousands of axons into the CNS of these animals most likely required insulating ensheathing of axons.

Modern Crustacean (e.g., prawns, shrimps and crabs) retain this arrangements having elaborated axonal ensheathment. In the prawns of the genus *Penaeus* (such as Japanese tiger shrimp, or Chinese white shrimp), axons are surrounded by glial membranes and by a large submyelin space positioned between the axonal membrane and the first layer of glial membrane. The ion currents thus are trapped in this space as if the normal axon is surrounded by a giant axon (the submyelin space acts in essence as a low-resistance pathway), this topography allows for an unprecedented speed of action potential propagation of up to 210 m/s [74, 152, 153]. The submyelin spaces are tightly sealed at the nodes thus allowing the saltatory conductance. The node (which in invertebrates is called the 'fenestration node') diameter and internodal distance is proportional to the axon diameter and, in prawns, vary between 5 and 50 μm and 3 and 12 μm, respectively. The thickness of the glial membranous sheath is ~10 μm; it is comprised of 10–60 stacked membrane layers separated by 8–9 nm. Like in vertebrates, voltage-gated sodium channels in prawns are concentrated at the nodes where their density can reach thousands of channels/μm^2. There is a fundamental difference between vertebrates and prawn axonal coverage. In vertebrates the single Schwann cell (peripheral nerves) or a process of an oligodendrocyte (CNS) spirals around the axon forming multiple membranous lamellae. In the prawns a single myelinating glial cell sends multiple processes forming multiple layers, with each process encircling the axon once, meeting itself on the opposite side in a seam [55]. Another difference is location of the nuclei of myelinating cell. In vertebrates it is located as a rule at the outer edge of myelin sheath, whereas in prawns the nuclei are randomly dispersed between membrane laminae [153].

Axons covered with multilayered glial membranes (although these glial cells do not produce myelin) are operative in some other invertebrates. In the earthworm *Lumbricus terrestrils* the central axons of 50–100 μm in diameter are enwrapped with 60–200 layers of cell membranes produced by many glial cells, nuclei of which are scattered along the axon [122]. In this structure the nodes are not clearly seen; nonetheless, the conductance velocity reaches ~20–45 m/s, which is higher compared to thicker giant axons of *Loligo* squid. The glial axonal coverage was also found in marine Annelida phoronids; in these animals axons are covered with 9–20 membranous layers [42]. Similar number of layers of glial membranes covers the axon of the aquatic sludge worm *Branchiura sowerbyi* [158].

Emergence of myelin is associated with relatively developed vertebrates. Compacted myelin sheaths are absent in lower vertebrates, such as hagfish and lampreys, and begun to develop in sharks and bony fish. There are some arguments indicating that the most ancient forms of myelin sheath emerged in placoderms (now extinct jawed armoured fish from the early Silurian period ~420 million years ago). These fish form the base for chondrichtyan and bony fishes. The fossil records indicate that the diameters of the foramina for oculomotor nerves in the jawed fish and in jawless Osteostraci fish (which do not have myelin) are the same (about 0.1 mm), whereas the length of nerve in placoderms was 10 times larger, that highlights the necessity for myelin to maintain the same speed of action potential-mediated signal transduction [155]. Further reasoning suggests the connection between appearance of the jaw in early Gnathostomata (jawed vertebrates, which embrace all higher vertebrates living today, including mammals) and myelination. By acquiring myelinated nerves, the jawed fishes arguably acquired better ability to hunt the prey, while keeping the axonal diameter the same or even smaller compared to their jawless predecessors [155].

Coverage of axons with glial membranes emerged in early evolutionary forms. In the beginning this coverage arguably supported axonal mechanical stability and provided energy support. At the same time glial membranous lamellae increased action potential propagation velocity, and once emerged, myelination provided obvious evolutionary advantages. One of the advantages was an increase in compactness of the nervous system and decrease in energy expenditure for restoring ion balances.

2.8 Evolution of Microglia

The evolutionary origins of microglia remain largely unexplored. It is possible to suggest, however, that appearance of innate immune and phagocytic cells in the nerve tissue coincided with the emergence of barriers separating early brains from the circulation. Such barriers restricted pathways for entry of immune/defence cells into the brain parenchyma, thus calling for a specialised intra-brain defence system. This problem was solved after immune cells found the way to migrate and retain in the nerve tissue; exposure to the specific neurochemical environment as well as epigenetic trends most likely stimulated acquisition of specific microglial phenotype. Evidence for phylogenetically early microglial cells is available for Annelida (leech), Mollusca (Bivalvia and snails) and some Arthropoda (insects) (see [68] for detailed review).

The nervous system of medicinal leech contains substantial numbers of microglial cells. The microglia of the leech is represented by small spindle-like shape cells. Insults to the leech nervous system trigger microglial activation; these cells migrate to the site of the insult and become phagocytes. The weak silver carbonate staining (a classical microglial staining technique developed by Pío del Río Hortega) is generally used to visualise activated microglia in leech. In response to infectious attack the leech microglial cells were found to produce and secrete antimicrobial peptides [125].

Well developed microglia were also found in nerve ganglia of molluscs. In the marine bivalve *Mytilus edulis* microglial cells can be activated and their migration can be instigated in response to various molecular signals including nitric oxide, opioids, cannabinoids and cytokines. Similarly, migrating microglia were observed in the snail *Planorbarius corneus* and in the insect *Leucophaea maderae*. The microglial cells (morphologically distinguished by phagocytic inclusions) of snail *Planorbis corneus*, were mainly concentrated in the neuropil of nerve ganglia while mechanical lesion increased the number of these phagocytic cells [110].

2.9 Conclusions

Most ancient glial cells developed as a supportive element of the sensory organs. The centralisation and increase in complexity of the nervous system created high demand for homeostatic support which was met by diversification of glial cells. Such a diversification resulted in multiple phenotypes in invertebrates, which are in possession of very specialised glial cells such as giant glial cells in the leech or cortex glia in *Drosophila*. These glial cells of invertebrates have performed many supportive functions from regulation of ion and neurotransmitter homeostasis to metabolic support and regulation of neuronal development. In the invertebrates glial cells formed brain to body barriers which stipulated the emergence of specialised immune and defence cells known as microglia. Increase in complexity of brain connectome and increase in axonal density were factors defining evolutionary benefits of the myelin sheath and development of myelinating cells. The evolution of myelination formed the basis for increased complexity of the nervous system that relies on interneuronal connections. In early Chordata radial glia become the main sub-type which was instrumental in formation of the layered brain. Subsequent increase in brain thickness promoted evolution of homeostatic astroglia. Finally, in the brain of primates and especially in the brain of humans, astrocytes become exceedingly complex and new types of astroglial cells involved in interlayer communication/integration have evolved.

Acknowledgments VP's work is supported by a grant from the National Institute of General Medical Sciences of the National Institutes of Health (R01GM123971). VP is an Honorary Professor at University of Rijeka, Croatia.

References

1. Altun ZF, Chen B, Wang ZW, Hall DH (2009) High resolution map of Caenorhabditis elegans gap junction proteins. Dev Dyn 238:1936–1950
2. Ari C, Kalman M (2008) Evolutionary changes of astroglia in Elasmobranchii comparing to amniotes: a study based on three immunohistochemical markers (GFAP, S-100, and glutamine synthetase). Brain Behav Evol 71:305–324
3. Awasaki T, Lai SL, Ito K, Lee T (2008) Organization and postembryonic development of glial cells in the adult central brain of Drosophila. J Neurosci 28:13742–13753

4. Bacaj T, Tevlin M, Lu Y, Shaham S (2008) Glia are essential for sensory organ function in *C. elegans*. Science 322:744–747
5. Bahney J, von Bartheld CS (2018) The cellular composition and glia-neuron ratio in the spinal cord of a human and a nonhuman primate: comparison with other species and brain regions. Anat Rec (Hoboken) 301:697–710
6. Bainton RJ, Tsai LT, Schwabe T, DeSalvo M, Gaul U, Heberlein U (2005) Moody encodes two GPCRs that regulate cocaine behaviors and blood-brain barrier permeability in Drosophila. Cell 123:145–156
7. Baumgart EV, Barbosa JS, Bally-Cuif L, Gotz M, Ninkovic J (2010) Stab wound injury of the zebrafish telencephalon: a model for comparative analysis of reactive gliosis. Glia 60:343–357
8. Baumgartner S, Littleton JT, Broadie K, Bhat MA, Harbecke R, Lengyel JA, Chiquet-Ehrismann R, Prokop A, Bellen HJ (1996) A Drosophila neurexin is required for septate junction and blood-nerve barrier formation and function. Cell 87:1059–1068
9. Beckervordersandforth RM, Rickert C, Altenhein B, Technau GM (2008) Subtypes of glial cells in the Drosophila embryonic ventral nerve cord as related to lineage and gene expression. Mech Dev 125:542–557
10. Bedini C, Lanfranchi A (1991) The central and peripheral nervous system of Acoela (plathelminthes). An electron microscopical study. Acta Zoologica (Stockholm) 72:101–106
11. Bery A, Cardona A, Martinez P, Hartenstein V (2010) Structure of the central nervous system of a juvenile acoel, Symsagittifera roscoffensis. Dev Genes Evol 220:61–76
12. Booth GE, Kinrade EF, Hidalgo A (2000) Glia maintain follower neuron survival during Drosophila CNS development. Development 127:237–244
13. Borycz J, Borycz JA, Loubani M, Meinertzhagen IA (2002) tan and ebony genes regulate a novel pathway for transmitter metabolism at fly photoreceptor terminals. J Neurosci 22:10549–10557
14. Buchanan RL, Benzer S (1993) Defective glia in the Drosophila brain degeneration mutant drop-dead. Neuron 10:839–850
15. Bushong EA, Martone ME, Jones YZ, Ellisman MH (2002) Protoplasmic astrocytes in CA1 stratum radiatum occupy separate anatomical domains. J Neurosci 22:183–192
16. Butler AB, Hodos W (2005) Vertebrate neuroanatomy: evolution and adaptation, 2nd edn. Wiley, New York
17. Cavalier-Smith T (1998) A revised six-kingdom system of life. Biol Rev Camb Philos Soc 73:203–266
18. Cavalier-Smith T (2009) Megaphylogeny, cell body plans, adaptive zones: causes and timing of eukaryote basal radiations. J Eukaryot Microbiol 56:26–33
19. Chaturvedi R, Reddig K, Li HS (2014) Long-distance mechanism of neurotransmitter recycling mediated by glial network facilitates visual function in Drosophila. Proc Natl Acad Sci USA 111:2812–2817
20. Chell JM, Brand AH (2010) Nutrition-responsive glia control exit of neural stem cells from quiescence. Cell 143:1161–1173
21. Christensen JR, Larsen KB, Lisanby SH, Scalia J, Arango V, Dwork AJ, Pakkenberg B (2007) Neocortical and hippocampal neuron and glial cell numbers in the rhesus monkey. Anat Rec (Hoboken) 290:330–340
22. Coggeshall RE, Fawcett DW (1964) The fine structure of the central nervous system of the leech, *Hirudo Medicinalis*. J Neurophysiol 27:229–289
23. Colombo JA, Reisin HD (2004) Interlaminar astroglia of the cerebral cortex: a marker of the primate brain. Brain Res 1006:126–131
24. Colombo JA, Sherwood CC, Hof PR (2004) Interlaminar astroglial processes in the cerebral cortex of great apes. Anat Embryol (Berl) 208:215–218
25. Colombo JA, Yanez A, Puissant V, Lipina S (1995) Long, interlaminar astroglial cell processes in the cortex of adult monkeys. J Neurosci Res 40:551–556
26. Csoknya M, Dénes V, Wilhelm M (2012) Glial cells in the central nervous system of earthworm, *Eisenia fetida*. Acta Biol Hung 63(Suppl. 1):114–128

27. De Pinto V, Caggese C, Prezioso G, Ritossa F (1987) Purification of the glutamine synthetase II isozyme of Drosophila melanogaster and structural and functional comparison of glutamine synthetases I and II. Biochem Genet 25:821–836
28. Deitmer JW (1991) Electrogenic sodium-dependent bicarbonate secretion by glial cells of the leech central nervous system. J Gen Physiol 98:637–655
29. Deitmer JW, Kristan WB Jr (1999) Glial responses during evoked behaviors in the leech. Glia 26:186–189
30. Deitmer JW, Rose CR, Munsch T, Schmidt J, Nett W, Schneider HP, Lohr C (1999) Leech giant glial cell: functional role in a simple nervous system. Glia 28:175–182
31. Deitmer JW, Schlue WR (1987) The regulation of intracellular pH by identified glial cells and neurones in the central nervous system of the leech. J Physiol 388:261–283
32. Deitmer JW, Schlue WR (1989) An inwardly directed electrogenic sodium-bicarbonate co-transport in leech glial cells. J Physiol 411:179–194
33. Deitmer JW, Schneider HP (1997) Intracellular acidification of the leech giant glial cell evoked by glutamate and aspartate. Glia 19:111–122
34. Deitmer JW, Verkhratsky AJ, Lohr C (1998) Calcium signalling in glial cells. Cell Calcium 24:405–416
35. DeSalvo MK, Mayer N, Mayer F, Bainton RJ (2011) Physiologic and anatomic characterization of the brain surface glia barrier of Drosophila. Glia 59:1322–1340
36. Doherty J, Logan MA, Tasdemir OE, Freeman MR (2009) Ensheathing glia function as phagocytes in the adult Drosophila brain. J Neurosci 29:4768–4781
37. Dykes IM, Freeman FM, Bacon JP, Davies JA (2004) Molecular basis of gap junctional communication in the CNS of the leech Hirudo medicinalis. J Neurosci 24:886–894
38. Ebens AJ, Garren H, Cheyette BN, Zipursky SL (1993) The Drosophila anachronism locus: a glycoprotein secreted by glia inhibits neuroblast proliferation. Cell 74:15–27
39. Edwards TN, Meinertzhagen IA (2010) The functional organisation of glia in the adult brain of Drosophila and other insects. Prog Neurobiol 90:471–497
40. Eriksen N, Pakkenberg B (2007) Total neocortical cell number in the mysticete brain. Anat Rec (Hoboken) 290:83–95
41. Felton CM, Johnson CM (2011) Modulation of dopamine-dependent behaviors by the Caenorhabditis elegans Olig homolog HLH-17. J Neurosci Res 89:1627–1636
42. Fernandez I, Pardos F, Benito J, Roldan C (1996) Ultrastructural observations on the phoronid nervous system. J Morphol 230:265–281
43. Freeman MR, Doherty J (2006) Glial cell biology in Drosophila and vertebrates. Trends Neurosci 29:82–90
44. Genova JL, Fehon RG (2003) Neuroglian, Gliotactin, and the Na+/K+ATPase are essential for septate junction function in Drosophila. J Cell Biol 161:979–989
45. Ghysen A (2003) The origin and evolution of the nervous system. Int J Dev Biol 47:555–562
46. Gobron S, Monnerie H, Meiniel R, Creveaux I, Lehmann W, Lamalle D, Dastugue B, Meiniel A (1996) SCO-spondin: a new member of the thrombospondin family secreted by the subcommissural organ is a candidate in the modulation of neuronal aggregation. J Cell Sci 109(Pt 5):1053–1061
47. Grosjean Y, Grillet M, Augustin H, Ferveur JF, Featherstone DE (2008) A glial amino-acid transporter controls synapse strength and courtship in Drosophila. Nat Neurosci 11:54–61
48. Grupp L, Wolburg H, Mack AF (2010) Astroglial structures in the zebrafish brain. J Comp Neurol 518:4277–4287
49. Hartenstein V (2011) Morphological diversity and development of glia in Drosophila. Glia 59:1237–1252
50. Hatan M, Shinder V, Israeli D, Schnorrer F, Volk T (2011) The Drosophila blood brain barrier is maintained by GPCR-dependent dynamic actin structures. J Cell Biol 192:307–319
51. Hawkins A, Olszewski J (1957) Glia/nerve cell index for cortex of the whale. Science 126:76–77
52. Hedgecock EM, Culotti JG, Hall DH (1990) The unc-5, unc-6, and unc-40 genes guide circumferential migrations of pioneer axons and mesodermal cells on the epidermis in C. elegans. Neuron 4:61–85

53. Heiman MG, Shaham S (2009) DEX-1 and DYF-7 establish sensory dendrite length by anchoring dendritic tips during cell migration. Cell 137:344–355
54. Helm C, Karl A, Beckers P, Kaul-Strehlow S, Ulbricht E, Kourtesis I, Kuhrt H, Hausen H, Bartolomaeus T, Reichenbach A, Bleidorn C (2017) Early evolution of radial glial cells in Bilateria. Proc Biol Sci 284
55. Heuser JE, Doggenweiler CF (1966) The fine structural organization of nerve fibers, sheaths, and glial cells in the prawn, Palaemonetes vulgaris. J Cell Biol 30:381–403
56. Hirth IC, Deitmer JW (2006) 5-Hydroxytryptamine-mediated increase in glutamate uptake by the leech giant glial cell. Glia 54:786–794
57. Hodgkin AL (1954) A note on conduction velocity. J Physiol 125:221–224
58. Holland ND (2003) Early central nervous system evolution: an era of skin brains? Nat Rev Neurosci 4:617–627
59. Holmgren E (1901) Beiträge zur Morphologie der Zelle: I. Nervenzellen. Anat Hefte 18:267–326
60. Hosoya T, Takizawa K, Nitta K, Hotta Y (1995) glial cells missing: a binary switch between neuronal and glial determination in Drosophila. Cell 82:1025–1036
61. Igaki T, Kanda H, Yamamoto-Goto Y, Kanuka H, Kuranaga E, Aigaki T, Miura M (2002) Eiger, a TNF superfamily ligand that triggers the Drosophila JNK pathway. EMBO J 21:3009–3018
62. Jelsing J, Nielsen R, Olsen AK, Grand N, Hemmingsen R, Pakkenberg B (2006) The postnatal development of neocortical neurons and glial cells in the Gottingen minipig and the domestic pig brain. J Exp Biol 209:1454–1462
63. Jones BW, Fetter RD, Tear G, Goodman CS (1995) glial cells missing: a genetic switch that controls glial versus neuronal fate. Cell 82:1013–1023
64. Kanda H, Igaki T, Kanuka H, Yagi T, Miura M (2002) Wengen, a member of the Drosophila tumor necrosis factor receptor superfamily, is required for Eiger signaling. J Biol Chem 277:28372–28375
65. Kandarian B, Sethi J, Wu A, Baker M, Yazdani N, Kym E, Sanchez A, Edsall L, Gaasterland T, Macagno E (2012) The medicinal leech genome encodes 21 innexin genes: different combinations are expressed by identified central neurons. Dev Genes Evol 222:29–44
66. Kawano T, Takuwa K, Kuniyoshi H, Juni N, Nakajima T, Yamamoto D, Kimura Y (1999) Cloning and characterization of a Drosophila melanogaster cDNA encoding a glutamate transporter. Biosci Biotechnol Biochem 63:2042–2044
67. Keller LC, Cheng L, Locke CJ, Muller M, Fetter RD, Davis GW (2011) Glial-derived prodegenerative signaling in the Drosophila neuromuscular system. Neuron 72:760–775
68. Kettenmann H, Hanisch UK, Noda M, Verkhratsky A (2011) Physiology of microglia. Physiol Rev 91:461–553
69. Kim J, Jones BW, Zock C, Chen Z, Wang H, Goodman CS, Anderson DJ (1998) Isolation and characterization of mammalian homologs of the Drosophila gene glial cells missing. Proc Natl Acad Sci USA 95:12364–12369
70. Kremer MC, Jung C, Batelli S, Rubin GM, Gaul U (2017) The glia of the adult Drosophila nervous system. Glia 65:606–638
71. Kretzschmar D, Hasan G, Sharma S, Heisenberg M, Benzer S (1997) The swiss cheese mutant causes glial hyperwrapping and brain degeneration in Drosophila. J Neurosci 17:7425–7432
72. Kurant E (2011) Keeping the CNS clear: glial phagocytic functions in Drosophila. Glia 59:1304–1311
73. Kurshan PT, Oztan A, Schwarz TL (2009) Presynaptic alpha2delta-3 is required for synaptic morphogenesis independent of its Ca2+-channel functions. Nat Neurosci 12:1415–1423
74. Kusano K (1966) Electrical activity and structural correlates of giant nerve fibers in Kuruma shrimp (*Penaeus japonicus*). J Cell Physiol 68:361–383
75. Labouesse M, Sookhareea S, Horvitz HR (1994) The Caenorhabditis elegans gene lin-26 is required to specify the fates of hypodermal cells and encodes a presumptive zinc-finger transcription factor. Development 120:2359–2368
76. Le Marrec-Croq F, Drago F, Vizioli J, Sautiere PE, Lefebvre C (2013) The leech nervous system: a valuable model to study the microglia involvement in regenerative processes. Clin Dev Immunol 2013:274019

77. Lidow MS, Song ZM (2001) Primates exposed to cocaine in utero display reduced density and number of cerebral cortical neurons. J Comp Neurol 435:263–275
78. Lievens JC, Rival T, Iche M, Chneiweiss H, Birman S (2005) Expanded polyglutamine peptides disrupt EGF receptor signaling and glutamate transporter expression in Drosophila. Hum Mol Genet 14:713–724
79. Lohr C, Deitmer JW (1997) Structural and physiological properties of leech giant glial cells as studied by confocal microscopy. Exp Biol Online 2:8
80. Lohr C, Deitmer JW (2006) Calcium signaling in invertebrate glial cells. Glia 54:642–649
81. MacDonald JM, Beach MG, Porpiglia E, Sheehan AE, Watts RJ, Freeman MR (2006) The Drosophila cell corpse engulfment receptor Draper mediates glial clearance of severed axons. Neuron 50:869–881
82. Magistretti PJ (2009) Neuroscience. Low-cost travel in neurons. Science 325:1349–1351
83. Mashanov VS, Zueva OR, Garcia-Arraras JE (2010) Organization of glial cells in the adult sea cucumber central nervous system. Glia 58:1581–1593
84. Mashanov VS, Zueva OR, Heinzeller T, Aschauer B, Naumann WW, Grondona JM, Cifuentes M, Garcia-Arraras JE (2009) The central nervous system of sea cucumbers (Echinodermata: Holothuroidea) shows positive immunostaining for a chordate glial secretion. Front Zool 6:11
85. Mayer F, Mayer N, Chinn L, Pinsonneault RL, Kroetz D, Bainton RJ (2009) Evolutionary conservation of vertebrate blood-brain barrier chemoprotective mechanisms in Drosophila. J Neurosci 29:3538–3550
86. McMiller TL, Johnson CM (2005) Molecular characterization of HLH-17, a C. elegans bHLH protein required for normal larval development. Gene 356:1–10
87. Meinertzhagen IA, O'Neil SD (1991) Synaptic organization of columnar elements in the lamina of the wild type in Drosophila melanogaster. J Comp Neurol 305:232–263
88. Melom JE, Littleton JT (2013) Mutation of a NCKX eliminates glial microdomain calcium oscillations and enhances seizure susceptibility. J Neurosci 33:1169–1178
89. Morante J, Vallejo DM, Desplan C, Dominguez M (2013) Conserved miR-8/miR-200 defines a glial niche that controls neuroepithelial expansion and neuroblast transition. Dev Cell 27:174–187
90. Morrison SJ, Spradling AC (2008) Stem cells and niches: mechanisms that promote stem cell maintenance throughout life. Cell 132:598–611
91. Mortensen HS, Pakkenberg B, Dam M, Dietz R, Sonne C, Mikkelsen B, Eriksen N (2014) Quantitative relationships in delphinid neocortex. Front Neuroanat 8:132
92. Muller M, Henrich A, Klockenhoff J, Dierkes PW, Schlue WR (2000) Effects of ATP and derivatives on neuropile glial cells of the leech central nervous system. Glia 29:191–201
93. Munsch T, Deitmer JW (1992) Calcium transients in identified leech glial cells in situ evoked by high potassium concentrations and 5-hydroxytryptamine. J Exp Biol 167:251–265
94. Nave KA, Werner HB (2014) Myelination of the nervous system: mechanisms and functions. Annu Rev Cell Dev Biol 30:503–533
95. Nedergaard M (1994) Direct signaling from astrocytes to neurons in cultures of mammalian brain cells. Science 263:1768–1771
96. Ng FS, Tangredi MM, Jackson FR (2011) Glial cells physiologically modulate clock neurons and circadian behavior in a calcium-dependent manner. Curr Biol 21:625–634
97. Nicholls JG, Kuffler SW (1964) Extracellular space as a pathway for exchange between blood and neurons in the central nervous system of the leech: ionic composition of glial cells and neurons. J Neurophysiol 27:645–671
98. Oberheim NA, Goldman SA, Nedergaard M (2012) Heterogeneity of astrocytic form and function. Methods Mol Biol 814:23–45
99. Oberheim NA, Takano T, Han X, He W, Lin JH, Wang F, Xu Q, Wyatt JD, Pilcher W, Ojemann JG, Ransom BR, Goldman SA, Nedergaard M (2009) Uniquely hominid features of adult human astrocytes. J Neurosci 29:3276–3287
100. Oberheim NA, Wang X, Goldman S, Nedergaard M (2006) Astrocytic complexity distinguishes the human brain. Trends Neurosci 29:547–553
101. Oikonomou G, Shaham S (2011) The glia of Caenorhabditis elegans. Glia 59:1253–1263

102. Oikonomou G, Shaham S (2012) On the morphogenesis of glial compartments in the sensory organs of Caenorhabditis elegans. Worm 1:51–55
103. Oland LA, Gibson NJ, Tolbert LP (2010) Localization of a GABA transporter to glial cells in the developing and adult olfactory pathway of the moth Manduca sexta. J Comp Neurol 518:815–838
104. Omoto JJ, Lovick JK, Hartenstein V (2016) Origins of glial cell populations in the insect nervous system. Curr Opin Insect Sci 18:96–104
105. Pakhotin P, Verkhratsky A (2005) Electrical synapses between Bergmann glial cells and Purkinje neurones in rat cerebellar slices. Mol Cell Neurosci 28:79–84
106. Pakkenberg B, Gundersen HJ (1988) Total number of neurons and glial cells in human brain nuclei estimated by the disector and the fractionator. J Microsc 150:1–20
107. Pakkenberg B, Gundersen HJ (1997) Neocortical neuron number in humans: effect of sex and age. J Comp Neurol 384:312–320
108. Parker RJ, Auld VJ (2006) Roles of glia in the Drosophila nervous system. Semin Cell Dev Biol 17:66–77
109. Paterson JR, Garcia-Bellido DC, Lee MS, Brock GA, Jago JB, Edgecombe GD (2011) Acute vision in the giant Cambrian predator Anomalocaris and the origin of compound eyes. Nature 480:237–240
110. Pentreath VW, Radojcic T, Seal LH, Winstanley EK (1985) The glial cells and glia-neuron relations in the buccal ganglia of Planorbis corneus (L.): cytological, qualitative and quantitative changes during growth and ageing. Philos Trans R Soc Lond B Biol Sci 307:399–455
111. Pereanu W, Spindler S, Cruz L, Hartenstein V (2007) Tracheal development in the Drosophila brain is constrained by glial cells. Dev Biol 302:169–180
112. Perens EA, Shaham S (2005) C. elegans daf-6 encodes a patched-related protein required for lumen formation. Dev Cell 8:893–906
113. Popescu LM, Faussone-Pellegrini MS (2010) TELOCYTES—a case of serendipity: the winding way from Interstitial Cells of Cajal (ICC), via Interstitial Cajal-Like Cells (ICLC) to TELOCYTES. J Cell Mol Med 14:729–740
114. Procko C, Shaham S (2010) Assisted morphogenesis: glial control of dendrite shapes. Curr Opin Cell Biol 22:560–565
115. Rabinowitch I, Chatzigeorgiou M, Schafer WR (2013) A gap junction circuit enhances processing of coincident mechanosensory inputs. Curr Biol 23:963–967
116. Rakic P (2003) Elusive radial glial cells: historical and evolutionary perspective. Glia 43:19–32
117. Reichenbach A (1989) Glia:neuron index: review and hypothesis to account for different values in various mammals. Glia 2:71–77
118. Reichenbach A, Neumann M, Bruckner G (1987) Cell length to diameter relation of rat fetal radial glia–does impaired K^+ transport capacity of long thin cells cause their perinatal transformation into multipolar astrocytes? Neurosci Lett 73:95–100
119. Reichenbach A, Pannicke T (2008) Neuroscience. A new glance at glia. Science 322:693–694
120. Ringstad N, Abe N, Horvitz HR (2009) Ligand-gated chloride channels are receptors for biogenic amines in C. elegans. Science 325:96–100
121. Rival T, Soustelle L, Strambi C, Besson MT, Iche M, Birman S (2004) Decreasing glutamate buffering capacity triggers oxidative stress and neuropil degeneration in the Drosophila brain. Curr Biol 14:599–605
122. Roots BI, Lane NJ (1983) Myelinating glia of earthworm giant axons: thermally induced intramembranous changes. Tissue Cell 15:695–709
123. Ryan TJ, Grant SG (2009) The origin and evolution of synapses. Nat Rev Neurosci 10:701–712
124. Saubermann AJ, Castiglia CM, Foster MC (1992) Preferential uptake of rubidium from extracellular space by glial cells compared to neurons in leech ganglia. Brain Res 577:64–72
125. Schikorski D, Cuvillier-Hot V, Leippe M, Boidin-Wichlacz C, Slomianny C, Macagno E, Salzet M, Tasiemski A (2008) Microbial challenge promotes the regenerative process of the injured central nervous system of the medicinal leech by inducing the synthesis of antimicrobial peptides in neurons and microglia. J Immunol 181:1083–1095

126. Schwabe T, Bainton RJ, Fetter RD, Heberlein U, Gaul U (2005) GPCR signaling is required for blood-brain barrier formation in drosophila. Cell 123:133–144
127. Seal RP, Daniels GM, Wolfgang WJ, Forte MA, Amara SG (1998) Identification and characterization of a cDNA encoding a neuronal glutamate transporter from Drosophila melanogaster. Recept Channels 6:51–64
128. Sousa-Nunes R, Yee LL, Gould AP (2011) Fat cells reactivate quiescent neuroblasts via TOR and glial insulin relays in Drosophila. Nature 471:508–512
129. Speder P, Liu J, Brand AH (2011) Nutrient control of neural stem cells. Curr Opin Cell Biol 23:724–729
130. Stenesen D, Moehlman AT, Kramer H (2015) The carcinine transporter CarT is required in Drosophila photoreceptor neurons to sustain histamine recycling. Elife 4:e10972
131. Stork T, Bernardos R, Freeman MR (2012) Analysis of glial cell development and function in Drosophila. Cold Spring Harb Protoc 2012:1–17
132. Stork T, Engelen D, Krudewig A, Silies M, Bainton RJ, Klambt C (2008) Organization and function of the blood-brain barrier in Drosophila. J Neurosci 28:587–597
133. Stout RF Jr, Parpura V (2011) Voltage-gated calcium channel types in cultured C. elegans CEPsh glial cells. Cell Calcium 50:98–108
134. Stout RF Jr, Verkhratsky A, Parpura V (2014) Caenorhabditis elegans glia modulate neuronal activity and behavior. Front Cell Neurosci 8:67
135. Sukhdeo SC, Sukhdeo MVK (1994) Mesehnchyme cells in Fasciola hepatica (Platyhelmintes): Primive glia? Tissue Cell 26:123–131
136. Sukhdeo SC, Sukhdeo MVK, Mettrick DF (1988) Neurocytology of the cerebral ganglion of *Fasciola hepatica* (Platyhelminthes). J Comp Neurol 278:337–343
137. Tasdemir-Yilmaz OE, Freeman MR (2014) Astrocytes engage unique molecular programs to engulf pruned neuronal debris from distinct subsets of neurons. Genes Dev 28:20–33
138. Thomas JH (1994) The mind of a worm. Science 264:1698–1699
139. Threadgold LT, Arme C (1974) Electron microscope studies of Fasciola hepatica. Expr Parasitol 35:389–405
140. True JR, Yeh SD, Hovemann BT, Kemme T, Meinertzhagen IA, Edwards TN, Liou SR, Han Q, Li J (2005) Drosophila tan encodes a novel hydrolase required in pigmentation and vision. PLoS Genet 1:e63
141. Tsacopoulos M, Veuthey AL, Saravelos SG, Perrottet P, Tsoupras G (1994) Glial cells transform glucose to alanine, which fuels the neurons in the honeybee retina. J Neurosci 14:1339–1351
142. Van Roy P, Briggs DE (2011) A giant Ordovician anomalocaridid. Nature 473:510–513
143. Verkhratsky A, Butt AM (2013) Glial physiology and pathophysiology. Wiley-Blackwell, Chichester
144. Verkhratsky A, Nedergaard M (2018) Physiology of astroglia. Physiol Rev 98:239–389
145. Viehweger E, Robitail S, Rohon MA, Jacquemier M, Jouve JL, Bollini G, Simeoni MC (2008) Measuring quality of life in cerebral palsy children. Ann Readapt Med Phys 51:119–137
146. Wadsworth WG, Bhatt H, Hedgecock EM (1996) Neuroglia and pioneer neurons express UNC-6 to provide global and local netrin cues for guiding migrations in C. elegans. Neuron 16:35–46
147. Ward S, Thomson N, White JG, Brenner S (1975) Electron microscopical reconstruction of the anterior sensory anatomy of the nematode Caenorhabditis elegans. J Comp Neurol 160:313–337
148. White JG, Southgate E, Thomson JN, Brenner S (1986) The structure of the nervous system of the nematode Caenorhabditis elegans. Philos Trans R Soc Lond B Biol Sci 314:1–340
149. Wuttke WA, Pentreath VW (1990) Evidence for the uptake of neuronally derived choline by glial cells in the leech central nervous system. J Physiol 420:387–408
150. Xiong WC, Montell C (1995) Defective glia induce neuronal apoptosis in the repo visual system of Drosophila. Neuron 14:581–590
151. Xiong WC, Okano H, Patel NH, Blendy JA, Montell C (1994) repo encodes a glial-specific homeo domain protein required in the Drosophila nervous system. Genes Dev 8:981–994

152. Xu K, Terakawa S (1993) Saltatory conduction and a novel type of excitable fenestra in shrimp myelinated nerve fibers. Jpn J Physiol 43(Suppl 1):S285–293
153. Xu K, Terakawa S (1999) Fenestration nodes and the wide submyelinic space form the basis for the unusually fast impulse conduction of shrimp myelinated axons. J Exp Biol 202:1979–1989
154. Yoshimura S, Murray JI, Lu Y, Waterston RH, Shaham S (2008) mls-2 and vab-3 Control glia development, hlh-17/Olig expression and glia-dependent neurite extension in C. elegans. Development 135:2263–2275
155. Zalc B, Goujet D, Colman D (2008) The origin of the myelination program in vertebrates. Curr Biol 18:R511–512
156. Zhang YV, Ormerod KG & Littleton JT. (2017). Astrocyte Ca^{2+} Influx Negatively Regulates Neuronal Activity. eNeuro 4
157. Ziegler AB, Brusselbach F, Hovemann BT (2013) Activity and coexpression of Drosophila black with ebony in fly optic lobes reveals putative cooperative tasks in vision that evade electroretinographic detection. J Comp Neurol 521:1207–1224
158. Zoran MJ, Drewes CD, Fourtner CR, Siegel AJ (1988) The lateral giant fibers of the tubificid worm, Branchiura sowerbyi: structural and functional asymmetry in a paired interneuronal system. J Comp Neurol 275:76–86

Chapter 3
Physiology of Astroglia

Alexei Verkhratsky, Vladimir Parpura, Nina Vardjan and Robert Zorec

Abstract Astrocytes are principal cells responsible for maintaining the brain homeostasis. Additionally, these glial cells are also involved in homocellular (astrocyte-astrocyte) and heterocellular (astrocyte-other cell types) signalling and metabolism. These astroglial functions require an expression of the assortment of molecules, be that transporters or pumps, to maintain ion concentration gradients across the plasmalemma and the membrane of the endoplasmic reticulum. Astrocytes sense and balance their neurochemical environment via variety of transmitter receptors and transporters. As they are electrically non-excitable, astrocytes display intracellular calcium and sodium fluctuations, which are not only used for operative signalling but can also affect metabolism. In this chapter we discuss the molecules that achieve ionic gradients and underlie astrocyte signalling.

Keywords Astrocytes · Brain homoeostasis · Neurotransmitter receptors · Ion channels · SLC transporters · Ca^{2+} signalling · Na^{+} signalling

---

A. Verkhratsky (✉)
Faculty of Biology, Medicine and Health, The University of Manchester, Manchester M13 9PT, UK
e-mail: Alexej.Verkhratsky@manchester.ac.uk

Faculty of Health and Medical Sciences, Center for Basic and Translational Neuroscience, University of Copenhagen, 2200 Copenhagen, Denmark

Achucarro Center for Neuroscience, IKERBASQUE, Basque Foundation for Science, 48011 Bilbao, Spain

V. Parpura
Department of Neurobiology, The University of Alabama at Birmingham, Birmingham, AL, USA

N. Vardjan · R. Zorec
Laboratory of Neuroendocrinology-Molecular Cell Physiology, Faculty of Medicine, Institute of Pathophysiology, University of Ljubljana, Ljubljana, Slovenia

Celica Biomedical, Ljubljana, Slovenia

A. Verkhratsky et al. (eds.), *Neuroglia in Neurodegenerative Diseases*,
Advances in Experimental Medicine and Biology 1175,
https://doi.org/10.1007/978-981-13-9913-8_3

3.1 Definition of Astroglia

Astroglia (also known as astrocytes) are a class of neural cells of ectodermal, neuroepithelial origin that sustain homeostasis and provide for defence of the central nervous system (CNS) (Fig. 3.1; [410]). The term astrocyte (αστρον κψτοσ; *astron, star and kytos, a hollow vessel*, later *cell*; that is a star-like cell) was introduced by Michael von Lenhossék in 1895 [213]; of note, he proposed to name all parenchymal glia spongiocytes, with only a subtype of these cells having characteristic morphology in Golgi's stained preparations being identified as astrocyte. On this matter Lenhossék wrote: 'I would suggest that all supporting cells be named spongiocytes. And the most common form in vertebrates be named spider cells or astrocytes, and use the term neuroglia only *cum grano salis* (with a grain of salt), at least until we have a clearer view'. The terms of protoplasmic (white matter) and fibrous (grey matter) glia were introduced by Albert von Kölliker and William Lloyd Andriezen [13, 195]).

Astrocytes demonstrate quite heterogeneous morphology across different brain structures (see Chap. 1). Nevertheless, the main physiological features of astroglial cells are somewhat similar, being specifically tailored for their homeostatic function. Astrocytes maintain homeostasis of the CNS at all levels of organisation [408–410] from molecular (ion and transmitter homeostasis, regulation of pH, metabolic energy support, Fig. 3.1), cellular (neurogenesis), network (synaptogenesis and synaptic maturation, maintenance and extinction), organ (regulation of the blood-brain barrier, operation of the glymphatic system) and systemic (chemosensing of oxygen, CO_2 and systemic Na^+ concentration).

In the CNS, astrocytes are integrated into cellular networks (known as syncytia), by gap junctions, which are specialised areas of apposing membranes of adjacent cells pierced by many hundreds of intercellular channels or connexons that form the conduit for intercellular transport of ions, second messengers and other biologically active molecules with a molecular weight lesser than 1000 Da. In the mammalian

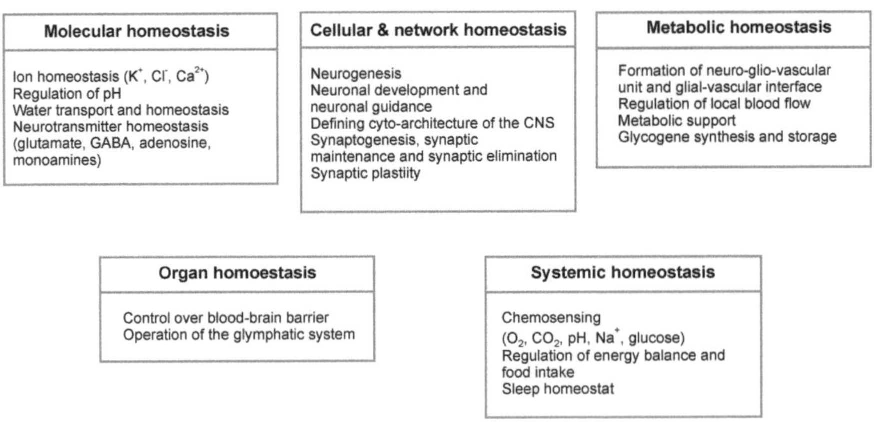

Fig. 3.1 Homoeostatic functions of astroglia

CNS, astroglial syncytia are anatomically segregated within different anatomical structures. In the sensory cortex, astroglial syncytia are confined to individual barrels and in the olfactory bulb to individual glomeruli [118, 148, 332]. Panglial syncytia that connect astrocytes and oligodendrocytes have been identified in the thalamus, neocortex and hippocampus [59, 123, 181, 296]. Whether astroglia form syncytia with neurones, remains an open issue.

3.2 Membrane Physiology and Ion Distribution

Astrocytes are electrically non-excitable cells, with a rather negative resting membrane potential (V_m) of about −80 mV. Disparity between cytosolic and extracellular ion concentrations (Fig. 3.2), together with the specific membrane ion permeability, define this negative V_m of astrocytes. At rest intra-astrocytic concentration of K^+ is between 120 and 140 mM and extracellular K^+ concentration is about 3 mM, which sets the equilibrium potential for K^+ (E_K) at −98 mV (at 37 °C). Concentration of cytosolic Na^+ in astrocytes (15–20 mM) is generally higher than in the majority of neurones (8–10 mM). With Na^+ concentration in the cerebrospinal fluid (CSF) around 145–155 mM, the corresponding E_{Na} ranges between +55 and +60 mV [192, 331]. Concentration of ionised Ca^{2+} in the cytosol of astrocytes ranges between 50 and 150 nM, which for extracellular [Ca^{2+}] of 1.4 mM sets the $E_{Ca}{}^{2+}$ at +120 to +

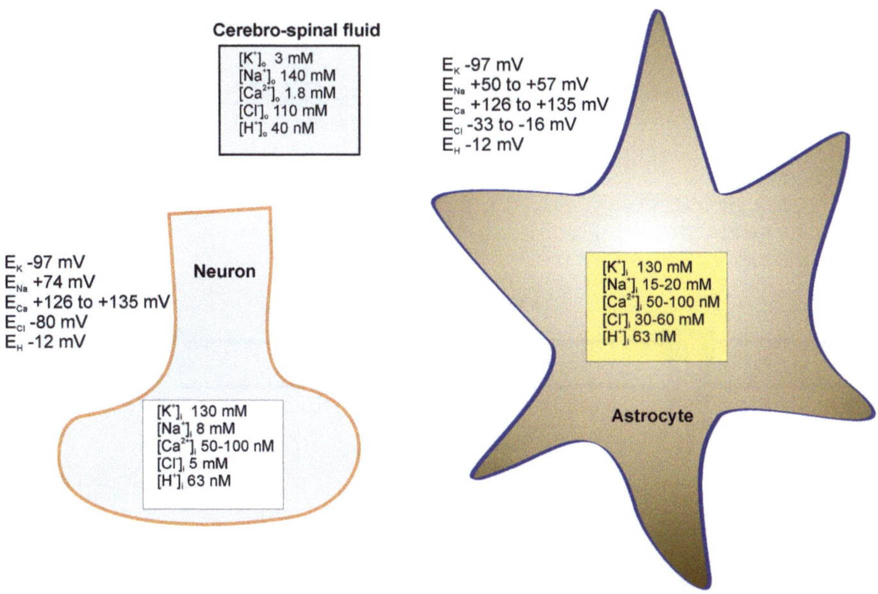

Fig. 3.2 Ion distribution (and corresponding values of equilibrium potentials for different ions) between the cerebrospinal fluid and cytosol of astrocytes and neurones. Modified from [413]

140 mV. Cytosolic concentration of free Mg^{2+} in cultured astrocytes measured with a fluorescent probe Mag-Fura-2 is around 125 μM [20]; the CSF Mg^{2+} has been determined at ~0.9 mM [384] giving $E_{Mg}{}^{2+}$ ~ +25 mV.

High cytosolic Cl^- concentration (30–50 mM) has been measured in cultured astrocytes and Bergmann glial cells in cerebellar slices [403]; this sets the $E_{Cl}{}^-$ around −35 mV (the $[Cl^-]_o$ is ~120 mM). The concentration of protons in astroglial cytosol is ~63 nM (pH 7.2), which, assuming the extracellular H^+ concentration to be ~40 nM (pH 7.4) sets the $E_H{}^+$ at ~−12 mV. The cytosol of astrocytes is rich in CO_2 (~1.2 mM) and $HCO_3{}^-$ (~17 mM).

The most characteristic electrophysiological signature of mature astrocytes is hyperpolarised resting potential (~−80 mV) and low input resistance (5–20 MΩ) indicative of high resting membrane permeability for K^+ [242, 243]; the current to voltage relationship of astroglial cells is nearly linear [2, 73, 161, 168]. Fluctuations of astroglial V_m generally reflect changes in extracellular K^+ concentration [11, 81].

3.3 Ion Channels

3.3.1 Potassium Channels

Glial membrane permeability is dominated by K^+ channels, several types of which are expressed in astrocytes (Fig. 3.3). These channels have distinct voltage-dependence, which covers the whole range of physiological membrane potentials, thus ensuring

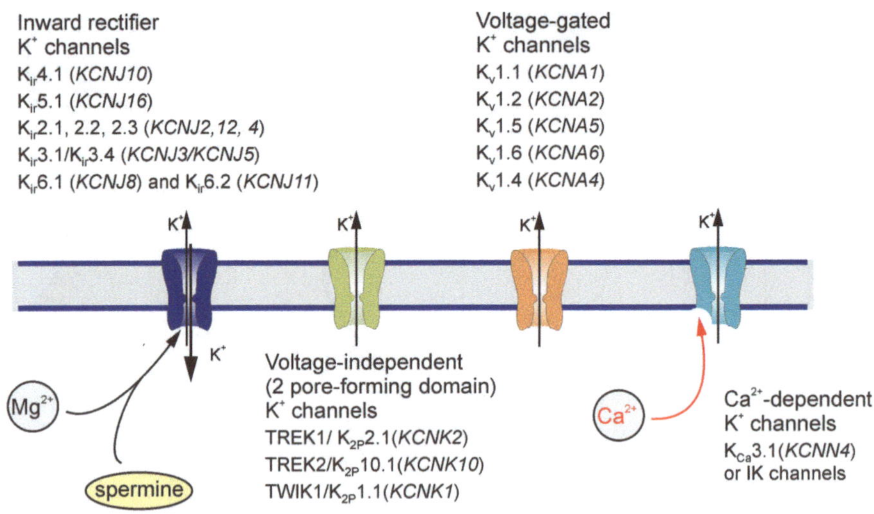

Fig. 3.3 Potassium channels in astroglia. Gene names for a given channels are shown in parentheses. Modified from [413]

the passive properties of astroglial plasma membrane. This prominent K^+ conductance of astrocytic plasmalemma defines the homeostatic capabilities of astrocytes [269], supporting movement of ions and providing electrical driving force for membrane transporters.

3.3.1.1 Inward Rectifier Potassium Channels, K_{ir}

Inward rectifying K^+ channels are so-called because they pass K^+ ions more easily into the cell (the inward direction) than out of the cell (the outward direction). The inward rectification occurs due to a voltage-dependent intracellular block by Mg^{2+} and polyamines [225]. These two transmembrane domain channels belong to a gene super-family represented by 16 subtypes (*KCNJ1–KCNJ18*), which are further divided into 7 families from $K_{ir}1.x$ to $K_{ir}7.x$ [139].

The main astroglial inward rectifying K^+ channels are represented by the $K_{ir}4.1$ subtype (product of *KCNJ10 gene*). These channels are detected in many types of astroglia including protoplasmic and fibrous astrocytes, from all brain areas including hippocampus, neocortex, optic nerve, cerebellum, spinal cord and retina [58, 140, 169], with some regional variations; astroglia such as Müller glia and Bergmann glia express them as well. High $K_{ir}4.1$ immunoreactivity was detected in hippocampal astrocytes, in astroglial cells in the cerebral cortex, in the deep cerebellar nuclei, in Bergmann glia and in Müller cells but not in astrocytes in white matter [309]. In the spinal cord, the expression of $K_{ir}4.1$ channels is the highest in astrocytes from the ventral horn and the lowest in astrocytes from the apex of the dorsal horn [273]. The $K_{ir}4.1$ channels are major contributors to the resting membrane potential of astroglial cells: functional inhibition or genetic deletion of K_{ir} currents markedly increases input resistance (up to 20-fold) and depolarises astrocytes (by ~20 mV) [273, 350].

The $K_{ir}4.1$ may co-assemble with $K_{ir}5.1$ channels forming heteromers, which were found in parenchymal and radial astrocytes, in the olfactory bulb, neocortex, cerebellum and retina [50, 138, 159, 251]. The $K_{ir}4.1/K_{ir}5.1$ heteromeric channels are concentrated in perisynaptic and perivascular processes of astrocytes and their processes close to the pia mater [138]. Some astrocytes in the hippocampus and cerebellum (including Bergmann glia) along with retinal Müller glia were found to express $K_{ir}2.1$, $K_{ir}2.2$ and $K_{ir}2.4$ channels [194, 214, 317, 380]. Astrocytes also express ATP-sensitive inward rectifying K^+ channels assembled from $K_{ir}6.x$ subunit and SUR1/2; these channels open upon intracellular ATP depletion [96, 174, 365, 395, 431].

3.3.1.2 Voltage-Independent K^+ Channels

Another type of K^+ channels, which contributes to the resting membrane permeability of astrocytes is represented by members of the two-pore-domain potassium channels (K_{2P}) family encoded by 15 KCNK genes [100]. Hippocampal astrocytes were found to express functional TREK1/$K_{2P}2.1$(*KCNK2*), TREK2/$K_{2P}10.1$(*KCNK10*) and TWIK1/$K_{2P}1.1$(*KCNK1*) channels [350, 432]. In cultured cortical astrocytes and astrocytes in hippocampal slices expression of TWIK-1/TREK-1 heterodimer (formed by disulphide bridge between cysteine-cysteine residuals of both subunits) has been demonstrated as the predominant channel type [156].

3.3.1.3 Voltage-Gated K^+ Channels, K_v

Astrocytes express delayed rectifying (K_D) and transient (K_A) voltage-gated K^+ channels. The delayed rectifying K^+ currents were identified in astrocytes throughout the CNS including the cortex, hippocampus, cerebellum and spinal cord [48]. At the molecular level, $K_v1.5$ (*KCNA5*), $K_v1.4$ (*KCNA4*) and $K_v11.1$/ERG1 (KCNH2) have been identified in astrocytes from the hippocampus and spinal cord [97, 98, 334]. Astrocytes are also in possession of fast (rapidly activating and inactivating) A-type K^+ currents mediated by K_v1, K_v3 and K_v4 channels [32].

3.3.1.4 Ca^{2+}-Dependent K^+ Channels, K_{Ca}

Several types of Ca^{2+}-dependent K^+ channels (K_{Ca}) were found in astrocytes in vitro and in situ. At the mRNA level SK (small conductance K_{Ca} 2.3/KCNN3) and IK (intermediate conductance $K_{Ca}3.1$/KCNN4) channels were detected in mouse cortical astrocytes in acutely isolated slices [222]. The $K_{Ca}2.3$ immunoreactivity was found in astroglial processes in the rat supraoptic nucleus [17]. BK (big conductance) channels ($K_{Ca}1.1$/KCNMA1) were identified in the perivascular astroglial endfeet in the hippocampus and cerebellum [315], whereas patch-clamp recordings obtained from endfeet revealed large-conductance (225 pS) BK-single channel currents [107].

3.3.2 *Sodium Channels*

3.3.2.1 Voltage-Gated Sodium Channels

Although being non-excitable cells, astrocytes express voltage-gated Na^+ channels (Na_v), which were detected in vitro and in situ, albeit at low density [26, 28, 42, 371–374]. Cultured astrocytes from the optic nerve, hippocampus and spinal cord were found to express fast tetrodotoxin (TTX)-sensitive and slow TTX-resistant Na^+ currents [373, 374]. At the molecular level astrocytes mainly express $Na_v1.5$ subunit which was identified both in vitro and in situ at mRNA and protein levels [40, 41, 286]; relatively low expression of $Na_v1.2$, $Na_v1.3$ [44] and $Na_v1.6$ [321, 345] were also detected; incidentally expression of $Na_v1.6$ channels was found to increase in reactive astroglia [434].

Distribution of Na^+ channels in astrocytes from different brain regions as well as their physiological role remain poorly understood; hitherto Na^+ currents have not been recorded from astroglial cells in vivo. Possibly voltage-gated Na^+ channels contribute to Na^+ signalling; it was also suggested that Na^+ influx mediated by these channels is needed for sustained activity of Na^+/K^+ pump (see [43, 285, 287] for further details).

3.3.2.2 $[Na^+]_O$—Regulated Na^+ Channels, Na_x

Specific type of Na^+ channels regulated by extracellular Na^+ concentration ($[Na^+]_o$) are named Na_x channels. They are expressed in astrocytes of the subfornical organ and organum vasculosum of the lamina terminalis (that are parts of circumventricular organs surrounding ventricles, structures where the blood-brain barrier is not as tight as in other parts of the brain) [267, 420]. These channels in vitro are opened following an increase in $[Na^+]_o$ to 150 mM. In vivo, in the presence of endothelin-3, which activates ET_B receptors expressed in astroglia, the threshold for Na_x channel activation is lowered to 140 mM of $[Na^+]_o$ [141]. The Na_x channels appear to operate as molecular sensors for $[Na^+]$ in the circulation (Fig. 3.4).

3.3.2.3 Epithelial Sodium Channel, ENaC

The epithelial Na^+ channels are non-voltage gated, amiloride-sensitive Na^+ channels widely expressed in the CNS [10, 238, 416]. Immunohistochemistry found strong expression of ENaCs in astrocytes in circumventricular organs, white matter and pia mater [237]. These channels, together with Na_x channels, may be involved in the regulation of systemic Na^+ homeostasis.

3.3.3 Calcium Channels

3.3.3.1 Voltage-Gated Ca^{2+} Channels

Early patch clamp recordings from astrocytes in culture revealed Ba^{2+} and Ca^{2+} currents sensitive to classic voltage-gated Ca^{2+} channel antagonists and enhanced by norepinephrine or by an increase in cytosolic 3′,5′-cyclic adenosine monophosphate, i.e. cAMP [27, 72, 228], the signalling by this second messenger complementing Ca^{2+} signalling in astrocytes [146]. Subsequently $Ca_v1.2$ and $Ca_v1.3$ channels were detected in the transcriptome of rodent cortical astrocytes [60, 430]. Several types of Ca^{2+} channel subunits (α1B or N-type, α1C/D or L-type, α1E or R-type and α1G or T-type) were detected at mRNA and protein levels in astrocytes in culture [209]. Pituicytes analysed immunohistochemically in situ were found to possess $Ca_v2.2$ (N-type) and $Ca_v2.3$ (R-type) channels [418]. Evidence for functional activity of

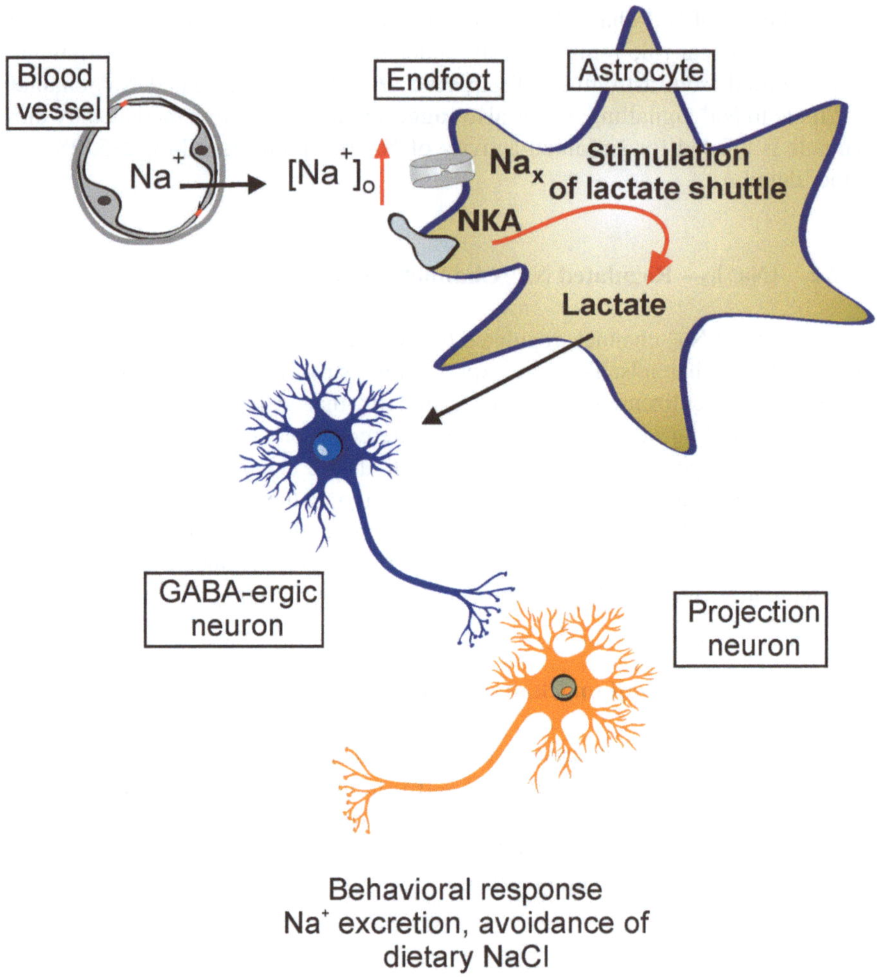

Fig. 3.4 Astroglial Na_x channels in systemic Na^+ regulation. Increases in blood Na^+ concentration activate Na_x sodium channels localised in astrocytes residing in the subfornical organ. This leads to an increase in cytosolic Na^+ concentration, which in turn increases astroglial production of lactate. Lactate released by astrocytes is accumulated by neighbouring neurones (release and uptake carried by MCT1 in astrocytes and MCT4 in neurones, respectively), thus increasing ATP production in neurones. Increased ATP in turn closes neuronal ATP-sensitive K^+ channels, which results in depolarisation and subsequent activation of neuronal networks responsible for systemic Na^+ homeostasis. NKA, sodium-potassium ATPase. Modified from [413]

voltage-gated Ca^{2+} channels in situ is rather thin and indirect (see for example [215, 294]) and no evidence from the in vivo experiments exist at all. There are some indications for increased expression of voltage-gated Ca^{2+} channels in pathological or reactive astrocytes [419, 421].

3.3.3.2 Orai or Ca^{2+}-Release Activated Ca^{2+} Channels

The Ca^{2+}-release activated Ca^{2+} channels of Orai family (Orai1,2,3) represent one of the main molecular pathways of the store-operated Ca^{2+} entry (SOCE) in non-excitable cells [289]. Activation of these plasmalemmal channels is controlled by the stromal interacting molecules, STIM1 and STIM2, which act as Ca^{2+} sensors of the endoplasmic reticulum (ER) [368]. In cultured astrocytes Orai1 and STIM1 were detected at a protein level; over-expression of Orai1 increased the amplitude of SOCE, whereas siRNA knock out decreased SOCE [248]. Electrophysiological recordings of I_{CRAC} performed on acutely dissociated Müller cells demonstrated sensitivity of the current to ORAI inhibitor Synta 66 [246].

3.3.3.3 Ca^{2+} Release Channels

Release of Ca^{2+} from the ER store in astrocytes is mainly mediated by Inositol 1,4,5-trisphosphate receptors ($InsP_3R$), of which type 2 predominates [133, 144, 354, 356, 414]. Genetic deletion of $InsP_3R$ type 2 ($InsP_3R2$) has been shown to significantly reduce or even completely abolish Ca^{2+} signalling in astroglial cells from the hippocampus and cortex [171, 302]. Other studies, however, reported $[Ca^{2+}]_i$ transients in $InsP_3R2^{-/-}$ mice astrocytes [130]. There is evidence for functional expression of $InsP_3Rs$ type 1 and 2 in astroglia [125, 327, 357].

Ryanodine receptors also have been identified in astrocytes [364, 414], although their functional role remains unclear. They supply Ca^{2+} necessary for Ca^{2+}-dependent glutamate release from cortical astrocytes in culture. Ca^{2+}-dependent glutamate release involves two classes of ER Ca^{2+} stores in astrocytes [150]. There is some evidence for astroglial expression of the two-pore channels (TPC) that release Ca^{2+}, and are activated by nicotinic acid adenine dinucleotide phosphate, i.e. NAADP [25, 301].

3.3.4 Transient Receptor Potential (TRP) Channels

Astrocytes express several types of cationic channels of TRP (transient receptor potential) family (Fig. 3.5). The 'ankyrin' channel TRPA1 was found in somata and processes of astrocytes in the brain stem in the rat trigeminal caudal nucleus using immunogold electron microscopy [211]. Functional expression of TRPA1 was also demonstrated in a sub-population of hippocampal astrocytes [360, 362]. The TRPC

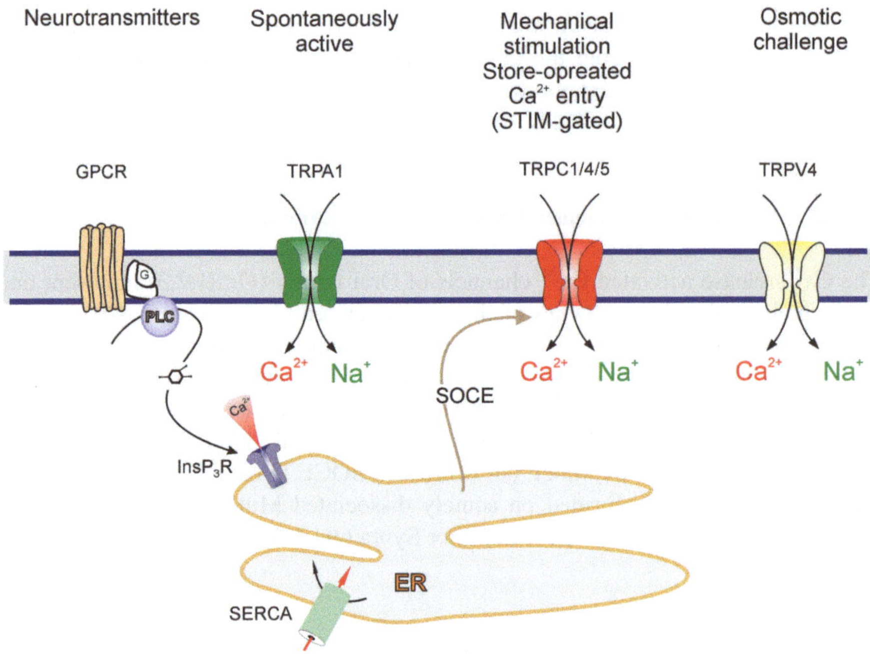

Fig. 3.5 Astroglial TRP channels. Activation of G-protein coupled receptors (GPCR), i.e. metabotropic stimulation, can lead to production of $InsP_3$ and release of Ca^{2+} from the ER store. The Ca^{2+} content of the ER store is refilled by Sarco(Endo)Plasmic Reticulum Ca^{2+} ATPase, i.e. SERCA. Depletion of the ER Ca^{2+} store activates (via STIM) TRPC channels in astrocytes which are therefore acting as a store-operated channel, contributing to capacitative Ca^{2+} entry. Activation of all TRP channels mediates Ca^{2+} and Na^+ influx. Modified from [413]

('canonical') channels were detected in freshly isolated and in primary cultured astrocytes, which were reported to express all subtypes of these channels from TRPC1 to TRPC6 [124, 244, 308]. These TRPC channels contribute to astroglial Ca^{2+} signalling induced by purinergic, glutamatergic and mechanical stimulation [231, 323, 324]. Astroglial TRPC channels represent a substantial pathway for store-operated Ca^{2+} entry in astroglia [412].

Rodent astrocytes from the brain and the spinal cord have been also reported to express TRPV1 channels [91, 151, 398]. Similarly TRPV4 channels have been identified in cortical and hippocampal astrocytes [22, 33, 57, 220]; these channels can be activated by hypo-osmotic stress and by cell swelling [33, 57] in the absence of addition of membrane by exocytosis [283].

3.3.5 *Hyperpolarisation-Activated Cyclic Nucleotide-Gated (HCN) Channels*

The cationic (Na^+/K^+) HCN channels are found in both healthy and reactive astrocytes in situ [145, 337].

3.3.6 *Acid-Sensitive Ion Channels*

The acid-sensitive ion channels (ASIC1-3) were described in reactive astrocytes in the context of chronic epilepsy, and their activation was claimed to contribute to seizure generation [423].

3.3.7 *Anion Channels*

Astrocytes express the following anion channels: (i) cystic fibrosis transmembrane conductance regulator or CFTR channels, (ii) voltage-dependent anion-selective channels or VDAC, (iii) Ca^{2+}-dependent Cl^- channels (iv) volume-regulated anion channels or VRAC and (v) ClC-1, -2 and -3 channels [186, 291, 410, 429]. The ClC-2 channels are concentrated in astroglial processes enwrapping GABAergic synapses [363], which may indicate their role for regulation of intra-cleft Cl^- concentration and hence of GABAergic transmission. Astrocytes also express an anion channel of Bestrophin (*Best*) family; these channels were suggested to contribute to the Ca^{2+}-dependent secretion of glutamate and GABA [290, 422].

3.3.8 *Aquaporins*

Three types of aquaporins, the AQP1, AQP4 and AQP9 were identified in astroglia although AQP4 is the most abundant [21, 256, 342]; the plasma membrane contains mainly the AQP4e isoform [219, 313]. Astroglial AQP4 channels are concentrated in the perivascular and subpial endfeet [256]. Genetic deletion of AQP4 affects olfaction [224] and hearing; it results in a decrease in astroglial water permeability [370], deficient K^+ buffering, compromised volume regulation [39, 219], and deficits in synaptic plasticity [346, 367], and memory [367, 427].

3.3.9 Connexons

Connexons form gap junctional channels that integrate astrocytes into functional syncytia, subject to regulation by G-protein coupled receptors [392]. Astrocyte-astrocyte homocellular gap junctions are composed of Cx26, Cx30 and Cx43 [117, 198, 258], of which Cx43 is the most abundant [257]. The Cx43 is expressed in astroglial cells in all CNS regions, whereas Cx30 is mostly expressed in the thalamus and leptomeninges [257, 369]. The Cx26 subtype has been detected in astrocytes in the hypothalamus, reticular thalamic and subthalamic nuclei [259]. Astrocyte-oligodendrocyte heterocellular gap junctions are composed of heterotypic channels represented, in vitro, by four complexes: Cx47/Cx43, Cx47/Cx30, Cx32/Cx30 or Cx32/Cx26 [230], although in situ Cx32/Cx30 and Cx47/Cx43 complexes appear to predominate [8, 277]. There are some sporadic reports of astrocyte-neuronal heterocellular contacts [9, 255, 279], and these contacts are arguably limited to developing brain.

Unpaired connexons, or hemichannels, have been identified in astrocytes in vitro and in vivo; all three major connexons (Cx26, Cx30 and Cx43) expressed in astrocytes can act as hemichannels [119]. The hemichannels are non-operational in healthy astrocytes, but can be activated by low external calcium concentration, by substantial depolarisation, by some specific intracellular Ca^{2+} signals, or by exposure to pro-inflammatory agents [274, 275]. The hemichannels can contribute to secretion of neurotransmitters and neuromodulators [293] and are subject to regulation via transmitters and G-protein coupled receptors.

3.3.10 Pannexons

Transcripts for pannexin 1 (Panx1) were identified in astroglia in vitro and in situ [153, 320], with Panx1 currents characterised in cultured cortical astrocytes [157]. Astroglial pannexons are activated by voltage and by activation of $P2X_7$ receptors; they are inhibited by broad-spectrum gap junction antagonists carbenoxolone and mefloquine, and they are permeable to fluorescent tracer YoPro [157]. Panx1 containing pannexons have been considered as a transmembrane conduit for ATP [78].

3.4 Receptors

Conceptually, astrocytes have been shown to express virtually any type of receptor found in the CNS. At the same time the pattern of receptors expressed by astrocytes in situ and in vivo is restrictive and is regulated by the local neurochemical environment [183, 410, 411].

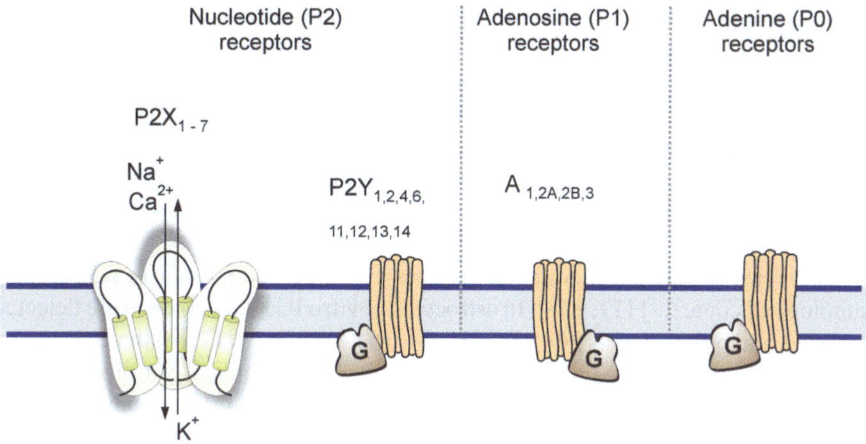

Fig. 3.6 Classes of purinoreceptors. ATP after being released from neurones and glia is rapidly degrading by ectonucleotidases into ADP, AMP and adenosine, which act on P1 metabotropic adenosine receptors, P2X ionotropic and P2Y metabotropic nucleotide receptors. Adenine stimulates A0 adenine metabotorpic receptors, which hitherto have not been detected in astrocytes. Modified from [406]

3.4.1 Purinoceptors

3.4.1.1 Adenosine Receptors

All four types of adenosine receptors (A_1, A_{2A}, A_{2B} and A_3) were found in astrocytes in the in vitro and in situ preparations [84]. These receptors are linked to intracellular second messenger systems including $InsP_3$, Ca^{2+} and cAMP cascades (Fig. 3.6). Activation of A_1 receptors triggered intracellular Ca^{2+} release as well as Ca^{2+} entry and potentiated histamine-induced Ca^{2+} mobilisation [297, 310]. Similarly, A_{2A} receptors were responsible for Ca^{2+} signalling in astrocytes from the olfactory bulb [90], whereas A_{2B} receptors triggered Ca^{2+} signals in cortical astroglia [306]. The A_3 receptors mediated adenosine- and guanosine-evoked $[Ca^{2+}]_i$ transients in cultured mouse astrocytes [68].

3.4.1.2 P2X Purinoceptors

Transcripts for all seven types of ionotropic purinoceptors ($P2X_1$–$P2X_7$, Fig. 3.6) have been detected in astrocytes in vitro and in tissue extracts [89, 109, 113, 162]. At the protein level, $P2X_{2,3,4}$ receptors were found in astrocytes from the nucleus accumbens [109]. $P2X_1$ and $P2X_2$ receptors were found in astroglial cells in the cerebellum [173, 221]; $P2X_4$ receptors were identified in astrocytes from the brainstem [18] and in Müller glia [142], whereas hippocampal astrocytes were immunoreactive for $P2X_{1\text{-}4}$, $P2X_6$ and $P2X_7$ receptors [197].

Ion currents mediated by heteromeric $P2X_{1/5}$ receptors were characterised in cortical mouse astrocytes [204]). These receptors contribute to 'glial synaptic currents' monitored in astrocytes in response to stimulation of neuronal afferents [200, 203]; in addition $P2X_{1/5}$ receptors produced spontaneous 'miniature' post-synaptic currents in astrocytes in cortical slices [203]. Astroglial $P2X_{1/5}$ receptors have intermediate Ca^{2+} permeability ($P_{Ca}{}^{2+}/P_{monovalent} \sim 2.2$), and their activation by endogenous agonists, or by synaptically released ATP, triggers transient cytoplasmic Ca^{2+} signals [281].

Astrocytes have been reported to express $P2X_7$ receptors both in healthy and pathological contexts [111, 158]. In astrocytes in vitro $P2X_7$ receptors were detected at mRNA and protein levels [89, 93, 113, 154, 164, 260, 282, 417]. Astroglial expression of $P2X_7$ receptors, as a rule, increases after brain injury of various aetiology [109, 110, 260]. In astrocytes in culture both $P2X_7$-mediated Ca^{2+} signals and membrane currents have been detected [93, 113, 266, 268, 335]. Astroglial $P2X_7$ currents were also characterised in rat and mouse cortical slices. Activation of $P2X_7$ receptors in cultured astrocytes may be associated with release of glutamate, GABA, ATP [24, 93, 94, 382]. The $P2X_7$-mediated release of glutamate was also identified in astrocytes in hippocampal slices [106].

3.4.1.3 P2Y Receptors

Astrocytes in cortex express transcripts for $P2Y_{1,2,4,6,12,13}$ and UDP-glucose $P2Y_{14}$ receptor [1, 36, 89, 113], whereas spinal cord astrocytes predominantly express mRNA for $P2Y_{1,2}$ receptors [104]. Stimulation of astroglial P2Y receptors triggers Ca^{2+} signals originating from $InsP_3$-induced ER Ca^{2+} release [38, 56, 160, 163, 175, 298, 299, 407].

3.4.2 *Glutamate Receptors*

3.4.2.1 Ionotropic Glutamate Receptors

Astrocytes from different regions of the brain express α-amino-3-hydroxy-5-methyl-isoxazole propionate (AMPA) receptors (Fig. 3.7), which have been characterised at expression and functional levels. All main subunits of AMPA receptors (GluA1–GluA4) have been detected in astroglial cells. In the hippocampus, AMPA receptors are assembled predominantly from GluA2 and GluA4 subunits, which are reflected by a linear I-V relation and low Ca^{2+} permeability [351]. In cortical astrocytes these receptors are composed of GluA1 and GluA4 subunits [75]. In Bergmann glial cells AMPA receptors do not contain GluA2 subunit, and accordingly they have a double-rectifying I–V relationship and low ($P_{Ca}{}^{2+}/P_{monovalent} \sim 1$) Ca^{2+} permeability [115, 253]. Conditional deletion of AMPA receptors, composed of GluA1 and GluA4

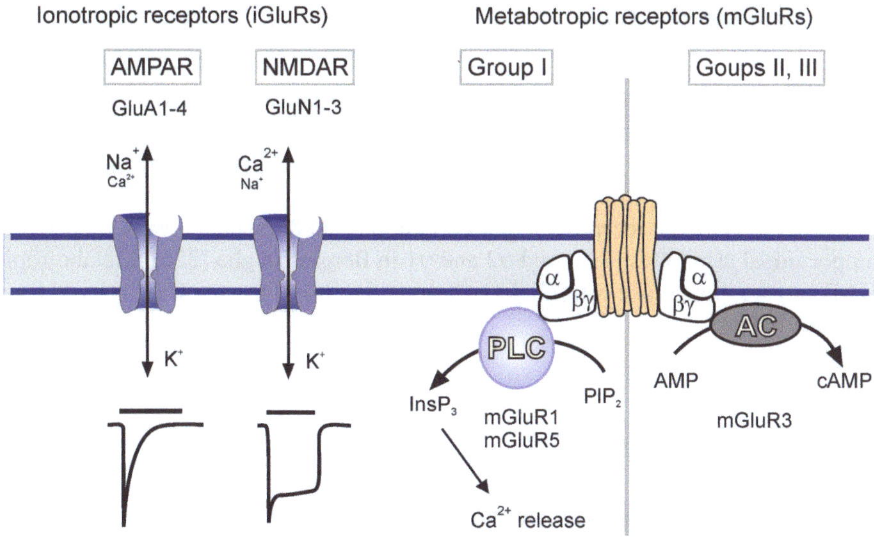

Fig. 3.7 Classes of glutamate receptors expressed in astrocytes. Current traces show a faster time course for AMPAR than NMDAR. AC, adenylyl cyclase; AMP, adenosine monophosphate; cAMP, cyclic AMP; $InsP_3$, inositol 1,4,5-trisphosphate; PIP_2, phosphatidylinositol 4,5-bisphosphate; PLC, phospholipase C. Modified from [406]

subunits, from Bergmann glia led to a retraction of glial perisynaptic processes and deficient fine motor coordination [338].

Astrocytes also possess functional N-methyl D-aspartate (NMDA) receptors. The transcriptome of human astroglia contains all 7 NMDA receptors subunits (GluN1, GluN2/A-D and GluN3A,B—[210]). In acute slices the NMDA-mediated currents and Ca^{2+} signals were found in astrocytes in the neocortex [202, 272, 348], in the spinal cord [437], and in some cells in the hippocampus [311, 376]. Astroglial NMDA receptors heterotetramers assembled from obligatory two GluN1 and additional subunits of each GluN2 C or D and GluN3; this composition underlies a weak Mg^{2+} block (which develops at ~−120 mV) and relatively low Ca^{2+} permeability ($P_{Ca}/P_{monovalent}$ ~ 3), as well as sensitivity to memantine and GluN2C/D subunit-selective antagonist UBP141 [95, 203, 280, 281].

3.4.2.2 Metabotropic Glutamate Receptors

The most abundant type of astroglial metabotropic receptors in mature CNS is represented by mGluR3, which inhibits adenylyl cyclase [385]. In younger animals astrocytes express mGluR1/5 receptors linked to Ca^{2+} signals and $[Ca^{2+}]_i$ oscillations [76, 191, 208, 312].

3.4.3 GABA Receptors

Astrocytes express both ionotropic $GABA_A$ and metabotropic $GABA_B$ receptors. The $GABA_A$ receptors mediated Cl^- currents have been characterised in astroglial cells in culture and in situ in the hippocampus, cerebellum, retina, hypothalamus, supraoptic nucleus and spinal cord [74, 179, 180, 182, 229, 252]. The subunit composition of astroglial $GABA_A$ receptors is not certain; $\alpha 1$ and $\beta 1$ subunits were detected in hippocampal astrocytes [112] and $\alpha 2$ and $\gamma 1$ in Bergmann glia [326]. Metabotropic $GABA_B$ receptors evoke astroglial Ca^{2+} signalling by triggering Ca^{2+} release from the ER [264].

3.4.4 Glycine Receptors

Glycine receptors mediate Cl^- currents in astrocytes in the spinal cord slices [295]. Single-cell RT-PCR performed on these astrocytes revealed expression of $\alpha 1$ and (in ~50% of cells) β-subunits of the receptor [188].

3.5 Acetylcholine Receptors

The ionotropic nicotinic acetylcholine receptors (nAChRs) have been characterised in astrocytes in culture and in acute slices. Activation of these receptors mediates Ca^{2+} influx and triggers Ca^{2+}-induced Ca^{2+} release [271, 353, 389]. Analysis of mRNA expression in rodent cortical astrocytes found $\alpha 4$, $\alpha 7$ and $\beta 2$ subunits, whereas in human astroglial cells from the hippocampus and entorhinal cortex $\alpha 3$, $\alpha 7$ and $\beta 4$ subunits were identified by immunocytochemistry [121, 390].

The metabotropic M_1 and M_2 AChRs mediate Ca^{2+} signalling in astrocytes from hippocampal slices [15].

3.6 Receptors for Monoamines

Astroglial cells express receptors for major monoamines including adrenoceptors, and receptors for serotonin, dopamine and histamine. Both α- and β-adrenoceptors have been identified and characterised in astrocytes in culture, in slices, and in vivo at transcript, protein and functional levels [14, 135]. The α_1-adrenoceptors are coupled to phospholipase-C (PLC) and $InsP_3$ signalling and hence to Ca^{2+} release from the ER [55, 193, 352]. Immunoreactivity for α_2-adrenoceptors was found in astrocytic processes in the brain tissue [14, 239]; activation of α_2-adrenoceptor also triggers Ca^{2+} signalling [265, 339]. β_1-, β_2- and β_3-adrenoceptors were characterised in astrocytes in vitro and in vivo [62, 232, 352]. β_1-adrenoceptors contribute to regulation of glycogen synthesis, while β_2-adrenoceptors coupled to adenylyl cyclase through G_s proteins together with possibly β_3-adrenoceptors regulate glucose uptake by modulating GLUT1 plasmalemmal glucose transporter [92, 155].

Astrocytes express 5-HT_{1A}, 5-HT_{2A}, 5-HT_{2B} and 5-HT_{5A} metabotropic serotonin receptors, [19, 61, 341]. 5-HT_2 receptors activate PLC/$InsP_3$/Ca^{2+} signalling cascade [341]. 5-HT_{2B} receptor seems rather abundant in astrocytes [196] with its expression as twice as large as in neurones [428]. Serotonin-specific reuptake inhibitors (major anti-depressant agents such as fluoxetine or sertraline) directly activate astroglial 5-HT_{2B} receptors [136, 428] with subsequent Ca^{2+} signalling [347] or phosphorylation of extracellular regulated kinases 1/2 (ERK1/2) or up-regulation of Ca^{2+}-dependent phospholipase A2 (cPLA2) [136, 428].

Dopamine D_1, D_2, D_4 and D_5 receptors have been detected in astroglia at transcript and protein levels [245] with higher expression of D_1 [425] or D_2 receptors [23]; astrocytes from the striatum were claimed to express D_5 receptors [52]. Strong presence of D_2 receptors was found in astroglial processes enwrapping cortical interneurones [184]. Activation of D_1 and D_2 receptors in astrocytes triggers Ca^{2+} signalling originating from the $InsP_3$-induced ER Ca^{2+} release [184, 322].

Astroglia express H_1, H_2 and H_3 histamine receptors [166]. Astrocytic H_1 receptors are coupled to PLC/$InsP_3$/Ca^{2+} signalling cascade [193, 355], regulation of glucose metabolism [16] and up-regulation of EAAT2/GLT-1 plasmalemmal glutamate transporter [105]; the latter also modulated by cytosolic Ca^{2+} [377].

3.7 Bradykinin Receptors

The B_2 bradykinin receptors were identified in cultured astrocytes [71]; stimulation of these receptors induces $InsP_3$ production, Ca^{2+} signalling and glutamate release [292, 378].

3.8 Cannabinoid Receptors

Astrocytes express cannabinoid CB_1 receptors, which are involved in regulation of cellular metabolism [46, 340]. The CB_1-mediated astroglial Ca^{2+} signalling was detected in response to neuronal release of endocannabinoids [261]. Activation of CB_1 receptors in astroglia was also claimed to regulate neuronal synaptic plasticity [262].

3.9 Neuropeptide Receptors

V_1 vasopressin receptors induced Ca^{2+} signalling was found in cultured astrocytes and in pituicytes [129, 167]. Oxytocin receptors linked to PLC/$InsP_3$/ER Ca^{2+} release were identified in rat embryonic cultured hippocampal astrocytes [87] and in hypothalamic astrocytes [199].

ET_A and ET_B endothelin receptors were initially described in cultured astrocytes [147, 405]. Stimulation of $ET_{A/B}$ receptors resulted in astroglial Ca^{2+} signalling [47, 233]. In mouse cerebellar Bergmann glial cells, endothelin evoked $[Ca^{2+}]_i$ transients sensitive to the selective ET_B receptor antagonist BQ-788 [399]. Activation of $ET_{A/B}$ receptors also suppresses astroglial gap junctions due to dephosphorylation of Cx43 [47, 116].

Receptors to atrial natriuretic peptide (NPR) were first identified in cultured mouse astrocytes [393]. Subsequently NPR-A [383], NPR-B [383, 438] and NPR-C [366, 383] receptors have been characterised. NPR-A and NPR-B increase intracellular cyclic guanosine monophosphate (cGMP), while NPR-C acts as a 'clearance receptor' that removes peptides from the extracellular space [314].

Astrocytes have been found to express δ and κ-opioid receptors [101, 102]; κ-opioid receptors induced Ca^{2+} signals sensitive to nifedipine [103], while δ-receptors, are linked to ER Ca^{2+} release [396]. Opioid receptors were claimed to regulate expression of plasmalemmal glutamate transporters [218], and astroglial growth [379].

3.10 Receptors for Leptin and Insulin

Expression of leptin receptors was detected in astrocytes in the subcommissural organ [79], in the nucleus tractus solitarius [80] and in the hypothalamus [149].

Insulin receptors have been characterised in astrocytes in vitro [131, 435]. Genetic deletion of insulin receptors from astrocytes affected brain glucose sensing, and reduced astroglial coverage of hypothalamic neurones [114]. Insulin may also act via the insulin-like growth factor 1 (IGF-1) receptor. Activation of insulin and IGF-1 receptors upregulates levels of glycogen in cultured rodent astrocytes [250].

3.11 Platelet-Activating Factor Receptor

Receptors for platelet-activating factor (1-*O*-alkyl-2-acetyl-*sn*-glycero-3-phosphocholine) were detected in astrocytes in vitro [53]. Activation of platelet-activating factor receptors stimulated production of $InsP_3$ [254, 304], induced secretion of nerve growth factor [54] and prostaglandin E_2 [391].

3.12 Protease-Activated Receptors (PAR)

Thrombin-activated PAR-1 and trypsin-activated PAR-2 receptors were found in cultured rat newborn astrocytes; activation of these receptors induced ER Ca^{2+} release [400, 401]. Astroglial PAR receptors are coupled to several signalling pathways, including stabilisation of hypoxia inducible factor-1α through ERK, JNK and PI3K/Akt cascades [436]. Activation of PAR-1 receptors (for example, by the selective peptide agonist TFLLR) is often used for selective stimulation of astroglia in situ [201, 359], although there are reports of neuronal Ca^{2+} signals triggered by the TFLLR-sensitive receptors [127].

3.13 Astroglial Membrane Transporters

3.13.1 ATP-Dependent Transporters

The most prominent and functionally important astroglial membrane P-type ATPase is represented by the Na^+-K^+ ATPase (NKA), that counter-transports Na^+ and K^+ with a stoichiometry of 3 Na^+ (expelled from the cell): 2 K^+ (imported into the cell). This defines electrogeneity of NKA. Astrocytes exclusively express α2 catalytic subunit [137, 165], which defines its peculiar properties. In contrast to neurones (which express α1 subunit) astroglial NKA is activated by physiological rises in $[K^+]_o$, while neuronal NKA is activated by an increase in $[Na^+]_i$ [134, 207, 329]. This stipulates the leading role of astroglial NKA in K^+ buffering. Astrocytes also express plasmalemmal Ca^{2+}-ATPases (PMCAs) and Sarco(Endo)Plasmic Reticulum Ca^{2+} ATP-ase (SERCA) both being responsible for Ca^{2+} homeostasis [410]. Astroglial plasma membrane and astroglial secretory vesicles possess the vacuolar V-type H^+ ATPase [288, 305].

3.13.2 Secondary Plasmalemmal Transporters of Solute Carrier (SLC) Family

3.13.2.1 Glutamate Transporters

Astrocytes represent the main sink for glutamate in the CNS [82, 410, 433]. Astrocytes express two types of plasmalemmal glutamate transporters: the excitatory amino acid transporters 1 and 2 (EAAT1/SLC1A6 and EAAT2/SLC1A2), which in rodent experiments are also referred to as GLAST1 (glutamate-aspartate transporter 1 [381]) and GLT-1 (glutamate transporter 1 [307]). The EAAT1 is predominantly expressed in the cerebellum [212], in the retina [319] and in circumventricular organs [37]; in all other parts of the brain the EAAT2 is the major type. The average density of EAAT1 is ~4700/μm^2 in Bergmann glia, and ~2300/μm^2 in the CA1 region of the hippocampus; the density of EAAT2 is ~8500/μm^2 in the hippocampus and ~740/μm^2 in the cerebellum [212]. Both transporters are concentrated in perisynaptic astroglial processes [66].

The stoichiometry of EAAT1 and EAAT2 is 3 Na^+, 1 H^+, 1 glutamate$^-$ in (glutamate is an anion at physiological conditions): 1 K^+ out [278, 426]. The equilibrium potential E_{EAAT} is the function of ions and glutamate concentration; the extracellular glutamate concentration varies between 25 nM at rest and 1 mM during synaptic transmission. The intracellular concentration of glutamate in astrocytes is ~0.3 mM due to high activity of glutamine synthetase [49, 132]. At the rest the E_{EAAT} is about +9 mV, whereas at 1 mM of glutamate in the cleft the transporter reverses at +145 mV [410]. The transporter is electrogenic and generates transmembrane current carried mainly by Na^+ ions [190, 404]. This Na^+ influx may elevate $[Na^+]_i$ by 10–30 mM [331].

Astrocytes also express the cystine/glutamate antiporter Sxc^-, which localises extrasynaptically [7, 70, 176]. This transporter is important for accumulation of cystine needed for production of glutathione.

3.13.2.2 Glutamine Transporters

The obligatory glutamate (and in proxy GABA) precursor glutamine is exported from astrocytes by SNAT3/SLC38A3 and SNAT5/SLC38A5 plasmalemmal glutamine transports, which are coupled with co-transport of 1 Na^+ and counter-transport of 1 H^+ [343]. Astroglial Na^+ signals stimulate glutamine efflux [397].

3.13.2.3 GABA Transporters

Astrocytes predominantly express GAT3 GABA plasmalemmal transporter with much lower expression of GAT1. The GAT3 are concentrated in astroglial processes [241, 325]. In the cerebellum GAT3 is localised in the perisynaptic processes of Bergmann glia, enwrapping inhibitory synapses [241]. In thalamic astrocytes GAT1 is concentrated in perisynaptic membranes, whereas GAT3 is localised more distantly being thus responsible for extrasynaptic GABA transport [31]. The stoichiometry of both transporters is 1 GABA: 2 Na^+: 1 Cl^-, being thus electrogenic [177, 223, 318]. The reversal potential E_{GAT} lies around −50 mV; hence, relatively small depolarisation and/or an increase in $[Na^+]_i$ favour the reverse mode operation of these transporters [402].

3.13.2.4 Glycine Transporters

Astroglial cells express GlyT1/SLC6A9 glycine plasmalemmal transporters [424]. Their stoichiometry is 1 glycine: 2 Na^+: 1 Cl^- [333]. The transporter can reverse at physiological membrane potentials [358]. This transporter-mediated glycine release was shown to be stimulated by dopamine [152].

3.13.2.5 Adenosine Transporters

Astrocytes express both equilibrative (i.e. controlled by adenosine transmembrane gradient, [187]) plasmalemmal transporters ENT-1/SLC29A1, ENT-2/SLC29A2, ENT-3/SLC29A3 and ENT-4/SLC29A4, and Na^+-dependent concentrative nucleoside plasmalemmal transporters CNT2/SLC28A2 and CNT3/SLC28A3, which co-transport adenosine together with 1 Na^+ [217, 300].

3.13.2.6 Transporters for Monoamines

It seems that the main astroglial monoamine plasmalemmal transporter is represented by norepinephrine transporter NET/SLC6A2 that couples monoamine transport with 2 Na^+ and 1 Cl^-, this transporter has higher affinity for dopamine than norepinephrine [349, 387].

3.13.2.7 D-Serine Transporters

Transmembrane transport of D-serine in astrocytes is mediated by a neutral amino acid transporter subtype ASCT2 (SLC1A5), which is an alanine-, serine-, cysteine-preferring neutral amino acid plasmalemmal transporter [234]. The ASCT2 is a Na^+-dependent with Na^+ to amino acid stoichiometry of 1:1 [235].

3.13.2.8 Sodium-Calcium Exchangers

All three known plasmalemmal sodium-calcium exchangers, NCX1/SLC8A1, NCX2/SLC8A2 and NCX3/SLC8A3, are expressed in astroglia [287]. The NCXs are mainly concentrated in astroglial perisynaptic processes, and co-localise with glutamate transporters and possibly with glutamate ionotropic receptors [45, 240]. The stoichiometry of astroglial NCX is 3 Na^+: 1 Ca^{2+}, and hence the equilibrium E_{NCX} lies at ~−85 to −90 mV at rest making it prone for fluctuating between forward and reverse modes [415]. Membrane depolarisation and increase in intracellular Na^+ concentration favour NCX to operate in the reverse mode, whereas increase in $[Ca^{2+}]_i$ promotes the forward mode. Operation of astroglial NCX was shown in vitro [120, 388] and in situ [189].

3.13.2.9 Sodium-Proton Exchanger, or NHE

Astrocytes express NHE1/SLC9A1 Na^+–H^+ exchanger [69, 85] with electroneutral stoichiometry 1 Na^+ (in): 1 H^+ (out) [276]. The NHE1 is primarily responsible for efflux of protons generated by cytoplasmic metabolism and accumulated by astrocytes through glutamate uptake (each glutamate brings a single H^+ ion) and Ca^{2+} extrusion (PMCA exchanges 2 H^+ for each Ca^{2+} ion expelled).

3.13.2.10 Sodium-Bicarbonate Co-transporter, NBC

The sodium-bicarbonate transporter NBCe1/SLC4A4 has been identified in astrocytes in culture [270] and in hippocampal slices [122]. The NBC stoichiometry is 1 Na^+: 2 HCO_3^- or 1 Na^+: 3 HCO_3^- [263], and this transporter can operate in both forward and reverse modes [394].

3.13.2.11 Sodium-Potassium-Chloride Co-transporter, NKCC1

The Na^+–K^+–Cl– co-transporter NKCC1/SLC12A2 has been detected in Bergmann glia [170], in astrocytes from the optic nerve [227] and spinal cord [316]. It has an electroneutral stoichiometry of 1 $Na^{+:}$ 1 K^+: 2 Cl^- [226]. Experiments in situ in hippocampal slices questioned the functional role of NKCC1 in protoplasmic astrocytes in the healthy brain [207].

3.13.2.12 Glucose Transporters

Astrocytes express the glucose transporter GLUT1/SLC2A1 [6], which is predominantly localised in endfeet and perisynaptic processes. Immunostaining revealed the presence of this transporter in grey matter astroglia [249]. It also contains GLUT4, a transporter sensitive to insulin in skeletal muscle, however the flux of glucose in astrocytes is not upregulated by insulin [250].

3.13.2.13 Monocarboxylate Transporters, MCT

Monocarboxylate transporters 1 and 4 (MCT1/SLC16A1, MCT4/SLC16A3) provide for export of lactate from astroglial cells [126]. They may, however, mediate both export or import of lactate depending on concentration gradients for monocarboxylate and H^+ [126].

3.14 Ionic Signalling in Astroglia

3.14.1 Calcium Signalling

Discovery of astroglial Ca^{2+} signals and propagating Ca^{2+} waves [63, 76, 83, 99, 108, 185, 236] led to the formulation of the concept of astrocytic ionic signalling as a basis for their excitability [411]. Astroglial Ca^{2+} signalling depends on both intracellular and extracellular sources (Fig. 3.8) [414]. Somatic Ca^{2+} signals almost entirely depend on Ca^{2+} release from the ER mediated by $InsP_3$ receptor type 2; deletion of this channel often substantially reduced or even eliminated somatic $[Ca^{2+}]_i$ transients [4, 171, 302, 303]. At the same time, Ca^{2+} signals in astroglial processes remain even in the $InsP_3R2^{-/-}$ mice [130, 172, 375]. These signals were mediated by plasmalemmal Ca^{2+} influx [336]. This Ca^{2+} influx may reflect upon Ca^{2+} entry through ionotropic receptors and plasmalemmal channels, or Ca^{2+} influx mediated by the reverse mode of NCX [29, 361, 414].

Mechanisms underlying Ca^{2+} signalling differ between astrocytes from different brain regions. Local Ca^{2+} microdomains in Bergmann glia and in the main processes of hippocampal astrocytes were mediated solely by $InsP_3Rs$ [86, 191]. Local $[Ca^{2+}]_i$ transients in hippocampal astrocytes in contrast are mediated by TRPA1 channels [360]. In neocortical astrocytes Ca^{2+} signals involve ryanodine receptor-mediated $[Ca^{2+}]_i$-induced Ca^{2+} release [284], which is not operative in hippocampal astroglia [30]. In astrocytes in vivo sensory stimulation triggers global synchronised Ca^{2+} signals in astrocytes in somato-sensory cortex, which depend entirely on $InsP_3R2$ [171]. In cortical astrocytes spontaneous local Ca^{2+} signals in fine processes originate from Ca^{2+} release from mitochondria [3].

Global Ca^{2+} signals in the mature astrocytes in vivo are mediated by α_1-adrenoceptors [88]. Similar global astroglial signalling is observed in attention and vigilance state, when widespread astrocytic responses are evoked by acetylcholine release from projection of the nucleus basalis of Meynert and are mediated through metabotropic cholinergic receptor-$InsP_3$ pathway [67, 386]. Global astroglial Ca^{2+} signals spreading through the entire cortex were observed in response to transcranial direct current stimulation; these signals were mediated through α_1-adrenoceptors [247].

Astroglial propagating Ca^{2+} wave is mediated either by intercellular diffusion of $InsP_3$ through gap junctions [5, 143, 216] or through regenerative paracrine ATP-mediated signalling [12, 77, 128] or through the combination of both [344]. Whether propagating Ca^{2+} waves develop in the in vivo brain in awake and behaving animals remains an open question.

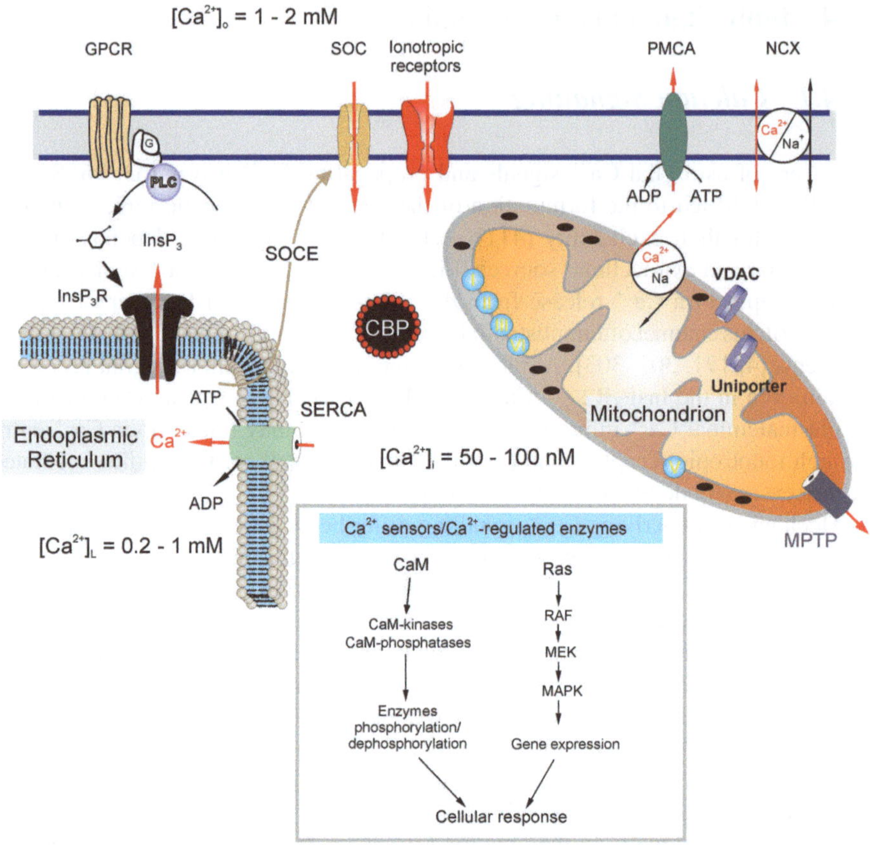

Fig. 3.8 Calcium distribution and calcium signalling cascades in intracellular compartments. Stimuli-induced increases in $[Ca^{2+}]_i$ could be caused by the entry of Ca^{2+} from the extracellular space through ionotropic receptors or store-operated channels (SOC). Plasmalemmal Ca^{2+} pumps/ATPases (PMCA) can extrude cytosolic Ca^{2+}, while the plasmalemmal sodium-calcium exchanger (NCX) can operate in both directions depending on intercellular Na^+ concentration and membrane potential. An additional source of Ca^{2+} is available from the ER internal store that possesses inositol 1,4,5 trisphosphate ($InsP_3$) receptors, which can be activated by the activity of metabotropic G-protein coupled receptors (GPCRs) and phospholipase C (PLC). The ER store is (re)filled by the activity of the store-specific Ca^{2+}-ATPase (SERCA). Cytosolic Ca^{2+} levels can be affected by a variety of cytosolic Ca^{2+}-binding proteins (CBPs) and by the action of mitochondria. A negative membrane potential exists across the inner mitochondrial membrane. Mitochondrial Ca^{2+} uptake occurs through voltage-dependent anion channels (VDACs) present in the outer membrane and by the uniporter in the inner membrane as the electrochemical gradient drives Ca^{2+} into the matrix, while free Ca^{2+} exits the mitochondrial matrix through the mitochondrial Na^+/Ca^{2+} exchanger and transient opening of the mitochondrial permeability transition pore (MPTP). Concentrations of free Ca^{2+} in different compartments are indicated on the scheme. Inset shows various Ca^{2+} effector molecules, sensors and enzymes. CaM, calmodulin; RAF, Rapidly Accelerated Fibrosarcoma, MAPK, mitogen-activated protein kinase (MAPK), MEK, MAPK kinase. Modified from [413]

3.15 Sodium Signalling

The concept of astroglial Na^+ signalling has been developed rather recently [192, 331]. Physiological stimulation triggers $[Na^+]_i$ transients in astrocytes in vitro [178, 329, 330] and in situ [189, 190, 205, 328]. Generation of Na^+ signals is accomplished through plasmalemmal Na^+ entry either through plasmalemmal channels or Na^+ coupled SLC transporters, whereas extrusion of Na^+ is primarily mediated by NKA [192, 331]. Resting $[Na^+]_i$ in astrocytes is higher than in neurones, being in the range of 15–20 mM.

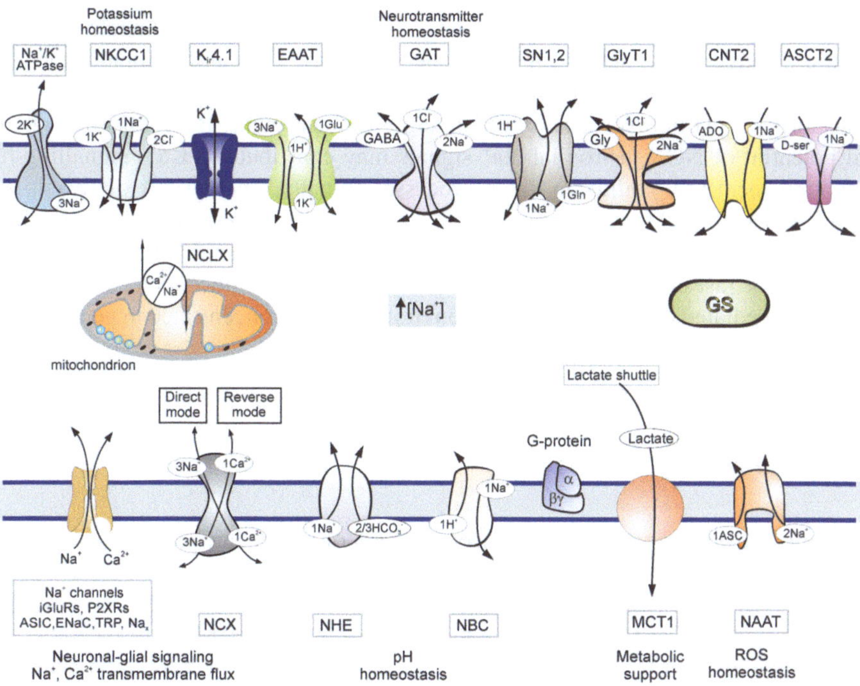

Fig. 3.9 Molecules of Na^+ homeostasis and targets of Na^+ signalling in astroglia. Schematic diagram showing receptors and transporters involved in and sensitive to changes in $[Na^+]_i$ and their relations to main homeostatic functions of astroglia. Abbreviations ASCT2, alanine-serine-cysteine transporter 2; ASIC—acid sensing ion channels; CNT2, concentrative nucleoside transporters; EAAT—excitatory amino acid transporters; ENaC—epithelial sodium channels; GAT—GABA transporters; GS—glutamine synthetase, GlyT1—glycine transporter. iGluRs—ionotropic glutamate receptors; Na_x—Na^+ channels activated by extracellular Na^+; NAAT—Na^+-dependent ascorbic acid transporter; NBC—Na^+/HCO_3^- (sodium-bicarbonate) co-transporter; NCX—Na^+/Ca^{2+} exchanger; NCLX—mitochondrial Na^+/Ca^{2+} exchanger; NHE—Na^+/H^+ exchanger; NKCC1—$Na^+/K^+/Cl^-$ cotransporter, MCT1—monocarboxylase transporter 1; P2XRs—ionotropic purinoceptors; SN1,2—sodium-coupled neutral amino acid transporters which underlie exit of glutamine; TRP—transient receptor potential channels. Reactive oxygen species (ROS). Modified from [413]

One of the main sources for Na^+ influx activated in response to neuronal activity is associated with operation of EAATs that co-transport 3 Na^+ with 1 glutamate; increase in extracellular glutamate may increase $[Na^+]_i$ by 10–30 mM [35, 65, 190]. Sodium influx may also be mediated by GABA transporters, ionotropic receptors, by TRP channels or by NCX operating in the forward mode [189, 205, 323]. Sodium entry may produce long-lasting $[Na^+]_i$ microdomains which, as per computational modelling, may be facilitated by fairly negative resting potential of astroglial plasmalemma [51]. Propagating Na^+ waves have been also detected in astrocytes in culture and in situ in hippocampal slices; these waves are propagating through gap junctions [206].

Astroglial Na^+ signals regulate multiple SLC transporters sensitive to transmembrane Na^+ gradients (Fig. 3.9); $[Na^+]_i$ also regulates glutamine-glutamate (GABA) shuttle through direct action on glutamine synthetase [34] and regulation of glutamine transporters [397]. Changes in $[Na^+]_i$ regulate K^+ buffering through the NKA transport and pH homeostasis by regulating NBC and NHE. By controlling reversal potential of NCX, astroglial Na^+ signals may contribute to Ca^{2+} signalling by initiating local Ca^{2+} influx in distal processes. Finally, fluctuations of $[Na^+]_i$ are coupled to astroglial metabolism, through controlling glycolysis and lactate production and possibly regulating ATP synthesis [64]. The sodium signalling system thus provides for fast coordination of neuronal activity with 'homeostatic' response of astroglia mediated through Na^+-dependent transporters, concentrated in perisynaptic processes.

3.16 Summary

Astroglial physiology is defined by a complement of ion channels, receptors for neurotransmitters, and neurohormones and membrane transporter systems. High expression of K^+ channels stabilises the membrane potential at negative level, thus ensuring electro-driving forces for operation of membrane transporters. Multiple receptors for neuroactive agents on astrocytes provide for input signals reflecting upon neuronal activity. Astrocytic ionic signalling regulates operation of transporters responsible for astroglial homeostatic response, central for astrocytic support of neuronal networks.

Acknowledgments VP's work is supported by a grant from the National Institute of General Medical Sciences of the National Institutes of Health (R01GM123971). VP is an Honorary Professor at University of Rijeka, Croatia.

References

1. Abbracchio MP, Ceruti S (2006) Roles of P2 receptors in glial cells: focus on astrocytes. Purinergic Signal 2:595–604
2. Adermark L, Lovinger DM (2008) Electrophysiological properties and gap junction coupling of striatal astrocytes. Neurochem Int 52:1365–1372

3. Agarwal A, Wu PH, Hughes EG, Fukaya M, Tischfield MA, Langseth AJ, Wirtz D, Bergles DE (2017) Transient opening of the mitochondrial permeability transition pore induces microdomain calcium transients in astrocyte processes. Neuron 93(587–605):e587
4. Agulhon C, Fiacco TA, McCarthy KD (2010) Hippocampal short- and long-term plasticity are not modulated by astrocyte Ca^{2+} signaling. Science 327:1250–1254
5. Allbritton NL, Meyer T, Stryer L (1992) Range of messenger action of calcium ion and inositol 1,4,5-trisphosphate. Science 258:1812–1815
6. Allen A, Messier C (2013) Plastic changes in the astrocyte GLUT1 glucose transporter and beta-tubulin microtubule protein following voluntary exercise in mice. Behav Brain Res 240:95–102
7. Allen JW, Shanker G, Aschner M (2001) Methylmercury inhibits the in vitro uptake of the glutathione precursor, cystine, in astrocytes, but not in neurons. Brain Res 894:131–140
8. Altevogt BM, Paul DL (2004) Four classes of intercellular channels between glial cells in the CNS. J Neurosci 24:4313–4323
9. Alvarez-Maubecin V, Garcia-Hernandez F, Williams JT, Van Bockstaele EJ (2000) Functional coupling between neurons and glia. J Neurosci 20:4091–4098
10. Amin MS, Wang HW, Reza E, Whitman SC, Tuana BS, Leenen FH (2005) Distribution of epithelial sodium channels and mineralocorticoid receptors in cardiovascular regulatory centers in rat brain. Am J Physiol Regul Integr Comp Physiol 289:R1787–R1797
11. Amzica F, Massimini M (2002) Glial and neuronal interactions during slow wave and paroxysmal activities in the neocortex. Cereb Cortex 12:1101–1113
12. Anderson CM, Bergher JP, Swanson RA (2004) ATP-induced ATP release from astrocytes. J Neurochem 88:246–256
13. Andriezen WL (1893) The neuroglia elements of the brain. Br Med J 2:227–230
14. Aoki C, Venkatesan C, Kurose H (1998) Noradrenergic modulation of the prefrontal cortex as revealed by electron microscopic immunocytochemistry. Adv Pharmacol 42:777–780
15. Araque A, Martin ED, Perea G, Arellano JI, Buno W (2002) Synaptically released acetylcholine evokes Ca^{2+} elevations in astrocytes in hippocampal slices. J Neurosci 22:2443–2450
16. Arbones L, Picatoste F, Garcia A (1990) Histamine stimulates glycogen breakdown and increases $^{45}Ca^{2+}$ permeability in rat astrocytes in primary culture. Mol Pharmacol 37:921–927
17. Armstrong WE, Rubrum A, Teruyama R, Bond CT, Adelman JP (2005) Immunocytochemical localization of small-conductance, calcium-dependent potassium channels in astrocytes of the rat supraoptic nucleus. J Comp Neurol 491:175–185
18. Ashour F, Deuchars J (2004) Electron microscopic localisation of $P2X_4$ receptor subunit immunoreactivity to pre- and post-synaptic neuronal elements and glial processes in the dorsal vagal complex of the rat. Brain Res 1026:44–55
19. Azmitia EC, Gannon PJ, Kheck NM, Whitaker-Azmitia PM (1996) Cellular localization of the $5\text{-}HT_{1A}$ receptor in primate brain neurons and glial cells. Neuropsychopharmacology 14:35–46
20. Babu AN, Cheng TP, Zhang A, Altura BT, Altura BM (1999) Low concentrations of ethanol deplete type-2 astrocytes of intracellular free magnesium. Brain Res Bull 50:59–62
21. Badaut J, Hirt L, Granziera C, Bogousslavsky J, Magistretti PJ, Regli L (2001) Astrocyte-specific expression of aquaporin-9 in mouse brain is increased after transient focal cerebral ischemia. J Cereb Blood Flow Metab 21:477–482
22. Bai JZ, Lipski J (2010) Differential expression of TRPM2 and TRPV4 channels and their potential role in oxidative stress-induced cell death in organotypic hippocampal culture. Neurotoxicology 31:204–214
23. Bal A, Bachelot T, Savasta M, Manier M, Verna JM, Benabid AL, Feuerstein C (1994) Evidence for dopamine D_2 receptor mRNA expression by striatal astrocytes in culture: in situ hybridization and polymerase chain reaction studies. Brain Res Mol Brain Res 23:204–212
24. Ballerini P, Rathbone MP, Di Iorio P, Renzetti A, Giuliani P, D'Alimonte I, Trubiani O, Caciagli F, Ciccarelli R (1996) Rat astroglial P2Z ($P2X_7$) receptors regulate intracellular calcium and purine release. NeuroReport 7:2533–2537

25. Barcelo-Torns M, Lewis AM, Gubern A, Barneda D, Bloor-Young D, Picatoste F, Churchill GC, Claro E, Masgrau R (2011) NAADP mediates ATP-induced Ca^{2+} signals in astrocytes. FEBS Lett 585:2300–2306
26. Barres BA, Chun LL, Corey DP (1988) Ion channel expression by white matter glia: I. Type 2 astrocytes and oligodendrocytes. Glia 1:10–30
27. Barres BA, Chun LL, Corey DP (1989) Calcium current in cortical astrocytes: induction by cAMP and neurotransmitters and permissive effect of serum factors. J Neurosci 9:3169–3175
28. Barres BA, Chun LL, Corey DP (1989) Glial and neuronal forms of the voltage-dependent sodium channel: characteristics and cell-type distribution. Neuron 2:1375–1388
29. Bazargani N, Attwell D (2016) Astrocyte calcium signaling: the third wave. Nat Neurosci 19:182–189
30. Beck A, Nieden RZ, Schneider HP, Deitmer JW (2004) Calcium release from intracellular stores in rodent astrocytes and neurons in situ. Cell Calcium 35:47–58
31. Beenhakker MP, Huguenard JR (2010) Astrocytes as gatekeepers of $GABA_B$ receptor function. J Neurosci 30:15262–15276
32. Bekar LK, Loewen ME, Cao K, Sun X, Leis J, Wang R, Forsyth GW, Walz W (2005) Complex expression and localization of inactivating K_v channels in cultured hippocampal astrocytes. J Neurophysiol 93:1699–1709
33. Benfenati V, Amiry-Moghaddam M, Caprini M, Mylonakou MN, Rapisarda C, Ottersen OP, Ferroni S (2007) Expression and functional characterization of transient receptor potential vanilloid-related channel 4 (TRPV4) in rat cortical astrocytes. Neuroscience 148:876–892
34. Benjamin AM (1987) Influence of Na^+, K^+, and Ca^{2+} on glutamine synthesis and distribution in rat brain cortex slices: a possible linkage of glutamine synthetase with cerebral transport processes and energetics in the astrocytes. J Neurochem 48:1157–1164
35. Bennay M, Langer J, Meier SD, Kafitz KW, Rose CR (2008) Sodium signals in cerebellar Purkinje neurons and Bergmann glial cells evoked by glutamatergic synaptic transmission. Glia 56:1138–1149
36. Bennett GC, Ford AP, Smith JA, Emmett CJ, Webb TE, Boarder MR (2003) P2Y receptor regulation of cultured rat cerebral cortical cells: calcium responses and mRNA expression in neurons and glia. Br J Pharmacol 139:279–288
37. Berger UV, Hediger MA (2000) Distribution of the glutamate transporters GLAST and GLT-1 in rat circumventricular organs, meninges, and dorsal root ganglia. J Comp Neurol 421:385–399
38. Bernstein M, Behnisch T, Balschun D, Reymann KG, Reiser G (1998) Pharmacological characterisation of metabotropic glutamatergic and purinergic receptors linked to Ca^{2+} signalling in hippocampal astrocytes. Neuropharmacology 37:169–178
39. Binder DK, Yao X, Zador Z, Sick TJ, Verkman AS, Manley GT (2006) Increased seizure duration and slowed potassium kinetics in mice lacking aquaporin-4 water channels. Glia 53:631–636
40. Black JA, Dib-Hajj S, Cohen S, Hinson AW, Waxman SG (1998) Glial cells have heart: rH1 Na^+ channel mRNA and protein in spinal cord astrocytes. Glia 23:200–208
41. Black JA, Newcombe J, Waxman SG (2010) Astrocytes within multiple sclerosis lesions upregulate sodium channel $Na_v1.5$. Brain 133:835–846
42. Black JA, Sontheimer H, Minturn JE, Ransom BR, Waxman SG (1992) The expression of sodium channels in astrocytes in situ and in vitro. Prog Brain Res 94:89–107
43. Black JA, Waxman SG (2013) Noncanonical roles of voltage-gated sodium channels. Neuron 80:280–291
44. Black JA, Westenbroek R, Minturn JE, Ransom BR, Catterall WA, Waxman SG (1995) Isoform-specific expression of sodium channels in astrocytes in vitro: immunocytochemical observations. Glia 14:133–144
45. Blaustein MP, Juhaszova M, Golovina VA, Church PJ, Stanley EF (2002) Na/Ca exchanger and PMCA localization in neurons and astrocytes: functional implications. Ann NY Acad Sci 976:356–366

46. Blazquez C, Sanchez C, Daza A, Galve-Roperh I, Guzman M (1999) The stimulation of ketogenesis by cannabinoids in cultured astrocytes defines carnitine palmitoyltransferase I as a new ceramide-activated enzyme. J Neurochem 72:1759–1768
47. Blomstrand F, Giaume C, Hansson E, Ronnback L (1999) Distinct pharmacological properties of ET-1 and ET-3 on astroglial gap junctions and $Ca^{(2+)}$ signaling. Am J Physiol 277:C616–C627
48. Bordey A, Sontheimer H (2000) Ion channel expression by astrocytes in situ: comparison of different CNS regions. Glia 30:27–38
49. Bramham CR, Torp R, Zhang N, Storm-Mathisen J, Ottersen OP (1990) Distribution of glutamate-like immunoreactivity in excitatory hippocampal pathways: a semiquantitative electron microscopic study in rats. Neuroscience 39:405–417
50. Brasko C, Hawkins V, De La Rocha IC, Butt AM (2016) Expression of $K_{ir}4.1$ and $K_{ir}5.1$ inwardly rectifying potassium channels in oligodendrocytes, the myelinating cells of the CNS. Brain Struct Funct
51. Breslin K, Wade JJ, Wong-Lin K, Harkin J, Flanagan B, Van Zalinge H, Hall S, Walker M, Verkhratsky A, McDaid L (2018) Potassium and sodium microdomains in thin astroglial processes: a computational model study. PLoS Comput Biol 14:e1006151
52. Brito V, Beyer C, Kuppers E (2004) BDNF-dependent stimulation of dopamine D_5 receptor expression in developing striatal astrocytes involves PI3-kinase signaling. Glia 46:284–295
53. Brodie C (1994) Functional PAF receptors in glia cells: binding parameters and regulation of expression. Int J Dev Neurosci 12:631–640
54. Brodie C (1995) Platelet activating factor induces nerve growth factor production by rat astrocytes. Neurosci Lett 186:5–8
55. Brune T, Deitmer JW (1995) Intracellular acidification and Ca^{2+} transients in cultured rat cerebellar astrocytes evoked by glutamate agonists and noradrenaline. Glia 14:153–161
56. Bruner G, Murphy S (1993) UTP activates multiple second messenger systems in cultured rat astrocytes. Neurosci Lett 162:105–108
57. Butenko O, Dzamba D, Benesova J, Honsa P, Benfenati V, Rusnakova V, Ferroni S, Anderova M (2012) The increased activity of TRPV4 channel in the astrocytes of the adult rat hippocampus after cerebral hypoxia/ischemia. PLoS ONE 7:e39959
58. Butt AM, Kalsi A (2006) Inwardly rectifying potassium channels (K_{ir}) in central nervous system glia: a special role for $K_{ir}4.1$ in glial functions. J Cell Mol Med 10:33–44
59. Butt AM, Ransom BR (1989) Visualization of oligodendrocytes and astrocytes in the intact rat optic nerve by intracellular injection of lucifer yellow and horseradish peroxidase. Glia 2:470–475
60. Cahoy JD, Emery B, Kaushal A, Foo LC, Zamanian JL, Christopherson KS, Xing Y, Lubischer JL, Krieg PA, Krupenko SA, Thompson WJ, Barres BA (2008) A transcriptome database for astrocytes, neurons, and oligodendrocytes: a new resource for understanding brain development and function. J Neurosci 28:264–278
61. Carson MJ, Thomas EA, Danielson PE, Sutcliffe JG (1996) The $5HT_{5A}$ serotonin receptor is expressed predominantly by astrocytes in which it inhibits cAMP accumulation: a mechanism for neuronal suppression of reactive astrocytes. Glia 17:317–326
62. Catus SL, Gibbs ME, Sato M, Summers RJ, Hutchinson DS (2011) Role of β-adrenoceptors in glucose uptake in astrocytes using beta-adrenoceptor knockout mice. Br J Pharmacol 162:1700–1715
63. Charles AC, Merrill JE, Dirksen ER, Sanderson MJ (1991) Intercellular signaling in glial cells: calcium waves and oscillations in response to mechanical stimulation and glutamate. Neuron 6:983–992
64. Chatton JY, Magistretti PJ, Barros LF (2016) Sodium signaling and astrocyte energy metabolism. Glia 64:1667–1676
65. Chatton JY, Pellerin L, Magistretti PJ (2003) GABA uptake into astrocytes is not associated with significant metabolic cost: implications for brain imaging of inhibitory transmission. Proc Natl Acad Sci USA 100:12456–12461

66. Chaudhry FA, Lehre KP, van Lookeren Campagne M, Ottersen OP, Danbolt NC, Storm-Mathisen J (1995) Glutamate transporters in glial plasma membranes: highly differentiated localizations revealed by quantitative ultrastructural immunocytochemistry. Neuron 15:711–720
67. Chen N, Sugihara H, Sharma J, Perea G, Petravicz J, Le C, Sur M (2012) Nucleus basalis-enabled stimulus-specific plasticity in the visual cortex is mediated by astrocytes. Proc Natl Acad Sci USA 109:E2832–E2841
68. Chen Y, Rathbone MP, Hertz L (2001) Guanosine-induced increase in free cytosolic calcium concentration in mouse astrocytes in primary cultures: does it act on an A_3 adenosine receptor? J Neurosci Res 65:184–189
69. Chesler M (2003) Regulation and modulation of pH in the brain. Physiol Rev 83:1183–1221
70. Cho Y, Bannai S (1990) Uptake of glutamate and cysteine in C-6 glioma cells and in cultured astrocytes. J Neurochem 55:2091–2097
71. Cholewinski AJ, Stevens G, McDermott AM, Wilkin GP (1991) Identification of B_2 bradykinin binding sites on cultured cortical astrocytes. J Neurochem 57:1456–1458
72. Chun LLY, Barres BA, Corey DP (1986) Induction of a calcium channel in astrocytes by cAMP. Soc Neurosci Abs 12:1346
73. Chvatal A, Pastor A, Mauch M, Sykova E, Kettenmann H (1995) Distinct populations of identified glial cells in the developing rat spinal cord slice: ion channel properties and cell morphology. Eur J Neurosci 7:129–142
74. Clark B, Mobbs P (1992) Transmitter-operated channels in rabbit retinal astrocytes studied in situ by whole-cell patch clamping. J Neurosci 12:664–673
75. Conti F, Minelli A, Brecha NC (1994) Cellular localization and laminar distribution of AMPA glutamate receptor subunits mRNAs and proteins in the rat cerebral cortex. J Comp Neurol 350:241–259
76. Cornell-Bell AH, Finkbeiner SM, Cooper MS, Smith SJ (1990) Glutamate induces calcium waves in cultured astrocytes: long-range glial signaling. Science 247:470–473
77. Cotrina ML, Lin JH, Lopez-Garcia JC, Naus CC, Nedergaard M (2000) ATP-mediated glia signaling. J Neurosci 20:2835–2844
78. Dahl G (2015) ATP release through pannexon channels. Philos Trans R Soc Lond B Biol Sci 370
79. Dall'Aglio C, Ceccarelli P, Pascucci L, Brecchia G, Boiti C (2006) Receptors for leptin and estrogen in the subcommissural organ of rabbits are differentially modulated by fasting. Brain Res 1124:62–69
80. Dallaporta M, Pecchi E, Pio J, Jean A, Horner KC, Troadec JD (2009) Expression of leptin receptor by glial cells of the nucleus tractus solitarius: possible involvement in energy homeostasis. J Neuroendocrinol 21:57–67
81. Dallerac G, Chever O, Rouach N (2013) How do astrocytes shape synaptic transmission? Insights from electrophysiology. Front Cell Neurosci 7:159
82. Danbolt NC (2001) Glutamate uptake. Prog Neurobiol 65:1–105
83. Dani JW, Chernjavsky A, Smith SJ (1992) Neuronal activity triggers calcium waves in hippocampal astrocyte networks. Neuron 8:429–440
84. Dare E, Schulte G, Karovic O, Hammarberg C, Fredholm BB (2007) Modulation of glial cell functions by adenosine receptors. Physiol Behav 92:15–20
85. Deitmer JW, Rose CR (1996) pH regulation and proton signalling by glial cells. Prog Neurobiol 48:73–103
86. Di Castro MA, Chuquet J, Liaudet N, Bhaukaurally K, Santello M, Bouvier D, Tiret P, Volterra A (2011) Local Ca^{2+} detection and modulation of synaptic release by astrocytes. Nat Neurosci 14:1276–1284
87. Di Scala-Guenot D, Mouginot D, Strosser MT (1994) Increase of intracellular calcium induced by oxytocin in hypothalamic cultured astrocytes. Glia 11:269–276
88. Ding F, O'Donnell J, Thrane AS, Zeppenfeld D, Kang H, Xie L, Wang F, Nedergaard M (2013) α_1-Adrenergic receptors mediate coordinated Ca^{2+} signaling of cortical astrocytes in awake, behaving mice. Cell Calcium 54:387–394

89. Dixon SJ, Yu R, Panupinthu N, Wilson JX (2004) Activation of P2 nucleotide receptors stimulates acid efflux from astrocytes. Glia 47:367–376
90. Doengi M, Deitmer JW, Lohr C (2008) New evidence for purinergic signaling in the olfactory bulb: A_{2A} and $P2Y_1$ receptors mediate intracellular calcium release in astrocytes. FASEB J 22:2368–2378
91. Doly S, Fischer J, Salio C, Conrath M (2004) The vanilloid receptor-1 is expressed in rat spinal dorsal horn astrocytes. Neurosci Lett 357:123–126
92. Dong JH, Chen X, Cui M, Yu X, Pang Q, Sun JP (2012) β_2-adrenergic receptor and astrocyte glucose metabolism. J Mol Neurosci 48:456–463
93. Duan S, Anderson CM, Keung EC, Chen Y, Swanson RA (2003) $P2X_7$ receptor-mediated release of excitatory amino acids from astrocytes. J Neurosci 23:1320–1328
94. Duan S, Neary JT (2006) $P2X_7$ receptors: properties and relevance to CNS function. Glia 54:738–746
95. Dzamba D, Honsa P, Valny M, Kriska J, Valihrach L, Novosadova V, Kubista M, Anderova M (2015) Quantitative analysis of glutamate receptors in glial cells from the cortex of GFAP/EGFP mice following ischemic injury: focus on NMDA receptors. Cell Mol Neurobiol 35:1187–1202
96. Eaton MJ, Skatchkov SN, Brune A, Biedermann B, Veh RW, Reichenbach A (2002) SURI and $K_{ir}6.1$ subunits of K_{ATP}-channels are co-localized in retinal glial (Muller) cells. NeuroReport 13:57–60
97. Edwards L, Nashmi R, Jones O, Backx P, Ackerley C, Becker L, Fehlings MG (2002) Upregulation of K_v 1.4 protein and gene expression after chronic spinal cord injury. J Comp Neurol 443:154–167
98. Emmi A, Wenzel HJ, Schwartzkroin PA, Taglialatela M, Castaldo P, Bianchi L, Nerbonne J, Robertson GA, Janigro D (2000) Do glia have heart? Expression and functional role for ether-a-go-go currents in hippocampal astrocytes. J Neurosci 20:3915–3925
99. Enkvist MO, Holopainen I, Akerman KE (1989) Glutamate receptor-linked changes in membrane potential and intracellular Ca^{2+} in primary rat astrocytes. Glia 2:397–402
100. Enyedi P, Czirjak G (2010) Molecular background of leak K^+ currents: two-pore domain potassium channels. Physiol Rev 90:559–605
101. Eriksson PS, Hansson E, Ronnback L (1990) δ and κ opiate receptors in primary astroglial cultures from rat cerebral cortex. Neurochem Res 15:1123–1126
102. Eriksson PS, Hansson E, Ronnback L (1992) δ and κ opiate receptors in primary astroglial cultures. Part II: Receptor sets in cultures from various brain regions and interactions with beta-receptor activated cyclic AMP. Neurochem Res 17:545–551
103. Eriksson PS, Nilsson M, Wagberg M, Hansson E, Ronnback L (1993) κ-opioid receptors on astrocytes stimulate L-type Ca^{2+} channels. Neuroscience 54:401–407
104. Fam SR, Gallagher CJ, Salter MW (2000) $P2Y_1$ purinoceptor-mediated Ca^{2+} signaling and Ca^{2+} wave propagation in dorsal spinal cord astrocytes. J Neurosci 20:2800–2808
105. Fang Q, Hu WW, Wang XF, Yang Y, Lou GD, Jin MM, Yan HJ, Zeng WZ, Shen Y, Zhang SH, Xu TL, Chen Z (2014) Histamine up-regulates astrocytic glutamate transporter 1 and protects neurons against ischemic injury. Neuropharmacology 77:156–166
106. Fellin T, Pozzan T, Carmignoto G (2006) Purinergic receptors mediate two distinct glutamate release pathways in hippocampal astrocytes. J Biol Chem 281:4274–4284
107. Filosa JA, Bonev AD, Straub SV, Meredith AL, Wilkerson MK, Aldrich RW, Nelson MT (2006) Local potassium signaling couples neuronal activity to vasodilation in the brain. Nat Neurosci 9:1397–1403
108. Finkbeiner S (1992) Calcium waves in astrocytes-filling in the gaps. Neuron 8:1101–1108
109. Franke H, Grosche J, Schadlich H, Krugel U, Allgaier C, Illes P (2001) P2X receptor expression on astrocytes in the nucleus accumbens of rats. Neuroscience 108:421–429
110. Franke H, Gunther A, Grosche J, Schmidt R, Rossner S, Reinhardt R, Faber-Zuschratter H, Schneider D, Illes P (2004) $P2X_7$ receptor expression after ischemia in the cerebral cortex of rats. J Neuropathol Exp Neurol 63:686–699

111. Franke H, Verkhratsky A, Burnstock G, Illes P (2012) Pathophysiology of astroglial purinergic signalling. Purinergic Signal 8:629–657
112. Fraser DD, Duffy S, Angelides KJ, Perez-Velazquez JL, Kettenmann H, MacVicar BA (1995) $GABA_A$/benzodiazepine receptors in acutely isolated hippocampal astrocytes. J Neurosci 15:2720–2732
113. Fumagalli M, Brambilla R, D'Ambrosi N, Volonte C, Matteoli M, Verderio C, Abbracchio MP (2003) Nucleotide-mediated calcium signaling in rat cortical astrocytes: role of P2X and P2Y receptors. Glia 43:218–230
114. Garcia-Caceres C, Quarta C, Varela L, Gao Y, Gruber T, Legutko B, Jastroch M, Johansson P, Ninkovic J, Yi CX, Le Thuc O, Szigeti-Buck K, Cai W, Meyer CW, Pfluger PT, Fernandez AM, Luquet S, Woods SC, Torres-Aleman I, Kahn CR, Gotz M, Horvath TL, Tschop MH (2016) Astrocytic insulin signaling couples brain glucose uptake with nutrient availability. Cell 166:867–880
115. Geiger JR, Melcher T, Koh DS, Sakmann B, Seeburg PH, Jonas P, Monyer H (1995) Relative abundance of subunit mRNAs determines gating and Ca^{2+} permeability of AMPA receptors in principal neurons and interneurons in rat CNS. Neuron 15:193–204
116. Giaume C, Cordier J, Glowinski J (1992) Endothelins inhibit junctional permeability in cultured mouse astrocytes. Eur J Neurosci 4:877–881
117. Giaume C, Fromaget C, el Aoumari A, Cordier J, Glowinski J, Gros D (1991) Gap junctions in cultured astrocytes: single-channel currents and characterization of channel-forming protein. Neuron 6:133–143
118. Giaume C, Koulakoff A, Roux L, Holcman D, Rouach N (2010) Astroglial networks: a step further in neuroglial and gliovascular interactions. Nat Rev Neurosci 11:87–99
119. Giaume C, Leybaert L, Naus CC, Saez JC (2013) Connexin and pannexin hemichannels in brain glial cells: properties, pharmacology, and roles. Front Pharmacol 4:88
120. Golovina VA, Bambrick LL, Yarowsky PJ, Krueger BK, Blaustein MP (1996) Modulation of two functionally distinct Ca^{2+} stores in astrocytes: role of the plasmalemmal Na/Ca exchanger. Glia 16:296–305
121. Graham AJ, Ray MA, Perry EK, Jaros E, Perry RH, Volsen SG, Bose S, Evans N, Lindstrom J, Court JA (2003) Differential nicotinic acetylcholine receptor subunit expression in the human hippocampus. J Chem Neuroanat 25:97–113
122. Grichtchenko II, Chesler M (1994) Depolarization-induced alkalinization of astrocytes in gliotic hippocampal slices. Neuroscience 62:1071–1078
123. Griemsmann S, Hoft SP, Bedner P, Zhang J, von Staden E, Beinhauer A, Degen J, Dublin P, Cope DW, Richter N, Crunelli V, Jabs R, Willecke K, Theis M, Seifert G, Kettenmann H, Steinhauser C (2015) Characterization of panglial gap junction networks in the thalamus, neocortex, and hippocampus reveals a unique population of glial cells. Cereb Cortex 25:3420–3433
124. Grimaldi M, Maratos M, Verma A (2003) Transient receptor potential channel activation causes a novel form of $[Ca^{2+}]_i$ oscillations and is not involved in capacitative Ca^{2+} entry in glial cells. J Neurosci 23:4737–4745
125. Grolla AA, Sim JA, Lim D, Rodriguez JJ, Genazzani AA, Verkhratsky A (2013) Amyloid-β and Alzheimer's disease type pathology differentially affects the calcium signalling toolkit in astrocytes from different brain regions. Cell Death Dis 4:e623
126. Halestrap AP (2012) The monocarboxylate transporter family—structure and functional characterization. IUBMB Life 64:1–9
127. Han KS, Mannaioni G, Hamill CE, Lee J, Junge CE, Lee CJ, Traynelis SF (2011) Activation of protease activated receptor 1 increases the excitability of the dentate granule neurons of hippocampus. Mol Brain. 4:32
128. Hassinger TD, Guthrie PB, Atkinson PB, Bennet MVL, Kater SB (1997) An extracellular signaling component in propagation of astrocytic calcium waves. Proc Natl Acad Sci USA 93:13268–13273
129. Hatton GI, Bicknell RJ, Hoyland J, Bunting R, Mason WT (1992) Arginine vasopressin mobilises intracellular calcium via V_1-receptor activation in astrocytes (pituicytes) cultured from adult rat neural lobes. Brain Res 588:75–83

130. Haustein MD, Kracun S, Lu XH, Shih T, Jackson-Weaver O, Tong X, Xu J, Yang XW, O'Dell TJ, Marvin JS, Ellisman MH, Bushong EA, Looger LL, Khakh BS (2014) Conditions and constraints for astrocyte calcium signaling in the hippocampal mossy fiber pathway. Neuron 82:413–429
131. Heni M, Hennige AM, Peter A, Siegel-Axel D, Ordelheide AM, Krebs N, Machicao F, Fritsche A, Haring HU, Staiger H (2011) Insulin promotes glycogen storage and cell proliferation in primary human astrocytes. PLoS ONE 6:e21594
132. Herman MA, Jahr CE (2007) Extracellular glutamate concentration in hippocampal slice. J Neurosci 27:9736–9741
133. Hertle DN, Yeckel MF (2007) Distribution of inositol-1,4,5-trisphosphate receptor isotypes and ryanodine receptor isotypes during maturation of the rat hippocampus. Neuroscience 150:625–638
134. Hertz L, Gerkau NJ, Xu J, Durry S, Song D, Rose CR, Peng L (2015) Roles of astrocytic Na^+, K^+-ATPase and glycogenolysis for K^+ homeostasis in mammalian brain. J Neurosci Res 93:1019–1030
135. Hertz L, Lovatt D, Goldman SA, Nedergaard M (2010) Adrenoceptors in brain: cellular gene expression and effects on astrocytic metabolism and $[Ca^{2+}]_i$. Neurochem Int 57:411–420
136. Hertz L, Rothman DL, Li B, Peng L (2015) Chronic SSRI stimulation of astrocytic 5-HT_{2B} receptors change multiple gene expressions/editings and metabolism of glutamate, glucose and glycogen: a potential paradigm shift. Front Behav Neurosci 9:25
137. Hertz L, Song D, Xu J, Peng L, Gibbs ME (2015) Role of the astrocytic Na^+, K^+-ATPase in K^+ homeostasis in brain: K^+ uptake, signaling pathways and substrate utilization. Neurochem Res 40:2505–2516
138. Hibino H, Fujita A, Iwai K, Yamada M, Kurachi Y (2004) Differential assembly of inwardly rectifying K^+ channel subunits, $K_{ir}4.1$ and $K_{ir}5.1$, in brain astrocytes. J Biol Chem 279:44065–44073
139. Hibino H, Inanobe A, Furutani K, Murakami S, Findlay I, Kurachi Y (2010) Inwardly rectifying potassium channels: their structure, function, and physiological roles. Physiol Rev 90:291–366
140. Higashi K, Fujita A, Inanobe A, Tanemoto M, Doi K, Kubo T, Kurachi Y (2001) An inwardly rectifying K^+ channel, $K_{ir}4.1$, expressed in astrocytes surrounds synapses and blood vessels in brain. Am J Physiol Cell Physiol 281:C922–C931
141. Hiyama TY, Yoshida M, Matsumoto M, Suzuki R, Matsuda T, Watanabe E, Noda M (2013) Endothelin-3 expression in the subfornical organ enhances the sensitivity of Na_x, the brain sodium-level sensor, to suppress salt intake. Cell Metab 17:507–519
142. Ho T, Vessey KA, Fletcher EL (2014) Immunolocalization of the $P2X_4$ receptor on neurons and glia in the mammalian retina. Neuroscience 277:55–71
143. Hofer T, Venance L, Giaume C (2002) Control and plasticity of intercellular calcium waves in astrocytes: a modeling approach. J Neurosci 22:4850–4859
144. Holtzclaw LA, Pandhit S, Bare DJ, Mignery GA, Russell JT (2002) Astrocytes in adult rat brain express type 2 inositol 1,4,5-trisphosphate receptors. Glia 39:69–84
145. Honsa P, Pivonkova H, Harantova L, Butenko O, Kriska J, Dzamba D, Rusnakova V, Valihrach L, Kubista M, Anderova M (2014) Increased expression of hyperpolarization-activated cyclic nucleotide-gated (HCN) channels in reactive astrocytes following ischemia. Glia 62:2004–2021
146. Horvat A, Zorec R, Vardjan N (2016) Adrenergic stimulation of single rat astrocytes results in distinct temporal changes in intracellular $Ca^{(2+)}$ and cAMP-dependent PKA responses. Cell Calcium 59:156–163
147. Hosli E, Hosli L (1991) Autoradiographic evidence for endothelin receptors on astrocytes in cultures of rat cerebellum, brainstem and spinal cord. Neurosci Lett 129:55–58
148. Houades V, Koulakoff A, Ezan P, Seif I, Giaume C (2008) Gap junction-mediated astrocytic networks in the mouse barrel cortex. J Neurosci 28:5207–5217
149. Hsuchou H, Pan W, Barnes MJ, Kastin AJ (2009) Leptin receptor mRNA in rat brain astrocytes. Peptides 30:2275–2280

150. Hua X, Malarkey EB, Sunjara V, Rosenwald SE, Li WH, Parpura V (2004) $Ca^{(2+)}$-dependent glutamate release involves two classes of endoplasmic reticulum $Ca^{(2+)}$ stores in astrocytes. J Neurosci Res 76:86–97
151. Huang C, Hu ZL, Wu WN, Yu DF, Xiong QJ, Song JR, Shu Q, Fu H, Wang F, Chen JG (2010) Existence and distinction of acid-evoked currents in rat astrocytes. Glia 58:1415–1424
152. Huang H, Barakat L, Wang D, Bordey A (2004) Bergmann glial GlyT1 mediates glycine uptake and release in mouse cerebellar slices. J Physiol 560:721–736
153. Huang Y, Grinspan JB, Abrams CK, Scherer SS (2007) Pannexin1 is expressed by neurons and glia but does not form functional gap junctions. Glia 55:46–56
154. Hung AC, Sun SH (2002) The $P2X_7$ receptor-mediated phospholipase D activation is regulated by both PKC-dependent and PKC-independent pathways in a rat brain-derived Type-2 astrocyte cell line, RBA-2. Cell Signal 14:83–92
155. Hutchinson DS, Summers RJ, Gibbs ME (2007) β_2- and β_3-adrenoceptors activate glucose uptake in chick astrocytes by distinct mechanisms: a mechanism for memory enhancement? J Neurochem 103:997–1008
156. Hwang EM, Kim E, Yarishkin O, Woo DH, Han KS, Park N, Bae Y, Woo J, Kim D, Park M, Lee CJ, Park JY (2014) A disulphide-linked heterodimer of TWIK-1 and TREK-1 mediates passive conductance in astrocytes. Nat Commun 5:3227
157. Iglesias R, Dahl G, Qiu F, Spray DC, Scemes E (2009) Pannexin 1: the molecular substrate of astrocyte "hemichannels". J Neurosci 29:7092–7097
158. Illes P, Verkhratsky A (2016) Purinergic neurone-glia signalling in cognitive-related pathologies. Neuropharmacology 104:62–75
159. Ishii M, Fujita A, Iwai K, Kusaka S, Higashi K, Inanobe A, Hibino H, Kurachi Y (2003) Differential expression and distribution of $K_{ir}5.1$ and $K_{ir}4.1$ inwardly rectifying K^+ channels in retina. Am J Physiol Cell Physiol 285:C260–C267
160. Ishimoto H, Nakahata N, Matsuoka I, Nakanishi H (1997) Effects of ATP on phosphoinositide hydrolysis and prostaglandin E2 generation in rabbit astrocytes. J Pharm Pharmacol 49:520–524
161. Isokawa M, McKhann GM 2nd (2005) Electrophysiological and morphological characterization of dentate astrocytes in the hippocampus. J Neurobiol 65:125–134
162. Jabs R, Guenther E, Marquordt K, Wheeler-Schilling TH (2000) Evidence for $P2X_3$, $P2X_4$, $P2X_5$ but not for $P2X_7$ containing purinergic receptors in Muller cells of the rat retina. Brain Res Mol Brain Res 76:205–210
163. Jimenez AI, Castro E, Communi D, Boeynaems JM, Delicado EG, Miras-Portugal MT (2000) Coexpression of several types of metabotropic nucleotide receptors in single cerebellar astrocytes. J Neurochem 75:2071–2079
164. John GR, Simpson JE, Woodroofe MN, Lee SC, Brosnan CF (2001) Extracellular nucleotides differentially regulate interleukin-1_b signaling in primary human astrocytes: implications for inflammatory gene expression. J Neurosci 21:4134–4142
165. Juhaszova M, Blaustein MP (1997) Na^+ pump low and high ouabain affinity α subunit isoforms are differently distributed in cells. Proc Natl Acad Sci USA 94:1800–1805
166. Juric DM, Mele T, Carman-Krzan M (2011) Involvement of histaminergic receptor mechanisms in the stimulation of NT-3 synthesis in astrocytes. Neuropharmacology 60:1309–1317
167. Jurzak M, Muller AR, Gerstberger R (1995) Characterization of vasopressin receptors in cultured cells derived from the region of rat brain circumventricular organs. Neuroscience 65:1145–1159
168. Kafitz KW, Meier SD, Stephan J, Rose CR (2008) Developmental profile and properties of sulforhodamine 101–Labeled glial cells in acute brain slices of rat hippocampus. J Neurosci Methods 169:84–92
169. Kalsi AS, Greenwood K, Wilkin G, Butt AM (2004) $K_{ir}4.1$ expression by astrocytes and oligodendrocytes in CNS white matter: a developmental study in the rat optic nerve. J Anat 204:475–485
170. Kanaka C, Ohno K, Okabe A, Kuriyama K, Itoh T, Fukuda A, Sato K (2001) The differential expression patterns of messenger RNAs encoding K-Cl cotransporters (KCC1,2) and Na-K-2Cl cotransporter (NKCC1) in the rat nervous system. Neuroscience 104:933–946

171. Kanemaru K, Kubota J, Sekiya H, Hirose K, Okubo Y, Iino M (2013) Calcium-dependent N-cadherin up-regulation mediates reactive astrogliosis and neuroprotection after brain injury. Proc Natl Acad Sci USA 110:11612–11617
172. Kanemaru K, Sekiya H, Xu M, Satoh K, Kitajima N, Yoshida K, Okubo Y, Sasaki T, Moritoh S, Hasuwa H, Mimura M, Horikawa K, Matsui K, Nagai T, Iino M, Tanaka KF (2014) In vivo visualization of subtle, transient, and local activity of astrocytes using an ultrasensitive $Ca^{(2+)}$ indicator. Cell Rep 8:311–318
173. Kanjhan R, Housley GD, Thorne PR, Christie DL, Palmer DJ, Luo L, Ryan AF (1996) Localization of ATP-gated ion channels in cerebellum using P2x2R subunit-specific antisera. NeuroReport 7:2665–2669
174. Karschin A, Brockhaus J, Ballanyi K (1998) KATP channel formation by the sulphonylurea receptors SUR1 with $K_{ir}6.2$ subunits in rat dorsal vagal neurons in situ. J Physiol 509(Pt 2):339–346
175. Kastritsis CH, Salm AK, McCarthy K (1992) Stimulation of the P2Y purinergic receptor on type 1 astroglia results in inositol phosphate formation and calcium mobilization. J Neurochem 58:1277–1284
176. Kato S, Ishita S, Sugawara K, Mawatari K (1993) Cystine/glutamate antiporter expression in retinal Muller glial cells: implications for DL-alpha-aminoadipate toxicity. Neuroscience 57:473–482
177. Kavanaugh MP, Arriza JL, North RA, Amara SG (1992) Electrogenic uptake of γ-aminobutyric acid by a cloned transporter expressed in Xenopus oocytes. J Biol Chem 267:22007–22009
178. Kelly T, Kafitz KW, Roderigo C, Rose CR (2009) Ammonium-evoked alterations in intracellular sodium and pH reduce glial glutamate transport activity. Glia 57:921–934
179. Kettenmann H, Backus KH, Schachner M (1984) Aspartate, glutamate and gamma-aminobutyric acid depolarize cultured astrocytes. Neurosci Lett 52:25–29
180. Kettenmann H, Backus KH, Schachner M (1987) γ-Aminobutyric acid opens Cl^- channels in cultured astrocytes. Brain Res 404:1–9
181. Kettenmann H, Ransom BR (1988) Electrical coupling between astrocytes and between oligodendrocytes studied in mammalian cell cultures. Glia 1:64–73
182. Kettenmann H, Schachner M (1985) Pharmacological properties of gamma-aminobutyric acid-, glutamate-, and aspartate-induced depolarizations in cultured astrocytes. J Neurosci 5:3295–3301
183. Kettenmann H, Zorec R (2013) Release of gliotransmitters and transmitter receptors in astrocytes. In: Kettenmann H, Ransom BR (eds) Neuroglia. Oxford University Press, New York, pp 197–211
184. Khan ZU, Koulen P, Rubinstein M, Grandy DK, Goldman-Rakic PS (2001) An astroglia-linked dopamine D_2-receptor action in prefrontal cortex. Proc Natl Acad Sci USA 98:1964–1969
185. Kim WT, Rioult MG, Cornell-Bell AH (1994) Glutamate-induced calcium signaling in astrocytes. Glia 11:173–184
186. Kimelberg HK, Macvicar BA, Sontheimer H (2006) Anion channels in astrocytes: biophysics, pharmacology, and function. Glia 54:747–757
187. King AE, Ackley MA, Cass CE, Young JD, Baldwin SA (2006) Nucleoside transporters: from scavengers to novel therapeutic targets. Trends Pharmacol Sci 27:416–425
188. Kirchhoff F, Mulhardt C, Pastor A, Becker CM, Kettenmann H (1996) Expression of glycine receptor subunits in glial cells of the rat spinal cord. J Neurochem 66:1383–1390
189. Kirischuk S, Kettenmann H, Verkhratsky A (1997) Na^+/Ca^{2+} exchanger modulates kainate-triggered Ca^{2+} signaling in Bergmann glial cells in situ. FASEB J 11:566–572
190. Kirischuk S, Kettenmann H, Verkhratsky A (2007) Membrane currents and cytoplasmic sodium transients generated by glutamate transport in Bergmann glial cells. Pflugers Arch 454:245–252
191. Kirischuk S, Kirchhoff F, Matyash V, Kettenmann H, Verkhratsky A (1999) Glutamate-triggered calcium signalling in mouse bergmann glial cells in situ: role of inositol-1,4,5-trisphosphate-mediated intracellular calcium release. Neuroscience 92:1051–1059

192. Kirischuk S, Parpura V, Verkhratsky A (2012) Sodium dynamics: another key to astroglial excitability? Trends Neurosci 35:497–506
193. Kirischuk S, Tuschick S, Verkhratsky A, Kettenmann H (1996) Calcium signalling in mouse Bergmann glial cells mediated by α_1-adrenoreceptors and H_1 histamine receptors. Eur J Neurosci 8:1198–1208
194. Kofuji P, Biedermann B, Siddharthan V, Raap M, Iandiev I, Milenkovic I, Thomzig A, Veh RW, Bringmann A, Reichenbach A (2002) K_{ir} potassium channel subunit expression in retinal glial cells: implications for spatial potassium buffering. Glia 39:292–303
195. von Kölliker A (1896) Handbuch der Gewebelehre des Menschen, 6 Aufl. Engelmann, Leipzig
196. Kong EK, Peng L, Chen Y, Yu AC, Hertz L (2002) Up-regulation of 5-HT_{2B} receptor density and receptor-mediated glycogenolysis in mouse astrocytes by long-term fluoxetine administration. Neurochem Res 27:113–120
197. Kukley M, Barden JA, Steinhauser C, Jabs R (2001) Distribution of P2X receptors on astrocytes in juvenile rat hippocampus. Glia 36:11–21
198. Kunzelmann P, Schroder W, Traub O, Steinhauser C, Dermietzel R, Willecke K (1999) Late onset and increasing expression of the gap junction protein connexin30 in adult murine brain and long-term cultured astrocytes. Glia 25:111–119
199. Kuo J, Hariri OR, Micevych P (2009) An interaction of oxytocin receptors with metabotropic glutamate receptors in hypothalamic astrocytes. J Neuroendocrinol 21:1001–1006
200. Lalo U, Palygin O, North RA, Verkhratsky A, Pankratov Y (2011) Age-dependent remodelling of ionotropic signalling in cortical astroglia. Aging Cell 10:392–402
201. Lalo U, Palygin O, Rasooli-Nejad S, Andrew J, Haydon PG, Pankratov Y (2014) Exocytosis of ATP from astrocytes modulates phasic and tonic inhibition in the neocortex. PLoS Biol 12:e1001747
202. Lalo U, Pankratov Y, Kirchhoff F, North RA, Verkhratsky A (2006) NMDA receptors mediate neuron-to-glia signaling in mouse cortical astrocytes. J Neurosci 26:2673–2683
203. Lalo U, Pankratov Y, Parpura V, Verkhratsky A (2011) Ionotropic receptors in neuronal-astroglial signalling: what is the role of "excitable" molecules in non-excitable cells. Biochim Biophys Acta 1813:992–1002
204. Lalo U, Pankratov Y, Wichert SP, Rossner MJ, North RA, Kirchhoff F, Verkhratsky A (2008) $P2X_1$ and $P2X_5$ subunits form the functional P2X receptor in mouse cortical astrocytes. J Neurosci 28:5473–5480
205. Langer J, Rose CR (2009) Synaptically induced sodium signals in hippocampal astrocytes in situ. J Physiol 587:5859–5877
206. Langer J, Stephan J, Theis M, Rose CR (2012) Gap junctions mediate intercellular spread of sodium between hippocampal astrocytes in situ. Glia 60:239–252
207. Larsen BR, Assentoft M, Cotrina ML, Hua SZ, Nedergaard M, Kaila K, Voipio J, MacAulay N (2014) Contributions of the Na^+/K^+-ATPase, NKCC1, and $K_{ir}4.1$ to hippocampal K^+ clearance and volume responses. Glia 62:608–622
208. Latour I, Gee CE, Robitaille R, Lacaille JC (2001) Differential mechanisms of Ca^{2+} responses in glial cells evoked by exogenous and endogenous glutamate in rat hippocampus. Hippocampus 11:132–145
209. Latour I, Hamid J, Beedle AM, Zamponi GW, Macvicar BA (2003) Expression of voltage-gated Ca^{2+} channel subtypes in cultured astrocytes. Glia 41:347–353
210. Lee MC, Ting KK, Adams S, Brew BJ, Chung R, Guillemin GJ (2010) Characterisation of the expression of NMDA receptors in human astrocytes. PLoS ONE 5:e14123
211. Lee SM, Cho YS, Kim TH, Jin MU, Ahn DK, Noguchi K, Bae YC (2012) An ultrastructural evidence for the expression of transient receptor potential ankyrin 1 (TRPA1) in astrocytes in the rat trigeminal caudal nucleus. J Chem Neuroanat 45:45–49
212. Lehre KP, Danbolt NC (1998) The number of glutamate transporter subtype molecules at glutamatergic synapses: chemical and stereological quantification in young adult rat brain. J Neurosci 18:8751–8757
213. von Lenhossék M (1895) Der feinere Bau des Nervensystems im Lichte neuester Forschung, 2nd edn. Fischer's Medicinische Buchhandlung H. Kornfield, Berlin

214. Leonoudakis D, Mailliard W, Wingerd K, Clegg D, Vandenberg C (2001) Inward rectifier potassium channel $K_{ir}2.2$ is associated with synapse-associated protein SAP97. J Cell Sci 114:987–998
215. Letellier M, Park YK, Chater TE, Chipman PH, Gautam SG, Oshima-Takago T, Goda Y (2016) Astrocytes regulate heterogeneity of presynaptic strengths in hippocampal networks. Proc Natl Acad Sci USA 113:E2685–E2694
216. Leybaert L, Sanderson MJ (2001) Intercellular calcium signaling and flash photolysis of caged compounds. A sensitive method to evaluate gap junctional coupling. Methods Mol Biol 154:407–430
217. Li B, Gu L, Hertz L, Peng L (2013) Expression of nucleoside transporter in freshly isolated neurons and astrocytes from mouse brain. Neurochem Res 38:2351–2358
218. Liang J, Chao D, Sandhu HK, Yu Y, Zhang L, Balboni G, Kim DH, Xia Y (2014) δ-Opioid receptors up-regulate excitatory amino acid transporters in mouse astrocytes. Br J Pharmacol 171:5417–5430
219. Lisjak M, Potokar M, Rituper B, Jorgacevski J, Zorec R (2017) AQP4e-based orthogonal arrays regulate rapid cell volume changes in astrocytes. J Neurosci 37:10748–10756
220. Liu X, Bandyopadhyay BC, Nakamoto T, Singh B, Liedtke W, Melvin JE, Ambudkar I (2006) A role for AQP5 in activation of TRPV4 by hypotonicity: concerted involvement of AQP5 and TRPV4 in regulation of cell volume recovery. J Biol Chem 281:15485–15495
221. Loesch A, Burnstock G (1998) Electron-immunocytochemical localization of $P2X_1$ receptors in the rat cerebellum. Cell Tissue Res 294:253–260
222. Longden TA, Dunn KM, Draheim HJ, Nelson MT, Weston AH, Edwards G (2011) Intermediate-conductance calcium-activated potassium channels participate in neurovascular coupling. Br J Pharmacol 164:922–933
223. Lu CC, Hilgemann DW (1999) GAT1 (GABA:Na^+:Cl^-) cotransport function. Steady state studies in giant Xenopus oocyte membrane patches. J Gen Physiol 114:429–444
224. Lu DC, Zhang H, Zador Z, Verkman AS (2008) Impaired olfaction in mice lacking aquaporin-4 water channels. FASEB J 22:3216–3223
225. Lu Z (2004) Mechanism of rectification in inward-rectifier K^+ channels. Annu Rev Physiol 66:103–129
226. Macaulay N, Zeuthen T (2012) Glial K^+ clearance and cell swelling: key roles for cotransporters and pumps. Neurochem Res 37:2299–2309
227. MacVicar BA, Feighan D, Brown A, Ransom B (2002) Intrinsic optical signals in the rat optic nerve: role for K^+ uptake via NKCC1 and swelling of astrocytes. Glia 37:114–123
228. MacVicar BA, Tse FW (1988) Norepinephrine and cyclic adenosine 3′:5′-cyclic monophosphate enhance a nifedipine-sensitive calcium current in cultured rat astrocytes. Glia 1:359–365
229. MacVicar BA, Tse FW, Crichton SA, Kettenmann H (1989) GABA-activated Cl^- channels in astrocytes of hippocampal slices. J Neurosci 9:3577–3583
230. Magnotti LM, Goodenough DA, Paul DL (2011) Functional heterotypic interactions between astrocyte and oligodendrocyte connexins. Glia 59:26–34
231. Malarkey EB, Ni Y, Parpura V (2008) Ca^{2+} entry through TRPC1 channels contributes to intracellular Ca^{2+} dynamics and consequent glutamate release from rat astrocytes. Glia 56:821–835
232. Mantyh PW, Rogers SD, Allen CJ, Catton MD, Ghilardi JR, Levin LA, Maggio JE, Vigna SR (1995) β_2-adrenergic receptors are expressed by glia in vivo in the normal and injured central nervous system in the rat, rabbit, and human. J Neurosci 15:152–164
233. Marsault R, Vigne P, Breittmayer JP, Frelin C (1990) Astrocytes are target cells for endothelins and sarafotoxin. J Neurochem 54:2142–2144
234. Martineau M, Parpura V, Mothet JP (2014) Cell-type specific mechanisms of D-serine uptake and release in the brain. Front Synaptic Neurosci 6:12
235. Maucler C, Pernot P, Vasylieva N, Pollegioni L, Marinesco S (2013) In vivo D-serine hetero-exchange through alanine-serine-cysteine (ASC) transporters detected by microelectrode biosensors. ACS Chem Neurosci. 4:772–781
236. McCarthy KD, Salm AK (1991) Pharmacologically-distinct subsets of astroglia can be identified by their calcium response to neuroligands. Neuroscience 41:325–333

237. Miller RL, Loewy AD (2013) ENaC γ-expressing astrocytes in the circumventricular organs, white matter, and ventral medullary surface: sites for Na^+ regulation by glial cells. J Chem Neuroanat 53:72–80
238. Miller RL, Wang MH, Gray PA, Salkoff LB, Loewy AD (2013) ENaC-expressing neurons in the sensory circumventricular organs become c-Fos activated following systemic sodium changes. Am J Physiol Regul Integr Comp Physiol 305:R1141–R1152
239. Milner TA, Lee A, Aicher SA, Rosin DL (1998) Hippocampal α_{2a}-adrenergic receptors are located predominantly presynaptically but are also found postsynaptically and in selective astrocytes. J Comp Neurol 395:310–327
240. Minelli A, Castaldo P, Gobbi P, Salucci S, Magi S, Amoroso S (2007) Cellular and subcellular localization of Na^+-Ca^{2+} exchanger protein isoforms, NCX1, NCX2, and NCX3 in cerebral cortex and hippocampus of adult rat. Cell Calcium 41:221–234
241. Minelli A, DeBiasi S, Brecha NC, Zuccarello LV, Conti F (1996) GAT-3, a high-affinity GABA plasma membrane transporter, is localized to astrocytic processes, and it is not confined to the vicinity of GABAergic synapses in the cerebral cortex. J Neurosci 16:6255–6264
242. Mishima T, Hirase H (2010) In vivo intracellular recording suggests that gray matter astrocytes in mature cerebral cortex and hippocampus are electrophysiologically homogeneous. J Neurosci 30:3093–3100
243. Mishima T, Sakatani S, Hirase H (2007) Intracellular labeling of single cortical astrocytes in vivo. J Neurosci Methods 166:32–40
244. Miyano K, Morioka N, Sugimoto T, Shiraishi S, Uezono Y, Nakata Y (2010) Activation of the neurokinin-1 receptor in rat spinal astrocytes induces Ca^{2+} release from IP_3-sensitive Ca^{2+} stores and extracellular Ca^{2+} influx through TRPC3. Neurochem Int 57:923–934
245. Miyazaki I, Asanuma M, Diaz-Corrales FJ, Miyoshi K, Ogawa N (2004) Direct evidence for expression of dopamine receptors in astrocytes from basal ganglia. Brain Res 1029:120–123
246. Molnar T, Yarishkin O, Iuso A, Barabas P, Jones B, Marc RE, Phuong TT, Krizaj D (2016) Store-operated calcium entry in muller glia is controlled by synergistic activation of TRPC and Orai channels. J Neurosci 36:3184–3198
247. Monai H, Ohkura M, Tanaka M, Oe Y, Konno A, Hirai H, Mikoshiba K, Itohara S, Nakai J, Iwai Y, Hirase H (2016) Calcium imaging reveals glial involvement in transcranial direct current stimulation-induced plasticity in mouse brain. Nat Commun 7:11100
248. Moreno C, Sampieri A, Vivas O, Pena-Segura C, Vaca L (2012) STIM1 and Orai1 mediate thrombin-induced Ca^{2+} influx in rat cortical astrocytes. Cell Calcium 52:457–467
249. Morgello S, Uson RR, Schwartz EJ, Haber RS (1995) The human blood-brain barrier glucose transporter (GLUT1) is a glucose transporter of gray matter astrocytes. Glia 14:43–54
250. Muhic M, Vardjan N, Chowdhury HH, Zorec R, Kreft M (2015) Insulin and insulin-like growth factor 1 (IGF-1) modulate cytoplasmic glucose and glycogen levels but not glucose transport across the membrane in astrocytes. J Biol Chem 290:11167–11176
251. Mulkey DK, Wenker IC (2011) Astrocyte chemoreceptors: mechanisms of H^+ sensing by astrocytes in the retrotrapezoid nucleus and their possible contribution to respiratory drive. Exp Physiol 96:400–406
252. Muller T, Fritschy JM, Grosche J, Pratt GD, Mohler H, Kettenmann H (1994) Developmental regulation of voltage-gated K^+ channel and GABAA receptor expression in Bergmann glial cells. J Neurosci 14:2503–2514
253. Muller T, Moller T, Berger T, Schnitzer J, Kettenmann H (1992) Calcium entry through kainate receptors and resulting potassium-channel blockade in Bergmann glial cells. Science 256:1563–1566
254. Murphy S, Welk G (1990) Hydrolysis of polyphosphoinositides in astrocytes by platelet-activating factor. Eur J Pharmacol 188:399–401
255. Nadarajah B, Thomaidou D, Evans WH, Parnavelas JG (1996) Gap junctions in the adult cerebral cortex: regional differences in their distribution and cellular expression of connexins. J Comp Neurol 376:326–342
256. Nagelhus EA, Ottersen OP (2013) Physiological roles of aquaporin-4 in brain. Physiol Rev 93:1543–1562

257. Nagy JI, Dudek FE, Rash JE (2004) Update on connexins and gap junctions in neurons and glia in the mammalian nervous system. Brain Res Brain Res Rev 47:191–215
258. Nagy JI, Li X, Rempel J, Stelmack G, Patel D, Staines WA, Yasumura T, Rash JE (2001) Connexin26 in adult rodent central nervous system: demonstration at astrocytic gap junctions and colocalization with connexin30 and connexin43. J Comp Neurol 441:302–323
259. Nagy JI, Lynn BD, Tress O, Willecke K, Rash JE (2011) Connexin26 expression in brain parenchymal cells demonstrated by targeted connexin ablation in transgenic mice. Eur J Neurosci 34:263–271
260. Narcisse L, Scemes E, Zhao Y, Lee SC, Brosnan CF (2005) The cytokine IL-1β transiently enhances $P2X_7$ receptor expression and function in human astrocytes. Glia. 49:245–258
261. Navarrete M, Araque A (2008) Endocannabinoids mediate neuron-astrocyte communication. Neuron 57:883–893
262. Navarrete M, Araque A (2010) Endocannabinoids potentiate synaptic transmission through stimulation of astrocytes. Neuron 68:113–126
263. Newman EA (1996) Acid efflux from retinal glial cells generated by sodium bicarbonate cotransport. J Neurosci 16:159–168
264. Nilsson M, Eriksson PS, Ronnback L, Hansson E (1993) GABA induces Ca^{2+} transients in astrocytes. Neuroscience 54:605–614
265. Nilsson M, Hansson E, Ronnback L (1991) Adrenergic and 5-HT_2 receptors on the same astroglial cell. A microspectrofluorimetric study on cytosolic Ca^{2+} responses in single cells in primary culture. Brain Res Dev Brain Res 63:33–41
266. Nobile M, Monaldi I, Alloisio S, Cugnoli C, Ferroni S (2003) ATP-induced, sustained calcium signalling in cultured rat cortical astrocytes: evidence for a non-capacitative, P2X7-like-mediated calcium entry. FEBS Lett 538:71–76
267. Noda M, Sakuta H (2013) Central regulation of body-fluid homeostasis. Trends Neurosci 36:661–673
268. Norenberg W, Schunk J, Fischer W, Sobottka H, Riedel T, Oliveira JF, Franke H, Illes P (2010) Electrophysiological classification of $P2X_7$ receptors in rat cultured neocortical astroglia. Br J Pharmacol 160:1941–1952
269. Nwaobi SE, Cuddapah VA, Patterson KC, Randolph AC, Olsen ML (2016) The role of glial-specific $K_{ir}4.1$ in normal and pathological states of the CNS. Acta Neuropathol 132:1–21
270. O'Connor ER, Sontheimer H, Ransom BR (1994) Rat hippocampal astrocytes exhibit electrogenic sodium-bicarbonate co-transport. J Neurophysiol 72:2580–2589
271. Oikawa H, Nakamichi N, Kambe Y, Ogura M, Yoneda Y (2005) An increase in intracellular free calcium ions by nicotinic acetylcholine receptors in a single cultured rat cortical astrocyte. J Neurosci Res 79:535–544
272. Oliveira JF, Riedel T, Leichsenring A, Heine C, Franke H, Krugel U, Norenberg W, Illes P (2011) Rodent cortical astroglia express in situ functional $P2X_7$ receptors sensing pathologically high ATP concentrations. Cereb Cortex 21:806–820
273. Olsen ML, Campbell SL, Sontheimer H (2007) Differential distribution of $K_{ir}4.1$ in spinal cord astrocytes suggests regional differences in K^+ homeostasis. J Neurophysiol 98:786–793
274. Orellana JA, Saez PJ, Cortes-Campos C, Elizondo RJ, Shoji KF, Contreras-Duarte S, Figueroa V, Velarde V, Jiang JX, Nualart F, Saez JC, Garcia MA (2012) Glucose increases intracellular free Ca^{2+} in tanycytes via ATP released through connexin 43 hemichannels. Glia 60:53–68
275. Orellana JA, Saez PJ, Shoji KF, Schalper KA, Palacios-Prado N, Velarde V, Giaume C, Bennett MV, Saez JC (2009) Modulation of brain hemichannels and gap junction channels by pro-inflammatory agents and their possible role in neurodegeneration. Antioxid Redox Signal 11:369–399
276. Orlowski J, Grinstein S (1997) Na^+/H^+ exchangers of mammalian cells. J Biol Chem 272:22373–22376
277. Orthmann-Murphy JL, Freidin M, Fischer E, Scherer SS, Abrams CK (2007) Two distinct heterotypic channels mediate gap junction coupling between astrocyte and oligodendrocyte connexins. J Neurosci 27:13949–13957

278. Owe SG, Marcaggi P, Attwell D (2006) The ionic stoichiometry of the GLAST glutamate transporter in salamander retinal glia. J Physiol 577:591–599
279. Pakhotin P, Verkhratsky A (2005) Electrical synapses between Bergmann glial cells and Purkinje neurones in rat cerebellar slices. Mol Cell Neurosci 28:79–84
280. Palygin O, Lalo U, Pankratov Y (2011) Distinct pharmacological and functional properties of NMDA receptors in mouse cortical astrocytes. Br J Pharmacol 163:1755–1766
281. Palygin O, Lalo U, Verkhratsky A, Pankratov Y (2010) Ionotropic NMDA and $P2X_{1/5}$ receptors mediate synaptically induced Ca^{2+} signalling in cortical astrocytes. Cell Calcium 48:225–231
282. Panenka W, Jijon H, Herx LM, Armstrong JN, Feighan D, Wei T, Yong VW, Ransohoff RM, MacVicar BA (2001) $P2X_7$-like receptor activation in astrocytes increases chemokine monocyte chemoattractant protein-1 expression via mitogen-activated protein kinase. J Neurosci 21:7135–7142
283. Pangrsic T, Potokar M, Haydon PG, Zorec R, Kreft M (2006) Astrocyte swelling leads to membrane unfolding, not membrane insertion. J Neurochem 99:514–523
284. Pankratov Y, Lalo U (2015) Role for astroglial α1-adrenoreceptors in gliotransmission and control of synaptic plasticity in the neocortex. Front Cell Neurosci 9:230
285. Pappalardo LW, Black JA, Waxman SG (2016) Sodium channels in astroglia and microglia. Glia
286. Pappalardo LW, Liu S, Black JA, Waxman SG (2014) Dynamics of sodium channel $Na_v1.5$ expression in astrocytes in mouse models of multiple sclerosis. NeuroReport 25:1208–1215
287. Pappalardo LW, Samad OA, Black JA, Waxman SG (2014) Voltage-gated sodium channel Na_v 1.5 contributes to astrogliosis in an in vitro model of glial injury via reverse Na^+/Ca^{2+} exchange. Glia 62:1162–1175
288. Pappas CA, Ransom BR (1993) A depolarization-stimulated, bafilomycin-inhibitable H^+ pump in hippocampal astrocytes. Glia 9:280–291
289. Parekh AB, Putney JW Jr (2005) Store-operated calcium channels. Physiol Rev 85:757–810
290. Park H, Han KS, Oh SJ, Jo S, Woo J, Yoon BE, Lee CJ (2013) High glutamate permeability and distal localization of Best1 channel in CA1 hippocampal astrocyte. Mol Brain 6:54
291. Parkerson KA, Sontheimer H (2004) Biophysical and pharmacological characterization of hypotonically activated chloride currents in cortical astrocytes. Glia 46:419–436
292. Parpura V, Basarsky TA, Liu F, Jeftinija K, Jeftinija S, Haydon PG (1994) Glutamate-mediated astrocyte-neuron signalling. Nature 369:744–747
293. Parpura V, Scemes E, Spray DC (2004) Mechanisms of glutamate release from astrocytes: gap junction “hemichannels”, purinergic receptors and exocytotic release. Neurochem Int 45:259–264
294. Parri HR, Gould TM, Crunelli V (2001) Spontaneous astrocytic Ca^{2+} oscillations in situ drive NMDAR-mediated neuronal excitation. Nat Neurosci 4:803–812
295. Pastor A, Chvatal A, Sykova E, Kettenmann H (1995) Glycine- and GABA-activated currents in identified glial cells of the developing rat spinal cord slice. Eur J Neurosci 7:1188–1198
296. Pastor A, Kremer M, Moller T, Kettenmann H, Dermietzel R (1998) Dye coupling between spinal cord oligodendrocytes: differences in coupling efficiency between gray and white matter. Glia 24:108–120
297. Peakman MC, Hill SJ (1995) Adenosine A_1 receptor-mediated changes in basal and histamine-stimulated levels of intracellular calcium in primary rat astrocytes. Br J Pharmacol 115:801–810
298. Pearce B, Langley D (1994) Purine- and pyrimidine-stimulated phosphoinositide breakdown and intracellular calcium mobilisation in astrocytes. Brain Res 660:329–332
299. Pearce B, Murphy S, Jeremy J, Morrow C, Dandona P (1989) ATP-evoked Ca^{2+} mobilisation and prostanoid release from astrocytes: P2-purinergic receptors linked to phosphoinositide hydrolysis. J Neurochem 52:971–977
300. Peng L, Huang R, Yu AC, Fung KY, Rathbone MP, Hertz L (2005) Nucleoside transporter expression and function in cultured mouse astrocytes. Glia 52:25–35

301. Pereira GJ, Hirata H, Fimia GM, do Carmo LG, Bincoletto C, Han SW, Stilhano RS, Ureshino RP, Bloor-Young D, Churchill G, Piacentini M, Patel S, Smaili SS (2011) Nicotinic acid adenine dinucleotide phosphate (NAADP) regulates autophagy in cultured astrocytes. J Biol Chem 286:27875–27881
302. Petravicz J, Boyt KM, McCarthy KD (2014) Astrocyte IP_3R2-dependent Ca^{2+} signaling is not a major modulator of neuronal pathways governing behavior. Front Behav Neurosci 8:384
303. Petravicz J, Fiacco TA, McCarthy KD (2008) Loss of IP_3 receptor-dependent Ca^{2+} increases in hippocampal astrocytes does not affect baseline CA1 pyramidal neuron synaptic activity. J Neurosci 28:4967–4973
304. Petroni A, Salami M, Blasevich M, Papini N, Galella G, Colombo C, Galli C (1994) Eicosanoid and inositol phosphate response to platelet-activating factor (PAF) and to a PAF antagonist in rat astroglial cells. Brain Res Dev Brain Res 78:169–174
305. Philippe JM, Dubois JM, Rouzaire-Dubois B, Cartron PF, Vallette F, Morel N (2002) Functional expression of V-ATPases in the plasma membrane of glial cells. Glia 37:365–373
306. Pilitsis JG, Kimelberg HK (1998) Adenosine receptor mediated stimulation of intracellular calcium in acutely isolated astrocytes. Brain Res 798:294–303
307. Pines G, Danbolt NC, Bjoras M, Zhang Y, Bendahan A, Eide L, Koepsell H, Storm-Mathisen J, Seeberg E, Kanner BI (1992) Cloning and expression of a rat brain L-glutamate transporter. Nature 360:464–467
308. Pizzo P, Burgo A, Pozzan T, Fasolato C (2001) Role of capacitative calcium entry on glutamate-induced calcium influx in type-I rat cortical astrocytes. J Neurochem 79:98–109
309. Poopalasundaram S, Knott C, Shamotienko OG, Foran PG, Dolly JO, Ghiani CA, Gallo V, Wilkin GP (2000) Glial heterogeneity in expression of the inwardly rectifying K^+ channel, $K_{ir}4.1$, in adult rat CNS. Glia. 30:362–372
310. Porter JT, McCarthy KD (1995) Adenosine receptors modulate $[Ca^{2+}]_i$ in hippocampal astrocytes in situ. J Neurochem 65:1515–1523
311. Porter JT, McCarthy KD (1995) GFAP-positive hippocampal astrocytes in situ respond to glutamatergic neuroligands with increases in $[Ca^{2+}]i$. Glia. 13:101–112
312. Porter JT, McCarthy KD (1996) Hippocampal astrocytes in situ respond to glutamate released from synaptic terminals. J Neurosci 16:5073–5081
313. Potokar M, Stenovec M, Jorgacevski J, Holen T, Kreft M, Ottersen OP, Zorec R (2013) Regulation of AQP4 surface expression via vesicle mobility in astrocytes. Glia 61:917–928
314. Potter LR, Abbey-Hosch S, Dickey DM (2006) Natriuretic peptides, their receptors, and cyclic guanosine monophosphate-dependent signaling functions. Endocr Rev 27:47–72
315. Price DL, Ludwig JW, Mi H, Schwarz TL, Ellisman MH (2002) Distribution of rSlo Ca^{2+}-activated K^+ channels in rat astrocyte perivascular endfeet. Brain Res 956:183–193
316. Price TJ, Hargreaves KM, Cervero F (2006) Protein expression and mRNA cellular distribution of the NKCC1 cotransporter in the dorsal root and trigeminal ganglia of the rat. Brain Res 1112:146–158
317. Raap M, Biedermann B, Braun P, Milenkovic I, Skatchkov SN, Bringmann A, Reichenbach A (2002) Diversity of K_{ir} channel subunit mRNA expressed by retinal glial cells of the guinea-pig. NeuroReport 13:1037–1040
318. Radian R, Kanner BI (1983) Stoichiometry of sodium- and chloride-coupled gamma-aminobutyric acid transport by synaptic plasma membrane vesicles isolated from rat brain. Biochemistry 22:1236–1241
319. Rauen T, Rothstein JD, Wassle H (1996) Differential expression of three glutamate transporter subtypes in the rat retina. Cell Tissue Res 286:325–336
320. Ray A, Zoidl G, Weickert S, Wahle P, Dermietzel R (2005) Site-specific and developmental expression of pannexin1 in the mouse nervous system. Eur J Neurosci 21:3277–3290
321. Reese KA, Caldwell JH (1999) Immunocytochemical localization of NaCh6 in cultured spinal cord astrocytes. Glia 26:92–96
322. Requardt RP, Hirrlinger PG, Wilhelm F, Winkler U, Besser S, Hirrlinger J (2012) Ca^{2+} signals of astrocytes are modulated by the NAD^+/NADH redox state. J Neurochem 120:1014–1025

323. Reyes RC, Verkhratsky A, Parpura V (2012) Plasmalemmal Na^+/Ca^{2+} exchanger modulates Ca^{2+}-dependent exocytotic release of glutamate from rat cortical astrocytes. ASN Neuro 4
324. Reyes RC, Verkhratsky A, Parpura V (2013) TRPC1-mediated Ca^{2+} and Na^+ signalling in astroglia: differential filtering of extracellular cations. Cell Calcium (in press)
325. Ribak CE, Tong WM, Brecha NC (1996) GABA plasma membrane transporters, GAT-1 and GAT-3, display different distributions in the rat hippocampus. J Comp Neurol 367:595–606
326. Riquelme R, Miralles CP, De Blas AL (2002) Bergmann glia $GABA_A$ receptors concentrate on the glial processes that wrap inhibitory synapses. J Neurosci 22:10720–10730
327. Ronco V, Grolla AA, Glasnov TN, Canonico PL, Verkhratsky A, Genazzani AA, Lim D (2014) Differential deregulation of astrocytic calcium signalling by amyloid-β, TNFα, IL-1β and LPS. Cell Calcium 55:219–229
328. Rose CR, Karus C (2013) Two sides of the same coin: sodium homeostasis and signaling in astrocytes under physiological and pathophysiological conditions. Glia 61:1191–1205
329. Rose CR, Ransom BR (1996) Intracellular sodium homeostasis in rat hippocampal astrocytes. J Physiol 491(Pt 2):291–305
330. Rose CR, Ransom BR (1997) Regulation of intracellular sodium in cultured rat hippocampal neurones. J Physiol 499(Pt 3):573–587
331. Rose CR, Verkhratsky A (2016) Principles of sodium homeostasis and sodium signalling in astroglia. Glia 64:1611–1627
332. Roux L, Benchenane K, Rothstein JD, Bonvento G, Giaume C (2011) Plasticity of astroglial networks in olfactory glomeruli. Proc Natl Acad Sci USA 108:18442–18446
333. Roux MJ, Supplisson S (2000) Neuronal and glial glycine transporters have different stoichiometries. Neuron 25:373–383
334. Roy ML, Saal D, Perney T, Sontheimer H, Waxman SG, Kaczmarek LK (1996) Manipulation of the delayed rectifier $K_V1.5$ potassium channel in glial cells by antisense oligodeoxynucleotides. Glia 18:177–184
335. Rubini P, Pagel G, Mehri S, Marquardt P, Riedel T, Illes P (2014) Functional $P2X_7$ receptors at cultured hippocampal astrocytes but not neurons. Naunyn Schmiedebergs Arch Pharmacol 387:943–954
336. Rungta RL, Bernier LP, Dissing-Olesen L, Groten CJ, LeDue JM, Ko R, Drissler S, MacVicar BA (2016) Ca^{2+} transients in astrocyte fine processes occur via Ca^{2+} influx in the adult mouse hippocampus. Glia 64:2093–2103
337. Rusnakova V, Honsa P, Dzamba D, Stahlberg A, Kubista M, Anderova M (2013) Heterogeneity of astrocytes: from development to injury—single cell gene expression. PLoS ONE 8:e69734
338. Saab AS, Neumeyer A, Jahn HM, Cupido A, Simek AA, Boele HJ, Scheller A, Le Meur K, Gotz M, Monyer H, Sprengel R, Rubio ME, Deitmer JW, De Zeeuw CI, Kirchhoff F (2012) Bergmann glial AMPA receptors are required for fine motor coordination. Science 337:749–753
339. Salm AK, McCarthy KD (1990) Norepinephrine-evoked calcium transients in cultured cerebral type 1 astroglia. Glia 3:529–538
340. Sanchez C, Galve-Roperh I, Rueda D, Guzman M (1998) Involvement of sphingomyelin hydrolysis and the mitogen-activated protein kinase cascade in the Delta9-tetrahydrocannabinol-induced stimulation of glucose metabolism in primary astrocytes. Mol Pharmacol 54:834–843
341. Sanden N, Thorlin T, Blomstrand F, Persson PA, Hansson E (2000) 5-Hydroxytryptamine$_{2B}$ receptors stimulate Ca^{2+} increases in cultured astrocytes from three different brain regions. Neurochem Int 36:427–434
342. Satoh J, Tabunoki H, Yamamura T, Arima K, Konno H (2007) Human astrocytes express aquaporin-1 and aquaporin-4 in vitro and in vivo. Neuropathology 27:245–256
343. Scalise M, Pochini L, Galluccio M, Indiveri C (2016) Glutamine transport. From energy supply to sensing and beyond. Biochim Biophys Acta
344. Scemes E, Giaume C (2006) Astrocyte calcium waves: what they are and what they do. Glia 54:716–725

345. Schaller KL, Krzemien DM, Yarowsky PJ, Krueger BK, Caldwell JH (1995) A novel, abundant sodium channel expressed in neurons and glia. J Neurosci 15:3231–3242
346. Scharfman HE, Binder DK (2013) Aquaporin-4 water channels and synaptic plasticity in the hippocampus. Neurochem Int 63:702–711
347. Schipke CG, Heuser I, Peters O (2011) Antidepressants act on glial cells: SSRIs and serotonin elicit astrocyte calcium signaling in the mouse prefrontal cortex. J Psychiatr Res 45:242–248
348. Schipke CG, Ohlemeyer C, Matyash M, Nolte C, Kettenmann H, Kirchhoff F (2001) Astrocytes of the mouse neocortex express functional N-methyl-D-aspartate receptors. FASEB J. 15:1270–1272
349. Schroeter S, Apparsundaram S, Wiley RG, Miner LH, Sesack SR, Blakely RD (2000) Immunolocalization of the cocaine- and antidepressant-sensitive l-norepinephrine transporter. J Comp Neurol 420:211–232
350. Seifert G, Huttmann K, Binder DK, Hartmann C, Wyczynski A, Neusch C, Steinhauser C (2009) Analysis of astroglial K^+ channel expression in the developing hippocampus reveals a predominant role of the $K_{ir}4.1$ subunit. J Neurosci 29:7474–7488
351. Seifert G, Steinhauser C (1995) Glial cells in the mouse hippocampus express AMPA receptors with an intermediate Ca^{2+} permeability. Eur J Neurosci 7:1872–1881
352. Shao Y, Sutin J (1992) Expression of adrenergic receptors in individual astrocytes and motor neurons isolated from the adult rat brain. Glia 6:108–117
353. Sharma G, Vijayaraghavan S (2001) Nicotinic cholinergic signaling in hippocampal astrocytes involves calcium-induced calcium release from intracellular stores. Proc Natl Acad Sci USA 98:4148–4153
354. Sharp AH, Nucifora FC Jr, Blondel O, Sheppard CA, Zhang C, Snyder SH, Russell JT, Ryugo DK, Ross CA (1999) Differential cellular expression of isoforms of inositol 1,4,5-triphosphate receptors in neurons and glia in brain. J Comp Neurol 406:207–220
355. Shelton MK, McCarthy KD (2000) Hippocampal astrocytes exhibit Ca^{2+}-elevating muscarinic cholinergic and histaminergic receptors in situ. J Neurochem 74:555–563
356. Sheppard CA, Simpson PB, Sharp AH, Nucifora FC, Ross CA, Lange GD, Russell JT (1997) Comparison of type 2 inositol 1,4,5-trisphosphate receptor distribution and subcellular Ca^{2+} release sites that support Ca^{2+} waves in cultured astrocytes. J Neurochem 68:2317–2327
357. Sherwood MW, Arizono M, Hisatsune C, Bannai H, Ebisui E, Sherwood JL, Panatier A, Oliet SH, Mikoshiba K (2017) Astrocytic IP_3 Rs: contribution to Ca^{2+} signalling and hippocampal LTP. Glia 65:502–513
358. Shibasaki K, Hosoi N, Kaneko R, Tominaga M, Yamada K (2016) Glycine release from astrocytes via functional reversal of GlyT1. J Neurochem
359. Shigetomi E, Bowser DN, Sofroniew MV, Khakh BS (2008) Two forms of astrocyte calcium excitability have distinct effects on NMDA receptor-mediated slow inward currents in pyramidal neurons. J Neurosci 28:6659–6663
360. Shigetomi E, Jackson-Weaver O, Huckstepp RT, O'Dell TJ, Khakh BS (2013) TRPA1 channels are regulators of astrocyte basal calcium levels and long-term potentiation via constitutive D-serine release. J Neurosci 33:10143–10153
361. Shigetomi E, Patel S, Khakh BS (2016) Probing the complexities of astrocyte calcium signaling. Trends Cell Biol 26:300–312
362. Shigetomi E, Tong X, Kwan KY, Corey DP, Khakh BS (2012) TRPA1 channels regulate astrocyte resting calcium and inhibitory synapse efficacy through GAT-3. Nat Neurosci 15:70–80
363. Sik A, Smith RL, Freund TF (2000) Distribution of chloride channel-2-immunoreactive neuronal and astrocytic processes in the hippocampus. Neuroscience 101:51–65
364. Simpson PB, Holtzclaw LA, Langley DB, Russell JT (1998) Characterization of ryanodine receptors in oligodendrocytes, type 2 astrocytes, and O-2A progenitors. J Neurosci Res 52:468–482
365. Skatchkov SN, Rojas L, Eaton MJ, Orkand RK, Biedermann B, Bringmann A, Pannicke T, Veh RW, Reichenbach A (2002) Functional expression of K_{ir} 6.1/SUR1-K_{ATP} channels in frog retinal Muller glial cells. Glia 38:256–267

366. Skowronska M, Zielinska M, Albrecht J (2010) Stimulation of natriuretic peptide receptor C attenuates accumulation of reactive oxygen species and nitric oxide synthesis in ammonia-treated astrocytes. J Neurochem 115:1068–1076
367. Skucas VA, Mathews IB, Yang J, Cheng Q, Treister A, Duffy AM, Verkman AS, Hempstead BL, Wood MA, Binder DK, Scharfman HE (2011) Impairment of select forms of spatial memory and neurotrophin-dependent synaptic plasticity by deletion of glial aquaporin-4. J Neurosci 31:6392–6397
368. Smyth JT, Dehaven WI, Jones BF, Mercer JC, Trebak M, Vazquez G, Putney JW Jr (2006) Emerging perspectives in store-operated Ca^{2+} entry: roles of Orai, Stim and TRP. Biochim Biophys Acta 1763:1147–1160
369. Sohl G, Odermatt B, Maxeiner S, Degen J, Willecke K (2004) New insights into the expression and function of neural connexins with transgenic mouse mutants. Brain Res Brain Res Rev 47:245–259
370. Solenov E, Watanabe H, Manley GT, Verkman AS (2004) Sevenfold-reduced osmotic water permeability in primary astrocyte cultures from AQP-4-deficient mice, measured by a fluorescence quenching method. Am J Physiol Cell Physiol 286:C426–C432
371. Sontheimer H, Black JA, Ransom BR, Waxman SG (1992) Ion channels in spinal cord astrocytes in vitro. I. Transient expression of high levels of Na^+ and K^+ channels. J Neurophysiol 68:985–1000
372. Sontheimer H, Fernandez-Marques E, Ullrich N, Pappas CA, Waxman SG (1994) Astrocyte Na^+ channels are required for maintenance of Na^+/K^+-ATPase activity. J Neurosci 14:2464–2475
373. Sontheimer H, Minturn JE, Black JA, Ransom BR, Waxman SG (1991) Two types of Na^+-currents in cultured rat optic nerve astrocytes: changes with time in culture and with age of culture derivation. J Neurosci Res 30:275–287
374. Sontheimer H, Waxman SG (1992) Ion channels in spinal cord astrocytes in vitro. II. Biophysical and pharmacological analysis of two Na^+ current types. J Neurophysiol 68:1001–1011
375. Srinivasan R, Huang BS, Venugopal S, Johnston AD, Chai H, Zeng H, Golshani P, Khakh BS (2015) Ca^{2+} signaling in astrocytes from $Ip3r2^{-/-}$ mice in brain slices and during startle responses in vivo. Nat Neurosci 18:708–717
376. Steinhauser C, Jabs R, Kettenmann H (1994) Properties of GABA and glutamate responses in identified glial cells of the mouse hippocampal slice. Hippocampus 4:19–35
377. Stenovec M, Kreft M, Grilc S, Pangrsic T, Zorec R (2008) EAAT2 density at the astrocyte plasma membrane and $Ca^{(2+)}$-regulated exocytosis. Mol Membr Biol 25:203–215
378. Stephens GJ, Cholewinski AJ, Wilkin GP, Djamgoz MB (1993) Calcium-mobilizing and electrophysiological effects of bradykinin on cortical astrocyte subtypes in culture. Glia 9:269–279
379. Stiene-Martin A, Gurwell JA, Hauser KF (1991) Morphine alters astrocyte growth in primary cultures of mouse glial cells: evidence for a direct effect of opiates on neural maturation. Brain Res Dev Brain Res 60:1–7
380. Stonehouse AH, Pringle JH, Norman RI, Stanfield PR, Conley EC, Brammar WJ (1999) Characterisation of $K_{ir}2.0$ proteins in the rat cerebellum and hippocampus by polyclonal antibodies. Histochem Cell Biol 112:457–465
381. Storck T, Schulte S, Hofmann K, Stoffel W (1992) Structure, expression, and functional analysis of a Na^+-dependent glutamate/aspartate transporter from rat brain. Proc Natl Acad Sci USA 89:10955–10959
382. Suadicani SO, Brosnan CF, Scemes E (2006) $P2X_7$ receptors mediate ATP release and amplification of astrocytic intercellular Ca^{2+} signaling. J Neurosci 26:1378–1385
383. Sumners C, Tang W, Paulding W, Raizada MK (1994) Peptide receptors in astroglia: focus on angiotensin II and atrial natriuretic peptide. Glia 11:110–116
384. Sun L, Kosugi Y, Kawakami E, Piao YS, Hashimoto T, Oyanagi K (2009) Magnesium concentration in the cerebrospinal fluid of mice and its response to changes in serum magnesium concentration. Magnes Res 22:266–272
385. Sun W, McConnell E, Pare JF, Xu Q, Chen M, Peng W, Lovatt D, Han X, Smith Y, Nedergaard M (2013) Glutamate-dependent neuroglial calcium signaling differs between young and adult brain. Science 339:197–200

386. Takata N, Mishima T, Hisatsune C, Nagai T, Ebisui E, Mikoshiba K, Hirase H (2011) Astrocyte calcium signaling transforms cholinergic modulation to cortical plasticity *in vivo*. J Neurosci 31:18155–18165
387. Takeda H, Inazu M, Matsumiya T (2002) Astroglial dopamine transport is mediated by norepinephrine transporter. Naunyn Schmiedebergs Arch Pharmacol. 366:620–623
388. Takuma K, Matsuda T, Hashimoto H, Kitanaka J, Asano S, Kishida Y, Baba A (1996) Role of Na^+-Ca^{2+} exchanger in agonist-induced Ca^{2+} signaling in cultured rat astrocytes. J Neurochem 67:1840–1845
389. Talantova M, Sanz-Blasco S, Zhang X, Xia P, Akhtar MW, Okamoto S, Dziewczapolski G, Nakamura T, Cao G, Pratt AE, Kang YJ, Tu S, Molokanova E, McKercher SR, Hires SA, Sason H, Stouffer DG, Buczynski MW, Solomon JP, Michael S, Powers ET, Kelly JW, Roberts A, Tong G, Fang-Newmeyer T, Parker J, Holland EA, Zhang D, Nakanishi N, Chen HS, Wolosker H, Wang Y, Parsons LH, Ambasudhan R, Masliah E, Heinemann SF, Pina-Crespo JC, Lipton SA (2013) Abeta induces astrocytic glutamate release, extrasynaptic NMDA receptor activation, and synaptic loss. Proc Natl Acad Sci USA 110:E2518–E2527
390. Teaktong T, Graham A, Court J, Perry R, Jaros E, Johnson M, Hall R, Perry E (2003) Alzheimer's disease is associated with a selective increase in α7 nicotinic acetylcholine receptor immunoreactivity in astrocytes. Glia. 41:207–211
391. Teather LA, Lee RK, Wurtman RJ (2002) Platelet-activating factor increases prostaglandin E_2 release from astrocyte-enriched cortical cell cultures. Brain Res 946:87–95
392. Tence M, Ezan P, Amigou E, Giaume C (2012) Increased interaction of connexin43 with zonula occludens-1 during inhibition of gap junctions by G protein-coupled receptor agonists. Cell Signal 24:86–98
393. Teoh R, Kum W, Cockram CS, Young JD, Nicholls MG (1989) Mouse astrocytes possess specific ANP receptors which are linked to cGMP production. Clin Exp Pharmacol Physiol 16:323–327
394. Theparambil SM, Naoshin Z, Thyssen A, Deitmer JW (2015) Reversed electrogenic sodium bicarbonate cotransporter 1 is the major acid loader during recovery from cytosolic alkalosis in mouse cortical astrocytes. J Physiol 593:3533–3547
395. Thomzig A, Wenzel M, Karschin C, Eaton MJ, Skatchkov SN, Karschin A, Veh RW (2001) $K_{ir}6.1$ is the principal pore-forming subunit of astrocyte but not neuronal plasma membrane K-ATP channels. Mol Cell Neurosci 18:671–690
396. Thorlin T, Eriksson PS, Persson PA, Aberg ND, Hansson E, Ronnback L (1998) δ-opioid receptors on astroglial cells in primary culture: mobilization of intracellular free calcium via a pertussis sensitive G protein. Neuropharmacology 37:299–311
397. Todd AC, Marx MC, Hulme SR, Broer S, Billups B (2017) SNAT3-mediated glutamine transport in perisynaptic astrocytes in situ is regulated by intracellular sodium. Glia
398. Toth A, Boczan J, Kedei N, Lizanecz E, Bagi Z, Papp Z, Edes I, Csiba L, Blumberg PM (2005) Expression and distribution of vanilloid receptor 1 (TRPV1) in the adult rat brain. Brain Res Mol Brain Res 135:162–168
399. Tuschick S, Kirischuk S, Kirchhoff F, Liefeldt L, Paul M, Verkhratsky A, Kettenmann H (1997) Bergmann glial cells in situ express endothelin$_B$ receptors linked to cytoplasmic calcium signals. Cell Calcium 21:409–419
400. Ubl JJ, Reiser G (1997) Characteristics of thrombin-induced calcium signals in rat astrocytes. Glia. 21:361–369
401. Ubl JJ, Vohringer C, Reiser G (1998) Co-existence of two types of $[Ca^{2+}]_i$-inducing protease-activated receptors (PAR-1 and PAR-2) in rat astrocytes and C6 glioma cells. Neuroscience 86:597–609
402. Unichenko P, Myakhar O, Kirischuk S (2012) Intracellular Na^+ concentration influences short-term plasticity of glutamate transporter-mediated currents in neocortical astrocytes. Glia 60:605–614
403. Untiet V, Kovermann P, Gerkau NJ, Gensch T, Rose CR, Fahlke C (2017) Glutamate transporter-associated anion channels adjust intracellular chloride concentrations during glial maturation. Glia 65:388–400

404. Vandenberg RJ, Ryan RM (2013) Mechanisms of glutamate transport. Physiol Rev 93:1621–1657
405. Venance L, Premont J, Glowinski J, Giaume C (1998) Gap junctional communication and pharmacological heterogeneity in astrocytes cultured from the rat striatum. J Physiol 510(Pt 2):429–440
406. Verkhratsky A, Butt AM (2013) Glial physiology and pathophysiology. Wiley-Blackwell, Chichester, p 560
407. Verkhratsky A, Kettenmann H (1996) Calcium signalling in glial cells. Trends Neurosci 19:346–352
408. Verkhratsky A, Nedergaard M (2014) Astroglial cradle in the life of the synapse. Philos Trans R Soc Lond B Biol Sci 369:20130595
409. Verkhratsky A, Nedergaard M (2016). The homeostatic astroglia emerges from evolutionary specialization of neural cells. Philos Trans R Soc Lond B Biol Sci 371
410. Verkhratsky A, Nedergaard M (2018) Physiology of astroglia. Physiol Rev 98:239–389
411. Verkhratsky A, Orkand RK, Kettenmann H (1998) Glial calcium: homeostasis and signaling function. Physiol Rev 78:99–141
412. Verkhratsky A, Parpura V (2014) Store-operated calcium entry in neuroglia. Neurosci Bull 30:125–133
413. Verkhratsky A, Parpura V (2015) Physiology of astroglia: channels, receptors, transporters, ion signaling and gliotransmission. Morgan & Claypool Publishers, 172 pp
414. Verkhratsky A, Rodriguez JJ, Parpura V (2012) Calcium signalling in astroglia. Mol Cell Endocrinol 353:45–56
415. Verkhratsky A, Trebak M, Perocchi F, Khananshvili D, Sekler I (2018) Crosslink between calcium and sodium signalling. Exp Physiol 103:157–169
416. Waldmann R, Champigny G, Bassilana F, Voilley N, Lazdunski M (1995) Molecular cloning and functional expression of a novel amiloride-sensitive Na^+ channel. J Biol Chem 270:27411–27414
417. Wang CM, Chang YY, Sun SH (2003) Activation of $P2X_7$ purinoceptor-stimulated TGF-β 1 mRNA expression involves PKC/MAPK signalling pathway in a rat brain-derived type-2 astrocyte cell line, RBA-2. Cell Signal 15:1129–1137
418. Wang D, Yan B, Rajapaksha WR, Fisher TE (2009) The expression of voltage-gated ca^{2+} channels in pituicytes and the up-regulation of L-type ca^{2+} channels during water deprivation. J Neuroendocrinol 21:858–866
419. Wang F, Du T, Liang C, Verkhratsky A, Peng L (2015) Ammonium increases Ca^{2+} signalling and upregulates expression of $Ca_v1.2$ gene in astrocytes in primary cultures and in the *in vivo* brain. Acta Physiol (Oxf) 214:261–274
420. Watanabe E, Hiyama TY, Shimizu H, Kodama R, Hayashi N, Miyata S, Yanagawa Y, Obata K, Noda M (2006) Sodium-level-sensitive sodium channel Na_x is expressed in glial laminate processes in the sensory circumventricular organs. Am J Physiol Regul Integr Comp Physiol 290:R568–R576
421. Westenbroek RE, Bausch SB, Lin RC, Franck JE, Noebels JL, Catterall WA (1998) Upregulation of L-type Ca^{2+} channels in reactive astrocytes after brain injury, hypomyelination, and ischemia. J Neurosci 18:2321–2334
422. Woo DH, Han KS, Shim JW, Yoon BE, Kim E, Bae JY, Oh SJ, Hwang EM, Marmorstein AD, Bae YC, Park JY, Lee CJ (2012) TREK-1 and Best1 channels mediate fast and slow glutamate release in astrocytes upon GPCR activation. Cell 151:25–40
423. Yang F, Sun X, Ding Y, Ma H, Yang TO, Ma Y, Wei D, Li W, Xu T, Jiang W (2016) Astrocytic acid-sensing ion channel 1a contributes to the development of chronic epileptogenesis. Sci Rep 6:31581
424. Zafra F, Aragon C, Olivares L, Danbolt NC, Gimenez C, Storm-Mathisen J (1995) Glycine transporters are differentially expressed among CNS cells. J Neurosci 15:3952–3969
425. Zanassi P, Paolillo M, Montecucco A, Avvedimento EV, Schinelli S (1999) Pharmacological and molecular evidence for dopamine D_1 receptor expression by striatal astrocytes in culture. J Neurosci Res 58:544–552

426. Zerangue N, Kavanaugh MP (1996) Flux coupling in a neuronal glutamate transporter. Nature 383:634–637
427. Zhang J, Li Y, Chen ZG, Dang H, Ding JH, Fan Y, Hu G (2013) Glia protein aquaporin-4 regulates aversive motivation of spatial memory in Morris water maze. CNS Neurosci Ther 19:937–944
428. Zhang S, Li B, Lovatt D, Xu J, Song D, Goldman SA, Nedergaard M, Hertz L, Peng L (2010) 5-HT_{2B} receptors are expressed on astrocytes from brain and in culture and are a chronic target for all five conventional 'serotonin-specific reuptake inhibitors'. Neuron Glia Biol 6:113–125
429. Zhang XD, Morishima S, Ando-Akatsuka Y, Takahashi N, Nabekura T, Inoue H, Shimizu T, Okada Y (2004) Expression of novel isoforms of the ClC-1 chloride channel in astrocytic glial cells in vitro. Glia. 47:46–57
430. Zhang Y, Chen K, Sloan SA, Bennett ML, Scholze AR, O'Keeffe S, Phatnani HP, Guarnieri P, Caneda C, Ruderisch N, Deng S, Liddelow SA, Zhang C, Daneman R, Maniatis T, Barres BA, Wu JQ (2014) An RNA-sequencing transcriptome and splicing database of glia, neurons, and vascular cells of the cerebral cortex. J Neurosci 34:11929–11947
431. Zhou M, Tanaka O, Suzuki M, Sekiguchi M, Takata K, Kawahara K, Abe H (2002) Localization of pore-forming subunit of the ATP-sensitive K^+-channel, $K_{ir}6.2$, in rat brain neurons and glial cells. Brain Res Mol Brain Res 101:23–32
432. Zhou M, Xu G, Xie M, Zhang X, Schools GP, Ma L, Kimelberg HK, Chen H (2009) TWIK-1 and TREK-1 are potassium channels contributing significantly to astrocyte passive conductance in rat hippocampal slices. J Neurosci 29:8551–8564
433. Zhou Y, Danbolt NC (2013) GABA and glutamate transporters in brain. Front Endocrinol (Lausanne) 4:165
434. Zhu H, Zhao Y, Wu H, Jiang N, Wang Z, Lin W, Jin J, Ji Y (2016) Remarkable alterations of Nav1.6 in reactive astrogliosis during epileptogenesis. Sci Rep 6:38108
435. Zhu SQ, Kum W, Ho SK, Young JD, Cockram CS (1990) Structure-function relationships of insulin receptor interactions in cultured mouse astrocytes. Brain Res 529:329–332
436. Zhu Z, Reiser G (2014) Signaling mechanism of protease activated receptor 1-induced proliferation of astrocytes: stabilization of hypoxia inducible factor-1alpha triggers glucose metabolism and accumulation of cyclin D1. Neurochem Int 79:20–32
437. Ziak D, Chvatal A, Sykova E (1998) Glutamate-, kainate- and NMDA-evoked membrane currents in identified glial cells in rat spinal cord slice. Physiol Res 47:365–375
438. Zielinska M, Fresko I, Konopacka A, Felipo V, Albrecht J (2007) Hyperammonemia inhibits the natriuretic peptide receptor 2 (NPR-2)-mediated cyclic GMP synthesis in the astrocytic compartment of rat cerebral cortex slices. Neurotoxicology 28:1260–1263

Chapter 4
Gliocrine System: Astroglia as Secretory Cells of the CNS

Nina Vardjan, Vladimir Parpura, Alexei Verkhratsky and Robert Zorec

Abstract Astrocytes are secretory cells, actively participating in cell-to-cell communication in the central nervous system (CNS). They sense signaling molecules in the extracellular space, around the nearby synapses and also those released at much farther locations in the CNS, by their cell surface receptors, get excited to then release their own signaling molecules. This contributes to the brain information processing, based on diffusion within the extracellular space around the synapses and on convection when locales relatively far away from the release sites are involved. These functions resemble secretion from endocrine cells, therefore astrocytes were termed to be a part of the gliocrine system in 2015. An important mechanism, by which astrocytes release signaling molecules is the merger of the vesicle membrane with the plasmalemma, i.e., exocytosis. Signaling molecules stored in astroglial secretory vesicles can be discharged into the extracellular space after the vesicle membrane fuses with the plasma membrane. This leads to a fusion pore formation, a channel that must widen to allow the exit of the Vesiclal cargo. Upon complete vesicle membrane fusion, this process also integrates other proteins, such as receptors, transporters

---

N. Vardjan (✉) · R. Zorec (✉)
Laboratory of Neuroendocrinology-Molecular Cell Physiology, Faculty of Medicine, Institute of Pathophysiology, University of Ljubljana, 1000 Ljubljana, Slovenia
e-mail: nina.vardjan@mf.uni-lj.si

R. Zorec
e-mail: robert.zorec@mf.uni-lj.si

Celica Biomedical, 1000 Ljubljana, Slovenia

V. Parpura
Department of Neurobiology, The University of Alabama at Birmingham, Birmingham, AL, USA

A. Verkhratsky
Faculty of Biology, Medicine and Health, The University of Manchester, Manchester M13 9PT, UK

Center for Basic and Translational Neuroscience, Faculty of Health and Medical Sciences, University of Copenhagen, 2200 Copenhagen, Denmark

Achucarro Center for Neuroscience, IKERBASQUE, Basque Foundation for Science, 48011 Bilbao, Spain

A. Verkhratsky et al. (eds.), *Neuroglia in Neurodegenerative Diseases*,
Advances in Experimental Medicine and Biology 1175,
https://doi.org/10.1007/978-981-13-9913-8_4

and channels into the plasma membrane, determining astroglial surface signaling landscape. Vesiclal cargo, together with the whole vesicle can also exit astrocytes by the fusion of multivesicular bodies with the plasma membrane (exosomes) or by budding of vesicles (ectosomes) from the plasma membrane into the extracellular space. These astroglia-derived extracellular vesicles can later interact with various target cells. Here, the characteristics of four types of astroglial secretory vesicles: synaptic-like microvesicles, dense-core vesicles, secretory lysosomes, and extracellular vesicles, are discussed. Then machinery for vesicle-based exocytosis, second messenger regulation and the kinetics of exocytotic vesicle content discharge or release of extracellular vesicles are considered. In comparison to rapidly responsive, electrically excitable neurons, the receptor-mediated cytosolic excitability-mediated astroglial exocytotic vesicle-based transmitter release is a relatively slow process.

Keywords Exocytosis · Astrocytes · Fusion pore · Gliocrine system · Secretory vesicles

4.1 Vesicular Network and Astroglial Secretion

Similarly to all eukaryotic cells, astrocytes (homoeostatic glial cells of the central nervous system, CNS), contain a complex cytoplasmic network of vesicles. Lysosome, a vesicular organelle discovered in 1955 [29], is present in astrocytes, and plays a prominent intermediate role in endo- and exocytotic vesicle pathways (Fig. 4.1) [152]. It has been hypothesized almost a century ago, that astrocytes act as secretory cells, when in 1910 Jean Nageotte, based on the microscopic observations, considered that astrocytes act as secretory cells [90]. In the last two decades, using a variety of experimental approaches (e.g., by optical and membrane capacitance measurements, electrochemical amperometry, and selective interference with proteins of the exocytotic machinery), it has been determined that astrocytes can release signaling molecules via a vesicle-based mechanism (i.e., exocytosis) and are thus actively involved in information processing in the brain [136, 139]. Although being electrically non-excitable, astrocytes, similarly to neurons, possess (i) exocytotic vesicles, (ii) express proteins for regulated SNARE (Soluble NSF Attachment protein REceptor)-dependent vesicular exocytosis and (iii) can respond to various extracellular stimuli with an increase in cytosolic second messengers triggering Vesiclal exocytosis. The SNARE components of exocytotic machinery in astrocytes are not identical to neurons, nor are the vesicle types, their fusion sites and regulation of exocytosis [77, 86, 136, 139, 152, 153].

4.2 Astroglial Secretory Vesicles

Astrocytes contain various different types of secretory vesicles loaded with different types of molecules (such as ATP, D-serine, glutamate, atrial natriuretic peptide

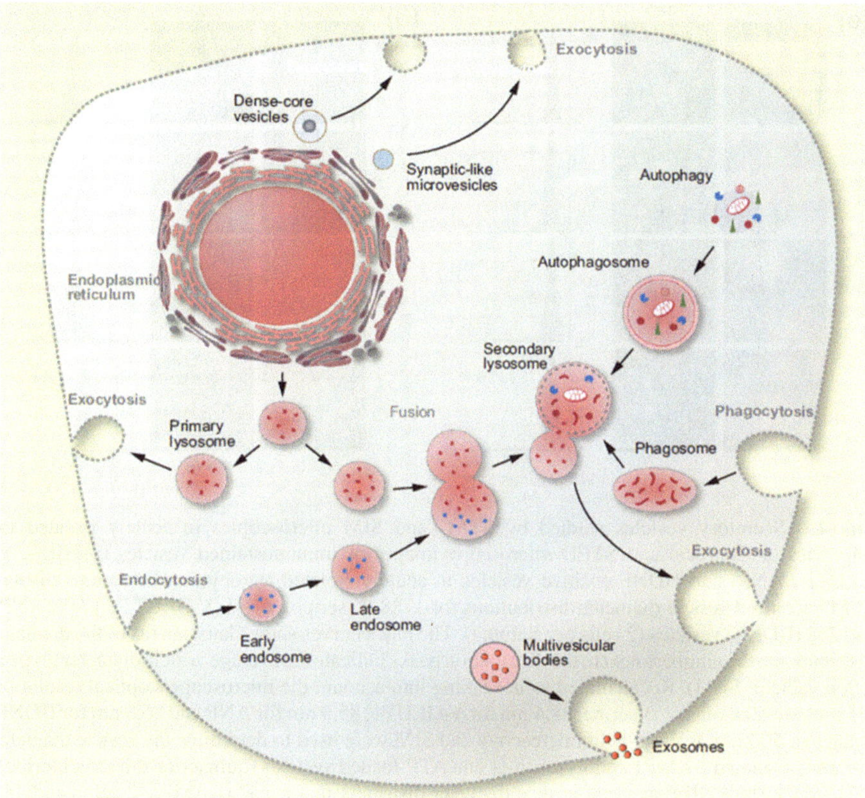

Fig. 4.1 The vesicle network in astrocytes. Lysosomes, first described in 1955, represent a central, prominent intermediate of endo- and exocytotic pathways in all eukaryotic cells, including astroglia. Intracellular secretory organelles (synaptic-like vesicles, dense-core vesicles and primary lysosomes) originate from the endoplasmic reticulum and Golgi complex. Primary lysosomes fuse with endosomes, phagosomes and autophagosomes and convert to secondary lysosomes that undergo exocytosis, thus expelling products of degradation. The multivesicular bodies contain exosomes that may carry various signaling factors. Modified with permission [152]

(ANP), brain-derived neurotrophic factor (BDNF), etc., Fig. 4.2) [39, 40, 101, 139]. These secretory vesicles are classified into synaptic-like microvesicles (SLMVs), [8, 12, 27, 58], dense-core vesicles (DCVs) [17], secretory lysosomes (SL) [71], and extracellular vesicles (EVs) [38].

4.2.1 Synaptic-Like Microvesicles

Astroglial SLMVs are clear electron-lucent vesicles, which are similar to neuronal synaptic vesicles [28, 60], their diameters range between 30 and 100 nm and these

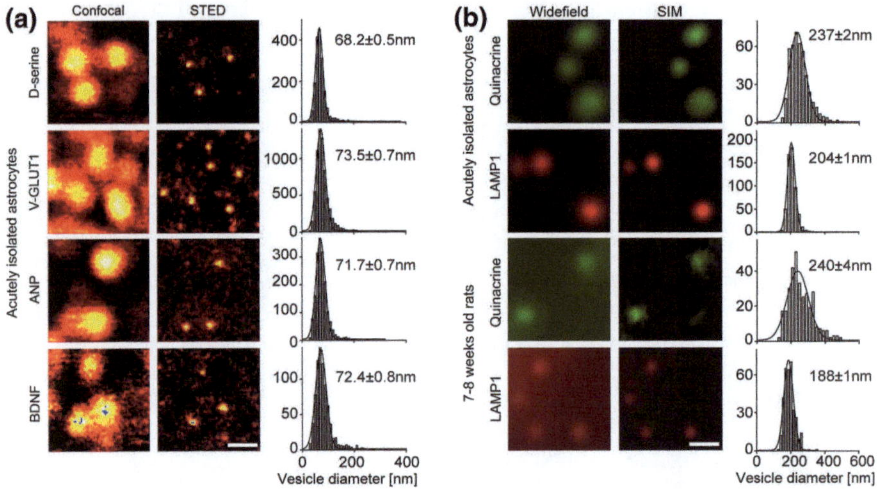

Fig. 4.2 Secretory vesicles studied by STED and SIM microscopies in acutely isolated rat astrocytes. **a** Confocal and STED microscopy images of immunostained vesicles D-serine-, V-GLUT1-, ANP- and BDNF-positive vesicles in acutely isolated astrocytes. Histograms display STED-acquired vesicle diameter distributions for 1788 (D-serine), 6787 (V-GLUT1), 1747 (ANP) and 798 (BDNF) vesicles (2 cells per staining). The black curves show Gaussian fits of the diameter distributions; the numbers next to the distribution peaks indicate the average vesicle diameter (expectation value ± SEM). Recalculated values taking into account the microscope's optical resolution (45 nm) are 80.8 nm for D-serine, 88.4 nm for V-GLUT1, 85.9 nm for ANP and 86.8 nm for BDNF. Scale bar, 500 nm. **b** Wide-field microscopy and SIM were used to determine the vesicle diameter of immunostained LAMP1 endolysosomes and ATP-loaded vesicles (quinacrine dihydrochloride). Histograms show SIM-acquired vesicle diameter distributions for 557 (LAMP1, 2 cells) and 445 (quinacrine, 2 cells) vesicles in acutely isolated astrocytes (upper two panels) and 338 (LAMP1, 3 cells) and 333 (quinacrine, 6 cells) vesicles in astrocytes isolated from 7- to 8-week-old rats (lower two panels). The black curves show Gaussian fits of the diameter distributions; the average vesicle diameter (expectation value ± SEM) is labeled next to the distribution peaks. Scale bar, 500 nm. Modified with permission [39]

SLMVs store low molecular weight signaling molecule glutamate (~147 Da) and in some astrocytes also D-serine (~105 Da). In hippocampal slices, larger SLMVs have been identified in astrocytes (1–3 μm in diameter), which may be generated by intracellular fusion of smaller vesicles or other organelles or both upon sustained Ca^{2+} or mechanical stimulation [59], perhaps a manifestation of a pathological status. D-serine has been recently proposed to reside preferentially in neurons, since biosynthetic enzyme of D-serine serine racemase is expressed almost entirely by neurons [97, 145], with astrocytes arguably being the source of L-serine, which cannot be synthesized in neurons. As revealed with electron microscopy, astrocytes with SLMVs lack the structurally organized active zones with clearly defined synaptic vesicle pools with hundreds to thousands of synaptic vesicles (SVs) per synapse (the readily releasable and the reserve vesicle pools), which are typically found in presynaptic neurons [12, 58]. However, SLMVs in astrocytes do organize in small spaced clusters (2–15 vesicles) located near the astrocytic plasma membrane of the perisynaptic astrocytic process. Endoplasmic reticulum appears located in close proximity to these clusters, suggesting that astrocytes contain functional nanodomains, where a local Ca^{2+} increase can trigger release of glutamate and/or D-serine [9, 12, 58, 82]. However, astrocytic perisynaptic processes are mainly devoid of subcellular organelles [105, 124].

Whether signaling molecules glutamate and D-serine are stored inside the same astroglial SLMVs is still a matter of debate. In cultured astrocytes vesicular SNARE protein vesicle-associated membrane protein 2 (VAMP2) and cellubrevin (VAMP3) were found colocalized with both vesicular glutamate transporters (VGluTs; [12, 14, 87]) and D-serine [80, 88], while studies on tissue astrocytes showed that glutamate and D-serine can be stored in distinct SLMVs within the same astrocyte [9]. Examination of immunopurified astroglial SLMVs showed that SLMVs can co-store both glutamate and D-serine [82]. The observation that isolated astroglial SLMVs and isolated neuronal SVs contain different signaling molecules (isolated SLMVs contain D-serine and glutamate [27, 82] and isolated neuronal SVs contain glutamate, glycine GABA and are devoid of D-serine [82, 125]) (although D-serine has been recently proposed to reside preferentially in neurons [97]) might indicate distinct physiological roles of SLMVs and SVs in the CNS.

SLMVs use VGluTs to move glutamate from the cell cytosol into vesicular lumen using a H^+ gradient, created by vacuolar-type H^+-ATPase (V-ATPase), with associated chloride flux. VGluTs 1, 2, and 3 were identified in the membrane of SLMVs in astrocytes in culture, and VGluTs 1 and 2 were shown to associate with SLMVs in tissue astrocytes of several brain areas of hippocampus (CA1), cerebral cortex, striatum, dentate-molecular layers [12, 82, 87, 92, 148], although VGluTs 1-3 were not identified in tissue astrocytes from mice grey matter, thalamic ventrobasal nucleus primary somatosensory cortex, hippocampus and cerebellum [70], suggesting that astrocytes from different brain regions may carry different vesicle types consistent with the regional and functional heterogeneity of astrocytes [91]. Vesicular D-serine transporters (VSerT) were identified in immunopurified astrocytic vesicles. They are likely D-serine/chloride co-transporters and use the H^+ gradient created by V-ATPase to refill the vesicles with D-serine [81, 82].

4.2.2 Dense-Core Vesicles

Astroglial dense-core vesicles (DCVs) are ultrastructurally similar to the large-dense core vesicles (LDCVs) that release neuropeptides and hormones from neuroendocrine cells [16] and neurons [60]. Although DCVs are not very abundant in astrocytes [27], both DCVs and SLMVs can coexist within the same astrocyte [94, 113]. Moreover, DCVs appear larger (100–600 nm; [17, 54, 109]) than SLMVs. The DCVs in cultured astrocytes may contain secretogranins II [17, 94, 109] and III [95], chromogranins [54], ANP [61, 94], neuropeptide Y [109, 113], and ATP [24, 96]. Secretogranins containing DCVs were identified also in astrocytes in human brain tissue [54], indicating the presence of DCVs in astrocytes in situ. Inositol-

1,4,5-triphosphate (IP_3) receptors (IP_3Rs), acting as IP_3-gated Ca^{2+} channels were detected on DCV membranes in astrocytes in brain tissue suggesting that DCVs also serve as IP_3-sensitive intracellular Ca^{2+} stores [54].

Using a super-resolution microscopy approach it was shown that the peptidergic ANP- and BDNF-containing vesicles have diameters less than 100 nm [39] and that the ANP-antibody retrieving vesicles do not exhibit a dense core [107]. Also, astrocytes contain fewer smaller and less dense secretory granules containing secretogranin II [17]. Thus, it appears that peptidergic granules in astrocytes are not uniform in morphological appearance.

Tissue-type plasminogen activator (tPA) is considered to be released by neurons but taken up by astrocytes, possibly into recycling vesicles as these vesicles can uptake ANP-antibodies [107]. Interestingly, tPA is constitutively endocytosed by astrocytes via the low-density lipoprotein-related protein receptor, and is then exocytosed in a regulated manner. Extracellular glutamate inhibits the exocytotic recycling of tPA by astrocytes and on the other hand, capturing extracellular tPA into astrocytes reduces the NMDA-mediated responses potentiated by tPA [20].

4.2.3 Secretory Lysosomes

Secretory lysosomes with diameters between 300 and 500 nm [23, 150] that store signaling molecule ATP, have been identified in cultured astrocytes [56, 71, 96, 150]. Secretory lysosomes in astrocytes as in other cell types are likely involved in membrane repair [3]. Astroglial secretory lysosomes express lysosomal-specific markers, including cathepsin D and lysosomal-associated membrane protein 1 (LAMP1 [150]), monomeric GTP-protein Rab 7, SNARE protein tetanus neurotoxin (TeNT)-insensitive VAMP (TI-VAMP/VAMP7), which contributes to TeNT-independent exocytotic release of ATP [138], and vesicular nucleotide transporter VNuT [120], which is involved in ATP storage [93] within secretory lysosomes in astrocytes and hence warranting ATP release [63] from these astrocytic secretory organelles. Secretory lysosomes in astrocytes can be specifically labeled with dextrans [55, 134], FM dyes, and by a fluorescent ATP analogue MANT-ATP [150]. They may coexist with SLMVs in the same astrocyte [72]. Fusion of secretory lysosomes is regulated and induced with slow, locally restricted Ca^{2+} elevations [71], which are distinct from Ca^{2+} spikes inducing SLMV fusion [138].

4.2.4 Extracellular Vesicles: Exosomes and Ectosomes

Exosomes and ectosomes are extracellular vesicles (EVs) released from cells to deliver signals to target cells (Fig. 4.3). EVs control different biological processes by transferring membrane proteins, lipids, signaling molecules, mRNAs, microRNAs (miRNAs), and activating receptors of recipient cells, possibly playing a role

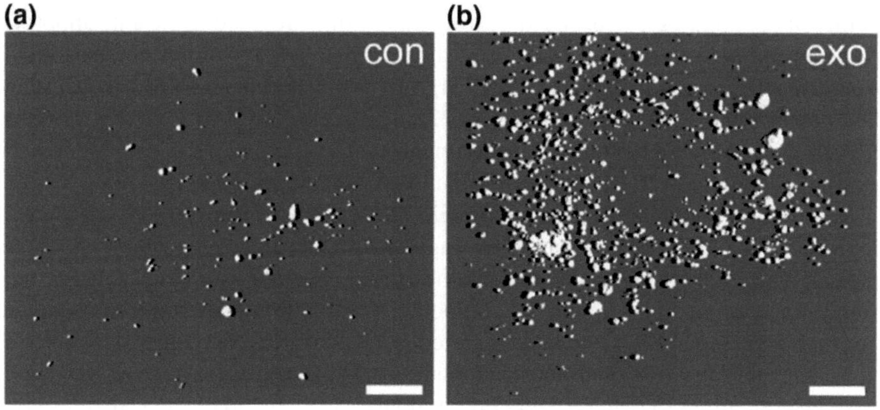

Fig. 4.3 Internalization of exosomes into astrocytes. Internalization of PKH26 nanoparticles and PKH26-positive particles of the exosome-containing samples into subcellular compartments of cultured astrocytes. **a, b** Representative three-dimensional shaded display of individual live cultured astrocytes that internalized PKH26 nanoparticles (**a**, con) and PKH26-positive particles present in the PKH26-labeled exosome-containing samples (**b**, exo) into intracellular compartments, observed as numerous bright fluorescent puncta. Scale bars, 10 μm. Modified with permission [112]

in autocrine regulation. Exosomes are released by exocytosis of multivesicular bodies (MVBs) containing intraluminal vesicles (ILVs) that are called exosomes when released from cells into the extracellular space. Ectosomes (also called microvesicles) are assembled by outward budding of the plasma membrane and released (shed) from the plasma membrane into the extracellular space. Exosomes are vesicles of 50–100 nm in diameter, while ectosomes are larger vesicles from 100 to >1,000 nm in diameter [25]. Astrocytes release both types of EVs [38].

Ectosomes carrying interleukin-1β (IL-1β) may shed from cultured astrocyte upon ATP stimulation through activation of ionotropic purinoreceptor $P2X_7$. This is associated with rapid activation of acid sphingomyelinase, which moves from luminal lysosomal compartment to the plasma membrane outer leaflet altering membrane structure/fluidity leading to vesicle blebbing and shedding 1–2 min after ATP stimulation [13]. Diameters of ectosomes shed by cultured astrocytes vary between 100 and 1,000 nm [13, 110]. Moreover, upon repetitive ATP stimulation cultured astrocytes release vesicles from the cell surface that can be from 1 up to 8 μm in diameter and express on their surface β1-integrin proteins and contain mitochondria and lipid droplets together with ATP [34]. Although it has not been directly demonstrated [34], these vesicles likely represent ectosomes due to their large size. In addition to interleukin-1β (IL-1β) [13], mitochondria, lipid droplets, and ATP [34], culture astrocyte-derived ectosomes may also carry fibroblast growth factor 2 and vascular endothelial growth factor [110], ectoenzyme nucleoside triphosphate diphosphohydrolases that hydrolyze extracellular nucleotides [21], and matrix metalloproteinases and their inhibitors [121]. Ectosomes shed from astrocytes in response to lipopolysaccharide-induced stress contain miRNA miR-34a that enhances the vulnerability of dopaminergic neurons to neurotoxins by downregulating the anti-apoptotic protein Bcl2 [78]. Recently, it has been shown that cytokines tumor necrosis factor α and IL-1β can modify the miRNA cargo of EVs shed from astrocytes to regulate neurotrophic signaling in neurons [22]. Astrocyte also shed EVs that promote transmigration of leukocytes into the brain through regulation of the peripheral acute cytokine response to IL-1β–induced inflammatory brain lesion [31].

Exosomes containing heat-shock protein 70 are released from cultured astrocytes in response to oxidative and heat stress, suggesting a mechanism by which astrocytes provide antioxidant protection to neurones [132]. Retinal astrocytes release exosomes that contain anti-angiogenic components that inhibit laser-induced choroidal neovascularization [43]. Exosomes secreted from astrocytes carrying synapsin I promote neurite outgrowth and neuronal survival [142]. Astrocyte-derived exosomes have been reported to contain mitochondrial DNA [41] and may carry also disease-specific cargo and promote neurological disorders by spreading pathology. Indeed, cultured astrocytes expressing mutant copper-zinc superoxide dismutase 1 (SOD1) secrete exosomes, which carry mutant SOD1. Astroglial derived mutant SOD1-positive exosomes can transfer mutant SOD1 to cultured neurons and induce motor neuron death. This suggests a role of EVs in the pathogenesis of amyotrophic lateral sclerosis [6]. Moreover, cultured astrocytes exposed to amyloid peptide release exosomes enriched with pro-apoptotic ceramide and prostate apoptosis response 4 (PAR4). These exosomes are taken up by astrocytes and promote their apoptosis suggesting that exosome-mediated astrocyte death may contribute to neurodegeneration in Alzheimer's disease. Exosome-mediated miRNA transfer from astrocytes to neurones has been suggested to participate in HIV-associated neurological disorders. Treatment of cultured astrocytes with pathogenic HIV trans-activator of transcription (Tat) protein and morphine triggers shuttling of miRNA miR29b via exosomes to neuronal cells, which results in decreased trophic factor platelet-derived growth factor (PDGF)-B expression and neuronal viability [52]. Nef (Negative Regulatory Factor), a protein encoded by primate lentiviruses such as HIV-1, has been shown to be released in EVs derived from astrocytes and human microglia and may accumulate in neighboring cells (Fig. 4.3) contributing to Nef-mediated neurotoxicity [118, 129]. Interestingly, this release appeared inhibited by elevated cytosolic calcium in human microglia [129]. Recently, it has been shown that reactive astrocytes release vimentin, an intermediate filament of the cytoskeleton, via exosomes. This promotes binding of exoenzyme *Clostridium botulinum* C3 transferase (that enzymatically inhibits small GTPases of the Rho family) to neuronal surface, which can be than internalized and promotes neuronal plasticity and growth [1].

As the field of studying EVs is still developing, it is important to note that when studying the internalization of EVs into cells, the methods and approaches have to be evaluated carefully. For example, when monitoring the internalization of EVs into astrocytes, EVs were labeled by a fluorescent dye PKH26, and it has been reported that a significant false-positive signal due to internalization of PKH26-nanoparticles was observed (Fig. 4.3), which can compromise the interpretation of EV internalization [112]. Thus, for EV uptake and functional studies it is critical to consider potential artifacts, since EVs are very small, often below the optical microscopy resolution.

4.3 SNARE and SNARE-Associated Proteins in Astrocytes

Astrocytes express vesicular R-SNARE and plasma membrane Q-SNARE proteins (Fig. 4.4). R-SNARE proteins synaptobrevin 2 (VAMP2), VAMP3 [27, 74, 80, 88, 99, 144], and TI-VAMP/VAMP7 [138] and Q-SNARE proteins SNAP23 and syntaxins 1, 2, 3, and 4 [48, 94, 148] have been identified in astrocytes as well as SNARE-associated proteins Munc18 [94] and synaptotagmin 4 [147]. In mammals, synaptotagmin 4 is not a Ca^{2+}-sensor for regulated exocytosis like synaptotagmin 1 in neurones is [143], but is important in modulating Ca^{2+}-evoked exocytosis [131]. SNARE proteins VAMP2 [144], VAMP3 [8, 12, 58, 123, 148], TI-VAMP/VAMP7 [138], SNAP23 [123], and syntaxin 1 [123] were confirmed also in brain tissue astrocytes using immunogold cytochemistry and confocal microscopy. Expression of synaptotagmins and other SNARE-associated proteins, such as Sec1/Munc18-like proteins, in brain tissue astrocytes still needs to be determined, although studies examining mRNA of astrocytes in brain tissue suggest expression of several synaptotagmin isoforms [84, 149] and SNARE-associated proteins [149].

The ternary SNARE fusion complex between vesicular and plasma membrane SNAREs [35, 36] in astrocytes is likely made of vesicular SNARE proteins VAMP2/3 (SLMVs) or TI-VAMP/VAMP7 (secretory lysosomes) and the plasma membrane SNAP23 and syntaxins [45, 85]. The formation of up to five SNARE complexes containing VAMP2 is believed to be sufficient to carry a single vesicle fusion in astrocytes [127].

Astroglial VAMP2 and VAMP3 colocalize with ATP [74] or D-serine [80]-storing vesicles. VAMP3 in astrocytes colocalizes also with the VGLUT1 and 2, vesicular glutamate transporters present on SLMVs that store glutamate [8, 12, 58, 148], and likely D-serine [82]. TI-VAMP/VAMP7 is present in the membrane of the astroglial late endocytic/lysosomal compartments [138] storing ATP [4, 24, 96, 150].

The functionality and physiological role of exocytotic apparatus in astrocytes consisting of aforementioned SNARE proteins has been addressed and confirmed in multiple studies. It has been shown that cleavage of SNARE proteins with tetanus (TeNT) and botulinum neurotoxins (BoNT) in cultured astrocytes attenuates exocytotic release of glutamate [4, 10, 11, 12, 53, 87, 104] as well as a reduction in membrane capacitance (C_m) increases [39, 61] and in amperometric spikes [23], implying the role of SNARE proteins in the release of glutamate from cultured astrocytes. The inactivation of VAMP2/VAMP3 in astrocytes by TeNT abolishes the release of glutamate or D-serine from astrocytes in brain tissue slices [47, 58, 106]. Additionally, in a mouse model in which a dominant negative SNARE transgene is expressed in astrocytes to interfere specifically with astroglial VAMP2/3 [44, 103] the synaptic transmission and plasticity in these animals were altered [49, 63, 89, 103, 133]. Furthermore, in mice with targeted expression of BoNT/B in Müller cells,

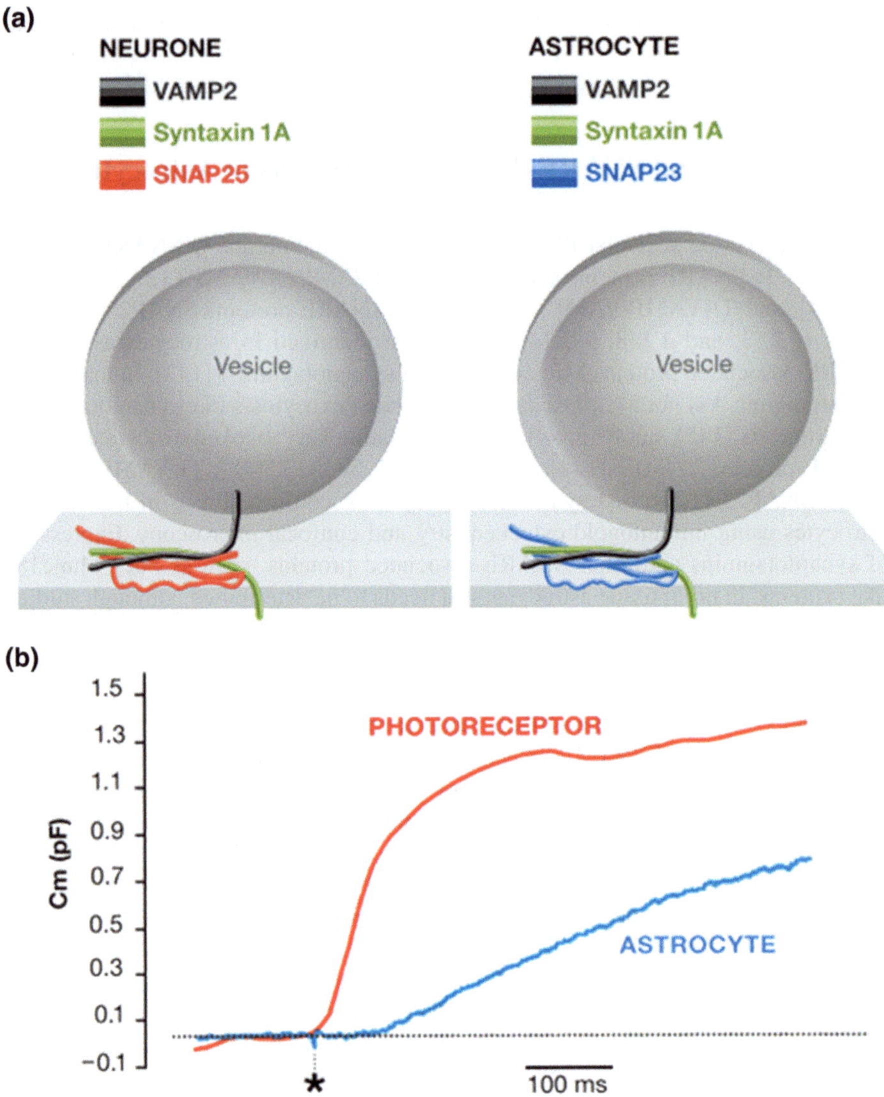

Fig. 4.4 Slowness of astroglial exocytosis. **a** Neuronal versus astrocytic SNAREs. Neurones and astrocytes alike express SNAREs VAMP2 and syntaxin 1; many astrocytes can also express VAMP3 in lieu of or in addition to VAMP2. Astrocytes express SNAP23, a homologue of neuronal SNAP25. At the plasma membrane, syntaxin 1A can form a binary cis complex with SNAP25B or SNAP23A, which then interacts with vesicular VAMP2 to form a ternary complex. A single ternary complex can tether the vesicle at the plasma membrane for a longer period of time, when it contains SNAP25B rather than SNAP23A, respectively. Of note, truncated syntaxin 1, lacking the N-terminal Habc domain and the linker region to the SNARE domain, is shown for simplicity. Drawings are not to scale. **b** Comparison of kinetics of neuronal and astroglial exocytosis. Time-dependent changes in membrane capacitance (Cm) recorded in a neuronal cell (trace in red, photoreceptor) and an astrocyte (trace in blue), elicited by a flash photolysis-induced increase in cytosolic Ca^{2+}. Note that the blue trace recorded in an astrocyte displays a significant delay between the stimulus (asterisk) and the response (trace components above the dotted line). Modified with permission [139]

a subtype of astroglia that expresses BoNT/B-sensitive VAMP2/3, the disruption of the Ca^{2+}-dependent vesicular glutamate release from Müller cells was observed [128]. Moreover, it has been shown in mice with inducible expression of TeNT in astrocytes that TeNT-sensitive vesicular release from astrocytes is necessary for sustaining gamma oscillations associated with recognition memory in mice [68]. Interestingly, the use of various botulinum toxins and dominant negative SNARE peptides has demonstrated that SNARE proteins determine the fusion frequency of individual vesicles monitored by the high-resolution membrane capacitance technique [39]. All these data clearly imply that SNARE-mediated exocytosis is present in astrocytes and essential for normal brain function.

4.4 Regulation and Kinetics of Secretory Vesicle Release in Astrocytes

4.4.1 GPCR-Mediated Regulation of Secretory Vesicle Release: Ca^{2+} and cAMP Signals

Neurones are electrically excitable and release neurotransmitters from synaptic vesicles in synaptic terminals in response to depolarization. In contrast to neurones, astrocytes are electrically silent and display only receptor-mediated cytosolic excitability. Astrocytes sense extracellular signalling molecules via plasma membrane receptors. They express a large number of various types of receptors and many of these receptors are metabotropic high affinity G-protein-coupled receptors (GPCRs) [2, 100, 140, 151]. Binding of signalling molecules to these receptors may increase cytosolic levels of free Ca^{2+} as well as other astrocytic cytosolic secondary messengers, including the cyclic adenosine monophosphate (cAMP). Such cytosolic excitability may lead to secretory vesicle release of signaling molecules from astrocytes (see Vesicular Network). These gliosignalling molecules can then interact with the receptors on neurons affecting neuronal excitability [17, 100] or affect receptors on other neighbouring cells.

Stimulation of astroglial GPCRs coupled to G_q protein leads to increases in intracellular levels of cytosolic Ca^{2+}. Activation of G_q GPCRs triggers IP^3 signaling cascade that releases Ca^{2+} into the cytosol from the IP^3-sensitive intracellular organelles acting as Ca^{2+} stores, such as endoplasmic reticulum (ER) [53, 69] and secretory vesicles [54]. Mitochondria can modulate those cytosolic calcium dynamics in astrocytes by taking up Ca^{2+} from the cytosol or releasing this ion into the cytosol at time of high or low Ca^{2+} cytosolic levels, respectively [115, 126]. Ca^{2+} can also partially enter astrocytes from the extracellular space through voltage-gated Ca^{2+} channels [67, 73, 102], ionotropic receptors [64], sodium-calcium exchanger [116] and through the transient receptor potential canonical type 1-containing channel [75]. G_q-induced cytosolic Ca^{2+} increases in astrocytes occur as oscillations or sustained elevations [100, 141, 151], spontaneously or in response to signaling molecules [26]. Astrocytes can intercellularly communicate through gap junction channels. They can propagate cytosolic Ca^{2+} excitability by diffusion of IP_3 or Ca^{2+} through gap junctions to neighboring unstimulated astrocytes in the form of intercellular Ca^{2+} waves [122]. They can also release glutamate or ATP in response to Ca^{2+} excitability [15, 26, 42].

Stimulation of astroglial GPCRs coupled to G_s proteins activates adenylyl cyclase (AC), an enzyme catalyzing the conversion of ATP to cAMP [114, 135]. cAMP activates a number of effectors in the cell, primarily cAMP-dependent protein kinase A, but signalling via cAMP-activated GTP-exchange protein [30], cAMP-gated ion channels, and Popeye domain-containing proteins [37] may also be triggered [7]. G_s protein activation induces persistent cAMP elevations [135, 137], which are at least in the case of adrenergic receptor activation 10-fold slower compared to G_q protein-triggered Ca^{2+} elevations [50, 51]. Whether G_s-induced cAMP excitability can be propagated via gap junctions needs to be evaluated [33]. It has been suggested that G_q- and G_s-mediated pathways in astrocytes can interact, since G_s-signaling pathway may enhance G_q-mediated Ca^{2+} responses and vice versa [5, 50, 51, 57].

GPCR G_q- and G_s-protein signalling pathways were shown to be involved in the regulation of secretory vesicle release of chemical messengers from astrocytes. Ca^{2+} elevations in astrocytes trigger the release of glutamate [10, 12, 98, 104, 148], ATP [4, 24], secretogranin II [17], ANP [62], and D-serine [88] from secretory vesicles. cAMP elevations can trigger the release of secretogranin II from astroglial peptidergic vesicles [17]. In astrocytes pretreated with the membrane-permeable cAMP analogue dibutyryl-cAMP the Ca^{2+}-triggered release of ANP from secretory vesicles was enhanced [94]. cAMP might trigger the fusion of secretory vesicles de novo or it may modulate the fusion pore dynamics of already pre-fused secretory vesicles by increasing the diameter and open time of a fusion pore between the vesicle and plasma membranes, which needs to be still determined. The latter mechanism has been observed in neuroendocrine cells [18].

4.4.2 *Kinetics of Secretory Vesicle Content Release in Astrocytes*

Temporal dynamics of secretory vesicle release from cultured astrocytes has been monitored using (i) electrophysiological techniques (amperometry [23] and membrane capacitance (C_m) measurements [61, 117] in combination with UV-flash photolysis-induced increases in cytosolic Ca^{2+} levels [61]), and (ii) optical techniques (real-time confocal microscopy and total internal reflector fluorescence microscopy, TIRFM) in combination with fluorescent markers of vesicular cycling/fusion, such as FM dyes [71, 72, 150], acridine orange [12, 32], quinacrine [96, 111], fluorescent dextrans [56], MANT-ATP [150], and genetically encoded chimeric proteins of specific membrane/luminal vesicle markers and green fluorescence proteins (GFP) or mCherry-derived proteins [77, 79].

4.4.2.1 Secretory Vesicle Fusion in Astrocytes Occurs with a Delay upon Stimulation

Compared to neurones it has been shown for all 4 secretory vesicles types described in astrocytes (Sect. 4.1) to fuse with the plasma membrane with a delay upon stimulation. As determined with C_m measurements the kinetics of secretory vesicle fusion in astrocytes is at least two orders of magnitude slower than that in neurons (Fig. 4.4b) [61], where secretory vesicle fusion occurs within <0.5 ms upon intracellular Ca^{2+} increase [131].

In respect to astroglial SLMVs the rise of cytosolic Ca^{2+} evoked by activation of metabotropic glutamatergic receptors [12, 19, 79] or purinergic receptors [119] triggers fusion events of SLMVs within hundreds of milliseconds after stimulation as determined in studies using fluorescently tagged VGluT1/2-containing vesicles (i.e., VGluT-pHluorin and VGluT-EGFP, which are chimeric proteins of VGluT and a pH-sensitive GFP protein ecliptic synapto-pHluorin (SpH; [83]) or EGFP. Ionomycin, a Ca^{2+} ionophore, triggers exocytotic fusion of SpH-labeled SLMVs within seconds [72]. In another study exocytotic bursts of SpH-labeled SLMVs occur within 6 s after mechanical stimulation of astrocytes, but other stimuli such as ATP, bradykinin, the Ca^{2+} ionophore 4-Br-A23187, α-latrotoxin, or hypertonicity cause fusion of SpH-labeled SLMVs following a delay of >1 min [77]. Secretion of astroglial peptidergic vesicles also occurs with a delay. Exocytosis of neuropeptide Y-positive peptidergic vesicles upon glutamate [113] or ionomycin stimulation [109] occurs with a delay of >1 min and exocytosis of emerald green-tagged AMP from peptidergic vesicles in 8-Br-cAMP-differentiated astrocytes occurred over a time scale of minutes upon ionomycin stimulation [94]. A similar time-course was observed, when exocytosis of FM-dye-labeled lysosomes was studied. FM-dye labeled lysosomes began to fuse with the plasma membrane with a delay of >1 min upon stimulation of astrocytes with Ca^{2+} ionophores A-23187 [71] and ionomycin, or upon ATP stimulation [150]. The exocytotic fusion of the majority of TI-VAMP positive quinacrine-labeled secretory vesicles [138], that likely represent secretory lysosomes, occurred with a delay of >2 min upon addition of different stimuli, including glutamate, ATP, ionomycin or upon stimulation with UV-induced Ca^{2+} uncaging [96, 111]. EGFP-LAMP1 (lysosomal-associated membrane protein 1)-labeled lysosomes and FITC-dextran-labeled lysosomes also undergo fusion with a delay of >40 s upon application of ionomycin [72], ATP and, a group I metabotropic glutamate receptor agonist (R/S)-3,5-dihydroxyphenylglycine [56]. Moreover, ectosomes carrying IL-1β start to bleb and shed from astroglial plasma membrane with a 1–2 min delay upon ATP stimulation [13].

The reason for such a loose excitation-secretion coupling in astrocytes [136] may be that (i) the major source of Ca^{2+} in astrocytes is not the extracellular space as in neurones, but intracellular IP_3-sensitive Ca^{2+}-storage organelles, which release Ca^{2+} only upon activation of receptor-mediated intracellular signalling cascades and production of IP_3, (ii) the lack of active zones in astrocytes and slower delivery of secretory vesicles to the plasma membrane fusion sites upon stimulus application compared to neurons, where there are active zones [108], or (iii) differences in exocytotic vesicle fusion machinery between astrocytes and neurons (Fig. 4.4) with astroglial machinery exhibiting a slower vesicle fusion dynamics compared to neuronal exocytotic machinery [85].

4.4.2.2 Modes of Astroglial Secretory Vesicles Fusion

Exocytotic fusion of secretory vesicles in astrocytes exists in two major forms [23], as observed in neurones and neuroendocrine cells [46]. Namely, amperometric studies revealed that dopamine-loaded astrocytic vesicles fuse with the plasma membrane either by transient (kiss-and-run) exocytosis, with vesicle content only partially released, or by full-fusion exocytosis [23]. In optical studies in which the exocytosis of SLMVs expressing spH [14, 77] or SLMVs co-expressing VGluT1-mCherry/VGluT1-pHluorin [79] was studied, both modes, the transient and the full-fusion, of SLMV exocytosis were shown to occur in the same astrocyte simultaneously under spontaneous or stimulated conditions. 50–60% of all spontaneous exocytotic events were the full-fusion events, while 40–50% were the transient fusion events. Depending on the type of a stimulus, the percentage of either type of event shifted toward transient or full-fusion modes of exocytosis upon stimulation. This indicates stimulus-dependent regulation of fusion pore opening [14, 23, 77]. Secretory lysosomes can also exhibit both transient and full-fusion modes of exocytosis. A rapid, total release of an FM dye was observed, followed by a slower and complete loss of EGFP-sialin (a lysosomal sialic acid transporter), from the same lysosomes upon mechanical stimulation, suggesting that secretory lysosomal fusion in astrocytes completes upon mechanical stimulation within seconds, without evidence for transient fusion [71]. Upon glutamate and ATP stimulation the release of FM dyes and MANT-ATP from LAMP1-positive lysosomes is only partial, implying the transient mode of secretory lysosomal exocytosis [150]. Vesicular nucleotide transporter mCherry has been shown to remain associated with the lysosomal membrane during the release of cathepsin D-Venus from the same lysosomes upon ATP, L-glutamate, and calcium ionophore A23187 stimulation, further suggesting that secretory lysosomes in astrocytes may not fully fuse with the plasma membrane [93]. Discrete increases in membrane capacitance, indicating single-vesicle fusion, revealed that astrocyte stimulation increases the frequency of predominantly transient fusion events in smaller vesicles (likely SLMVs and peptidergic vesicles), whereas larger vesicles (likely secretory lysosomes) transitioned to full fusion suggesting that vesicles with different diameters in astrocytes exhibit different capacities to discharge their cargo, due to distinct fusion pore properties [39].

The underlying molecular mechanisms controlling the fusion pore state are not clearly known, but may among others involve SNARE proteins and dynamin. Dynamin, a multidomain GTPase involved in vesicle scission from the plasmalemma during endocytosis, has been shown to be involved in the regulation of a fusion pore during spontaneous exocytosis in astrocytes, since activators of dynamin Ryngo™-1-23 promoted fusion pore closure by prolonging closed and by shortening open fusion pore dwell times [66]. DnSNARE (dominant-negative domain of synaptobrevin 2 protein) peptide, which interferes with endogenous VAMP2 expression and thus prevents VAMP2-mediated membrane fusion, has been shown to stabilize the fusion-pore diameter to narrow, release-unproductive diameters regardless of vesicle diameter, implying the regulatory role of SNAREs in governing vesicle fusion in astrocytes [39]. The fusion pore can alone be a subject of regulation by ketamine, an anesthetic that exhibits analgesic, psychotomimetic, and rapid antidepressant effects. It has been shown recently, using high-resolution cell-attached membrane capacitance measurements, that ketamine evokes long-lasting flickering of a narrow fusion pore that is incapable of transiting to full fission [65]. Furthermore, ketamine treatment also suppressed ATP-triggered vesicle fusion and BDNF secretion by increasing the probability of a narrow fusion pore open state and/or by reducing astrocytic Ca^{2+} excitability [130].

4.5 Non-vesicular Astroglial Secretion

Astrocytes can also release signalling molecules by a non-vesicle-based mechanisms (i) through plasmalemmal channels (e.g., volume-regulated anion channels, connexons/pannexons (hemichannels), ionotropic pore-forming $P2X_7$ purinergic receptors, the two-pore-domain potassium channel Trek-1, or Bestrophin-1 channels [146], and (ii) through plasmalemmal transporters (e.g., reversal uptake by plasma membrane excitatory amino acid (glutamate) transporters, (hetero)exchange via the cystine–glutamate antiporter or organic anion transporters) [76]. With the exception of Bestrophin-1, these non-vesicular release mechanisms are Ca^{2+}-independent and might be activated only under pathological conditions [2, 45].

4.6 Concluding Remarks

Astrocytes are involved in many processes in the CNS through sensing extracellular signaling molecules by surface GPCRs, responding to this with cytosolic excitation, which then stimulates the release of their own astroglial chemical messengers, gliosignaling molecules. Many studies support the existence of vesicule-based secretion of transmitters from astrocytes, in response to GPCR-mediated stimulation. These studies have shown that astrocytes possess various types of secretory vesicles. The exocytotic fusion of these vesicles is regulated at the level of a single fusion pore and it occurs in two modes, as transient and full-fusion exocytosis. Moreover, astrocytes, which are electrical silent, but exhibit GPCR-mediated cytosolic excitability, respond to stimulation with a delay in exocytosis compared to fast responsive electrically excitable neurones. Such slow release kinetics of vesicle signaling apparatus suggests that astrocytes are acting as integrators of information, modulating neuronal activity in a slow-time domain. Although the physiological relevance of astroglial exocytosis in vivo is still not clear, it is predicted that astrocytes participate in information processing in the brain by exocytotic release of signaling molecules.

Acknowledgements NV's and RZ's work was supported by grants from the Slovenian Research Agency (P3-0310, J3-9266, J3-7605) and COST Action CM1207 (GLISTEN), COST Action BM1402 (MouseAGE), COST Action CA15214 (EuroCellNet), COST Action CA18133 (ERNEST). VP's work is supported by a grant from the National Institute of General Medical Sciences of the National Institutes of Health (R01GM123971). VP is an Honorary Professor at University of Rijeka, Croatia.

References

1. Adolf A, Rohrbeck A, Münster-Wandowski A, Johansson M, Kuhn HG, Kopp MA, Brommer B, Schwab JM, Just I, Ahnert-Hilger G, Höltje M (2018) Release of astroglial vimentin by extracellular vesicles: modulation of binding and internalization of C3 transferase in astrocytes and neurons. Glia
2. Agulhon C, Petravicz J, McMullen AB, Sweger EJ, Minton SK, Taves SR, Casper KB, Fiacco TA, McCarthy KD (2008) What is the role of astrocyte calcium in neurophysiology? Neuron 59:932–946
3. Andrews NW, Chakrabarti S (2005) There's more to life than neurotransmission: the regulation of exocytosis by synaptotagmin VII. Trends Cell Biol 15:626–631
4. Bal-Price A, Moneer Z, Brown GC (2002) Nitric oxide induces rapid, calcium-dependent release of vesicular glutamate and ATP from cultured rat astrocytes. Glia 40:312–323
5. Balázs R, Miller S, Chun Y, O'Toole J, Cotman CW (1998) Metabotropic glutamate receptor agonists potentiate cyclic AMP formation induced by forskolin or beta-adrenergic receptor activation in cerebral cortical astrocytes in culture. J Neurochem 70:2446–2458
6. Basso M, Pozzi S, Tortarolo M, Fiordaliso F, Bisighini C, Pasetto L, Spaltro G, Lidonnici D, Gensano F, Battaglia E, Bendotti C, Bonetto V (2013) Mutant copper-zinc superoxide dismutase (SOD1) induces protein secretion pathway alterations and exosome release in astrocytes: implications for disease spreading and motor neuron pathology in amyotrophic lateral sclerosis. J Biol Chem 288:15699–15711
7. Beavo JA, Brunton LL (2002) Cyclic nucleotide research—still expanding after half a century. Nat Rev Mol Cell Biol 3:710–718
8. Bergersen LH, Gundersen V (2009) Morphological evidence for vesicular glutamate release from astrocytes. Neuroscience 158:260–265
9. Bergersen LH, Morland C, Ormel L, Rinholm JE, Larsson M, Wold JF, Røe AT, Stranna A, Santello M, Bouvier D, Ottersen OP, Volterra A, Gundersen V (2012) Immunogold detection of L-glutamate and D-serine in small synaptic-like microvesicles in adult hippocampal astrocytes. Cereb Cortex 22:1690–1697
10. Bezzi P, Carmignoto G, Pasti L, Vesce S, Rossi D, Rizzini BL, Pozzan T, Volterra A (1998) Prostaglandins stimulate calcium-dependent glutamate release in astrocytes. Nature 391:281–285
11. Bezzi P, Domercq M, Brambilla L, Galli R, Schols D, De Clercq E, Vescovi A, Bagetta G, Kollias G, Meldolesi J, Volterra A (2001) CXCR11-activated astrocyte glutamate release via TNFalpha: amplification by microglia triggers neurotoxicity. Nat Neurosci 4:702–710
12. Bezzi P, Gundersen V, Galbete JL, Seifert G, Steinhäuser C, Pilati E, Volterra A (2004) Astrocytes contain a vesicular compartment that is competent for regulated exocytosis of glutamate. Nat Neurosci 7:613–620
13. Bianco F, Perrotta C, Novellino L, Francolini M, Riganti L, Menna E, Saglietti L, Schuchman EH, Furlan R, Clementi E, Matteoli M, Verderio C (2009) Acid sphingomyelinase activity triggers microparticle release from glial cells. EMBO J 28:1043–1054

14. Bowser DN, Khakh BS (2007) Two forms of single-vesicle astrocyte exocytosis imaged with total internal reflection fluorescence microscopy. Proc Natl Acad Sci U S A. 104:4212–4217
15. Bowser DN, Khakh BS (2007) Vesicular ATP is the predominant cause of intercellular calcium waves in astrocytes. J Gen Physiol 129:485–491
16. Burgoyne RD, Morgan A (2003) Secretory granule exocytosis. Physiol Rev 83:581–632
17. Calegari F, Coco S, Taverna E, Bassetti M, Verderio C, Corradi N, Matteoli M, Rosa P (1999) A regulated secretory pathway in cultured hippocampal astrocytes. J Biol Chem 274:22539–22547
18. Calejo AI, Jorgacevski J, Kucka M, Kreft M, Goncalves PP, Stojilkovic SS, Zorec R (2013) cAMP-mediated stabilization of fusion pores in cultured rat pituitary lactotrophs. J Neurosci 33:8068–8078
19. Calì C, Marchaland J, Regazzi R, Bezzi P (2008) SDF 1-alpha (CXCL12) triggers glutamate exocytosis from astrocytes on a millisecond time scale: imaging analysis at the single-vesicle level with TIRF microscopy. J Neuroimmunol 198:82–91
20. Casse F, Bardou I, Danglot L, Briens A, Montagne A, Parcq J, Alahari A, Galli T, Vivien D, Docagne F (2012) Glutamate controls tPA recycling by astrocytes, which in turn influences glutamatergic signals. J Neurosci 32:5186–5199
21. Ceruti S, Colombo L, Magni G, Viganò F, Boccazzi M, Deli MA, Sperlágh B, Abbracchio MP, Kittel A (2011) Oxygen-glucose deprivation increases the enzymatic activity and the microvesicle-mediated release of ectonucleotidases in the cells composing the blood-brain barrier. Neurochem Int 59:259–271
22. Chaudhuri AD, Dastgheyb RM, Yoo SW, Trout A, Talbot CC, Hao H, Witwer KW, Haughey NJ (2018) TNFα and IL-1β modify the miRNA cargo of astrocyte shed extracellular vesicles to regulate neurotrophic signaling in neurons. Cell Death Dis 9:363
23. Chen X, Wang L, Zhou Y, Zheng LH, Zhou Z (2005) “Kiss-and-run” glutamate secretion in cultured and freshly isolated rat hippocampal astrocytes. J Neurosci 25:9236–9243
24. Coco S, Calegari F, Pravettoni E, Pozzi D, Taverna E, Rosa P, Matteoli M, Verderio C (2003) Storage and release of ATP from astrocytes in culture. J Biol Chem 278:1354–1362
25. Cocucci E, Meldolesi J (2015) Ectosomes and exosomes: shedding the confusion between extracellular vesicles. Trends Cell Biol 25:364–372
26. Cornell-Bell AH, Finkbeiner SM, Cooper MS, Smith SJ (1990) Glutamate induces calcium waves in cultured astrocytes: long-range glial signaling. Science 247:470–473
27. Crippa D, Schenk U, Francolini M, Rosa P, Verderio C, Zonta M, Pozzan T, Matteoli M, Carmignoto G (2006) Synaptobrevin2-expressing vesicles in rat astrocytes: insights into molecular characterization, dynamics and exocytosis. J Physiol 570:567–582
28. de Camilli P, Navone F (1987) Regulated secretory pathways of neurons and their relation to the regulated secretory pathway of endocrine cells. Ann N Y Acad Sci 493:461–479
29. de Duve C (2005) The lysosome turns fifty. Nat Cell Biol 7:847–849
30. de Rooij J, Zwartkruis FJ, Verheijen MH, Cool RH, Nijman SM, Wittinghofer A, Bos JL (1998) Epac is a Rap1 guanine-nucleotide-exchange factor directly activated by cyclic AMP. Nature 396:474–477
31. Dickens AM, Tovar-Y-Romo LB, Yoo SW, Trout AL, Bae M, Kanmogne M, Megra B, Williams DW, Witwer KW, Gacias M, Tabatadze N, Cole RN, Casaccia P, Berman JW, Anthony DC, Haughey NJ (2017) Astrocyte-shed extracellular vesicles regulate the peripheral leukocyte response to inflammatory brain lesions. Sci Signal 10
32. Domercq M, Brambilla L, Pilati E, Marchaland J, Volterra A, Bezzi P (2006) P2Y1 receptor-evoked glutamate exocytosis from astrocytes: control by tumor necrosis factor-alpha and prostaglandins. J Biol Chem 281:30684–30696
33. Scemes E, Stout Jr RF, Spray DC (2017) Adrenergic receptors on astrocytes modulate gap junctions. In: Nina V, Robert Z (eds) Noradrenergic signaling and astroglia. Academic Press, United Kingdom, 321
34. Falchi AM, Sogos V, Saba F, Piras M, Congiu T, Piludu M (2013) Astrocytes shed large membrane vesicles that contain mitochondria, lipid droplets and ATP. Histochem Cell Biol 139:221–231

35. Fasshauer D, Antonin W, Margittai M, Pabst S, Jahn R (1999) Mixed and non-cognate SNARE complexes. Characterization of assembly and biophysical properties. J Biol Chem 274:15440–15446
36. Fasshauer D, Sutton RB, Brunger AT, Jahn R (1998) Conserved structural features of the synaptic fusion complex: SNARE proteins reclassified as Q- and R-SNAREs. Proc Natl Acad Sci U S A 95:15781–15786
37. Froese A, Breher SS, Waldeyer C, Schindler RF, Nikolaev VO, Rinné S, Wischmeyer E, Schlueter J, Becher J, Simrick S, Vauti F, Kuhtz J, Meister P, Kreissl S, Torlopp A, Liebig SK, Laakmann S, Müller TD, Neumann J, Stieber J, Ludwig A, Maier SK, Decher N, Arnold HH, Kirchhof P, Fabritz L, Brand T (2012) Popeye domain containing proteins are essential for stress-mediated modulation of cardiac pacemaking in mice. J Clin Invest 122:1119–1130
38. Frühbeis C, Fröhlich D, Kuo WP, Krämer-Albers EM (2013) Extracellular vesicles as mediators of neuron-glia communication. Front Cell Neurosci 7:182
39. Gucek A, Jorgacevski J, Singh P, Geisler C, Lisjak M, Vardjan N, Kreft M, Egner A, Zorec R (2016) Dominant negative SNARE peptides stabilize the fusion pore in a narrow, release-unproductive state. Cell Mol Life Sci 73:3719–3731
40. Gucek A, Vardjan N, Zorec R (2012) Exocytosis in astrocytes: transmitter release and membrane signal regulation. Neurochem Res 37:2351–2363
41. Guescini M, Genedani S, Stocchi V, Agnati LF (2010) Astrocytes and Glioblastoma cells release exosomes carrying mtDNA. J Neural Transm (Vienna) 117:1–4
42. Guthrie PB, Knappenberger J, Segal M, Bennett MV, Charles AC, Kater SB (1999) ATP released from astrocytes mediates glial calcium waves. J Neurosci 19:520–528
43. Hajrasouliha AR, Jiang G, Lu Q, Lu H, Kaplan HJ, Zhang HG, Shao H (2013) Exosomes from retinal astrocytes contain antiangiogenic components that inhibit laser-induced choroidal neovascularization. J Biol Chem 288:28058–28067
44. Halassa MM, Florian C, Fellin T, Munoz JR, Lee SY, Abel T, Haydon PG, Frank MG (2009) Astrocytic modulation of sleep homeostasis and cognitive consequences of sleep loss. Neuron 61:213–219
45. Hamilton NB, Attwell D (2010) Do astrocytes really exocytose neurotransmitters? Nat Rev Neurosci 11:227–238
46. Harata N, Aravanis A, Tsien R (2006) Kiss-and-run and full-collapse fusion as modes of exo-endocytosis in neurosecretion. J Neurochem 97:1546–1570
47. Henneberger C, Papouin T, Oliet SH, Rusakov DA (2010) Long-term potentiation depends on release of D-serine from astrocytes. Nature 463:232–236
48. Hepp R, Perraut M, Chasserot-Golaz S, Galli T, Aunis D, Langley K, Grant NJ (1999) Cultured glial cells express the SNAP-25 analogue SNAP-23. Glia 27:181–187
49. Hines DJ, Haydon PG (2013) Inhibition of a SNARE-sensitive pathway in astrocytes attenuates damage following stroke. J Neurosci 33:4234–4240
50. Horvat A, Vardjan N (2019) Astroglial cAMP signalling in space and time. Neurosci Lett 689:5–10
51. Horvat A, Zorec R, Vardjan N (2016) Adrenergic stimulation of single rat astrocytes results in distinct temporal changes in intracellular Ca^{2+} and cAMP-dependent PKA responses. Cell Calcium 59:156–163
52. Hu G, Yao H, Chaudhuri AD, Duan M, Yelamanchili SV, Wen H, Cheney PD, Fox HS, Buch S (2012) Exosome-mediated shuttling of microRNA-29 regulates HIV Tat and morphine-mediated neuronal dysfunction. Cell Death Dis 3:e381
53. Hua X, Malarkey EB, Sunjara V, Rosenwald SE, Li WH, Parpura V (2004) Ca^{2+}-dependent glutamate release involves two classes of endoplasmic reticulum Ca^{2+} stores in astrocytes. J Neurosci Res 76:86–97
54. Hur YS, Kim KD, Paek SH, Yoo SH (2010) Evidence for the existence of secretory granule (dense-core vesicle)-based inositol 1,4,5-trisphosphate-dependent Ca^{2+} signaling system in astrocytes. PLoS ONE 5:e11973
55. Jaiswal JK, Andrews NW, Simon SM (2002) Membrane proximal lysosomes are the major vesicles responsible for calcium-dependent exocytosis in nonsecretory cells. J Cell Biol 159:625–635

56. Jaiswal JK, Fix M, Takano T, Nedergaard M, Simon SM (2007) Resolving vesicle fusion from lysis to monitor calcium-triggered lysosomal exocytosis in astrocytes. Proc Natl Acad Sci U S A 104:14151–14156
57. Jiménez AI, Castro E, Mirabet M, Franco R, Delicado EG, Miras-Portugal MT (1999) Potentiation of ATP calcium responses by A2B receptor stimulation and other signals coupled to Gs proteins in type-1 cerebellar astrocytes. Glia 26:119–128
58. Jourdain P, Bergersen LH, Bhaukaurally K, Bezzi P, Santello M, Domercq M, Matute C, Tonello F, Gundersen V, Volterra A (2007) Glutamate exocytosis from astrocytes controls synaptic strength. Nat Neurosci 10:331–339
59. Kang N, Peng H, Yu Y, Stanton PK, Guilarte TR, Kang J (2013) Astrocytes release D-serine by a large vesicle. Neuroscience 240:243–257
60. Klyachko V, Jackson M (2002) Capacitance steps and fusion pores of small and large-dense-core vesicles in nerve terminals. Nature 418:89–92
61. Kreft M, Stenovec M, Rupnik M, Grilc S, Krzan M, Potokar M, Pangrsic T, Haydon PG, Zorec R (2004) Properties of Ca^{2+}-dependent exocytosis in cultured astrocytes. Glia 46:437–445
62. Krzan M, Stenovec M, Kreft M, Pangrsic T, Grilc S, Haydon PG, Zorec R (2003) Calcium-dependent exocytosis of atrial natriuretic peptide from astrocytes. J Neurosci 23:1580–1583
63. Lalo U, Palygin O, Rasooli-Nejad S, Andrew J, Haydon PG, Pankratov Y (2014) Exocytosis of ATP from astrocytes modulates phasic and tonic inhibition in the neocortex. PLoS Biol 12:e1001747
64. Lalo U, Pankratov Y, Parpura V, Verkhratsky A (2011) Ionotropic receptors in neuronal-astroglial signalling: what is the role of "excitable" molecules in non-excitable cells. Biochim Biophys Acta 1813:992–1002
65. Lasic E, Rituper B, Jorgacevski J, Kreft M, Stenovec M, Zorec R (2016) Subanesthetic doses of ketamine stabilize the fusion pore in a narrow flickering state in astrocytes. J Neurochem 138:909–917
66. Lasic E, Stenovec M, Kreft M, Robinson PJ, Zorec R (2017) Dynamin regulates the fusion pore of endo- and exocytotic vesicles as revealed by membrane capacitance measurements. Biochim Biophys Acta 1861:2293–2303
67. Latour I, Hamid J, Beedle AM, Zamponi GW, Macvicar BA (2003) Expression of voltage-gated Ca^{2+} channel subtypes in cultured astrocytes. Glia 41:347–353
68. Lee HS, Ghetti A, Pinto-Duarte A, Wang X, Dziewczapolski G, Galimi F, Huitron-Resendiz S, Piña-Crespo JC, Roberts AJ, Verma IM, Sejnowski TJ, Heinemann SF (2014) Astrocytes contribute to gamma oscillations and recognition memory. Proc Natl Acad Sci U S A. 111:E3343–3352
69. Leybaert L, Sanderson MJ (2012) Intercellular Ca^{2+} waves: mechanisms and function. Physiol Rev 92:1359–1392
70. Li D, Hérault K, Silm K, Evrard A, Wojcik S, Oheim M, Herzog E, Ropert N (2013) Lack of evidence for vesicular glutamate transporter expression in mouse astrocytes. J Neurosci 33:4434–4455
71. Li D, Ropert N, Koulakoff A, Giaume C, Oheim M (2008) Lysosomes are the major vesicular compartment undergoing Ca^{2+}-regulated exocytosis from cortical astrocytes. J Neurosci 28:7648–7658
72. Liu T, Sun L, Xiong Y, Shang S, Guo N, Teng S, Wang Y, Liu B, Wang C, Wang L, Zheng L, Zhang CX, Han W, Zhou Z (2011) Calcium triggers exocytosis from two types of organelles in a single astrocyte. J Neurosci 31:10593–10601
73. MacVicar BA (1984) Voltage-dependent calcium channels in glial cells. Science 226:1345–1347
74. Maienschein V, Marxen M, Volknandt W, Zimmermann H (1999) A plethora of presynaptic proteins associated with ATP-storing organelles in cultured astrocytes. Glia 26:233–244
75. Malarkey EB, Ni Y, Parpura V (2008) Ca^{2+} entry through TRPC1 channels contributes to intracellular Ca^{2+} dynamics and consequent glutamate release from rat astrocytes. Glia 56:821–835
76. Malarkey EB, Parpura V (2008) Mechanisms of glutamate release from astrocytes. Neurochem Int 52:142–154

77. Malarkey EB, Parpura V (2011) Temporal characteristics of vesicular fusion in astrocytes: examination of synaptobrevin 2-laden vesicles at single vesicle resolution. J Physiol 589:4271–4300
78. Mao S, Sun Q, Xiao H, Zhang C, Li L (2015) Secreted miR-34a in astrocytic shedding vesicles enhanced the vulnerability of dopaminergic neurons to neurotoxins by targeting Bcl-2. Protein Cell 6:529–540
79. Marchaland J, Calì C, Voglmaier SM, Li H, Regazzi R, Edwards RH, Bezzi P (2008) Fast subplasma membrane Ca^{2+} transients control exo-endocytosis of synaptic-like microvesicles in astrocytes. J Neurosci 28:9122–9132
80. Martineau M, Galli T, Baux G, Mothet JP (2008) Confocal imaging and tracking of the exocytotic routes for D-serine-mediated gliotransmission. Glia 56:1271–1284
81. Martineau M, Parpura V, Mothet JP (2014) Cell-type specific mechanisms of D-serine uptake and release in the brain. Front Synaptic Neurosci 6:12
82. Martineau M, Shi T, Puyal J, Knolhoff AM, Dulong J, Gasnier B, Klingauf J, Sweedler JV, Jahn R, Mothet JP (2013) Storage and Uptake of D-Serine into Astrocytic Synaptic-Like Vesicles Specify Gliotransmission. J Neurosci 33:3413–3423
83. Miesenböck G, De Angelis D, Rothman J (1998) Visualizing secretion and synaptic transmission with pH-sensitive green fluorescent proteins. Nature 394:192–195
84. Mittelsteadt T, Seifert G, Alvárez-Barón E, Steinhäuser C, Becker AJ, Schoch S (2009) Differential mRNA expression patterns of the synaptotagmin gene family in the rodent brain. J Comp Neurol 512:514–528
85. Montana V, Liu W, Mohideen U, Parpura V (2009) Single molecule measurements of mechanical interactions within ternary SNARE complexes and dynamics of their disassembly: SNAP25 vs. SNAP23. J Physiol 587:1943–1960
86. Montana V, Malarkey EB, Verderio C, Matteoli M, Parpura V (2006) Vesicular transmitter release from astrocytes. Glia 54:700–715
87. Montana V, Ni Y, Sunjara V, Hua X, Parpura V (2004) Vesicular glutamate transporter-dependent glutamate release from astrocytes. J Neurosci 24:2633–2642
88. Mothet JP, Pollegioni L, Ouanounou G, Martineau M, Fossier P, Baux G (2005) Glutamate receptor activation triggers a calcium-dependent and SNARE protein-dependent release of the gliotransmitter D-serine. Proc Natl Acad Sci U S A 102:5606–5611
89. Nadjar A, Blutstein T, Aubert A, Laye S, Haydon PG (2013) Astrocyte-derived adenosine modulates increased sleep pressure during inflammatory response. Glia 61:724–731
90. Nageotte J (1910) Phenomenes de secretion dans le protoplasma des cellules nevrogliques de la substance grise, Paris
91. Oberheim NA, Goldman SA, Nedergaard M (2012) Heterogeneity of astrocytic form and function. Methods Mol Biol 814:23–45
92. Ormel L, Stensrud MJ, Bergersen LH, Gundersen V (2012) VGLUT1 is localized in astrocytic processes in several brain regions. Glia 60:229–238
93. Oya M, Kitaguchi T, Yanagihara Y, Numano R, Kakeyama M, Ikematsu K, Tsuboi T (2013) Vesicular nucleotide transporter is involved in ATP storage of secretory lysosomes in astrocytes. Biochem Biophys Res Commun 438:145–151
94. Paco S, Margeli MA, Olkkonen VM, Imai A, Blasi J, Fischer-Colbrie R, Aguado F (2009) Regulation of exocytotic protein expression and Ca^{2+}-dependent peptide secretion in astrocytes. J Neurochem 110:143–156
95. Paco S, Pozas E, Aguado F (2010) Secretogranin III is an astrocyte granin that is overexpressed in reactive glia. Cereb Cortex 20:1386–1397
96. Pangrsic T, Potokar M, Stenovec M, Kreft M, Fabbretti E, Nistri A, Pryazhnikov E, Khiroug L, Giniatullin R, Zorec R (2007) Exocytotic release of ATP from cultured astrocytes. J Biol Chem 282:28749–28758
97. Papouin T, Henneberger C, Rusakov DA, Oliet SHR (2017) Astroglial versus neuronal d-serine: fact checking. Trends Neurosci 40:517–520
98. Parpura V, Basarsky TA, Liu F, Jeftinija K, Jeftinija S, Haydon PG (1994) Glutamate-mediated astrocyte-neuron signalling. Nature 369:744–747

99. Parpura V, Fang Y, Basarsky T, Jahn R, Haydon PG (1995) Expression of synaptobrevin II, cellubrevin and syntaxin but not SNAP-25 in cultured astrocytes. FEBS Lett 377:489–492
100. Parpura V, Verkhratsky A (2012) The astrocyte excitability brief: from receptors to gliotransmission. Neurochem Int 61:610–621
101. Parpura V, Zorec R (2010) Gliotransmission: exocytotic release from astrocytes. Brain Res Rev 63:83–92
102. Parri HR, Crunelli V (2001) Pacemaker calcium oscillations in thalamic astrocytes in situ. NeuroReport 12:3897–3900
103. Pascual O, Casper KB, Kubera C, Zhang J, Revilla-Sanchez R, Sul JY, Takano H, Moss SJ, McCarthy K, Haydon PG (2005) Astrocytic purinergic signaling coordinates synaptic networks. Science 310:113–116
104. Pasti L, Zonta M, Pozzan T, Vicini S, Carmignoto G (2001) Cytosolic calcium oscillations in astrocytes may regulate exocytotic release of glutamate. J Neurosci 21:477–484
105. Patrushev I, Gavrilov N, Turlapov V, Semyanov A (2013) Subcellular location of astrocytic calcium stores favors extrasynaptic neuron-astrocyte communication. Cell Calcium 54:343–349
106. Perea G, Araque A (2007) Astrocytes potentiate transmitter release at single hippocampal synapses. Science 317:1083–1086
107. Potokar M, Stenovec M, Kreft M, Kreft ME, Zorec R (2008) Stimulation inhibits the mobility of recycling peptidergic vesicles in astrocytes. Glia 56:135–144
108. Potokar M, Vardjan N, Stenovec M, Gabrijel M, Trkov S, Jorgačevski J, Kreft M, Zorec R (2013) Astrocytic vesicle mobility in health and disease. Int J Mol Sci 14:11238–11258
109. Prada I, Marchaland J, Podini P, Magrassi L, D'Alessandro R, Bezzi P, Meldolesi J (2011) REST/NRSF governs the expression of dense-core vesicle gliosecretion in astrocytes. J Cell Biol 193:537–549
110. Proia P, Schiera G, Mineo M, Ingrassia AM, Santoro G, Savettieri G, Di Liegro I (2008) Astrocytes shed extracellular vesicles that contain fibroblast growth factor-2 and vascular endothelial growth factor. Int J Mol Med 21:63–67
111. Pryazhnikov E, Khiroug L (2008) Sub-micromolar increase in $[Ca^{2+}](i)$ triggers delayed exocytosis of ATP in cultured astrocytes. Glia 56:38–49
112. Puzar Dominkus P, Stenovec M, Sitar S, Lasic E, Zorec R, Plemenitas A, Zagar E, Kreft M, Lenassi M (2018) PKH26 labeling of extracellular vesicles: Characterization and cellular internalization of contaminating PKH26 nanoparticles. Biochim Biophys Acta Biomembr 1860:1350–1361
113. Ramamoorthy P, Whim MD (2008) Trafficking and fusion of neuropeptide Y-containing dense-core granules in astrocytes. J Neurosci 28:13815–13827
114. Rathbone MP, Middlemiss PJ, DeLuca B, Jovetich M (1991) Extracellular guanosine increases astrocyte cAMP: inhibition by adenosine A2 antagonists. NeuroReport 2:661–664
115. Reyes RC, Parpura V (2008) Mitochondria modulate Ca^{2+}-dependent glutamate release from rat cortical astrocytes. J Neurosci 28:9682–9691
116. Reyes RC, Verkhratsky A, Parpura V (2012) Plasmalemmal Na^{+}/Ca^{2+} exchanger modulates Ca^{2+}-dependent exocytotic release of glutamate from rat cortical astrocytes. ASN Neuro 4
117. Rituper B, Gucek A, Jorgacevski J, Flasker A, Kreft M, Zorec R (2013) High-resolution membrane capacitance measurements for the study of exocytosis and endocytosis. Nat Protoc 8:1169–1183
118. Sami Saribas A, Cicalese S, Ahooyi TM, Khalili K, Amini S, Sariyer IK (2017) HIV-1 Nef is released in extracellular vesicles derived from astrocytes: evidence for Nef-mediated neurotoxicity. Cell Death Dis 8:e2542
119. Santello M, Bezzi P, Volterra A (2011) TNFα controls glutamatergic gliotransmission in the hippocampal dentate gyrus. Neuron 69:988–1001
120. Sawada K, Echigo N, Juge N, Miyaji T, Otsuka M, Omote H, Yamamoto A, Moriyama Y (2008) Identification of a vesicular nucleotide transporter. Proc Natl Acad Sci U S A 105:5683–5686
121. Sbai O, Ould-Yahoui A, Ferhat L, Gueye Y, Bernard A, Charrat E, Mehanna A, Risso JJ, Chauvin JP, Fenouillet E, Rivera S, Khrestchatisky M (2010) Differential vesicular distribution and trafficking of MMP-2, MMP-9, and their inhibitors in astrocytes. Glia 58:344–366

122. Scemes E, Suadicani SO, Spray DC (2000) Intercellular communication in spinal cord astrocytes: fine tuning between gap junctions and P2 nucleotide receptors in calcium wave propagation. J Neurosci 20:1435–1445
123. Schubert V, Bouvier D, Volterra A (2011) SNARE protein expression in synaptic terminals and astrocytes in the adult hippocampus: a comparative analysis. Glia 59:1472–1488
124. Semyanov A (2018) Spatiotemporal pattern of calcium activity in astrocytic network. Cell Calcium 78:15–25
125. Sild M, Van Horn MR (2013) Astrocytes use a novel transporter to fill gliotransmitter vesicles with D-serine: evidence for vesicular synergy. J Neurosci 33:10193–10194
126. Simpson PB, Russell JT (1998) Role of mitochondrial Ca^{2+} regulation in neuronal and glial cell signalling. Brain Res Rev 26:72–81
127. Singh P, Jorgačevski J, Kreft M, Grubišić V, Stout RF, Potokar M, Parpura V, Zorec R (2014) Single-vesicle architecture of synaptobrevin2 in astrocytes. Nat Commun 5:3780
128. Slezak M, Grosche A, Niemiec A, Tanimoto N, Pannicke T, Münch TA, Crocker B, Isope P, Härtig W, Beck SC, Huber G, Ferracci G, Perraut M, Reber M, Miehe M, Demais V, Lévêque C, Metzger D, Szklarczyk K, Przewlocki R, Seeliger MW, Sage-Ciocca D, Hirrlinger J, Reichenbach A, Reibel S, Pfrieger FW (2012) Relevance of exocytotic glutamate release from retinal glia. Neuron 74:504–516
129. Stenovec M, Lasic E, Dominkus PP, Bobnar ST, Zorec R, Lenassi M, Kreft M (2018) Slow release of HIV-1 Protein Nef from vesicle-like structures is inhibited by cytosolic calcium elevation in single human microglia. Mol Neurobiol
130. Stenovec M, Lasič E, Božić M, Bobnar ST, Stout RF, Grubišić V, Parpura V, Zorec R (2016) Ketamine inhibits ATP-evoked exocytotic release of brain-derived neurotrophic factor from vesicles in cultured rat astrocytes. Mol Neurobiol 53:6882–6896
131. Südhof TC (2012) Calcium control of neurotransmitter release. Cold Spring Harb Perspect Biol 4:a011353
132. Taylor AR, Robinson MB, Gifondorwa DJ, Tytell M, Milligan CE (2007) Regulation of heat shock protein 70 release in astrocytes: role of signaling kinases. Dev Neurobiol 67:1815–1829
133. Turner JR, Ecke LE, Briand LA, Haydon PG, Blendy JA (2013) Cocaine-related behaviors in mice with deficient gliotransmission. Psychopharmacology 226:167–176
134. Vardjan N, Gabrijel M, Potokar M, Svajger U, Kreft M, Jeras M, de Pablo Y, Faiz M, Pekny M, Zorec R (2012) IFN-γ-induced increase in the mobility of MHC class II compartments in astrocytes depends on intermediate filaments. J Neuroinflamm 9:144
135. Vardjan N, Kreft M, Zorec R (2014) Dynamics of β-adrenergic/cAMP signaling and morphological changes in cultured astrocytes. Glia
136. Vardjan N, Parpura V, Zorec R (2016) Loose excitation-secretion coupling in astrocytes. Glia 64:655–667
137. Vardjan N, Zorec R (2015) Excitable Astrocytes: Ca^{2+}- and cAMP-regulated exocytosis. Neurochem Res
138. Verderio C, Cagnoli C, Bergami M, Francolini M, Schenk U, Colombo A, Riganti L, Frassoni C, Zuccaro E, Danglot L, Wilhelm C, Galli T, Canossa M, Matteoli M (2012) TI-VAMP/VAMP7 is the SNARE of secretory lysosomes contributing to ATP secretion from astrocytes. Biol Cell 104:213–228
139. Verkhratsky A, Matteoli M, Parpura V, Mothet JP, Zorec R (2016) Astrocytes as secretory cells of the central nervous system: idiosyncrasies of vesicular secretion. EMBO J
140. Verkhratsky A, Nedergaard M (2018) Physiology of astroglia. Physiol Rev 98:239–389
141. Volterra A, Liaudet N, Savtchouk I (2014) Astrocyte Ca^{2+} signalling: an unexpected complexity. Nat Rev Neurosci 15:327–335
142. Wang S, Cesca F, Loers G, Schweizer M, Buck F, Benfenati F, Schachner M, Kleene R (2011) Synapsin I is an oligomannose-carrying glycoprotein, acts as an oligomannose-binding lectin, and promotes neurite outgrowth and neuronal survival when released via glia-derived exosomes. J Neurosci 31:7275–7290
143. Wang Z, Chapman ER (2010) Rat and Drosophila synaptotagmin 4 have opposite effects during SNARE-catalyzed membrane fusion. J Biol Chem 285:30759–30766

144. Wilhelm A, Volknandt W, Langer D, Nolte C, Kettenmann H, Zimmermann H (2004) Localization of SNARE proteins and secretory organelle proteins in astrocytes in vitro and in situ. Neurosci Res 48:249–257
145. Wolosker H, Balu DT, Coyle JT (2016) The rise and fall of the d-Serine-mediated gliotransmission hypothesis. Trends Neurosci 39:712–721
146. Woo DH, Han KS, Shim JW, Yoon BE, Kim E, Bae JY, Oh SJ, Hwang EM, Marmorstein AD, Bae YC, Park JY, Lee CJ (2012) TREK-1 and Best1 channels mediate fast and slow glutamate release in astrocytes upon GPCR activation. Cell 151(1):25–40
147. Zhang Q, Fukuda M, Van Bockstaele E, Pascual O, Haydon PG (2004) Synaptotagmin IV regulates glial glutamate release. Proc Natl Acad Sci U S A 101:9441–9446
148. Zhang Q, Pangrsic T, Kreft M, Krzan M, Li N, Sul JY, Halassa M, Van Bockstaele E, Zorec R, Haydon PG (2004) Fusion-related release of glutamate from astrocytes. J Biol Chem 279:12724–12733
149. Zhang Y, Chen K, Sloan SA, Bennett ML, Scholze AR, O'Keeffe S, Phatnani HP, Guarnieri P, Caneda C, Ruderisch N, Deng S, Liddelow SA, Zhang C, Daneman R, Maniatis T, Barres BA, Wu JQ (2014) An RNA-sequencing transcriptome and splicing database of glia, neurons, and vascular cells of the cerebral cortex. J Neurosci 34:11929–11947
150. Zhang Z, Chen G, Zhou W, Song A, Xu T, Luo Q, Wang W, Gu XS, Duan S (2007) Regulated ATP release from astrocytes through lysosome exocytosis. Nat Cell Biol 9:945–953
151. Zorec R, Araque A, Carmignoto G, Haydon PG, Verkhratsky A, Parpura V (2012) Astroglial excitability and gliotransmission: an appraisal of Ca^{2+} as a signalling route. ASN Neuro 4
152. Zorec R, Parpura V, Verkhratsky A (2018) Astroglial vesicular network: evolutionary trends, physiology and pathophysiology. Acta Physiol (Oxf) 222
153. Zorec R, Verkhratsky A, Rodriguez JJ, Parpura V (2016) Astrocytic vesicles and gliotransmitters: Slowness of vesicular release and synaptobrevin2-laden vesicle nanoarchitecture. Neuroscience 323:67–75

Chapter 5
Physiology of Oligodendroglia

Arthur M. Butt, Maria Papanikolaou and Andrea Rivera

Abstract Oligodendrocytes are the myelinating cells of the CNS, producing the insulating myelin sheath that facilitates rapid electrical conduction of axonal action potentials. Oligodendrocytes arise from oligodendrocyte progenitor cells (OPCs) under the control of multiple factors, including neurotransmitters and other neuron-derived factors. A significant population of OPCs persists in the adult CNS, where they are often referred to as NG2-glia, because they are identified by their expression of the NG2 chondroitin sulphate proteoglycan (CSPG4). In the adult brain, the primary function of NG2-glia is the life-long generation of oligodendrocytes to replace myelin lost through natural 'wear and tear' and pathology, as well as to provide new oligodendrocytes to myelinate new connections formed in response to new life experiences. NG2-glia contact synapses and respond to neurotransmitters and potassium released during neuronal transmission; to this end, NG2-glia (OPCs) express multiple neurotransmitter receptors and ion channels, with prominent roles being identified for glutamatergic signalling and potassium channels in oligodendrocyte differentiation. Myelinating oligodendrocytes also express a wide range of neurotransmitter receptors and ion channels, together with transporters and gap junctions; together, these have critical functions in cellular ion and water homeostasis, as well as metabolism, which is essential for maintaining myelin and axon integrity. An overriding theme is that oligodendrocyte function and myelination is not only essential for rapid axonal conduction, but is essential for learning and the long-term integrity of axons and neurones. Hence, myelination underpins cognitive function and the massive computing power of the human brain and myelin loss has devastating effects on CNS function. This chapter focuses on normal oligodendrocyte physiology.

Keywords Oligodendrocyte · Oligodendrocyte precursor cell · OPC · NG2-glia · Myelin · Axon

---

A. M. Butt (✉) · M. Papanikolaou · A. Rivera
School of Pharmacy and Biomedical Science, University of Portsmouth, St Michael's Building, White Sawn Road, Portsmouth PO1 2DT, UK
e-mail: Arthur.butt@port.ac.uk

A. Verkhratsky et al. (eds.), *Neuroglia in Neurodegenerative Diseases*, Advances in Experimental Medicine and Biology 1175,
https://doi.org/10.1007/978-981-13-9913-8_5

5.1 Introduction

Individual oligodendrocytes can myelinate 30 or more axons and each myelin sheath extends along the axon for over 50–100 μm to form internodes that are interrupted by the nodes of Ranvier [12, 55]. In this way, the axonal membrane is divided into alternating nonconductive portions underneath the myelin sheath and the conductive nodes of Ranvier, where the sodium and potassium channels that mediate the action potentials are localised [76]. The myelin sheath is formed by concentric lamellae of the oligodendroglial plasmalemma wrapped around the axon; compacted myelin is formed by fused phospholipid bilayers, which gives the myelin its insulating properties [65]. The compacted myelin is surrounded by a cytoplasmic ridge that forms a conduit through which proteins and other chemicals are transported from the oligodendrocyte cell body [47]. At the node of Ranvier, the cytoplasmic ridges form the paranodal loops that establish complex adhesion junctions with the axon, which is essential for the separation of axonal potassium and sodium channels that generate the action potential [25]. The loss or disruption of myelin, such as occurs in demyelinating diseases and other pathologies, results in disruption of action potential propagation or conduction block, and ultimately loss of axonal integrity and neuronal death [58].

5.2 Myelin

The main constituents of myelin are lipids (70% of its dry weight) and proteins (30% of the dry weight), many of which are specific to myelin and are used to identify oligodendrocytes [47]. Cholesterol is a major component of myelin (27%) and is essential for myelination [60]; the blood–brain barrier prevents dietary cholesterol from entering the brain and astrocytes are proposed to be the main source of cholesterol in the brain, although oligodendrocytes are capable of de novo synthesis [35]. In addition, myelin contains phospholipids that are rich in glycosphingolipids, in particular, galactocerebroside (GalC), and sulphatides [34]. The major myelin proteins are myelin basic protein (MBP) and proteolipid protein (PLP), which constitute about 80% of CNS protein, together with numerous proteins that make up a small but significant fractions of myelin, including 2′, 3′-cyclic nucleotide-3′-phosphodiesterase (CNP), myelin-associated glycoprotein (MAG), and myelin oligodendrocyte glycoprotein (MOG); notably, the absence of PLP or CNP result in axonal degeneration [48]. In addition, myelin contains gap junctions, predominantly formed by connexins Cx32 and Cx47, the latter being specific to oligodendrocytes in the CNS, together with Cx29, which are crucial for ion homeostasis, myelination and axonal metabolism and integrity [70]. A consistent theme is that ablation of individual oligodendrocyte genes, ranging from myelin genes to connexins, results in axonal demise, demonstrating that oligodendrocytes and axons are interdependent functional units.

5.3 Oligodendrocyte Differentiation

Oligodendrocytes are generated from OPCs that arise from multipotent neural stem cells (NSCs) in the subventricular zone (SVZ) [5]. From these focal sources, OPCs migrate to populate the entire CNS, where they undergo local proliferation and pass through a number of intermediate stages to differentiate into myelinating oligodendrocytes [68]. A significant population of OPCs persists in the adult CNS, where they are often referred to as NG2-glia, which have the stem cell-like property of self-maintaining and the capacity to generate oligodendrocytes throughout life [2]. OPCs are identified by their expression of PDGFRα and NG2 (cspg4) [56] and, as they differentiate, they exit cell cycle and lose Pdgra and NG2, and transiently express GPR17, before expressing the transcription factor myelin gene regulatory factor (MRF) and myelin-related proteins, such as MBP and PLP [19, 23]; GPR17 and MRF appear in oligodendrocytes shortly before the onset of myelination and, respectively, negatively and positively regulate terminal differentiation into myelinating oligodendrocytes. The specification, migration, proliferation and differentiation of OPCs are regulated by a highly complex interplay between intrinsic and extrinsic factors that both negatively and positively influence oligodendrocyte generation and myelination [18]. These include key growth factors, such as platelet-derived growth factor (PDGF) and fibroblast growth factor (FGF), which act, respectively, via PDGFRα and FGF receptors-1, -2 and -3 (FGFR1-3), and insulin-like growth factor (IGF-1) [54]. In addition, Wnt signalling drives OPC expansion throughout life [5], and oestrogen positively regulates oligodendrocyte differentiation and myelination [72]. Cytokines and chemokines also act on oligodendrocytes and OPCs through a wide range of receptors, such as interleukins (e.g., IL-1β and IL-6) and the CXCL12/CXCR4/CXCR7 axis, which regulate OPC proliferation, migration, and differentiation [35, 54]. Furthermore, OPC expresses a range of neurotransmitter receptors and ion channels that regulate their migration, proliferation and differentiation [27]. The disruption or loss of OPCs, or their capacity for generating myelinating oligodendrocytes, has devastating effects on CNS function and ultimately leads to death. Hence, the key feature of OPC is that their self-maintenance and differentiation into oligodendrocytes are regulated by multifarious factors, helping to ensure that disruption of any single factor does not result in their loss of function. Most of these factors are also involved in oligodendrocyte pathologies and are important in regeneration and remyelination in diseases such as multiple sclerosis (MS).

5.4 Neurotransmitter Receptors

Oligodendrocytes and OPCs express ligand-gated ion channels and G protein-coupled receptors (GPCR) for a wide range of neurotransmitters (Fig. 5.1) [43, 71]. In oligodendrocytes, neurotransmitter receptors mediate intercellular communication with the neurons/axons they myelinate [37]; in OPCs, they are generally con-

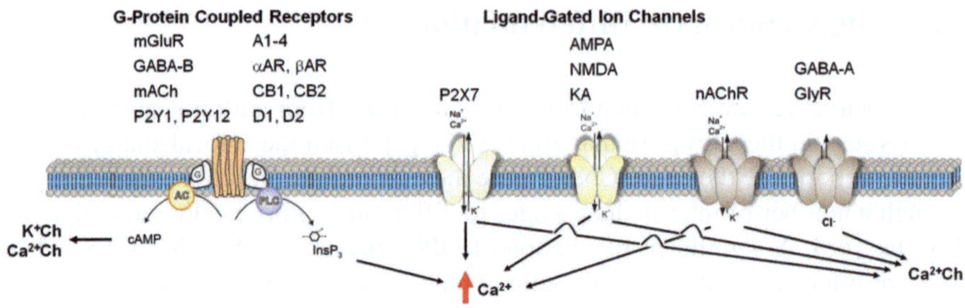

Fig. 5.1 Oligodendrocyte lineage cells express ligand-gated ion channels and G-Protein-coupled receptors (GPCR) for multiple neurotransmitters. A prominent feature of oligodendrocytes at all stages of differentiation is the expression of glutamate receptors, which play key roles in OPC proliferation and differentiation, and in homeostasis and metabolism in oligodendrocytes. A general characteristic is that most receptors can mediate increased intracellular Ca^{2+}, which has diverse effects on OPCs and oligodenrocytes (see text for further details)

sidered to regulate differentiation and many mediate their effects through changes in intracellular Ca^{2+} [10, 11].

Glutamate receptors are highly expressed by oligodendrocytes and OPCs [37]. Ionotropic glutamate receptors (iGluRs) of the AMPA, kainate and NMDA types are abundantly expressed throughout the oligodendrocyte lineage, which allows the flow of K^+, Na^+ and Ca^{2+}. In OPCs, AMPAR is permeable to Ca^{2+} [7, 26, 28], and Ca^{2+} permeability may be downregulated during oligodendrocyte development [33]. AMPAR activation in oligodendrocyte lineage cells mediates signalling from axons that regulate OPC differentiation and myelination [15]. Oligodendrocytes also express functional NMDARs, although their physiological importance is unclear [36]. All three groups of mGluR have been demonstrated in OPCs, but maybe downregulated as they differentiate into mature oligodendrocytes [13].

GABA receptors have been demonstrated to be functional in OPCs and are proposed to regulate their differentiation into oligodendrocytes [27]. Intracellular [Cl^-] is maintained high in glia and activation of $GABA_AR$ leads to Cl^- efflux and cell depolarization, which activates voltage-operated calcium channels (VOCC) and increased intracellular Ca^{2+} in OPC and oligodendrocytes [4]. Metabotropic $GABA_BR$ is also expressed by OPCs and stimulate proliferation and migration [41].

Purine receptors are widely expressed by OPCs and oligodendrocytes, where they are important in development, myelin maintenance and pathology [11, 57]. Adenosine receptors (AR) are GPCR are of four subtypes (A1-4), all of which have been identified in OPCs, where they regulate migration, proliferation, and differentiation, and appear to be downregulated in oligodendrocytes [17]; A2A-R activation decreases cAMP and stimulates outward rectifying potassium channels (Kv) and OPC differentiation (see below). P2XR are expressed by OPCs and oligodendrocytes, with most robust evidence for the $P2X_7$ subtype. P2X7R mediate a rise in intracellular Ca^{2+}, as well as activating multiple intracellular pathways, including MAPK, PKC, and PI3 K, all of which regulate OPC proliferation, differentiation and myelination. In addition, $P2X_7R$ is implicated in the loss of oligodendrocytes and myelin in ischemia and demyelination. P2YR are GPCR and a key feature in OPCs and OLs are the prominent expression of $P2Y_1R$ that mediate raised intracellular Ca^{2+}, which regulates migration, proliferation and differentiation in OPCs. In addition, $P2Y_{12}R$ are enriched in oligodendrocytes and are implicated in demyelination in MS [3].

Numerous other neurotransmitter receptors are reported in OPCs and oligodendrocytes that are implicated in the regulation of OPC differentiation and myelination [43]; these include acetylcholine receptors (AChR), both nicotinic and muscarinic [21], as well as cannabinoid receptors CB_1 and CB_2 [32].

5.5 Ion Channels and Transporters

Oligodendrocyte lineage cells express diverse ion channels (Fig. 5.2) [39, 64]. Kv (Kv1.3, Kv1.4, Kv1.5, Kv1.6) are prominent in OPCs and regulate their proliferation and differentiation, and are generally downregulated during differentiation [69]. OPCs and oligodendrocytes also express inward rectifying potassium channels (Kir), with a prominent role for Kir4.1, as homomers and as heteromers with Kir5.1 [9]; Kir4.1 facilitate clearance of K^+ released during axonal firing and selective deletion of Kir4.1 from OPCs or mature oligodendrocytes results in profound functional

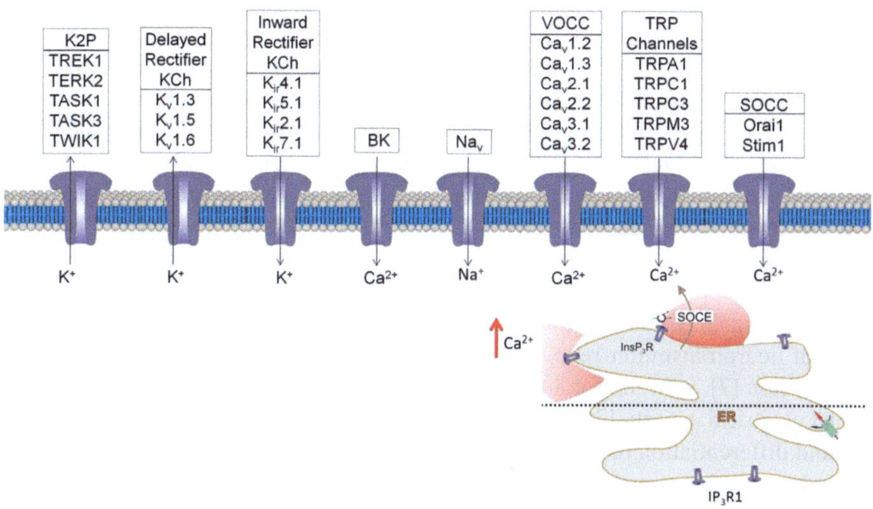

Fig. 5.2 Oligodendrocyte lineage cells express diverse ion channels. A key feature is the expression of potassium channels: delayed rectifier Kv are prominent in OPCs, where they regulate proliferation and differentiation, and are generally downregulated in myelinating oligodendrocytes; inward rectifier channels are expressed throughout the oligodendrocyte lineage and play an important role in ion and water homeostasis. In addition, oligodendrocyte lineage cells express multiple ion channels that mediate Ca^{2+} flux, which is important for myelination and are implicated in pathology

impairments and axon degeneration [38, 62]. Other potassium channels may also be important in maintaining oligodendroglial function and integrity, including Kir2.1, Kir7.1 and TASK1 channels [8, 31, 52].

A key feature of oligodendrocyte lineage cells is the important role for calcium in regulating OPCs and myelination, by influx through ion channels or by activation of receptors [71]. OPCs express voltage-operated calcium channels (VOCC) that regulate OPC maturation and myelination [51, 61]. In addition, store-operated calcium channels (SOCC) and TRP channels are another important mechanism of calcium influx [52]; TRPA1, TRPM3 and ASIC are expressed by oligodendrocytes [20, 30, 52], whilst TRPV4 are expressed by OPC [49]. OPC also express BK channels that mediate Ca^{2+} influx [14]. Oligodendrocytes express the major cation and anion transporters, including Ca-ATPase and Na–Ca exchangers (NCX), which are important in regulating intracellular calcium, together with Na-K-pumps and a variety of anion transporters proteins [53]; Na–K–Cl and K–Cl co-transporters in oligodendrocytes can promote pathological Na^+ entry into oligodendrocytes, which then triggers reverse Na–Ca exchange resulting in Ca^{2+} entry and injury. The physiological role of these diverse mechanisms appears to be regulation of proliferation and differentiation in OPCs and homeostasis and maintenance of cellular and myelin integrity in oligodendrocytes.

5.6 NG2-Glia

As noted above, a significant population of OPCs persists in the adult CNS, where they are often referred to as NG2-glia, since they are identified using antibodies against NG2. Genetic fate mapping has demonstrated that NG2-glia generate oligodendrocytes in the adult brain, which is essential for functional mtyelin repair [50] and for myelination of new connections formed in response to new life experiences [44, 75]. It is evident that neuronal activity enhances myelin formation, termed adaptive myelination, and this is important for nervous system plasticity and repair [16, 22, 50, 74]. This supports the concept that neurotransmission may drive differentiation of NG2-glia (Fig. 5.3), consistent with abundant evidence that NG2-glia express a wide range of neurotransmitter receptors and ion channels [39] and respond to synaptic transmission [7], as well as contacting axons at nodes of Ranvier and responding to axonal electrical activity [28]. In this context, regulation of NG2-glial cell proliferation and differentiation have been indicated for glutamatergic, GABAergic, purinergic and potassium signalling [27, 37, 57, 69]; these may act via changes in intracellular Ca^{2+} and calcium-dependent intracellular signalling pathways, including ERKs and CREB [63]. Adenosine and ATP have been shown to mediate axonal control of differentiation and myelination via raised $[Ca^{2+}]_i$ in NG2-glia [66]. Similarly, glutamate released from electrically active axons acting on AMPAR promotes proliferation and differentiation of NG2-glia [15, 73], whereas GABA acting on GABAAR finely tunes OPC self-maintenance capacity and negatively regulate the generation of oligodendrocytes [6, 29]. NG2-glia also sense changes in extracellular K^+ dur-

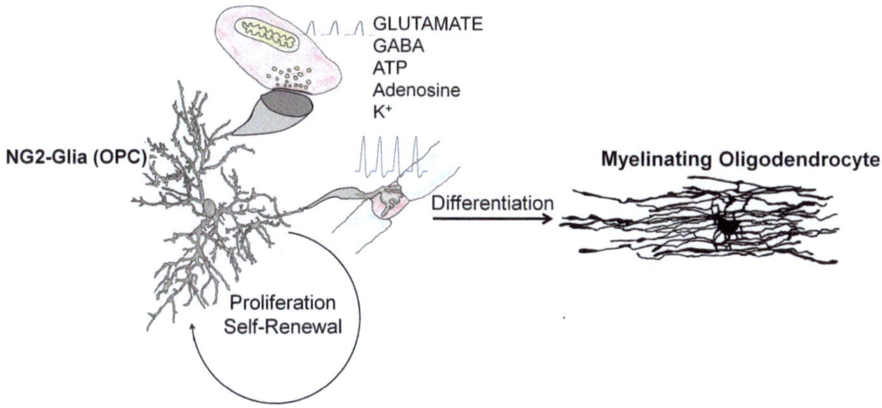

Fig. 5.3 Relationships of NG2-glia with neurons. NG2-glial cells function as adult OPCs. They are multiprocessed cells that extend processes to contact neurons at their sites of electrical and chemical activity at synapses, the sites of chemical transmission between neurons, and nodes of Ranvier, the sites of action potential propagation along axons. Neurotransmitters and K^+ released from synapses and axons act on ion channels and neurotransmitter receptors on NG2-glia to regulate their self-renewal and differentiation into myelinating oligodendrocytes. In this way, neuronal activity can drive myelination during development and in the adult

ing neuronal activity [42], and Kv regulate their proliferation and differentiation [69]; in contrast, selective deletion of Kir4.1 in OPCs did not appear to impair their development [38]. The balance of evidence is that neuronal activity drives adaptive myelination and that NG2-glia are the source of newly generated oligodendrocytes. However, blocking or stimulating synaptic signalling directly has only subtle effects on NG2-glia, indicating neurotransmitters alone do not drive oligodendrogenesis.

5.7 Oligodendrocyte–Axon Interactions and Metabolism

In addition to enabling rapid electrical conduction, myelin is required for axonal integrity, and an important mechanism is metabolic support (Fig. 5.4) [1]. Oligodendrocytes may provide support in the form of glucose [45], but, in general, it appears oligodendrocytes deliver lactate to axons, which they release through MCT1 into the periaxonal space, from where it is taken up by axons via MCT2 [24, 40]. A recent study provides a mechanism by which metabolic support is coupled to axonal activity: action potentials trigger axonal release of glutamate, which activates oligodendroglial NMDAR to evoke a rise in intracellular Ca^{2+} [46]; this stimulates oligodendroglial expression of the glucose transporter GLUT1 and glucose uptake, which is metabolised to lactate and released to axons [59]. The physiological importance of oligodendrocyte–axon metabolic support is unclear and may be more critical under conditions of glucose deprivation and high-frequency activity [67].

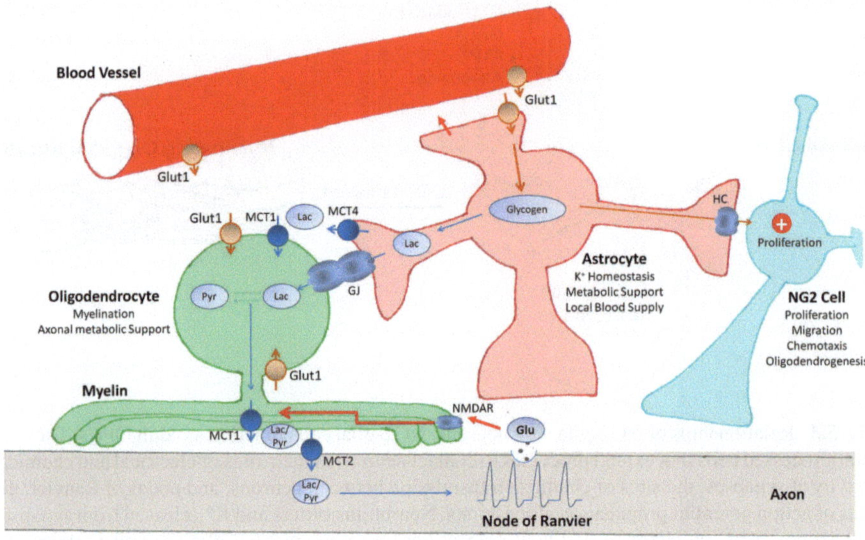

Fig. 5.4 Oligodendrocytes provide metabolic support for axons. Glutamate released during action potential propagation can act directly on oligodendrocyte NMDA-type receptors to stimulate their metabolic support of axons. NMDAR raise intracellular Ca^{2+} which increases activity of glucose transporters (Glut1), which is converted to lactate and released into the periaxonal space via monocarboxylic transporters (MCT1), which are highly expressed by oligodendrocytes, and taken up by the axons via the neuronal MCT2. Astrocytes are another source of lactate for oligodendrocytes, via gap junctions (GJ) and by release through monocarboxylate transporters (MCT4) and subsequent uptake by oligodendrocytes (MCT1). Astrocytes also release glucose via hemichannels to provide metabolic support for NG2-glia, which is necessary for their proliferation and regeneration of oligodendrocytes

5.8 Concluding Remarks

Oligodendrocytes are defined by their myelinating function in the CNS. Myelination provides rapid nervous transmission, without which the brain could not achieve its massive computing power. Myelination is also essential for axonal integrity: oligodendrocytes and the axons they myelinate are completely interdependent functional units and dysfunction in one results in loss of function of the other. The underlying mechanisms are not fully resolved, but recent findings indicate oligodendrocytes are an important source of metabolic support for axons. In addition, new studies demonstrate the importance of adaptive myelination for neural circuit plasticity and learning. The wide range of neurotransmitter receptors and ion channels expressed by oligodendrocytes and OPCs play key roles in these functions and are increasingly recognised as being important in both oligodendrocyte and axonal integrity.

Acknowledgements The authors are supported by grants from the BBSRC and MRC.

References

1. Alexandra IM, Constanze D, Klaus-Armin N (2018) An emerging role of dysfunctional axon-oligodendrocyte coupling in neurodegenerative diseases. Dialogues Clin Neurosci 20:283–292
2. Almeida RG, Lyons DA (2017) On myelinated axon plasticity and neuronal circuit formation and function. J Neurosci 37:10023–10034
3. Amadio S, Montilli C, Magliozzi R, Bernardi G, Reynolds R, Volonte C (2010) P2Y12 receptor protein in cortical gray matter lesions in multiple sclerosis. Cereb Cortex 20:1263–1273
4. Angulo MC, le Meur K, Kozlov AS, Charpak S, Audinat E (2008) GABA, a forgotten gliotransmitter. Prog Neurobiol 86:297–303
5. Azim K, Angonin D, Marcy G, Pieropan F, Rivera A, Donega V, Cantu C, Williams G, Berninger B, Butt AM, Raineteau O (2017) Pharmacogenomic identification of small molecules for lineage specific manipulation of subventricular zone germinal activity. PLoS Biol 15:e2000698
6. Balia M, Benamer N, Angulo MC (2017) A specific GABAergic synapse onto oligodendrocyte precursors does not regulate cortical oligodendrogenesis. Glia 65:1821–1832
7. Bergles DE, Roberts JD, Somogyi P, Jahr CE (2000) Glutamatergic synapses on oligodendrocyte precursor cells in the hippocampus. Nature 405:187–191
8. Brasko C, Butt AM (2018) Expression of Kir2.1 inward rectifying potassium channels in optic nerve glia: evidence for heteromeric association with Kir4.1 and Kir5.1. Neuroglia 1:176–187
9. Brasko C, Hawkins V, de la Rocha IC, Butt AM (2017) Expression of Kir4.1 and Kir5.1 inwardly rectifying potassium channels in oligodendrocytes, the myelinating cells of the CNS. Brain Struct Funct 222:41–59
10. Butt AM (2006) Neurotransmitter-mediated calcium signalling in oligodendrocyte physiology and pathology. Glia 54:666–675
11. Butt AM, Fern RF, Matute C (2014) Neurotransmitter signaling in white matter. Glia 62:1762–1779
12. Butt AM, Ransom BR (1989) Visualization of oligodendrocytes and astrocytes in the intact rat optic nerve by intracellular injection of lucifer yellow and horseradish peroxidase. Glia 2:470–475
13. Butt AM, Vanzulli I, Papanikolaou M, de la Rocha IC, Hawkins VE (2017) Metabotropic glutamate receptors protect oligodendrocytes from acute ischemia in the mouse optic nerve. Neurochem Res 42:2468–2478
14. Buttigieg J, Eftekharpour E, Karimi-Abdolrezaee S, Fehlings MG (2011) Molecular and electrophysiological evidence for the expression of BK channels in oligodendroglial precursor cells. Eur J Neurosci 34:538–547
15. Chen TJ, Kula B, Nagy B, Barzan R, Gall A, Ehrlich I, Kukley M (2018) In vivo regulation of oligodendrocyte precursor cell proliferation and differentiation by the AMPA-receptor subunit GluA2. Cell Rep 25:852–861.e7
16. Chorghay Z, Karadottir RT, Ruthazer ES (2018) White matter plasticity keeps the brain in tune: axons conduct while glia wrap. Front Cell Neurosci 12:428
17. Coppi E, Cellai L, Maraula G, Dettori I, Melani A, Pugliese AM, Pedata F (2015) Role of adenosine in oligodendrocyte precursor maturation. Front Cell Neurosci 9:155
18. Elbaz B, Popko B (2019) Molecular control of oligodendrocyte development. Trends Neurosci 42:263–277
19. Emery B, Agalliu D, Cahoy JD, Watkins TA, Dugas JC, Mulinyawe SB, Ibrahim A, Ligon KL, Rowitch DH, Barres BA (2009) Myelin gene regulatory factor is a critical transcriptional regulator required for CNS myelination. Cell 138:172–185
20. Feldman DH, Horiuchi M, Keachie K, McCauley E, Bannerman P, Itoh A, Itoh T, Pleasure D (2008) Characterization of acid-sensing ion channel expression in oligodendrocyte-lineage cells. Glia 56:1238–1249
21. Fields RD, Dutta DJ, Belgrad J, Robnett M (2017) Cholinergic signaling in myelination. Glia 65:687–698
22. Foster AY, Bujalka H, Emery B (2019) Axoglial interactions in myelin plasticity: evaluating the relationship between neuronal activity and oligodendrocyte dynamics. Glia

23. Fumagalli M, Lecca D, Coppolino GT, Parravicini C, Abbracchio MP (2017) Pharmacological properties and biological functions of the GPR17 receptor, a potential target for neuro-regenerative medicine. Adv Exp Med Biol 1051:169–192
24. Funfschilling U, Supplie LM, Mahad D, Boretius S, Saab AS, Edgar J, Brinkmann BG, Kassmann CM, Tzvetanova ID, Mobius W, Diaz F, Meijer D, Suter U, Hamprecht B, Sereda MW, Moraes CT, Frahm J, Goebbels S, Nave KA (2012) Glycolytic oligodendrocytes maintain myelin and long-term axonal integrity. Nature 485:517–521
25. Ghosh A, Sherman DL, Brophy PJ (2018) The axonal cytoskeleton and the assembly of nodes of ranvier. Neuroscientist 24:104–110
26. Haberlandt C, Derouiche A, Wyczynski A, Haseleu J, Pohle J, Karram K, Trotter J, Seifert G, Frotscher M, Steinhauser C, Jabs R (2011) Gray matter NG2 cells display multiple Ca2 + -signaling pathways and highly motile processes. PLoS ONE 6:e17575
27. Habermacher C, Angulo MC, Benamer N (2019) Glutamate versus GABA in neuron-oligodendroglia communication. Glia
28. Hamilton N, Vayro S, Wigley R, Butt AM (2010) Axons and astrocytes release ATP and glutamate to evoke calcium signals in NG2-glia. Glia 58:66–79
29. Hamilton NB, Clarke LE, Arancibia-Carcamo IL, Kougioumtzidou E, Matthey M, Karadottir R, Whiteley L, Bergersen LH, Richardson WD, Attwell D (2017) Endogenous GABA controls oligodendrocyte lineage cell number, myelination, and CNS internode length. Glia 65:309–321
30. Hamilton NB, Kolodziejczyk K, Kougioumtzidou E, Attwell D (2016) Proton-gated Ca(2 +)-permeable TRP channels damage myelin in conditions mimicking ischaemia. Nature 529:523–527
31. Hawkins V, Butt A (2013) TASK-1 channels in oligodendrocytes: a role in ischemia mediated disruption. Neurobiol Dis 55:87–94
32. Ilyasov AA, Milligan CE, Pharr EP, Howlett AC (2018) The endocannabinoid system and oligodendrocytes in health and disease. Front Neurosci 12:733
33. Itoh T, Beesley J, Itoh A, Cohen AS, Kavanaugh B, Coulter DA, Grinspan JB, Pleasure D (2002) AMPA glutamate receptor-mediated calcium signaling is transiently enhanced during development of oligodendrocytes. J Neurochem 81:390–402
34. Jackman N, Ishii A, Bansal R (2009) Oligodendrocyte development and myelin biogenesis: parsing out the roles of glycosphingolipids. Physiol (Bethesda) 24:290–297
35. Kiray H, Lindsay SL, Hosseinzadeh S, Barnett SC (2016) The multifaceted role of astrocytes in regulating myelination. Exp Neurol 283:541–549
36. Krasnow AM, Attwell D (2016) NMDA receptors: power switches for oligodendrocytes. Neuron 91:3–5
37. Kula B, Chen TJ, Kukley M (2019) Glutamatergic signaling between neurons and oligodendrocyte lineage cells: Is it synaptic or non-synaptic? Glia
38. Larson VA, Mironova Y, Vanderpool KG, Waisman A, Rash JE, Agarwal A, Bergles DE (2018) Oligodendrocytes control potassium accumulation in white matter and seizure susceptibility. Elife, 7
39. Larson VA, Zhang Y, Bergles DE (2016) Electrophysiological properties of NG2(+) cells: matching physiological studies with gene expression profiles. Brain Res 1638:138–160
40. Lee Y, Morrison BM, Li Y, Lengacher S, Farah MH, Hoffman PN, Liu Y, Tsingalia A, Jin L, Zhang PW, Pellerin L, Magistretti PJ, Rothstein JD (2012) Oligodendroglia metabolically support axons and contribute to neurodegeneration. Nature 487:443–448
41. Luyt K, Slade TP, Dorward JJ, Durant CF, Wu Y, Shigemoto R, Mundell SJ, Varadi A, Molnar E (2007) Developing oligodendrocytes express functional GABA(B) receptors that stimulate cell proliferation and migration. J Neurochem 100:822–840
42. Maldonado PP, Velez-Fort M, Levavasseur F, Angulo MC (2013) Oligodendrocyte precursor cells are accurate sensors of local K + in mature gray matter. J Neurosci 33:2432–2442
43. Marinelli C, Bertalot T, Zusso M, Skaper SD, Giusti P (2016) Systematic review of pharmacological properties of the oligodendrocyte lineage. Front Cell Neurosci 10:27
44. McKenzie IA, Ohayon D, Li H, de Faria JP, Emery B, Tohyama K, Richardson WD (2014) Motor skill learning requires active central myelination. Science 346:318–322

45. Meyer N, Richter N, Fan Z, Siemonsmeier G, Pivneva T, Jordan P, Steinhauser C, Semtner M, Nolte C, Kettenmann H (2018) Oligodendrocytes in the mouse corpus callosum maintain axonal function by delivery of glucose. Cell Rep 22:2383–2394
46. Micu I, Plemel JR, Caprariello AV, Nave KA, Stys PK (2018) Axo-myelinic neurotransmission: a novel mode of cell signalling in the central nervous system. Nat Rev Neurosci 19:49–58
47. Muller C, Bauer NM, Schafer I, White R (2013) Making myelin basic protein -from mRNA transport to localized translation. Front Cell Neurosci 7:169
48. Nave KA, Trapp BD (2008) Axon-glial signaling and the glial support of axon function. Annu Rev Neurosci 31:535–561
49. Ohashi K, Deyashiki A, Miyake T, Nagayasu K, Shibasaki K, Shirakawa H, Kaneko S (2018) TRPV4 is functionally expressed in oligodendrocyte precursor cells and increases their proliferation. Pflugers Arch 470:705–716
50. Ortiz FC, Habermacher C, Graciarena M, Houry PY, Nishiyama A, Oumesmar BN, Angulo MC (2019) Neuronal activity in vivo enhances functional myelin repair. JCI Insight, 5
51. Paez PM, Fulton D, Colwell CS, Campagnoni AT (2009) Voltage-operated Ca(2 +) and Na(+) channels in the oligodendrocyte lineage. J Neurosci Res 87:3259–3266
52. Papanikolaou M, Lewis A, Butt AM (2017) Store-operated calcium entry is essential for glial calcium signalling in CNS white matter. Brain Struct Funct 222:2993–3005
53. Parpura V, Sekler I, Fern R (2016) Plasmalemmal and mitochondrial Na(+) -Ca(2 +) exchange in neuroglia. Glia 64:1646–1654
54. Patel JR, Klein RS (2011) Mediators of oligodendrocyte differentiation during remyelination. FEBS Lett 585:3730–3737
55. Ransom BR, Butt AM, Black JA (1991) Ultrastructural identification of HRP-injected oligodendrocytes in the intact rat optic nerve. Glia 4:37–45
56. Rivera A, Vanzuli I, Arellano JJ, Butt A (2016) Decreased regenerative capacity of oligodendrocyte progenitor cells (ng2-glia) in the ageing brain: a vicious cycle of synaptic dysfunction, myelin loss and neuronal disruption? Curr Alzheimer Res 13:413–418
57. Rivera A, Vanzulli I, Butt AM (2016) A central role for ATP signalling in glial interactions in the CNS. Curr Drug Targets 17:1829–1833
58. Saab AS, Nave KA (2017) Myelin dynamics: protecting and shaping neuronal functions. Curr Opin Neurobiol 47:104–112
59. Saab AS, Tzvetavona ID, Trevisiol A, Baltan S, Dibaj P, Kusch K, Mobius W, Goetze B, Jahn HM, Huang W, Steffens H, Schomburg ED, Perez-Samartin A, Perez-Cerda F, Bakhtiari D, Matute C, Lowel S, Griesinger C, Hirrlinger J, Kirchhoff F, Nave KA (2016) Oligodendroglial NMDA receptors regulate glucose import and axonal energy metabolism. Neuron 91:119–132
60. Saher G, Stumpf SK (2015) Cholesterol in myelin biogenesis and hypomyelinating disorders. Biochim Biophys Acta 1851:1083–1094
61. Santiago Gonzalez DA, Cheli VT, Zamora NN, Lama TN, Spreuer V, Murphy GG, Paez PM (2017) Conditional deletion of the l-type calcium channel Cav1.2 in NG2-positive cells impairs remyelination in mice. J Neurosci, 37, 10038–10051
62. Schirmer L, Möbius W, Zhao C, Cruz-Herranz A, Haim LB, Cordano C, Shiow LR, Kelley KW, Sadowski B, Timmons G, Pröbstel AK (2018) Oligodendrocyte-encoded Kir4.1 function is required for axonal integrity. Elife, 7
63. Soliven B (2001) Calcium signalling in cells of oligodendroglial lineage. Microsc Res Tech 52:672–679
64. Spitzer SO, Sitnikov S, Kamen Y, Evans KA, Kronenberg-Versteeg D, Dietmann S, de Faria Jr O, Agathou S, Karadottir RT (2019) Oligodendrocyte progenitor cells become regionally diverse and heterogeneous with age. Neuron, 101, 459–471.e5
65. Stassart RM, Mobius W, Nave KA, Edgar JM (2018) The axon-myelin unit in development and degenerative disease. Front Neurosci 12:467
66. Stevens B, Porta S, Haak LL, Gallo V, Fields RD (2002) Adenosine: a neuron-glial transmitter promoting myelination in the CNS in response to action potentials. Neuron 36:855–868
67. Trevisiol A, Saab AS, Winkler U, Marx G, Imamura H, Mobius W, Kusch K, Nave KA, Hirrlinger J (2017) Monitoring ATP dynamics in electrically active white matter tracts. Elife, 6

68. van Bruggen D, Agirre E, Castelo-Branco G (2017) Single-cell transcriptomic analysis of oligodendrocyte lineage cells. Curr Opin Neurobiol 47:168–175
69. Vautier F, Belachew S, Chittajallu R, Gallo V (2004) Shaker-type potassium channel subunits differentially control oligodendrocyte progenitor proliferation. Glia 48:337–345
70. Vejar S, Oyarzun JE, Retamal MA, Ortiz FC, Orellana JA (2019) Connexin and pannexin-based channels in oligodendrocytes: implications in brain health and disease. Front Cell Neurosci 13:3
71. Verkhratskiĭ AN, Butt A (2013) Glial physiology and pathophysiology. Chichester, West Sussex, UK; Hoboken, NJ, USA, Wiley-Blackwell
72. Voskuhl RR, Itoh N, Tassoni A, Matsukawa MA, Ren E, Tse V, Jang E, Suen TT, Itoh Y (2019) Gene expression in oligodendrocytes during remyelination reveals cholesterol homeostasis as a therapeutic target in multiple sclerosis. Proc Natl Acad Sci U S A 116:10130–10139
73. Wake H, Lee PR, Fields RD (2011) Control of local protein synthesis and initial events in myelination by action potentials. Science 333:1647–1651
74. Williamson JM, Lyons DA (2018) Myelin dynamics throughout life: an ever-changing landscape? Front Cell Neurosci 12:424
75. Xiao L, Ohayon D, McKenzie IA, Sinclair-Wilson A, Wright JL, Fudge AD, Emery B, Li H, Richardson WD (2016) Rapid production of new oligodendrocytes is required in the earliest stages of motor-skill learning. Nat Neurosci 19:1210–1217
76. Zhang C, Rasband MN (2016) Cytoskeletal control of axon domain assembly and function. Curr Opin Neurobiol 39:116–121

Chapter 6
Physiology of Microglia

Tuan Leng Tay, Micaël Carrier and Marie-Ève Tremblay

Abstract Microglia constitute the major immune cells that permanently reside in the central nervous system (CNS) alongside neurons and other glial cells. These resident immune cells are critical for proper brain development, actively maintain brain health throughout the lifespan and rapidly adapt their function to the physiological or pathophysiological needs of the organism. Cutting-edge fate mapping and imaging techniques applied to animal models enabled a revolution in our understanding of their roles during normal physiological conditions. Here, we highlight studies that demonstrate the embryonic yolk sac origin of microglia and describe factors, including crosstalk with the periphery and external environment, that regulate their differentiation, homeostasis and function in the context of healthy CNS. The diversity of microglial phenotypes observed across the lifespan, between brain compartments and between sexes is also discussed. Understanding what defines specific microglial phenotypes is critical for the development of innovative therapies to modulate their effector functions and improve clinical outcomes.

Keywords Microglia · Origin · Development · Homeostasis · Physiological roles · Periphery · Environment

6.1 Introduction

Microglia are a prominent type of glia in the central nervous system (CNS) originating from a single mesodermal source in contrast to all other brain parenchymal cells

T. L. Tay (✉)
Institute of Biology I, University of Freiburg, Hauptstr. 1, 79104 Freiburg, Germany
e-mail: tuan.leng.tay@biologie.uni-freiburg.de

Cluster of Excellence BrainLinks-BrainTools, University of Freiburg, Freiburg, Germany

Institute of Biology III, University of Freiburg, Schänzlestr. 1, 79104 Freiburg, Germany

M. Carrier · M.-È. Tremblay (✉)
Axe Neurosciences, Centre de Recherche du CHU de Québec, 2705, Boulevard Laurier, Québec, QC G1V 4G2, Canada
e-mail: tremblay.marie-eve@crchudequebec.ulaval.ca

A. Verkhratsky et al. (eds.), *Neuroglia in Neurodegenerative Diseases*, Advances in Experimental Medicine and Biology 1175,
https://doi.org/10.1007/978-981-13-9913-8_6

that have multiple neuroectodermal lineages. In the past, studies on microglia centred on their function as resident macrophages of the brain and mediators of injury, inflammation and neurodegeneration [22, 113, 114]. Advances in mouse genetics enabled fate mapping of mammalian microglia across conditions of health and disease [51, 52, 110, 143, 160, 166]. Importantly, in vivo imaging techniques revealed real-time microglial activities in the brain milieu [32, 86, 104, 147, 155]. Adult microglia appear morphologically uniform at steady state, but they are functionally heterogeneous in their physiological responses, which may be attributed to their local environment including neuronal activity [4, 26, 33, 58, 86, 124].

Numerous studies have shifted the field's attention to the physiological functions of microglia in brain development, activity and plasticity. Microglia are crucial regulators of CNS development and homeostasis via neuronal–microglial interactions, scavenging of cellular debris, secretion of trophic factors and synaptic modelling [145]. Broadly speaking, physiological microglial functions are required for learning, memory and cognition through the modulation of neuronal numbers and neural connectivity [158]. Microglia regulate neuronal density through coordinated control of neurogenesis, oligodendrogenesis, as well as neuronal survival and turnover. These processes mainly take place during perinatal development [3, 62, 132, 149, 162], but also persist during adolescence and adulthood [18, 128, 133]. Microglia provide neurotrophic support to neurons, notably through the secretion of insulin-like growth factor 1 for cortical layer V neuronal survival [149], the maintenance of embryonic forebrain basal progenitors [3] and the facilitation of neuroblast survival and migration to the adult olfactory bulb [116]. As scavenging phagocytes, microglia accumulate in regions containing high densities of neural precursors or apoptotic neurons, to facilitate neuronal turnover during developmental cell death [6, 29, 90, 111, 141].

In the prenatal mouse brain, microglia additionally regulate the wiring of forebrain dopaminergic circuits [136]. Multiple electrophysiological and high resolution microscopy studies in zebrafish [86] and mouse [7, 10, 27, 72, 74, 89, 118, 120, 121, 164] demonstrated microglial modulation of activity-dependent synaptic maturation, activity and plasticity [145]. Once neuronal circuits are established, microglia contribute to the refinement of synaptic connections across adolescence and adulthood [11, 147, 155]. Microglia–synapse interactions were characterised in the thalamus, cerebral cortex, amygdala and hippocampus, in postnatal development, adolescence, adulthood and normal ageing, [2, 95, 109, 124, 147, 148]. For instance, correlative light and electron microscopy of postnatal day (P) 15 mouse hippocampus confirmed that microglia mediate both the elimination and formation of synaptic elements [159]. The evidence that microglial processes selectively and partially phagocytose presynaptic structures and induce postsynaptic spine head filopodia supports the hypothesis of a microglia-dependent mechanism for the remodelling and maturation of synaptic circuits [159].

The plurality of microglial functions suggests that environmental conditions impairing microglia during brain development could compromise essential processes including neural connectivity. During adulthood, impaired microglial remodelling of neuronal circuits could severely impair learning and memory functions [158]. Dys-

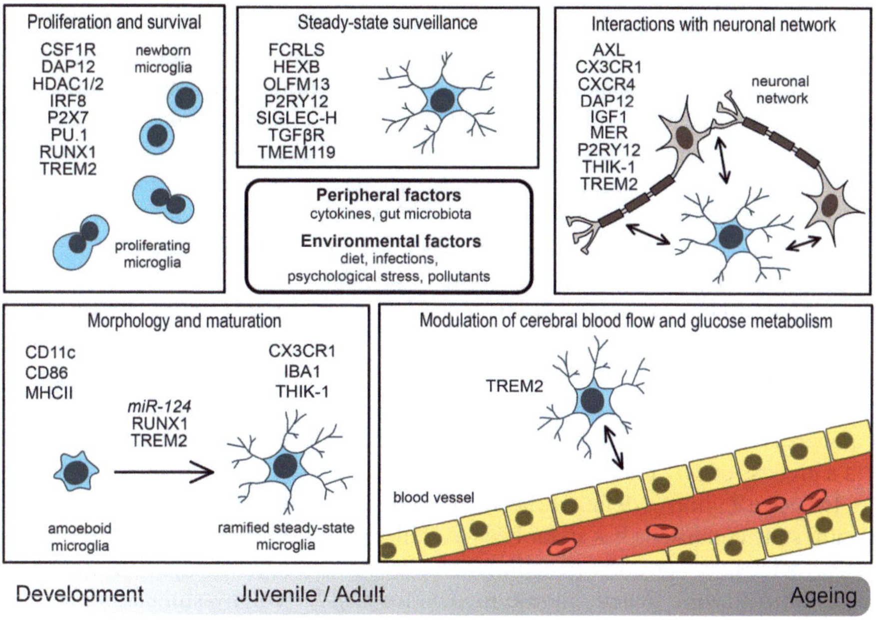

Fig. 6.1 Factors that impact physiological microglial development, homeostasis and function across the lifespan

functional or perturbed microglial homeostasis could have direct consequences on the onset of severe neurodegenerative or neuropsychiatric disorders at an early age or later in adulthood [142]. Here, we examine the innate conditions and external circumstances that support the proper maturation, homeostasis and function of physiological microglia (Fig. 6.1).

6.2 Origin and Maintenance of Microglia

Microglial colonisation of the CNS occurs before neuron and glial cells such as astrocytes and oligodendrocytes develop, and is conserved across vertebrate species [125, 141, 153]. Unlike neuroectodermal cells, microglia have a myeloid origin. Several fate mapping studies have established that microglia arise solely from yolk sac (YS) primitive macrophages [51, 54, 70, 71, 77, 126]. The microglial fate of YS precursors is specified between embryonic day (E) 7.0 and E7.5 [51]. Multi-lineage c-kit^{+} erythromyeloid YS precursor cells reside in blood islands of the proximal YS where they mature from A1 (CD45^{+} c-kit^{lo} CX_3CR1^{-} F4/80^{-}) into A2 (CD45^{+} c-kit^{-} CX_3CR1^{+} F4/80^{hi}) amoeboid macrophages [77]. They thereafter adopt a phenotype of mature macrophages in the neuroepithelium at E10.5 [97, 121]. Microglia seed the rudimentary brain via the leptomeninges and lateral ventricles before E9.5

and distribute themselves throughout the cortical wall from both directions at different speeds with varying rates of proliferation and maturation, depending on the brain region and developmental stage [4, 51, 141]. Normal blood circulation is necessary for YS macrophage recruitment [51]. Human amoeboid microglia penetrate the developing cerebral cortex at 4.5 gestational weeks (gw) via the pial surface, ventricles and choroid plexus [98, 153]. Microglia migrate radially and tangentially from the periphery towards the putative white matter, subplate and cortical plate layers, while pial cells populate the prospective cortical layer I. At 12–13 gw, the second wave of microglia invade only the white matter via the vasculature [98, 153]. Microglial cell numbers rise steadily in the first two postnatal weeks of mouse development, followed by a gradual decline by 50% from week three to six, after which their density stabilises [103]. The decline in the rate of proliferation is concomitant with an increase in apoptosis, contributing to the overall reduction in microglial cell numbers [103]. Mature microglia maintain their numbers in the brain throughout life via a process of self-renewal [5, 51, 66, 115, 145].

The contribution of definitive haematopoiesis as an origin to brain microglia under steady-state conditions has so far been excluded by specific targeting of *Kit*-positive cells, foetal liver- or aorta-gonad-mesonephros-derived haematopoietic stem cells in myeloid-specific genetic mouse models for *Csf-1r* (colony-stimulating factor-1 receptor), *Flt3* (Fms-like tyrosine kinase 3), *Myb* (myeloblastosis), *Runx1* (runt-related transcription factor 1) and *Tie2* (angiopoietin receptor) [51, 54, 70, 126, 131]. In contrast, pulse labelling of zebrafish macrophage precursors demonstrated that adult microglia derived from the ventral walls of the dorsal aorta in zebrafish [165]. Moreover, the existence of a significant number of *Hoxb8* lineage of haematopoietic mononuclear cells in the newborn and adult mouse brain under homeostasis [24] remains unexplained and challenges the single YS source of microglia. Mice lacking *Hoxb8* expression in microglia display obsessive–compulsive-like over-grooming behaviour. A plausible point of entry for these cells is via the brain ventricular choroid plexus [130], since *Hoxb8*-expressing cells distribute in a gradient from the pial surface and ventricular lining to the parenchyma during the first two postnatal weeks [24]. Even though genetic tracing studies predominantly support the single YS origin view, physiological microglia exhibit heterogeneous gene expression profiles, self-maintenance rates and morphological phenotypes across CNS compartments, as will be discussed later. We next examine the intrinsic and external factors that impact microglial development, identity and homeostasis.

6.3 Factors Required for Microglial Development and Homeostasis

Multiple studies have focused on elucidating the origin and renewal of microglia, as well as identifying factors which may affect their maturation, proliferation and apoptosis under normal physiological conditions (Fig. 6.1). Understanding these attributes is important to expand the available possibilities to target and control the effector functions of microglia in CNS diseases.

6.3.1 Cell Signalling Pathways Required for Microglial Proliferation and Maturation

The transcription factors Pu.1, a member of the Ets family [119], and interferon regulatory factor Irf8 both function as heterodimers in the determination of brain macrophage phenotype during the development of YS microglial precursors [8, 77, 96]. Another transcription factor that regulates the differentiation of myeloid cells is Runx1, which is expressed in a subpopulation of amoeboid microglia restricted to the ventricles during early postnatal forebrain development [169]. Runx1 has been described to mediate microglial proliferation. However, by P10, colocalisation of the proliferation marker Ki67 with Runx1 was no longer observed, with microglia shifting towards a ramified morphology, suggesting its role as a maturation factor [169]. It is still unknown how Runx1 regulates these events in a spatiotemporal dependent manner. The microRNA miR-124, which binds the mRNA of transcription factor C/EBPalpha and which in turn downregulates Pu.1, was specifically identified on microglia among myeloid cell populations [112]. miR-124 is functionally conserved in zebrafish and mouse. It controls the motility and phagocytic activities of microglia by promoting ramified surveillant phenotypes over reactive amoeboid ones [112, 140]. In the E13.5 mouse spinal cord, the purinergic ionotropic receptor P2X7 was also shown to mediate the proliferation of embryonic microglia, consequently regulating microglial density [117].

An essential pathway that defines microglial cell number involves Csf-1R (CD115). Csf-1R knockout mouse embryos are depleted of microglia and display impaired brain architecture [42]. Neuron-derived interleukin (IL)-34, the second ligand for Csf-1R, is more critical for regulating microglial cell density than Csf-1 in the adult brain; however its requirement during perinatal development is still controversial [59, 156]. The upregulation of brain Csf-1 levels did not mitigate the physiological decline in microglial numbers after the third postnatal week in mouse [103], suggesting that alternative microglial responses, to still unknown developmental signals, could favour microglial cell death during CNS maturation. Evidence for the requirement of Csf-1R signalling in microglial survival was provided by the near complete loss of mature microglia observed in adult mice treated with Csf-1R inhibitors [41]. Similarly, mice deficient for the Csf-1R adaptor protein DAP12 (DNAX activation protein of 12 kDa) reportedly have reduced microglial cell numbers at adulthood [107], without overt impact on their density during development [77]. While young DAP12-deficient mice do not show neurological deficits up to four weeks of age, synaptic function and plasticity are impaired, suggesting that DAP12 contributes to microglial physiology and communication with neurons [120]. Loss of signalling through the triggering receptor expressed on myeloid cells 2 (TREM2), another interaction partner of DAP12, has been implicated in CNS pathology [151]. Dissecting the impact of TREM2 knockout or mutation on microglial biology revealed that it plays key roles in regulating microglial proliferation, survival, clustering, autophagy, metabolism, as well as phagocytosis [150, 168], with consequences on CNS development and health maintenance. For instance, TREM2 knockout mice display enhanced excitatory neurotransmission, reduced functional brain connectivity, as well as repetitive behaviour and altered sociability [46]. In addition, TREM2 loss-of-function mutation in mouse reduced cerebral blood flow and glucose metabolism, thus uncovering new roles for microglia in regulating brain metabolism [79]. Moreover, the expression of *histone deacetylases* (*Hdac*) *1* and *2* and their target genes was shown to be developmentally regulated in prenatal murine microglia [31]. CX_3CR1-targeted ablation of *Hdac 1* and *2* during microglial precursor stages led to cellular malformation, impaired cell proliferation and the induction of apoptosis. Altered microglial cell density was observed up to 6 weeks of age while the morphological deficits persisted. Both enzymes are, however, redundant for the maintenance of adult microglia [31].

6.3.2 Cell Signalling Pathways Required for Microglial Homeostasis

Microglia utilise a vast number of surface receptors for cytokines, chemokines, purines, hormones and neurotransmitters in order to quickly react to changes in brain homeostasis. Similar to other tissue-resident macrophages, steady-state microglia express high levels of common markers including the fractalkine receptor CX_3CR1, Csf-1R, the integrin CD11b, surface glycoproteins F4/80 and CD68, ionised calcium-binding adapter molecule 1 (IBA1) and intermediate levels of pan-haematopoietic CD45 [60]. While earlier bulk transcriptomic studies have identified several genes to distinguish microglia from other cell types in the CNS or myeloid cells located in the periphery [20, 21, 25, 50, 129], groundbreaking single-cell transcriptomic analyses of microglia are unveiling the heterogeneity of their total pool in healthy and disease conditions [63, 75, 76, 85, 91, 92, 144]. Microglia 'signature' factors are now known to comprise *P2ry12*, *Fcrls*, *Tmem119*, *Olfml3*, *Hexb*, *Tgfbr1* and *Siglec-H*. It remains unknown, however, whether these factors contribute to specifying microglial motility, morphology and functions [20, 25, 50, 69]. Among these factors, the transforming growth factor β (TGFβ) appears critical for mediating microglial survival and phenotypic differentiation, since microglial density is drastically reduced in TGFβ receptor-deficient mice [20]. TGFβ signalling induces microglia to adopt a ramified morphology concomitant with reduced levels of CD86, MHC class II and CD11c and upregulation of CX_3CR1 and IBA1 in vitro [1]. The transmembrane protein TMEM119 was also described to distinguish resident $IBA1^+$ microglia from CNS-associated macrophages of the choroid plexus, meninges and $CD163^+$ perivascular cells and IBA^+ $CD68^+$ infiltrating macrophages in the postnatal mammalian brain, but its function remains unknown [9, 122]. The surveillant state of microglia is widely believed to be maintained by signalling between the neuron-secreted fractalkine (CX_3CL1) and its microglial CX_3CR1 receptor [12]. Several studies have shown the requirement of fractalkine signalling for the recruitment of microglial cells into the early postnatal hippocampus and cerebral cortex [72, 108, 149]. A transient reduction of microglial numbers was detected in CX_3CR1 knockout mice during the early postnatal period. Concomitant deficits in microglia-mediated synaptic pruning, weakened synaptic transmission and decreased functional brain connectivity were reported [109, 167], with consequences on cognition both early in life and into adulthood [158].

6.3.3 *Modulation of Microglial Process Remodelling and Surveillance*

Microglia constantly remodel the structure of their processes as required for surveilling the CNS [32, 86, 104, 147, 155], even following the death of the organism [36]. Studies that focused on elucidating the mechanisms regulating microglial cell migration and process motility revealed that purinergic signalling through microglial P2RY12 drives process response to laser injury in the cerebral cortex in vivo [32, 67], process remodelling in retinal explants [47] and filopodia extension in mouse models of status epilepticus [45] and neuropathic pain [61]. In addition, ATP release triggered by dendritic neuronal NMDA receptor activation induces the outgrowth of microglial processes in acute mouse hippocampal slices, suggesting a purine-mediated form of neuron–microglia communication [38]. Extracellular calcium reduction induced microglial process convergence onto neuronal dendrites independently from action potential firing in mouse brain slices and in vivo. This process is mediated by microglial P2RY12, suggesting that microglial interactions with neurons are guided by dendritic calcium reduction in the healthy brain [44]. Furthermore, the inhibition of the two-pore domain K^+ channel THIK-1 revealed its necessity for microglial membrane potential, process ramification, dynamic surveillance and release of pro-inflammatory IL-1β, using pharmacological or gene knockout interventions, in mouse brain slices. However, blocking P2RY12 did not affect microglial membrane potential, ramification or surveillance ex vivo [88]. Lastly, genetic deletion of the TAM receptor tyrosine kinases Mer and Axl revealed their involvement in the regulation of microglial physiology, including their process motility and response to laser injury in vivo, as well as phagocytic elimination of apoptotic newborn neurons generated during adult neurogenesis [48]. Whether and how these different pathways intersect still remains elusive.

6.3.4 *Impact of the Periphery*

Despite being residents of the so-called immune-privileged CNS, numerous studies point to an interrelationship between microglia and the periphery in the absence of pathology. Microglia are constantly modulated by blood cytokine diffusion and transport into the brain during infection, immune-related molecules secreted from the endothelial cells of the blood–brain barrier, and peripheral immune signals from the autonomic nervous system [37]. In particular, the gut microbiota was implicated in the normal development and maintenance of microglial cell homeostasis. Rats exposed to prenatal helminth infection have aberrant microglial tiling and reduced microglial response to early-life immune challenges [161]. Mice bred in germ-free conditions have increased microglial cell density and delayed maturation, with the effects regulated by short-chain fatty acids derived from bacterial fermentation by-products of microbiota [43]. Rodents raised without gut microbiota also display increased blood–brain barrier permeability [17]. Introduction of short-chain fatty acids into germ-free mice could potentially allow signalling to peripheral splenic macrophages which subsequently traffic into the CNS to promote maturation of microglia [99]. The microglial response to gut microbiota is subject to sex-specific effects, namely, germ-free male embryonic and female adult microglia were more severely perturbed than other groups [146]. Altered microglial transcriptome, chromatin accessibility and colonisation of the developing neocortex were further observed in the absence of the microbiota [146]. Altogether, maintaining the physiological CNS immune system depends on complex yet little known interactions between the peripheral and central nervous systems, apart from microglial responses to their immediate surroundings.

6.3.5 Impact of the External Environment

Microglial phenotype is constantly shaped by the exposure to various environmental factors, acting on both the brain and periphery, across the lifespan. Inflammation associated with psychological stress, infection, dietary imbalance and environmental pollutants alters physiological microglial phenotype and function [13, 23, 142, 158]. Brief exposure to these external insults (particularly during CNS development and maturation) can exert long-lasting impacts on microglia through 'priming', a process which increases their sensitivity to later challenges. This increased sensitivity accompanied by exacerbated phagocytosis and cytokines release impacts cognitive processes such as learning and memory. The environmental influence on microglia and its implications in brain health and disease are discussed in greater detail elsewhere [64, 142, 158, 163].

One example of an environmental impact on microglial brain colonisation, maturation and/or function is prenatal maternal immune activation. It can be induced, for instance, using the viral mimic polyriboinosinic-polyribocytidylic acid (poly I:C), which exerts different results dependent on the timing. Exposure to poly I:C at E14.5 accelerated the transcriptomic maturation profile of early postnatal mouse microglia towards an adult signature [91]. Complex changes in hippocampal microglial gene signature and decreased phagocytic activity were further induced in mice treated with poly I:C at E15 and correlated with schizophrenic-like behavioural abnormalities at P60 [93]. Furthermore, poly I:C treatment at E9.5 led to more profound changes in microglial inflammatory states and cell density in male than female mice, as well as produced schizophrenia-associated behavioural deficits at P80-90 [73].

Hippocampal microglial molecular profile, density, morphology and process motility were all affected in juvenile mice exposed to early-life stressors such as separation anxiety [35] and dietary deficiency in the essential n-3 polyunsaturated fatty acid [87]. High-fat diet in adult rodents also alters microglial reactivity and activity in selected brain milieus with adverse implications in the CNS inflammatory status and function that were, however, reversible with a low-fat diet [15, 65, 152]. To intervene in neuropsychiatric disorders and progressive neurodegeneration, the variable susceptibility of normal microglia in young and adult brains to their environments needs to be better understood.

6.4 Heterogeneity of Microglia

The increasing accessibility of single-cell transcriptomic analysis has revealed the highest diversity of microglial cellular states during normal physiological development and ageing [63, 85]. Significant distinctions in bulk microglia gene expression, morphology, ultrastructure, distribution, bioenergetics, immunophenotype and cellular properties compared across regions have been reported in healthy rodent brains [33, 34, 39, 58, 82, 123]. Some of these differences correlate with functional requirements for steady-state CNS development or surveillance. The immediate surroundings of a microglial cell influence its identity through a selection pressure for exclusive gene enhancers [55, 56, 81]. For example, reciprocal influence on cell densities exerted by microglia and neuronal progenitors in the ventricular and subventricular zone (SVZ) of the developing cortex was shown to depend on signalling between microglial receptor CXCR4 and CXCL12 secreted by basal progenitors [3]. Microglia in the adult SVZ and rostral migratory stream also comprise a morphologically and antigenically distinct phenotype, distinguishable by lower expression of purinergic receptor P2RY12 and its lack of ATP-driven chemotaxis [116]. Under healthy conditions, specific upregulation of the ubiquitin-specific protease 18 within white matter microglia, but not grey matter, mitigates tissue destruction via tonic interferon signalling [53]. Elevated neuronal death in the adult mouse cerebellum correlates with an enriched cell clearance phenotype which can be epigenetically suppressed in cortical or striatal microglia by the polycomb repressive complex 2 [6]. Evidence of varying microglial turnover rates across brain regions in mouse and human was provided by lineage tracing studies using genetic, thymidine analog or carbon dating [5, 83, 115, 143]. Taken together, the specific interactions between microglial cells and neuronal progenitors or neurons in each microenvironment may be prerequisites for the proper recruitment of particular microglial phenotypes that are required for the maturation, maintenance and plasticity of local neural circuits.

6.5 Sexual Dimorphism of Microglia

To elucidate the control of microglial phenotype and homeostasis, it has become increasingly important to examine the influence of sex. After all, sexual dimorphism in immune responses has been well documented in the peripheral innate and adaptive immune systems across the animal kingdom [78]. Physiological microglial density differs between males and females across stages of the lifespan and brain regions comprising the preoptic area (POA), the hippocampus, parietal cortex and amygdala [84, 101, 102, 127]. Sex-specific microglial response to neuropathic pain [135], chronic stress [16] and gut microbiota [146] were also reported in rodent and human foetal microglia. Isolated microglia from postnatal or adult mice of both sexes express glucocorticoid receptor, mineralocorticoid receptor and oestrogen receptor alpha together with 17β-hydroxysteroid dehydrogenase type 1 and 5α-reductase type 1 involved in steroid hormones metabolism [28, 57, 134]. It is thus conceivable that microglia contribute to sex-dependent brain regulation via their interaction with the endocrine system during CNS development. Steroid hormones act on their receptors, which belong to the superfamily of nuclear receptor transcription factors, to recruit enzymes and other protein components for histone modifications that exert epigenetic effects on the developing brain, thus determining sex differences in brain and behaviour [94]. At adulthood, female microglia grafted in male mice keep their sex-specific gene expressions and remain more immunoprotective than male microglia, in a context of ischemia, suggesting differences that may emerge during development [154]. Reduced DNA methylation in rodent POA, which is associated with male-specific behaviours, enables the expression of masculinising genes during a small perinatal time window [105]. While the POA in male rats reportedly displays more reactive microglial phenotype and a higher dendritic spine density compared to females [84], a direct relationship between microglia and brain sexualisation has not been established due to cellular heterogeneity in the POA. Considering that sexual dimorphic microglial cell density, (phagocytic) function and response may significantly modulate synaptogenesis and shape neuronal circuitry differently, existing data based on a single sex or mixed sex cohorts should be interpreted with care.

6.6 Microglia in Normal Ageing

Several phenotypes of non-steady state microglia become more common in physiologically aged brains absent of overt pathology compared with younger ones. Aged microglia are prone to upregulate pro-inflammatory genes and antigen-presenting markers, while anti-inflammatory cytokines and microglial activation inhibitory factors are down-regulated [100]. Genome-wide analysis of bulk isolated microglia from discrete brain areas revealed regional variability in ageing [58]. Dystrophic or senescent microglial cells with altered morphology, diminished reactivity and motility and reduced phagocytic capacity have been described in normally aged human and rodent brains [30, 68, 80, 137–139, 148, 157]. In addition, the prevalence of 'dark' microglia characterised by augmented signs of oxidative stress at the ultra-structural level is associated with ageing and several disease models in mice [14]. The reciprocal effects of age-associated microglial phenotypes to their environment are under active investigation [19, 40, 49, 106].

6.7 Conclusion

Microglia are active contributors to normal brain development and physiology. Of note, microglial dysfunctional or loss has been implicated in normal ageing and diseases, as discussed in other chapters. Within this context, it is necessary to understand the intrinsic factors and external conditions that shape the maturation, homeostasis and functions of these long-residing YS-derived tissue-resident macrophages of the CNS. Intrinsic cellular, molecular and epigenetic mechanisms in addition to external peripheral responses and environmental changes determine microglial phenotype, as well as dynamics, phagocytic behaviour, synaptic interactions and release of various mediators that modulate cognition. A better understanding of the mechanisms that govern microglial effector functions during normal physiological conditions will contribute significantly to the strategic development of better targeted treatments for CNS diseases across the lifespan.

Acknowledgements This work was supported by a grant from the Natural Sciences and Engineering Research Council of Canada (NSERC) RGPIN-2014-05308 to M.E.T., who is a Canada Research Chair of *Neuroimmune plasticity in health and therapy*. T.L.T. was supported by the German Research Foundation (DFG, TA 1029/1-1 and EXC 1086), Ministry of Science, Research and the Arts of Baden-Württemberg (7532.21/2.1.6), Klaus Tschira Boost Fund (KT10), and Wissenschaftliche Gesellschaft Freiburg (Helmut-Holzer Prize). M.C. is a recipient of the Master training awards from Université Laval and Fonds de recherche du Québec—Santé (FRQS). We are grateful to Julie C. Savage for critical comments on the manuscript and Maude Bordeleau for preparing the figure.

Conflict of interest The authors declare no competing financial interests.

References

1. Abutbul S, Shapiro J, Szaingurten-Solodkin I et al (2012) TGF-β signaling through SMAD2/3 induces the quiescent microglial phenotype within the CNS environment. Glia 60:1160–1171. https://doi.org/10.1002/glia.22343
2. Acharjee S, Verbeek M, Gomez CD et al (2018) Reduced Microglial Activity and Enhanced Glutamate Transmission in the Basolateral Amygdala in Early CNS Autoimmunity. J Neurosci 38:9019–9033. https://doi.org/10.1523/JNEUROSCI.0398-18.2018
3. Arnò B, Grassivaro F, Rossi C et al (2014) Neural progenitor cells orchestrate microglia migration and positioning into the developing cortex. Nat Commun 5:5611. https://doi.org/10.1038/ncomms6611

4. Arnoux I, Hoshiko M, Mandavy L et al (2013) Adaptive phenotype of microglial cells during the normal postnatal development of the somatosensory "Barrel" cortex. Glia 61:1582–1594. https://doi.org/10.1002/glia.22503
5. Askew K, Li K, Olmos-Alonso A et al (2017) Coupled Proliferation and Apoptosis Maintain the Rapid Turnover of Microglia in the Adult Brain. Cell Rep 18:391–405. https://doi.org/10.1016/j.celrep.2016.12.041
6. Ayata P, Badimon A, Strasburger HJ et al (2018) Epigenetic regulation of brain region-specific microglia clearance activity. Nat Neurosci 21:1049–1060. https://doi.org/10.1038/s41593-018-0192-3
7. Béchade C, Cantaut-Belarif Y, Bessis A (2013) Microglial control of neuronal activity. Front Cell Neurosci 7:32. https://doi.org/10.3389/fncel.2013.00032
8. Beers DR, Henkel JS, Xiao Q et al (2006) Wild-type microglia extend survival in PU.1 knockout mice with familial amyotrophic lateral sclerosis. Proc Natl Acad Sci USA 103:16021–16026. https://doi.org/10.1073/pnas.0607423103
9. Bennett ML, Bennett FC, Liddelow SA et al (2016) New tools for studying microglia in the mouse and human CNS. Proc Natl Acad Sci USA 113:E1738–E1746. https://doi.org/10.1073/pnas.1525528113
10. Bessis A, Béchade C, Bernard D, Roumier A (2007) Microglial control of neuronal death and synaptic properties. Glia 55:233–238. https://doi.org/10.1002/glia.20459
11. Bialas AR, Stevens B (2013) TGF-β signaling regulates neuronal C1q expression and developmental synaptic refinement. Nat Neurosci 16:1773–1782. https://doi.org/10.1038/nn.3560
12. Biber K, Neumann H, Inoue K, Boddeke HWGM (2007) Neuronal "On" and "Off" signals control microglia. Trends Neurosci 30:596–602. https://doi.org/10.1016/j.tins.2007.08.007
13. Bilbo SD (2013) Frank A. Beach award: programming of neuroendocrine function by early-life experience: a critical role for the immune system. Horm Behav 63:684–691. https://doi.org/10.1016/j.yhbeh.2013.02.017
14. Bisht K, Sharma KP, Lecours C et al (2016) Dark microglia: A new phenotype predominantly associated with pathological states. Glia 64:826–839. https://doi.org/10.1002/glia.22966
15. Bocarsly ME, Fasolino M, Kane GA et al (2015) Obesity diminishes synaptic markers, alters microglial morphology, and impairs cognitive function. Proc Natl Acad Sci USA 112:15731–15736. https://doi.org/10.1073/pnas.1511593112
16. Bollinger JL, Bergeon Burns CM, Wellman CL (2016) Differential effects of stress on microglial cell activation in male and female medial prefrontal cortex. Brain Behav Immun 52:88–97. https://doi.org/10.1016/j.bbi.2015.10.003
17. Braniste V, Al-Asmakh M, Kowal C et al (2014) The gut microbiota influences blood-brain barrier permeability in mice. Sci Transl Med 6:263ra158. https://doi.org/10.1126/scitranslmed.3009759
18. Brown GC, Neher JJ (2014) Microglial phagocytosis of live neurons. Nat Rev Neurosci 15:209–216. https://doi.org/10.1038/nrn3710
19. Bussian TJ, Aziz A, Meyer CF et al (2018) Clearance of senescent glial cells prevents tau-dependent pathology and cognitive decline. Nature 562:578–582. https://doi.org/10.1038/s41586-018-0543-y
20. Butovsky O, Jedrychowski MP, Moore CS et al (2014) Identification of a unique TGF-β-dependent molecular and functional signature in microglia. Nat Neurosci 17:131–143. https://doi.org/10.1038/nn.3599
21. Butovsky O, Siddiqui S, Gabriely G et al (2012) Modulating inflammatory monocytes with a unique microRNA gene signature ameliorates murine ALS. J Clin Invest 122:3063–3087. https://doi.org/10.1172/JCI62636
22. Cartier N, Lewis C-A, Zhang R, Rossi FMV (2014) The role of microglia in human disease: therapeutic tool or target? Acta Neuropathol 128:363–380. https://doi.org/10.1007/s00401-014-1330-y
23. Castanon N, Luheshi G, Layé S (2015) Role of neuroinflammation in the emotional and cognitive alterations displayed by animal models of obesity. Front Neurosci 9:229. https://doi.org/10.3389/fnins.2015.00229

24. Chen S-K, Tvrdik P, Peden E et al (2010) Hematopoietic origin of pathological grooming in Hoxb8 mutant mice. Cell 141:775–785. https://doi.org/10.1016/j.cell.2010.03.055
25. Chiu IM, Morimoto ETA, Goodarzi H et al (2013) A neurodegeneration-specific gene-expression signature of acutely isolated microglia from an amyotrophic lateral sclerosis mouse model. Cell Rep 4:385–401. https://doi.org/10.1016/j.celrep.2013.06.018
26. Clark AK, Gruber-Schoffnegger D, Drdla-Schutting R et al (2015) Selective activation of microglia facilitates synaptic strength. J Neurosci 35:4552–4570. https://doi.org/10.1523/JNEUROSCI.2061-14.2015
27. Costello DA, Lyons A, Denieffe S et al (2011) Long term potentiation is impaired in membrane glycoprotein CD200-deficient mice: a role for Toll-like receptor activation. J Biol Chem 286:34722–34732. https://doi.org/10.1074/jbc.M111.280826
28. Crain JM, Nikodemova M, Watters JJ (2013) Microglia express distinct M1 and M2 phenotypic markers in the postnatal and adult central nervous system in male and female mice. J Neurosci Res 91:1143–1151. https://doi.org/10.1002/jnr.23242
29. Cunningham CL, Martínez-Cerdeño V, Noctor SC (2013) Microglia regulate the number of neural precursor cells in the developing cerebral cortex. J Neurosci 33:4216–4233. https://doi.org/10.1523/JNEUROSCI.3441-12.2013
30. Damani MR, Zhao L, Fontainhas AM et al (2011) Age-related alterations in the dynamic behavior of microglia. Aging Cell 10:263–276. https://doi.org/10.1111/j.1474-9726.2010.00660.x
31. Datta M, Staszewski O, Raschi E et al (2018) Histone Deacetylases 1 and 2 Regulate Microglia Function during Development, Homeostasis, and Neurodegeneration in a Context-Dependent Manner. Immunity 48:514–529.e6. https://doi.org/10.1016/j.immuni.2018.02.016
32. Davalos D, Grutzendler J, Yang G et al (2005) ATP mediates rapid microglial response to local brain injury in vivo. Nat Neurosci 8:752–758. https://doi.org/10.1038/nn1472
33. De Biase LM, Schuebel KE, Fusfeld ZH et al (2017) Local Cues Establish and Maintain Region-Specific Phenotypes of Basal Ganglia Microglia. Neuron 95:341–356.e6. https://doi.org/10.1016/j.neuron.2017.06.020
34. de Haas AH, Boddeke HWGM, Biber K (2008) Region-specific expression of immunoregulatory proteins on microglia in the healthy CNS. Glia 56:888–894. https://doi.org/10.1002/glia.20663
35. Delpech J-C, Wei L, Hao J et al (2016) Early life stress perturbs the maturation of microglia in the developing hippocampus. Brain Behav Immun 57:79–93. https://doi.org/10.1016/j.bbi.2016.06.006
36. Dibaj P, Steffens H, Nadrigny F et al (2010) Long-lasting post-mortem activity of spinal microglia in situ in mice. J Neurosci Res 88:2431–2440. https://doi.org/10.1002/jnr.22402
37. Dilger RN, Johnson RW (2008) Aging, microglial cell priming, and the discordant central inflammatory response to signals from the peripheral immune system. J Leukoc Biol 84:932–939. https://doi.org/10.1189/jlb.0208108
38. Dissing-Olesen L, LeDue JM, Rungta RL et al (2014) Activation of neuronal NMDA receptors triggers transient ATP-mediated microglial process outgrowth. J Neurosci 34:10511–10527. https://doi.org/10.1523/JNEUROSCI.0405-14.2014
39. Doorn KJ, Brevé JJP, Drukarch B et al (2015) Brain region-specific gene expression profiles in freshly isolated rat microglia. Front Cell Neurosci 9:84. https://doi.org/10.3389/fncel.2015.00084
40. Elmore MRP, Hohsfield LA, Kramár EA et al (2018) Replacement of microglia in the aged brain reverses cognitive, synaptic, and neuronal deficits in mice. Aging Cell 17:e12832. https://doi.org/10.1111/acel.12832
41. Elmore MRP, Najafi AR, Koike MA et al (2014) Colony-stimulating factor 1 receptor signaling is necessary for microglia viability, unmasking a microglia progenitor cell in the adult brain. Neuron 82:380–397. https://doi.org/10.1016/j.neuron.2014.02.040
42. Erblich B, Zhu L, Etgen AM et al (2011) Absence of colony stimulation factor-1 receptor results in loss of microglia, disrupted brain development and olfactory deficits. PLoS ONE 6:e26317. https://doi.org/10.1371/journal.pone.0026317

43. Erny D, Hrabě de Angelis AL, Jaitin D et al (2015) Host microbiota constantly control maturation and function of microglia in the CNS. Nat Neurosci 18:965–977. https://doi.org/10.1038/nn.4030
44. Eyo UB, Gu N, De S et al (2015) Modulation of microglial process convergence toward neuronal dendrites by extracellular calcium. J Neurosci 35:2417–2422. https://doi.org/10.1523/JNEUROSCI.3279-14.2015
45. Eyo UB, Peng J, Swiatkowski P et al (2014) Neuronal hyperactivity recruits microglial processes via neuronal NMDA receptors and microglial P2Y12 receptors after status epilepticus. J Neurosci 34:10528–10540. https://doi.org/10.1523/JNEUROSCI.0416-14.2014
46. Filipello F, Morini R, Corradini I et al (2018) The Microglial Innate Immune Receptor TREM2 Is Required for Synapse Elimination and Normal Brain Connectivity. Immunity 48:979–991.e8. https://doi.org/10.1016/j.immuni.2018.04.016
47. Fontainhas AM, Wang M, Liang KJ et al (2011) Microglial morphology and dynamic behavior is regulated by ionotropic glutamatergic and GABAergic neurotransmission. PLoS ONE 6:e15973. https://doi.org/10.1371/journal.pone.0015973
48. Fourgeaud L, Través PG, Tufail Y et al (2016) TAM receptors regulate multiple features of microglial physiology. Nature 532:240–244. https://doi.org/10.1038/nature17630
49. Füger P, Hefendehl JK, Veeraraghavalu K et al (2017) Microglia turnover with aging and in an Alzheimer's model via long-term in vivo single-cell imaging. Nat Neurosci 20:1371–1376. https://doi.org/10.1038/nn.4631
50. Gautier EL, Shay T, Miller J et al (2012) Gene-expression profiles and transcriptional regulatory pathways that underlie the identity and diversity of mouse tissue macrophages. Nat Immunol 13:1118–1128. https://doi.org/10.1038/ni.2419
51. Ginhoux F, Greter M, Leboeuf M et al (2010) Fate Mapping Analysis Reveals That Adult Microglia Derive from Primitive Macrophages. Science 330:841–845. https://doi.org/10.1126/science.1194637
52. Goldmann T, Wieghofer P, Müller PF et al (2013) A new type of microglia gene targeting shows TAK1 to be pivotal in CNS autoimmune inflammation. Nat Neurosci 16:1618–1626. https://doi.org/10.1038/nn.3531
53. Goldmann T, Zeller N, Raasch J et al (2015) USP18 lack in microglia causes destructive interferonopathy of the mouse brain. EMBO J 34:1612–1629. https://doi.org/10.15252/embj.201490791
54. Gomez Perdiguero E, Klapproth K, Schulz C et al (2015) Tissue-resident macrophages originate from yolk-sac-derived erythro-myeloid progenitors. Nature 518:547–551. https://doi.org/10.1038/nature13989
55. Gosselin D, Link VM, Romanoski CE et al (2014) Environment drives selection and function of enhancers controlling tissue-specific macrophage identities. Cell 159:1327–1340. https://doi.org/10.1016/j.cell.2014.11.023
56. Gosselin D, Skola D, Coufal NG, et al (2017) An environment-dependent transcriptional network specifies human microglia identity. Science 356:. https://doi.org/10.1126/science.aal3222
57. Gottfried-Blackmore A, Sierra A, Jellinck PH et al (2008) Brain microglia express steroid-converting enzymes in the mouse. J Steroid Biochem Mol Biol 109:96–107. https://doi.org/10.1016/j.jsbmb.2007.12.013
58. Grabert K, Michoel T, Karavolos MH et al (2016) Microglial brain region-dependent diversity and selective regional sensitivities to aging. Nat Neurosci 19:504–516. https://doi.org/10.1038/nn.4222
59. Greter M, Lelios I, Pelczar P et al (2012) Stroma-derived interleukin-34 controls the development and maintenance of langerhans cells and the maintenance of microglia. Immunity 37:1050–1060. https://doi.org/10.1016/j.immuni.2012.11.001
60. Greter M, Merad M (2013) Regulation of microglia development and homeostasis. Glia 61:121–127. https://doi.org/10.1002/glia.22408
61. Gu N, Eyo UB, Murugan M et al (2016) Microglial P2Y12 receptors regulate microglial activation and surveillance during neuropathic pain. Brain Behav Immun 55:82–92. https://doi.org/10.1016/j.bbi.2015.11.007

62. Hagemeyer N, Hanft K-M, Akriditou M-A et al (2017) Microglia contribute to normal myelinogenesis and to oligodendrocyte progenitor maintenance during adulthood. Acta Neuropathol 134:441–458. https://doi.org/10.1007/s00401-017-1747-1
63. Hammond TR, Dufort C, Dissing-Olesen L et al (2019) Single-Cell RNA Sequencing of Microglia throughout the Mouse Lifespan and in the Injured Brain Reveals Complex Cell-State Changes. Immunity 50:253–271.e6. https://doi.org/10.1016/j.immuni.2018.11.004
64. Hanamsagar R, Bilbo SD (2017) Environment matters: microglia function and dysfunction in a changing world. Curr Opin Neurobiol 47:146–155. https://doi.org/10.1016/j.conb.2017.10.007
65. Hao S, Dey A, Yu X, Stranahan AM (2016) Dietary obesity reversibly induces synaptic stripping by microglia and impairs hippocampal plasticity. Brain Behav Immun 51:230–239. https://doi.org/10.1016/j.bbi.2015.08.023
66. Hashimoto D, Chow A, Noizat C et al (2013) Tissue resident macrophages self-maintain locally throughout adult life with minimal contribution from circulating monocytes. Immunity 38:. https://doi.org/10.1016/j.immuni.2013.04.004
67. Haynes SE, Hollopeter G, Yang G et al (2006) The P2Y12 receptor regulates microglial activation by extracellular nucleotides. Nat Neurosci 9:1512–1519. https://doi.org/10.1038/nn1805
68. Hefendehl JK, Neher JJ, Sühs RB et al (2014) Homeostatic and injury-induced microglia behavior in the aging brain. Aging Cell 13:60–69. https://doi.org/10.1111/acel.12149
69. Hickman SE, Kingery ND, Ohsumi TK et al (2013) The microglial sensome revealed by direct RNA sequencing. Nat Neurosci 16:1896–1905. https://doi.org/10.1038/nn.3554
70. Hoeffel G, Chen J, Lavin Y et al (2015) C-Myb(+) erythro-myeloid progenitor-derived fetal monocytes give rise to adult tissue-resident macrophages. Immunity 42:665–678. https://doi.org/10.1016/j.immuni.2015.03.011
71. Hoeffel G, Wang Y, Greter M et al (2012) Adult Langerhans cells derive predominantly from embryonic fetal liver monocytes with a minor contribution of yolk sac-derived macrophages. J Exp Med 209:1167–1181. https://doi.org/10.1084/jem.20120340
72. Hoshiko M, Arnoux I, Avignone E et al (2012) Deficiency of the microglial receptor CX3CR72 impairs postnatal functional development of thalamocortical synapses in the barrel cortex. J Neurosci 32:15106–15111. https://doi.org/10.1523/JNEUROSCI.1167-12.2012
73. Hui CW, St-Pierre A, El Hajj H et al (2018) Prenatal Immune Challenge in Mice Leads to Partly Sex-Dependent Behavioral, Microglial, and Molecular Abnormalities Associated with Schizophrenia. Front Mol Neurosci 11:13. https://doi.org/10.3389/fnmol.2018.00013
74. Ji K, Akgul G, Wollmuth LP, Tsirka SE (2013) Microglia actively regulate the number of functional synapses. PLoS ONE 8:e56293. https://doi.org/10.1371/journal.pone.0056293
75. Jordão MJC, Sankowski R, Brendecke SM et al (2019) Single-cell profiling identifies myeloid cell subsets with distinct fates during neuroinflammation. Science 363:eaat7554. https://doi.org/10.1126/science.aat7554
76. Keren-Shaul H, Spinrad A, Weiner A et al (2017) A unique microglia type associated with restricting development of Alzheimer's disease. Cell 169:1276–1290.e17. https://doi.org/10.1016/j.cell.2017.05.018
77. Kierdorf K, Erny D, Goldmann T et al (2013) Microglia emerge from erythromyeloid precursors via Pu.1- and Irf8-dependent pathways. Nat Neurosci 16:273–280. https://doi.org/10.1038/nn.3318
78. Klein SL, Flanagan KL (2016) Sex differences in immune responses. Nat Rev Immunol 16:626–638. https://doi.org/10.1038/nri.2016.90
79. Kleinberger G, Brendel M, Mracsko E et al (2017) The FTD-like syndrome causing TREM2 T66M mutation impairs microglia function, brain perfusion, and glucose metabolism. EMBO J 36:1837–1853. https://doi.org/10.15252/embj.201796516
80. Kreisel T, Frank MG, Licht T et al (2014) Dynamic microglial alterations underlie stress-induced depressive-like behavior and suppressed neurogenesis. Mol Psychiatry 19:699–709. https://doi.org/10.1038/mp.2013.155

81. Lavin Y, Winter D, Blecher-Gonen R et al (2014) Tissue-resident macrophage enhancer landscapes are shaped by the local microenvironment. Cell 159:1312–1326. https://doi.org/10.1016/j.cell.2014.11.018
82. Lawson LJ, Perry VH, Dri P, Gordon S (1990) Heterogeneity in the distribution and morphology of microglia in the normal adult mouse brain. Neuroscience 39:151–170
83. Lawson LJ, Perry VH, Gordon S (1992) Turnover of resident microglia in the normal adult mouse brain. Neuroscience 48:405–415
84. Lenz KM, Nugent BM, Haliyur R, McCarthy MM (2013) Microglia are essential to masculinization of brain and behavior. J Neurosci 33:2761–2772. https://doi.org/10.1523/JNEUROSCI.1268-12.2013
85. Li Q, Cheng Z, Zhou L et al (2019) Developmental Heterogeneity of Microglia and Brain Myeloid Cells Revealed by Deep Single-Cell RNA Sequencing. Neuron 101:207–223.e10. https://doi.org/10.1016/j.neuron.2018.12.006
86. Li Y, Du X-F, Liu C-S et al (2012) Reciprocal regulation between resting microglial dynamics and neuronal activity in vivo. Dev Cell 23:1189–1202. https://doi.org/10.1016/j.devcel.2012.10.027
87. Madore C, Nadjar A, Delpech J-C et al (2014) Nutritional n-3 PUFAs deficiency during perinatal periods alters brain innate immune system and neuronal plasticity-associated genes. Brain Behav Immun 41:22–31. https://doi.org/10.1016/j.bbi.2014.03.021
88. Madry C, Kyrargyri V, Arancibia-Cárcamo IL et al (2018) Microglial Ramification, Surveillance, and Interleukin-1β Release Are Regulated by the Two-Pore Domain K + Channel THIK-1. Neuron 97:299–312.e6. https://doi.org/10.1016/j.neuron.2017.12.002
89. Maggi L, Trettel F, Scianni M et al (2009) LTP impairment by fractalkine/CX3CL1 in mouse hippocampus is mediated through the activity of adenosine receptor type 3 (A3R). J Neuroimmunol 215:36–42. https://doi.org/10.1016/j.jneuroim.2009.07.016
90. Marín-Teva JL, Dusart I, Colin C et al (2004) Microglia promote the death of developing Purkinje cells. Neuron 41:535–547
91. Matcovitch-Natan O, Winter DR, Giladi A et al (2016) Microglia development follows a stepwise program to regulate brain homeostasis. Science 353:aad8670. https://doi.org/10.1126/science.aad8670
92. Mathys H, Adaikkan C, Gao F et al (2017) Temporal tracking of microglia activation in neurodegeneration at single-cell resolution. Cell Rep 21:366–380. https://doi.org/10.1016/j.celrep.2017.09.039
93. Mattei D, Ivanov A, Ferrai C et al (2017) Maternal immune activation results in complex microglial transcriptome signature in the adult offspring that is reversed by minocycline treatment. Transl Psychiatry 7:e1120. https://doi.org/10.1038/tp.2017.80
94. McCarthy MM, Auger AP, Bale TL et al (2009) The epigenetics of sex differences in the brain. J Neurosci 29:12815–12823. https://doi.org/10.1523/JNEUROSCI.3331-09.2009
95. Milior G, Lecours C, Samson L et al (2016) Fractalkine receptor deficiency impairs microglial and neuronal responsiveness to chronic stress. Brain Behav Immun 55:114–125. https://doi.org/10.1016/j.bbi.2015.07.024
96. Minten C, Terry R, Deffrasnes C et al (2012) IFN regulatory factor 8 is a key constitutive determinant of the morphological and molecular properties of microglia in the CNS. PLoS ONE 7:e49851. https://doi.org/10.1371/journal.pone.0049851
97. Mizutani M, Pino PA, Saederup N et al (2012) The fractalkine receptor but not CCR97 is present on microglia from embryonic development throughout adulthood. J Immunol 188:29–36. https://doi.org/10.4049/jimmunol.1100421
98. Monier A, Adle-Biassette H, Delezoide A-L et al (2007) Entry and distribution of microglial cells in human embryonic and fetal cerebral cortex. J Neuropathol Exp Neurol 66:372–382. https://doi.org/10.1097/nen.0b013e3180517b46
99. Mosher KI, Wyss-Coray T (2015) Go with your gut: microbiota meet microglia. Nat Neurosci 18:930–931. https://doi.org/10.1038/nn.4051
100. Mosher KI, Wyss-Coray T (2014) Microglial dysfunction in brain aging and Alzheimer's disease. Biochem Pharmacol 88:594–604. https://doi.org/10.1016/j.bcp.2014.01.008

101. Mouton PR, Long JM, Lei D-L et al (2002) Age and gender effects on microglia and astrocyte numbers in brains of mice. Brain Res 956:30–35
102. Nelson LH, Warden S, Lenz KM (2017) Sex differences in microglial phagocytosis in the neonatal hippocampus. Brain Behav Immun 64:11–22. https://doi.org/10.1016/j.bbi.2017.03.010
103. Nikodemova M, Kimyon RS, De I et al (2015) Microglial numbers attain adult levels after undergoing a rapid decrease in cell number in the third postnatal week. J Neuroimmunol 278:280–288. https://doi.org/10.1016/j.jneuroim.2014.11.018
104. Nimmerjahn A, Kirchhoff F, Helmchen F (2005) Resting microglial cells are highly dynamic surveillants of brain parenchyma in vivo. Science 308:1314–1318. https://doi.org/10.1126/science.1110647
105. Nugent BM, Wright CL, Shetty AC et al (2015) Brain feminization requires active repression of masculinization via DNA methylation. Nat Neurosci 18:690–697. https://doi.org/10.1038/nn.3988
106. Olah M, Patrick E, Villani A-C et al (2018) A transcriptomic atlas of aged human microglia. Nat Commun 9:539. https://doi.org/10.1038/s41467-018-02926-5
107. Otero K, Turnbull IR, Poliani PL et al (2009) Macrophage colony-stimulating factor induces the proliferation and survival of macrophages via a pathway involving DAP12 and beta-catenin. Nat Immunol 10:734–743. https://doi.org/10.1038/ni.1744
108. Paolicelli RC, Bisht K, Tremblay M-È (2014) Fractalkine regulation of microglial physiology and consequences on the brain and behavior. Front Cell Neurosci 8:129. https://doi.org/10.3389/fncel.2014.00129
109. Paolicelli RC, Bolasco G, Pagani F et al (2011) Synaptic pruning by microglia is necessary for normal brain development. Science 333:1456–1458. https://doi.org/10.1126/science.1202529
110. Parkhurst CN, Yang G, Ninan I et al (2013) Microglia promote learning-dependent synapse formation through brain-derived neurotrophic factor. Cell 155:1596–1609. https://doi.org/10.1016/j.cell.2013.11.030
111. Peri F, Nüsslein-Volhard C (2008) Live imaging of neuronal degradation by microglia reveals a role for v0-ATPase a1 in phagosomal fusion in vivo. Cell 133:916–927. https://doi.org/10.1016/j.cell.2008.04.037
112. Ponomarev ED, Veremeyko T, Barteneva N et al (2011) MicroRNA-124 promotes microglia quiescence and suppresses EAE by deactivating macrophages via the C/EBP-α-PU.1 pathway. Nat Med 17:64–70. https://doi.org/10.1038/nm.2266
113. Prinz M, Priller J (2014) Microglia and brain macrophages in the molecular age: from origin to neuropsychiatric disease. Nat Rev Neurosci 15:300–312. https://doi.org/10.1038/nrn3722
114. Prinz M, Priller J, Sisodia SS, Ransohoff RM (2011) Heterogeneity of CNS myeloid cells and their roles in neurodegeneration. Nat Neurosci 14:1227–1235. https://doi.org/10.1038/nn.2923
115. Réu P, Khosravi A, Bernard S et al (2017) The lifespan and turnover of microglia in the human brain. Cell Rep 20:779–784. https://doi.org/10.1016/j.celrep.2017.07.004
116. Ribeiro Xavier AL, Kress BT, Goldman SA et al (2015) A distinct population of microglia supports adult neurogenesis in the subventricular zone. J Neurosci 35:11848–11861. https://doi.org/10.1523/JNEUROSCI.1217-15.2015
117. Rigato C, Swinnen N, Buckinx R et al (2012) Microglia proliferation is controlled by P2X7 receptors in a Pannexin-1-independent manner during early embryonic spinal cord invasion. J Neurosci 32:11559–11573. https://doi.org/10.1523/JNEUROSCI.1042-12.2012
118. Rogers JT, Morganti JM, Bachstetter AD et al (2011) CX3CR118 deficiency leads to impairment of hippocampal cognitive function and synaptic plasticity. J Neurosci 31:16241–16250. https://doi.org/10.1523/JNEUROSCI.3667-11.2011
119. Rosenbauer F, Tenen DG (2007) Transcription factors in myeloid development: balancing differentiation with transformation. Nat Rev Immunol 7:105–117. https://doi.org/10.1038/nri2024
120. Roumier A, Béchade C, Poncer J-C et al (2004) Impaired synaptic function in the microglial KARAP/DAP12-deficient mouse. J Neurosci 24:11421–11428. https://doi.org/10.1523/JNEUROSCI.2251-04.2004

121. Roumier A, Pascual O, Béchade C et al (2008) Prenatal activation of microglia induces delayed impairment of glutamatergic synaptic function. PLoS ONE 3:e2595. https://doi.org/10.1371/journal.pone.0002595
122. Satoh J, Kino Y, Asahina N et al (2016) TMEM119 marks a subset of microglia in the human brain. Neuropathology 36:39–49. https://doi.org/10.1111/neup.12235
123. Savage JC, Picard K, González-Ibáñez F, Tremblay M-È (2018) A brief history of microglial ultrastructure: distinctive features, phenotypes, and functions discovered over the past 60 years by electron microscopy. Front Immunol 9:803. https://doi.org/10.3389/fimmu.2018.00803
124. Schafer DP, Lehrman EK, Kautzman AG et al (2012) Microglia sculpt postnatal neural circuits in an activity and complement-dependent manner. Neuron 74:691. https://doi.org/10.1016/j.neuron.2012.03.026
125. Schlegelmilch T, Henke K, Peri F (2011) Microglia in the developing brain: from immunity to behaviour. Curr Opin Neurobiol 21:5–10. https://doi.org/10.1016/j.conb.2010.08.004
126. Schulz C, Gomez Perdiguero E, Chorro L et al (2012) A lineage of myeloid cells independent of Myb and hematopoietic stem cells. Science 336:86–90. https://doi.org/10.1126/science.1219179
127. Schwarz JM, Sholar PW, Bilbo SD (2012) Sex differences in microglial colonization of the developing rat brain. J Neurochem 120:948–963. https://doi.org/10.1111/j.1471-4159.2011.07630.x
128. Sellner S, Paricio-Montesinos R, Spieß A et al (2016) Microglial CX3CR128 promotes adult neurogenesis by inhibiting Sirt 1/p65 signaling independent of CX3CL1. Acta Neuropathol Commun 4:102. https://doi.org/10.1186/s40478-016-0374-8
129. Serrats J, Schiltz JC, García-Bueno B et al (2010) Dual roles for perivascular macrophages in immune-to-brain signaling. Neuron 65:94–106. https://doi.org/10.1016/j.neuron.2009.11.032
130. Shechter R, London A, Schwartz M (2013) Orchestrated leukocyte recruitment to immune-privileged sites: absolute barriers versus educational gates. Nat Rev Immunol 13:206–218. https://doi.org/10.1038/nri3391
131. Sheng J, Ruedl C, Karjalainen K (2015) Most tissue-resident macrophages except microglia are derived from fetal hematopoietic stem cells. Immunity 43:382–393. https://doi.org/10.1016/j.immuni.2015.07.016
132. Shigemoto-Mogami Y, Hoshikawa K, Goldman JE et al (2014) Microglia enhance neurogenesis and oligodendrogenesis in the early postnatal subventricular zone. J Neurosci 34:2231–2243. https://doi.org/10.1523/JNEUROSCI.1619-13.2014
133. Sierra A, Encinas JM, Deudero JJP et al (2010) Microglia shape adult hippocampal neurogenesis through apoptosis-coupled phagocytosis. Cell Stem Cell 7:483–495. https://doi.org/10.1016/j.stem.2010.08.014
134. Sierra A, Gottfried-Blackmore A, Milner TA et al (2008) Steroid hormone receptor expression and function in microglia. Glia 56:659–674. https://doi.org/10.1002/glia.20644
135. Sorge RE, Mapplebeck JCS, Rosen S et al (2015) Different immune cells mediate mechanical pain hypersensitivity in male and female mice. Nat Neurosci 18:1081–1083. https://doi.org/10.1038/nn.4053
136. Squarzoni P, Oller G, Hoeffel G et al (2014) Microglia modulate wiring of the embryonic forebrain. Cell Rep 8:1271–1279. https://doi.org/10.1016/j.celrep.2014.07.042
137. Streit WJ (2006) Microglial senescence: does the brain's immune system have an expiration date? Trends Neurosci 29:506–510. https://doi.org/10.1016/j.tins.2006.07.001
138. Streit WJ, Braak H, Xue Q-S, Bechmann I (2009) Dystrophic (senescent) rather than activated microglial cells are associated with tau pathology and likely precede neurodegeneration in Alzheimer's disease. Acta Neuropathol 118:475–485. https://doi.org/10.1007/s00401-009-0556-6
139. Streit WJ, Sammons NW, Kuhns AJ, Sparks DL (2004) Dystrophic microglia in the aging human brain. Glia 45:208–212. https://doi.org/10.1002/glia.10319
140. Svahn AJ, Giacomotto J, Graeber MB et al (2016) miR-124 contributes to the functional maturity of microglia. Dev Neurobiol 76:507–518. https://doi.org/10.1002/dneu.22328

141. Swinnen N, Smolders S, Avila A et al (2013) Complex invasion pattern of the cerebral cortex by microglial cells during development of the mouse embryo. Glia 61:150–163. https://doi.org/10.1002/glia.22421
142. Tay TL, Béchade C, D'Andrea I et al (2018a) Microglia gone rogue: impacts on psychiatric disorders across the lifespan. Front Mol Neurosci 10. https://doi.org/10.3389/fnmol.2017.00421
143. Tay TL, Mai D, Dautzenberg J et al (2017) A new fate mapping system reveals context-dependent random or clonal expansion of microglia. Nat Neurosci 20:793–803. https://doi.org/10.1038/nn.4547
144. Tay TL, Sagar, Dautzenberg J et al (2018) Unique microglia recovery population revealed by single-cell RNAseq following neurodegeneration. Acta Neuropathol Commun 6:87. https://doi.org/10.1186/s40478-018-0584-3
145. Tay TL, Savage JC, Hui CW et al (2017) Microglia across the lifespan: from origin to function in brain development, plasticity and cognition. J Physiol (Lond) 595:1929–1945. https://doi.org/10.1113/JP272134
146. Thion MS, Low D, Silvin A et al (2018) Microbiome influences prenatal and adult microglia in a sex-specific manner. Cell 172:500–516.e16. https://doi.org/10.1016/j.cell.2017.11.042
147. Tremblay M-È, Lowery RL, Majewska AK (2010) Microglial interactions with synapses are modulated by visual experience. PLoS Biol 8:e1000527. https://doi.org/10.1371/journal.pbio.1000527
148. Tremblay M-È, Zettel ML, Ison JR et al (2012) Effects of aging and sensory loss on glial cells in mouse visual and auditory cortices. Glia 60:541–558. https://doi.org/10.1002/glia.22287
149. Ueno M, Fujita Y, Tanaka T et al (2013) Layer V cortical neurons require microglial support for survival during postnatal development. Nat Neurosci 16:543–551. https://doi.org/10.1038/nn.3358
150. Ulland TK, Song WM, Huang SC-C et al (2017) TREM2 maintains microglial metabolic fitness in Alzheimer's disease. Cell 170:649–663.e13. https://doi.org/10.1016/j.cell.2017.07.023
151. Ulrich JD, Ulland TK, Colonna M, Holtzman DM (2017) Elucidating the role of TREM2 in Alzheimer's disease. Neuron 94:237–248. https://doi.org/10.1016/j.neuron.2017.02.042
152. Valdearcos M, Robblee MM, Benjamin DI et al (2014) Microglia dictate the impact of saturated fat consumption on hypothalamic inflammation and neuronal function. Cell Rep 9:2124–2138. https://doi.org/10.1016/j.celrep.2014.11.018
153. Verney C, Monier A, Fallet-Bianco C, Gressens P (2010) Early microglial colonization of the human forebrain and possible involvement in periventricular white-matter injury of preterm infants. J Anat 217:436–448. https://doi.org/10.1111/j.1469-7580.2010.01245.x
154. Villa A, Gelosa P, Castiglioni L et al (2018) Sex-specific features of microglia from adult mice. Cell Rep 23:3501–3511. https://doi.org/10.1016/j.celrep.2018.05.048
155. Wake H, Moorhouse AJ, Jinno S et al (2009) Resting microglia directly monitor the functional state of synapses in vivo and determine the fate of ischemic terminals. J Neurosci 29:3974–3980. https://doi.org/10.1523/JNEUROSCI.4363-08.2009
156. Wang Y, Szretter KJ, Vermi W et al (2012) IL-34 is a tissue-restricted ligand of CSF1R required for the development of Langerhans cells and microglia. Nat Immunol 13:753–760. https://doi.org/10.1038/ni.2360
157. Wasserman JK, Yang H, Schlichter LC (2008) Glial responses, neuron death and lesion resolution after intracerebral hemorrhage in young versus aged rats. Eur J Neurosci 28:1316–1328. https://doi.org/10.1111/j.1460-9568.2008.06442.x
158. Weinhard L, d'Errico P, Tay TL (2018) Headmasters: microglial regulation of learning and memory in health and disease. Molecular, vol 5, pp 63–89. https://doi.org/10.3934/molsci.2018.1.63
159. Weinhard L, di Bartolomei G, Bolasco G et al (2018) Microglia remodel synapses by presynaptic trogocytosis and spine head filopodia induction. Nat Commun 9. https://doi.org/10.1038/s41467-018-03566-5

160. Wieghofer P, Knobeloch K-P, Prinz M (2015) Genetic targeting of microglia. Glia 63:1–22. https://doi.org/10.1002/glia.22727
161. Williamson LL, McKenney EA, Holzknecht ZE et al (2016) Got worms? Perinatal exposure to helminths prevents persistent immune sensitization and cognitive dysfunction induced by early-life infection. Brain Behav Immun 51:14–28. https://doi.org/10.1016/j.bbi.2015.07.006
162. Wlodarczyk A, Holtman IR, Krueger M et al (2017) A novel microglial subset plays a key role in myelinogenesis in developing brain. EMBO J 36:3292–3308. https://doi.org/10.15252/embj.201696056
163. Wolf SA, Boddeke HWGM, Kettenmann H (2017) Microglia in physiology and disease. Annu Rev Phys 79:619–643. https://doi.org/10.1146/annurev-physiol-022516-034406
164. Wu Y, Dissing-Olesen L, MacVicar BA, Stevens B (2015) Microglia: dynamic mediators of synapse development and plasticity. Trends Immunol 36:605–613. https://doi.org/10.1016/j.it.2015.08.008
165. Xu J, Zhu L, He S et al (2015) Temporal-spatial resolution fate mapping reveals distinct origins for embryonic and adult microglia in zebrafish. Dev Cell 34:632–641. https://doi.org/10.1016/j.devcel.2015.08.018
166. Yona S, Kim K-W, Wolf Y et al (2013) Fate mapping reveals origins and dynamics of monocytes and tissue macrophages under homeostasis. Immunity 38:79–91. https://doi.org/10.1016/j.immuni.2012.12.001
167. Zhan Y, Paolicelli RC, Sforazzini F et al (2014) Deficient neuron-microglia signaling results in impaired functional brain connectivity and social behavior. Nat Neurosci 17:400–406. https://doi.org/10.1038/nn.3641
168. Zheng H, Jia L, Liu C-C et al (2017) TREM2 promotes microglial survival by activating Wnt/β-catenin pathway. J Neurosci 37:1772–1784. https://doi.org/10.1523/JNEUROSCI.2459-16.2017
169. Zusso M, Methot L, Lo R et al (2012) Regulation of postnatal forebrain amoeboid microglial cell proliferation and development by the transcription factor Runx1. J Neurosci 32:11285–11298. https://doi.org/10.1523/JNEUROSCI.6182-11.2012

Chapter 7
General Pathophysiology of Astroglia

Alexei Verkhratsky, Margaret S. Ho, Nina Vardjan, Robert Zorec and Vladimir Parpura

Abstract Astroglial cells are involved in most if not in all pathologies of the brain. These cells can change the morpho-functional properties in response to pathology or innate changes of these cells can lead to pathologies. Overall pathological changes in astroglia are complex and diverse and often vary with different disease stages. We classify astrogliopathologies into reactive astrogliosis, astrodegeneration with astroglial atrophy and loss of function, and pathological remodelling of astrocytes. Such changes can occur in neurological, neurodevelopmental, metabolic and psychiatric disorders as well as in infection and toxic insults. Mutation in astrocyte-specific genes leads to specific pathologies, such as Alexander disease, which is a leukodystrophy. We discuss changes in astroglia in the pathological context and identify some molecular entities underlying pathology. These entities within astroglia may repent targets for novel therapeutic intervention in the management of brain pathologies.

Keywords Astrocyte · Neuropathology · Alexander disease · Stroke · Psychiatric diseases · Metabolic diseases · Neurotrauma · Infectious diseases · Systemic

A. Verkhratsky (✉)
Faculty of Biology, Medicine and Health, The University of Manchester, Oxford Road, Manchester M13 9PT, UK
e-mail: Alexej.Verkhratsky@manchester.ac.uk

Center for Basic and Translational Neuroscience, Faculty of Health and Medical Sciences, University of Copenhagen, 2200 Copenhagen, Denmark

Achucarro Center for Neuroscience, IKERBASQUE, Basque Foundation for Science, 48011 Bilbao, Spain

M. S. Ho
School of Life Science and Technology, ShanghaiTech University, 201210 Shanghai, China

N. Vardjan · R. Zorec
Laboratory of Neuroendocrinology-Molecular Cell Physiology, Faculty of Medicine, Institute of Pathophysiology, University of Ljubljana, Ljubljana, Slovenia

Celica BIOMEDICAL, Ljubljana, Slovenia

V. Parpura
Department of Neurobiology, The University of Alabama at Birmingham, Birmingham, AL, USA

A. Verkhratsky et al. (eds.), *Neuroglia in Neurodegenerative Diseases*,
Advances in Experimental Medicine and Biology 1175,
https://doi.org/10.1007/978-981-13-9913-8_7

inflammation · Sepsis-associated encephalopathy · Toxic encephalopathy · Hepatic encephalopathy · Autistic spectrum disorders · Epilepsy

7.1 Prologue: Neuroglia in Neurological Diseases

The role of neuroglia in neurological disorders have been widely accepted by leading neuroanatomists and neurologists of the nineteenth century, from Rudolf Virchow (who indicated that the 'interstitial tissue of the brain and spinal marrow are one of the most frequent seats of morbid change' [222]) to Carl Fromann, Alois Alzheimer and Nicolas Achucarro [1, 6, 8, 77], to name but a few. The neuropathological philosophy of the twentieth century was dominated by neurono-centric views, while the last decade witnessed the resurgence of neurogliopathology. Recently, the pathological potential of neuroglia in general, and astrocytes in particular, has been extensively studied and the fundamental principles of astrogliopathology have been defined [33, 74, 79, 152, 156, 173, 191, 192, 216, 219, 220, 234].

Neuroglial cells are primary homeostatic and defensive cells of the nervous system; and naturally, all types of glia are contributing to neuropathological developments. Astrocytes are a part of neural networks; they interact with neurones, with other glia and with blood vessels, thus, maintaining the structural and functional integrity of the neural tissue. Astrocytes are indispensable for maintaining neuronal functional and neuronal survival both in physiology and in pathology [214]. Therefore, astroglial failure creates a disease-permissive landscape and underlies neuronal malfunction, neuronal death and neurological deficits.

7.2 Principles of Astrogliopathology

Pathological changes in astroglia in neurological diseases are complex and diverse (Fig. 7.1). These changes can be generic or disease-specific. They often vary at different disease stages. In the context of human pathology, changes are affected by age and comorbidity. Astrogliopathology is classified into (i) reactive astrogliosis; (ii) astrodegeneration with astroglial atrophy and loss of function; and (iii) pathological remodelling of astrocytes (Fig. 7.1, [74, 156, 220]); all these pathological reactions occur together or in isolation.

7.2.1 Reactive Astrogliosis

Reactive astrogliosis is observed in many neurological disorders. Until very recently, astroglial reactivity was considered the sole manifestation of astrogliopathology. From histopathological point of view, astroglial reactivity is characterised by mor-

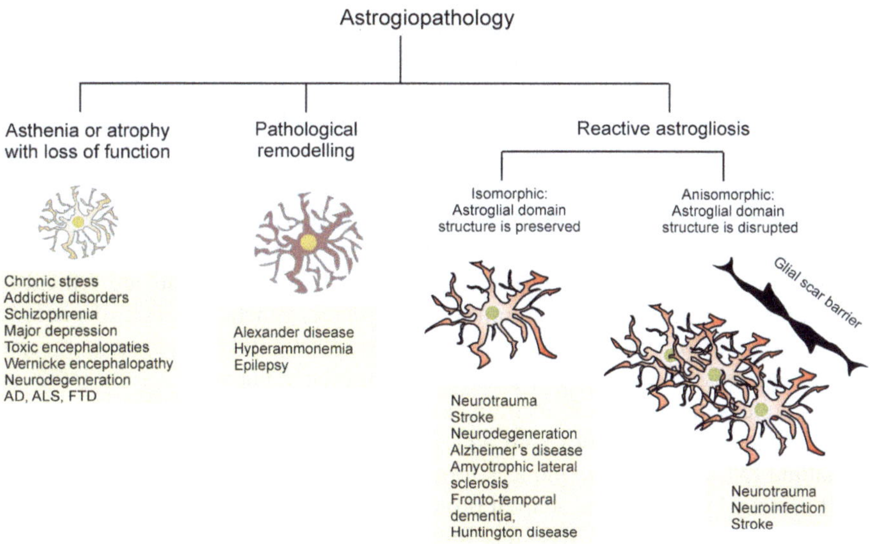

Fig. 7.1 Principles of astrogliopathology. Astrocytes undergo several types of morpho-functional changes in the brain pathology (see text for details). AD, Alzheimer's disease; ALS, amyotrophic lateral sclerosis; FTD, fronto-temporal dementia

phological hypertrophy and up-regulation of two major cytoskeletal intermediate filaments/proteins, glial fibrillary acidic protein (GFAP) and vimentin [95, 155, 191]. Reactive astrocytes undergo a variety of substantial modifications, showing multiple phenotypes with both neuroprotective and neurotoxic features. These phenotypes arguably are disease-specific, although they all can share some common properties [120, 121, 156]. Transcriptomes of reactive astrocytes in ischemia and endotoxin activation, for example, show significant differences [233].

Conceptually, reactive astrogliosis represents an evolutionary conserved (the first manifestations of astroglial reactivity are observed in many invertebrates including annelids and insects) defensive reprogramming of astroglia aimed at: (i) increased neuroprotection and trophic support of the nervous tissue; (ii) isolation of the lesioned area; (iii) reconstruction of the compromised blood–brain barrier; and (iv) facilitating the post-lesion regeneration of the nervous tissue [7, 156, 191]. The astrogliotic programme, therefore, has a high degree of flexibility and tailors functional and biochemical reprogramming of astrocytes to the nature and strength of the insult. Even within the same lesioned regions, astrocytes demonstrate a degree of heterogeneity in expression of transcription factors, inflammatory agents and signalling molecules [78, 92].

The initiation of astrogliosis is regulated mainly by damage-associated molecular patterns (DAMPs) or pathogen-associated molecular patterns (PAMPs). The DAMPs are endogenous molecules released from damaged or dying cells (ATP being the most prominent example), blood-borne factors that infiltrate brain parenchyma, etc. The PAMPs are exogenous molecules associated with infectious invaders such as bacteria or viruses; they mostly act through Toll-like receptors (TLRs) widely expressed in astrocytes [99, 203]. Astroglial cells express a wide range of receptors to both DAMPs and PAMPs: $P2X_7$ purinoceptors, TLRs, nucleotide-binding oligomerisation domains (NOD)-like receptors (NLRs), double-stranded RNA-dependent protein kinase, scavenger receptors, mannose receptor and receptors for complement components and mediators, such as CXCL10, CCL2, interleukin-6 and B-cell-activating factor of the tumour necrosis factor (TNF) family [70]. Often, exposure of astrocytes to DAMPs and PAMPs evokes cytosolic Ca^{2+} increases due to its release from the endoplasmic reticulum (ER) intracellular store. These Ca^{2+} signals are critical for instigating the astrogliotic programme. For instance, genetic deletion of predominant astroglial Ca^{2+} release channel of the ER, inositol 1,4,5-triphospate receptor type 2, suppresses astrogliotic response [101]. Similarly, pharmacological inhibition of Ca^{2+} release from the ER restrains astroglial reactivity triggered by amyloid-β [2]. Stimulation of astrocytes with ATP (a classical DAMP) not only triggers Ca^{2+} signalling [34, 211] but also induces formation of inflammasomes comprised of the NLR protein-1 or -2 LR, the adaptor protein apoptosis-associated speck-like protein containing a C-terminal caspase recruitment domain (ASC) and caspase-1. Activation of these inflammasomes leads to the processing of inflammatory caspase-1 and interleukin-1β (IL-1β) [137].

Reactive astrogliosis is classified according to the morphological properties and severity (Fig. 7.1). From the morphological perspective, astrogliosis is divided into isomorphic and anisomorphic astrogliosis. The isomorphic astrogliosis preserves astroglial territorial domains and it is fully reversible, whereas anisomorphic astrogliosis proceeds with violation of territorial domains, cell migration and territorial overlap, formation of astroglial palisades and ultimately the scar formation [156]. According to the severity, astrogliosis is classified into (i) mild to moderate astrogliosis; (ii) severe diffuse astrogliosis; and (iii) severe astrogliosis with compact scar formation [190, 191]. Fundamentally, astrogliosis provides for defence of the nervous tissue; it increases neuroprotection and is ultimately important for post-lesion regeneration. Even scar formation carries clear definitive function isolating the damaged area of the CNS and saving the whole at the expense of its part [156, 210]. Suppression of reactive astrogliosis usually exacerbates the course of pathology [156]. Inhibition of astroglial reactivity enlarges the size of the traumatic lesions and aggravates neurological deficit [148]. Deletion of GFAP and vimentin, both being critical for the execution of astrogliotic programme, facilitates the development of ischaemic infarcts [116] and exacerbates post-traumatic synaptic loss [154]. Ablation of astroglial reactivity increased the accumulation of β-amyloid and reduced microglial association with senile plaques in the animal model of the Alzheimer's disease [157]. Nonetheless, in conditions of prolonged stress or severe damage, reactive astrocytes may acquire neurotoxic potential and astrogliosis as a process can become maladaptive [155].

7.2.2 *Astroglial Atrophy*

Astrodegeneration is a widespread class of astrocytopathy, which is represented by morphological atrophy, increased astroglial death and hence decrease in astroglial density and an impairment of homeostatic functions. Astrodegeneration has been observed in various types of neuropathologies [90, 212, 221]. Astrodegeneration is particularly prominent in major psychiatric diseases. For instance, schizophrenia, major depressive disorder, Wernicke–Korsakoff encephalopathy, and addictive disorders are all accompanied with a reduction in the packing density of astrocytes and a failure of their homeostatic cascades, the latter most notably associated with glutamate homeostasis and glutamate–glutamine shuttle, which are both impaired [50, 53, 54, 134, 163, 165, 178, 218]. Aberrant astroglial glutamate transport and catabolism are arguably responsible for abnormal neurotransmission as well as for excitotoxic neuronal death, both resulting in psychotic symptoms. In amyotrophic lateral sclerosis, insufficient astroglial glutamate clearance from the extracellular space instigates excitotoxic death of large motor neurones [177, 207], whereas in Alzheimer's disease, reduced astroglial synaptic coverage contributes to early synaptic extinction and cognitive deficiency [215].

7.2.3 *Pathological Remodelling of Astrocytes*

Pathological remodelling of astrocytes covers abnormalities associated with an acquisition of abnormal molecular cascades or functional properties, which drive pathology [74, 156]. Pathological remodelling of astroglia contributes to various leukodystrophies, most notably to Alexander disease, megalencephalic leukoencephalopathy with subcortical cysts or vanishing white matter syndrome, in all of which the astrocytopathy initiates lesions of the white matter [113]. In Alexander disease, astroglial expression of sporadically mutated GFAP gene results in early and severe leukomalacia [131]. Pathological remodelling in astroglia has been also described in mesial temporal lobe epilepsy, in which astrocytes acquire aberrant morphology, reduce gap junctional coupling and down-regulate expression of $K_{ir}4.1$ channels; all these changes compromise K^+ homeostasis thus contributing to the initiation of seizures [19].

7.3 Astrogliopathology in Neurological Diseases

7.3.1 *Neurotrauma*

Traumatic injury of the brain and of the spinal cord are classified according to their nature (penetrating wounds or concussions; the later when occurring in the cervical spinal cord is known medically as cervical cord neurapraxia), their severity (mild, moderate or severe), volume (focal or diffuse), outcome (death, vegetative state, severe disability, moderate disability and good recovery) and anatomical localisation. According to its very nature, a traumatic injury to the CNS has a complex pathophysiology associated not only with direct damage to neural cells, but also with a damage to the whole organ with destruction of the blood–brain barrier and blood vessels, ischaemic insults, opening the way for secondary infection, etc. Neurotrauma predominantly triggers astrogliotic response; reactive phenotypes, however, very much depend on the pathological context [32, 33] with the severity of the damage and its anatomical localisation affecting astroglial activation.

In the healthy brain, astrocytes form numerous barriers with blood vessels and with cerebrospinal fluid; endfeet of astroglial cells together with the parenchymal basement membrane create *glia limitans*, which physically separates the brain parenchyma from blood vessels, perivascular spaces and the meninges. In response to a neurotrauma, an astrogliotic scar barrier is formed that delineates and isolates the areas of a focal damage from the healthy brain. Suppression of astrogliosis with consequent malformation of an astroglial scar markedly exacerbates tissue damage and neurological deficit [189]. The heterogeneity of reactive astroglial phenotypes very much depends on the distance to the lesion core. Close to the lesion astrocytes lose their domains, their processes overlap and the astroglial palisades are formed, reflecting anisomorphic astrogliosis. Astrocytes gather around the damaged sites and form the scar [32]. Astrocytes distant to the lesion core undergo isomorphic gliosis; they become hypertrophic and arguably neuroprotective. Contribution of astrocytes to tissue pathology in neurotrauma is multifaceted. Besides forming a protective scar astrocytes regulate inflammatory response, provide for homeostatic protection of the nervous tissue through the removal of extracellular glutamate, buffering K^+ or releasing scavengers of reactive oxygen species and regulate post-traumatic remodelling of synaptic networks. Reactive astrocytes are indispensable in remodelling as suppression of astrogliosis down-regulates post-traumatic regeneration of synaptic connectivity and neuronal networks [7].

7.4 Infectious Diseases

7.4.1 Infection of Nervous Tissue

Infections of the CNS caused by bacteria, viruses, fungi and parasites are classified into meningitis, encephalitis or brain abscess. Not every infectious agent can invade the CNS. Rather, only certain neurotropic viruses, bacteria, fungi and parasites can penetrate into the brain and the spinal cord with relative ease. Furthermore, most of the pathogens are effectively stopped by the brain barriers [110]. Infectious agents may cross the blood–brain barrier using the paracellular route, via transcytotic mechanism, inside entering monocytes (the Trojan horse hypothesis) as well as by other mechanisms such as, for example, hijacking of β-adrenergic receptors as shown for *Neisseria meningitidis* (meningococcus) [47].

Neuroglial contribution to the infectious lesions of the CNS is of fundamental importance. Neuroprotective activation of astrocytes and microglia to a large extent defines the spread of infection through the nervous tissue and hence determines the outcome of the disease. The glial response, in turn, depends on the nature of an infectious agent. For example, contact of astrocytes with the Gram-positive bacteria such as *Pneumococcus* or *Staphylococcus* triggers rapid astrogliotic activation with marked cellular hypertrophy and up-regulation of GFAP expression [96] accompanied with synthesis and secretion of pro-inflammatory agents such as TNF-α, ILs and macrophage inflammatory protein 1α [122]. Activation of astrocytes by infectious agents (see for example [70, 197, 228]) is mediated mainly through pattern recognition receptors (PRRs), which are represented by TLR 2, 3, 4, 9 [65], NLRs, retinoic acid-inducible gene (RIG)-like receptors (RLRs) and cytokine receptors [108]. The NOD2 receptor, operational in astrocytes, recognises a minimal motif present in all bacterial peptidoglycans and it is required for astroglial reactive reprogramming in response to *N. meningitidis* and *Borrelia burgdorferi* [43].

Activation of astrocytes is also linked to TLR receptors [67]. Distinct TLR subtypes recognise and respond to different PAMPs. Lipopolysaccharide (LPS), for example, signals through TLR4; TLR3 is activated by double-stranded RNA; peptidoglycans interacts with TLR2, while TLR9 senses CpG DNA [40]. Activated TLRs interact with adaptor proteins myeloid differentiation factor 88 (MyD88) and/or a TIR-containing adaptor molecule, Toll/interferon-1 receptor domain-containing adaptor inducing interferon-γ (TRIF), which acts as a part of relevant signalling cascade [40]. Bacterial infection of the nervous tissue down-regulates expression of connexins hence decreasing gap junctional connectivity of astroglial syncytia [66]. Direct interaction of several bacteria such as *Streptococcus pneumonia*, *B. burgdorferi* and *N. meningitides* triggers astroglial reactivity as well as increases the production of pro-inflammatory cytokines and chemokines such as IL-6, TNF-α, IL-8, CXCL-1 and CXCL-10 [240]. Besides activation, astrocytes may undergo pathological remodelling and act as a reservoir for infection. Furthermore, astrocytes can promote apoptotic death of their uninfected neighbours through gap junction route [68, 240]. Astroglial reactivity, that includes overexpression of GFAP, keeps infectious process at bay. Indeed, genetic deletion of GFAP associated with suppressed astrogliosis, significantly exacerbated the neurological damage induced by intra-brain injection of *S. aureus* [197].

Astrocytes are fundamental players not only in bacterial but also in viral infections of the CNS. First and foremost, astrocytes can be directly infected by a virus. For example, astrocytes accumulate human immunodeficiency virus-1 (HIV-1) in a cluster of differentiation 81 (CD81)-lined vesicles. Inside these vesicles, the virus is protected from degradation [83]. The very same vesicles contributed to the secondary *trans*-infection of T-cells [83]. In the dementia caused by HIV brain infection, astrocytes undergo both reactive remodelling and astroglial degeneration and astroglial death. These processes may reduce homeostatic support and hence contribute to cognitive deficit [46]. The astroglial infection with HIV-1 (similarly to bacterial infection) decreased expression of connexins and syncytial connectivity [150]. In a similar manner, astrogliotic response is mounted in response to infection with the herpes simplex virus 1 (HSV1). Here, activation of astrocytes is mediated by TLR3 and it is neuroprotective. Deletion of TLR3 suppressed astrogliosis and exacerbated HSV pathology in mice [168] as well as in humans [91]. Infection with cytomegalovirus (CMV) was associated with astroglial homeostatic failure. The CMV infected astrocytes showed decreased released of thrombospondins and deficient glutamate uptake, possibly linked to an increased excitotoxicity [235, 236]. Neurotropic viruses of the family of Flaviviridaem represented by Zika virus and tick-borne encephalitis virus (TBEV) invade astrocytes by endocytosis [159, 240]. Astroglial infection with TBEV does not visibly affect their survival or function, and it is generally believed that astroglial cells act as a reservoir for this type of virus [240]. Astrocytes also represent the cellular target for some protozoan parasite, most notably for *Toxoplasma gondii*. Astrocytes infected with *T. gondii* undergo biochemical remodelling associated with up-regulated synthesis of kynurenic acid that in turn may be linked to some forms of schizophrenia, which will be discussed in appropriate section below. In addition, infection of astrocytes with this protozoan results in the loss of gap junctions [37].

7.4.2 Systemic Infections and the Brain: Sepsis-Associated Encephalopathy

Systemic inflammation frequently accompanies various infectious and non-infectious diseases including degenerative and metabolic disorders. This systemic inflammation often is manifested in the form of sepsis. Sepsis (and in particular abdominal sepsis) is frequently accompanied with an acute brain dysfunction, generally defined as sepsis-associated encephalopathy or SAE [187]. From the clinical perspective, the SAE is regarded as a sign of the severity of a septic state, which potentially worsens the prognosis [158]. The SAE is defined as a clinical syndrome associated with the general brain dysfunction that develops in sepsis in the absence of primary infection of the nervous tissue. The histopathological signs of the SAE include infarctions, petechial and small focal haemorrhages, septic-embolic abscesses and septicopyemic microabscesses, disseminated intravascular coagulation (DIC) syndrome with fibrinous microthrombi, multifocal necrotising leukoencephalopathy, necrotic or apoptotic neuronal death, perivascular and cytotoxic oedema, damage of the blood–brain barrier and reactive neuroinflammation [89, 186, 187]. Sepsis is often associated with the formation of abscesses and microabscesses in the brain parenchyma, which can be regarded as directly associated with the SAE. The SAE, especially at the early stages is often associated with 'sickness behaviour', the syndrome accompanying system inflammation. The symptomatology of sickness behaviour syndrome includes anxiety, anorexia, anhedonia, depression, cognitive changes, including decreased concentration, learning and memory [56].

At the neurochemical level, the leading pathological changes in an SAE are represented by aberrant neurotransmission, which is responsible for cognitive and psychotic symptoms. Substantial changes in expression of main neurotransmitter receptors including receptors for γ-aminobutyric acid (GABA), serotonin, dopamine and noradrenaline have been observed in the brain in systemic infections [100, 208]. Changes in neurotransmitter homeostasis in sepsis are arguably related to alterations of amino acids levels in the blood characterised by a decrease in branched chain amino acids together with relative increase in aromatic amino acids [15]. In addition, compromised brain barriers allow a substantial influx of aromatic amino acids, such as tyrosine, phenylalanine and tryptophan, which may act as false neurotransmitters and alter biosynthesis of *true* neurotransmitters (e.g., dopamine, noradrenaline and serotonin—[188]).

Astrocytes, endfeet of which form *glia limitans* and hence can be considered as the parenchymal portion of brain barriers (the blood–brain and the blood–cerebrospinal fluid), define, to a very large extent, the resistance of the nervous tissue to the systemic inflammation. Intimate contacts of astrocytes with all other cellular elements of the nervous tissue allow them to regulate the relationship between the CNS and systemic physiology and pathology. In the context of SAE, astroglial reactivity is the principal mechanism that limits the propagation of pathological agents through the nervous tissue [41, 193]. Inhibition of astrogliotic response compromises astroglial barrier function and aggravates encephalopathy in the context of systemic inflammation or infectious lesion to the brain. For example, in transgenic mice with deleted gene for GFAP (this intervention suppresses astrogliosis), brain abscesses caused by *Staphylococcus aureus* or *Toxoplasma gondii* were much larger. Lesions become poorly demarcated, bacterial penetration significantly increased, neuronal death was much exacerbated and severe brain oedema developed [197]. Suppression of astroglial reactivity by activation of NF-κB signalling cascade in retinal ischemia or in spinal cord injury, is associated with an increased neuronal damage [27, 63]. Finally, inability of astrocytes to acquire reactive phenotype results in swelling, cytotoxic oedema and spread of damage in infectious abscesses [187].

Astrocytes contribute to the pathology of the blood–brain barrier, which is classified as disruptive and non-disruptive alterations, with both variants present in systemic inflammation [187]. The non-disruptive BBB pathology develops at the molecular and cellular levels when BBB permeability is affected following up- or down-regulation of receptors and transporters expressed in endothelial cells and astrocytes [209]. Disruptive BBB alterations develop through anatomical changes, which include degradation of glycocalyx, a loss of integrity of tight junctions, mitochondrial damage, appearance of fenestrae between endothelial cells, endothelial cells death, collapse of *glia limitans* and astrocytopathy. Disruption of BBB in systemic inflammation is mediated by blood-derived metalloproteinases, prostanoids, nitric oxide and reactive oxygen species [38, 136]. The switch between non-disruptive and disruptive BBB pathology depends on the severity of systemic inflammation. At the early stages of SAE the non-disruptive changes prevail, whereas in severe sepsis, both non-disruptive and disruptive changes occur [209].

7.5 Toxic Damage of the CNS

7.5.1 Heavy Metal Toxic Encephalopathies

Heavy metals, which cause severe brain damage with cognitive deficits, target primarily astrocytes. This is because heavy metals (such as manganese, lead, aluminium or mercury, in the form of methylmercury) mainly accumulate into astrocytes through different plasmalemmal transporters. In general, heavy metals down-regulate astroglial expression of glutamate transporters which decrease glutamate clearance and trigger excitotoxicity [198, 199, 217, 232].

Poisoning by methylmercury is known as Minamata disease named after the city of Minamata in Japan where the disease was first described [129]. The symptoms of Minamata disease include visual abnormalities, sensory lesions, cerebellar ataxia, hearing loss, weakness, tremor and cognitive decline. Methylmercury primarily accumulates in astroglia, where it inhibits glutamate and cystine uptake [5]. Suppression of glutamate uptake instigates exocytotic neuronal death, whereas inhibition of cystine transport limits astroglial synthesis of glutathione hence reducing astroglial capacity to counteract the accumulation of reactive oxygen species; both these processes contribute to neurotoxicity and neuronal death [57, 232].

Exposure to toxic concentrations of lead similarly causes neurodegeneration. Lead primarily accumulates in astroglia, where it down-regulates expression of EAAT-2 glutamate transporter, increases astroglial production of vascular endothelial growth factor, and impairs astroglia-associated water homeostasis by increasing the water permeability of aquaporin 4 [85]. Arguably, these mechanisms contribute to cytotoxic and vascular brain oedema observed in patients with lead poisoning.

Aluminium toxic encephalopathy is manifested by cognitive impairments, speech alterations, seizures and flapping wrist tremor (asterixis). Treatment of cultured astrocytes with aluminium led to swelling, destruction of the cytoskeleton, reduction in gap junctional connectivity, inhibition of glutamate uptake and increased astroglial apoptosis. Loss of astroglial glutamate uptake triggered neuronal death in neuronal–glial co-cultures [198, 199].

The main symptom of acute manganese neurotoxicity is an acute psychosis, whereas chronic manganese poisoning leads to parkinsonism. Astrocytes possess the high capacity manganese transport system; treatment of primary cultured astrocytes with manganese suppresses glutamate uptake and promotes apoptosis [57].

7.5.2 Hyperammonemia and Hepatic Encephalopathy

Increase in blood ammonium accompanies several diseases. The most frequent cause of hyperammonemia is, however, associated with an acute or chronic liver failure (the liver being the main organ for ammonia clearance). Hyperammonemia affects the brain and triggers a condition generally known as hepatic encephalopathy, manifested by cognitive and behavioural impairment; symptoms include confusion, forgetfulness, irritability and alterations of consciousness, such as lethargy and somnolence. Severe hyperammonemia provokes brain oedema, coma and death [31, 35, 73]. In the CNS ammonia is detoxified by glutamine synthetase localised exclusively in astrocytes; this enzyme catabolises ammonium reaction with glutamate and produces glutamine [3, 143, 175]. This reaction is central for glutamate-glutamine shuttle; it also fixes ammonium, which is liberated during physiological neuronal activity [126]. Ammonium overload occludes this pathway and blocks glutamine synthetase hence causing major disturbances of glutamatergic and GABAergic (as glutamate is the precursor to GABA) neurotransmission, which underlie all the symptoms outlined above [31, 35].

Hyperammonemia also affects homeostatic astroglial functions. Exposure of astrocytes to ammonium results in a down-regulation of expression of inward rectifying $K_{ir}4.1$ channels, an event mediated through astrocytic NMDA receptors by a yet uncharacterised mechanism. Decrease in the density of $K_{ir}4.1$ channels, in turn, affects astroglial K^+ buffering which may impair neuronal excitability [144, 166]. Exposure of astrocytes to ammonium also produces aberrant Ca^{2+} signalling by increasing expression of Ca^{2+}-permeable TRPC1 channels, up-regulating expression of $Ca_v1.2$ voltage-gated Ca^{2+} channels and facilitating Ca^{2+} release from the intercellular stores [86, 119, 223]. Increased Ca^{2+} load of astroglial cytosol, in turn, triggers the exocytotic secretion of glutamate which further exacerbates excitotoxic damage of the nervous tissue [82, 139]. Finally, increased ammonium compromises astroglial transport of Na^+ and H^+ which contributes to aberrant pH regulation in the CNS [105, 106]. All these molecular changes result in impaired synaptic transmission, synaptic plasticity and cognitive capabilities [45].

7.6 Astrogliopathology in Stroke

A disruption of the blood flow results either from a blood vessel rupture (that causes a haemorrhage), or by a restriction of blood supply to the brain or parts of the brain, because of a vascular occlusion (thrombosis or embolism), or to a systemic decrease in blood supply (resulting, for example, from a heart failure). This status is generally referred to as brain ischaemia. As a consequence, brain ischaemia can be either global, or focal, the latter corresponding to a stroke.

Astrogliopathology in stroke is complex and multifaceted, with astrocytes being both neuroprotective and neurotoxic [81, 239]. Focal ischaemia results in the infarction of nervous tissue creating a zone of pan-necrosis or an infarction core. At this core, all cells, neurones, glia and other non-neuronal cells undergo rapid necrosis. The size of the core is determined by anatomical location and duration of the ischemic attack. Quite frequently the focal ischemia is transient, as the blood flow can be restored when the vessel blockage is removed. In this case, restored blood flow results in reperfusion of the damaged area, which itself is potentially damaging because of the production of reactive oxygen species and secondary ion imbalances.

The infarction core is surrounded by the ischemic penumbra, which contains viable cells, although with compromised metabolism and function. The infarction core is formed rapidly, within minutes to hours after initiation of the stroke. This is followed by a much slower process of expansion of the infarction zone through the penumbra, which develops over many hours and days. The final neurological deficit is often defined by the limits of the infarction expansion, which in turn depends on astroglial response.

Astrocytes support neurones in the ischaemic penumbra through several homeostatic pathways. First and foremost, astrocytes maintain homeostasis of glutamate in the ischaemic zones. They also feed neurones with metabolic substrates such as lactate. Energising astroglial mitochondria, for example, increase neuroprotection in the ischaemic context [180]. Taming glutamate excitotoxicity, which always follows stroke, almost solely falls onto astroglial cells. Down-regulation of expression of the astroglial glutamate transporter GLT-1/EAAT1 with siRNA increases the size of the infarct [167], whereas targeted overexpression of GLT-1 in astrocytes limits the infarction volume and alleviates neurological deficit [88]. Similarly, stimulation of glutamate uptake with pharmacological agents such as tamoxifen or riluzole decreased infarction volume in animal models [227, 238]. Of note, astroglial glutamate transporters are Na^+ dependent, and hence maintenance of Na^+ transmembrane gradients is critical for glutamate clearance [109]. Another important component of astroglia-dependent neuroprotection in the ischaemic penumbra is associated with antioxidant defence. Astrocytes are critical for both glutathione and ascorbic acid systems, which are the most powerful scavengers of reactive oxygen species [61, 62, 125]. Progression of cell death through the ischaemic penumbra is mediated by spreading depolarisation, which stresses astroglial ionostatic capacity. Furthermore, astrocytes may propagate death signal, triggering distant neuronal death [115, 142].

An important component of astroglial response to stroke is associated with reactive astrogliosis. Ischaemic damage to the brain tissue rapidly instigates astroglial activation through the release of DAMPs; the severity of astrogliotic remodelling and reactive phenotypes depend on the distance to the ischemic core [33]. Astrocytes close to the ischaemic core undergo anisomorphic gliosis, form astroglial palisades and produce astroglial scar that limits the damage to the nervous tissue. In parallel, distantly to the core astrocytes undergo isomorphic, neuroprotective astrogliotic remodelling, which is critically important for post-lesion regeneration The main outcome of astrogliosis in the immediate vicinity of the necrotic area is the formation of an astroglial scar, whereas more peripheral reactive astrocytes are important for post-lesion regeneration [81].

7.7 Metabolic Disorders

7.7.1 Congenital Glutamine Deficiency with Glutamine Synthetase Mutations

Congenital glutamine synthetase deficiency, a rather rare recessive inborn disease, results from mutations to the gene *GLUL* that encodes astroglia-specific glutamine synthetase, thus, this disorder can be considered as a specific astrogliopathy. This disease is characterised by pronounced malformation of the brain with severe white matter deficiency and abnormal gyri formation. Functionally, this deficiency is manifested as epileptic encephalopathy. The deficit in glutamine synthetase in the liver promotes chronic hyperammonemia. In addition, levels of glutamine are reduced in the brain as well as in other organism fluids. The disease results in prenatal malformation of various organs and is generally incompatible with life. Most of the infants die shortly after birth. The leading pathophysiological mechanism is associated with impaired ability of astrocytes to produce glutamine, which affects excitatory and inhibitory transmission; in addition, deficient glutamine synthetase cannot properly detoxify ammonium [195].

7.7.2 Pyruvate Carboxylase Deficiency

Pyruvate carboxylase is an enzyme of gluconeogenesis and it also contributes to anaplerotic metabolic pathways (i.e. producing intermediates for metabolic chains such as the Krebs cycle). In the CNS, pyruvate carboxylase is predominantly expressed in astrocytes. Pyruvate carboxylase deficiency is an autosomal recessive disease associated with impaired metabolism. The symptoms include retardation of mental development, recurrent seizures and metabolic acidosis [127]. There are three clinically distinct forms: type A, or the infantile form, in which children die in the early years; type B, or the severe neonatal form, with many neurological signs including pyramidal symptoms, in which babies die within 3 months after birth; type C or the benign form, which is characterised by mild neurological developmental deficits. The cellular pathogenesis remains largely unknown, but it is probably linked to reduced astroglial homeostatic function, such as glutamate buffering and regulation of angiogenesis [57]

7.7.3 Niemann–Pick Type C Disease

Niemann–Pick disease type C is a progressive neurodegenerative disease associated with hepatosplenomegaly. It is characterised as an autosomal recessive lysosomal storage disease, which results from loss-of-function mutations of genes encoding

NPC-1 or NPC-2 proteins [176]. These proteins are localised in astroglial perisynaptic processes and may be involved in the regulation of cholesterol transport and, hence, synaptogenesis or synaptic maintenance [153]. Astroglia-specific genetic deletion of *Npc1* from mice resulted in reduced neuronal cholesterol, which was associated with decreased neuronal and glial death and three times increase in the life span [237]. There is also evidence of a possible contribution for NPC-1 protein in calcium homeostasis and signalling.

7.7.4 Aceruloplasminemia

The enzyme ceruloplasmin (also known as ferroxidase) is a part of iron metabolism. In the CNS this enzyme is expressed almost exclusively in perivascular astrocytes. Ceruloplasmin is an important component of protection of the nervous tissue against iron-associated lipid peroxidation and formation of hydroxyl radicals. Mutation of the ceruloplasmin gene with loss-of-function causes the autosomal recessive disease known as aceruloplasminemia, which can be defined as an inherited neurodegenerative disorder with systemic iron-overload syndrome [138]. This disease is characterised by primary lesions to astrocytes, which affects their morphology and results in an appearance of foamy spheroid bodies at the vascular endfeet [147]. Aceruloplasminemia is also associated with neuronal death and the appearance of iron deposition.

7.8 Alexander Disease

Alexander disease (AxD), named after William Stewart Alexander, a neuropathologist who described it for the first time [4], is a rare, chronic and usually fatal neurodegenerative disorder. Clinically, AxD may be defined as a severe leukodystrophy; pathophysiologically, it is a primary genetic astrogliopathology [131]. The AxD results from a dominant gain-of-function mutation of the gene encoding GFAP. This leads to astroglial pathology that, in turn, results in a severe damage to the developing white matter. The histopathological hallmark of AxD is an accumulation of protein aggregates, known as Rosenthal fibres, around astroglial nuclei and endfeet [131]. AxD is subclassified into: (i) Type I, characterised with an early onset and severe mental and physical disabilities, megalencephaly, seizures, spasticity, difficulty speaking and swallowing, and (ii) Type II, with a later onset and somewhat different and milder clinical manifestations with normal development and head size, with rare occurrence of seizures, but with ataxia, visual and autonomic abnormalities, troubles in sleeping patterns, hyperreflexia, difficulty speaking and swallowing [160].

Astrocytes in AxD demonstrate reactive morphology. These glial cells also remodel their biochemistry and secretome. In particular, astrocytes start to release pro-inflammatory factors TNF-α and IL-1β. In addition, astrocytes in AxD have reduced expression of glutamate plasmalemmal transporters, decreased activity of proteasomes, increased autophagy and increased activity of stress-activated protein kinase/c-Jun N-terminal kinase (JNK) pathway [131]. Multiple mechanisms by which pathological mutation of GFAP affects cellular functions have been considered. These include: (i) mutated GFAP through positive feedback loop inhibits proteasome function which activates JNK, and activated JNK directly further inhibits proteasome [204]; (ii) mutated GFAP inactivates one or more proteins by degradation of the Rosenthal fibres, where fragments of the small stress proteins, HSP27, αB crystalline, the 20S proteasome subunit, p-JNK, p62 and plectin, have been detected [131]. So far the AxD remains incurable, although several therapeutic strategies aimed at reducing GFAP expression are in development.

7.9 Neurodevelopmental Disorders

7.9.1 Autism Spectrum Disorders (ASD)

The class of autistics spectrum disorders (ASD) embraces numerous pathological conditions of heterogeneous clinical presentation and pathophysiology. They all, however, are manifested by deficits in social interactions and restrictive patterns of behaviours. Some of the autistic diseases are associated with intellectual deficits [162]. The underlying mechanism of ASDs is most likely associated with malformation of neuronal networks and aberrant neurotransmission in embryonic development caused by environmental and/or intrinsic factors [42, 80, 130, 174]. Formation of neuronal ensembles, synaptogenesis and synaptic elimination all critically depend on the performance of the astroglial cradle, which controls birth, life and death of synapses [213]. Astrocytes are responsible for neuroprotection and detoxification of harmful agents, including reactive oxygen species (through secreting antioxidants such as glutathione and ascorbic acid [30, 229]). Astrocytes tame excitotoxicity through glutamate uptake and they also control neurotransmitters catabolism and supply of neurotransmitter precursors [214]. In parallel, astrocytes are the main target for neurotoxic factors, such as heavy metals, which are linked to the aetiology of ASD [234]. Astrogliopathology in ASD has not been investigated in great details; there are some indications for astrogliosis [234], increased expression of connexin 43 and decreased expression of aquaporin 4 [71].

7.9.2 Down Syndrome

Down syndrome (DS), which is linked to the trisomy of chromosome 21, is characterised by mental retardation. In DS, the density of astrocytes is significantly reduced in the cortex [102] with decreased ability to properly support synaptogenesis and neuronal maturation [44].

7.9.3 Fragile X Syndrome

Expression of Fragile X mental retardation protein (FXMRP) causes a specific form of a neurodevelopmental disease manifested in ASD symptoms and mental disability, Fragile X syndrome that is also known as Martin–Bell syndrome or Escalante's syndrome [107]. Expression of FXMRP in astrocytes weakens their homeostatic function and neuroprotection in the in vitro experiments. Co-culturing healthy neurones with astrocytes harbouring FXMRP leads to abnormal neuronal dendritic morphology and reduced synaptic connectivity. In contrast, co-culturing FXMRP expressing neurones with healthy astrocytes prevents the development of abnormal dendritic morphology [97, 98].

7.9.4 Costello Syndrome

Costello syndrome (named so after its discoverer Jack Costello [48]) belongs to the family of the so-called RASopathies (where RAS stands for rat sarcoma) characterised by aberrant Ras signalling [205]. In this pathology, astroglial cells expressing a mutated *HRAS* (Harvey rat sarcoma viral oncogene homolog) gene demonstrate hyperactive Ras signalling, which accelerates differentiation and maturation of astrocytes, and leads to astroglial hypertrophy. This is also associated with pathological extracellular matrix and abnormal formation of neuronal networks that in turn causes cognitive and behavioural abnormalities [112].

7.10 Major Neuropsychiatric Diseases

7.10.1 Schizophrenia

In schizophrenia the wide spectrum of astroglial abnormalities is present. Conceptually, schizophrenia is associated with astroglial asthenia, atrophy, loss of homeostatic capabilities and arguably pathological remodelling, while reactive changes are not characteristic. Decrease in astroglial numbers, as well as dystrophic or swollen astroglial profiles, appear in various brain regions, including cortical and hippocampal structures [69, 163, 181, 226]. Astrocytes derived from human induced pluripotent stem cells obtained from schizophrenic patients and injected into mice, demonstrated atrophic morphology and loss of homeostatic functions [230].

Astrocytes in schizophrenia are characterised by a significant down-regulation of expression of several astroglia-specific molecules such as deiodinase type II, aquaporin-4, S100β, glutamine synthetase, plasmalemmal glutamate transporters and thrombospondin. These changes were the most prominent in the deep layers of the anterior cingulate gyrus, suggesting that a subset of astrocytes localised to specific cortical layers can be affected [231]. In the prefrontal cortex and hippocampus, a decrease in the expression of EAAT1/2 plasmalemmal glutamate transporters has been detected [16, 17, 146, 185], which may be linked to abnormalities in glutamatergic transmission. Genetic deletion of EAAT1 glutamate transporter promoted appearance of schizophrenia-like phenotypes manifested by locomotor hyperactivity and abnormal social behaviour [103, 104]. Astrocytes from rodent phencyclidine model of schizophrenia demonstrated a decrease in the expression of plasmalemmal cystine–glutamate exchanger Sxc^- [12], which modulates extrasynaptic concentration of glutamate and contributes to the biosynthesis of glutathione. Astrocytes may promote aberrant neurotransmission through synthesis and release of kynurenic acid that acts as an endogenous inhibitor of the NMDA receptor glycine binding site; kynurenic acid also blocks acetylcholine nicotinic receptors. The astroglial production of kynurenic acid is significantly up-regulated following brain infection with *Toxoplasma gondii*, which increases the risk of schizophrenia [183].

7.10.2 Mood Disorders

Astrogliopathology seems to be rather prominent in mood disorders [165, 178, 218]. The total number of glial cells and of astrocytes, in particular, is decreased in the orbitofrontal area and anterior cingulate, prefrontal, entorhinal and subgenual cortices, as well as the amygdala of the brains obtained from patients with major depression or bipolar disorder. [26, 49, 50, 149, 164]. In animals subjected to chronic stress, which instigates depressive phenotypes, GFAP expression and number of GFAP positive cells were reduced [28, 55]. Similarly, in models of attention deficit disorder and depression other astroglial markers, including aquaporin 4, astroglial connexins, astroglial plasmalemmal glutamate transporters and glutamine synthetase were all down-regulated [14, 21, 184].

Ablation of astrocytes in the medial prefrontal cortex of mice with the neuroglial toxin L-α-aminoadipic acid triggered an emergence of a depressive phenotype similar to that induced by chronic stress [13]. Exposure to chronic stress led to a down-regulation of astroglial expression of connexin 43 along with the reduction of gap junctional coupling in astrocytic syncytia. Pharmacological inhibition of astroglial connexon-based channels in the prefrontal cortex induced depressive behaviour manifested by anhedonia [201]. A similar phenotype was observed after inhibition of astroglial plasmalemmal glutamate transporters [18]. Chronic treatment with antidepressants directly affected astroglia, by increasing expression of a variety of receptors and transporters responsible for CNS homeostasis and limiting glutamate release [53, 60, 123, 171]. In conclusion, mood disorders are associated with astroglial degeneration and astroglial asthenia, which compromise brain homeostatic reserve and arguably synaptic transmission.

7.10.3 *Addictive Disorders*

Various nosological forms of addictive disorders are associated with astrogliopathies. Post-mortem analysis of the human brain samples revealed both astroglial reactivity with astroglial degeneration, and astroglial cell death with astroglial atrophy [9, 36, 72, 134, 145, 200, 225]. In the animal models of addiction with cocaine, methamphetamine and morphine, astroglial activation and increase in GFAP expression have been identified [25, 76, 84, 194]. In contrast, in the model of chronic alcoholism a decrease in GFAP expression and morphological atrophy of astrocytes were detected [75, 172]. In post-mortem tissues isolated from alcoholic sufferers, both hypertrophic GFAP positive astrocytes as well as areas with decreased GFAP expression and decreased density of astrocytes were described [52, 133].

The number of astrocytes is decreased in the prefrontal cortex of alcoholics [135]. A similar decrease in astroglial density and GFAP expression was detected in the prelimbic cortex of ethanol-preferring chronically alcoholic rats [133]. Additionally, a decrease in astrocyte density was observed in response to acute binge drinking in male (but not female) adult rats [111]. Ablation of astroglia with L-α-aminoadipic acid or uncoupling astroglial syncytia using a pharmacological inhibitor of connexin channels in the prefrontal cortex increases alcohol preference [132].

Addictive disorders are linked to astroglial plasmalemmal glutamate transport. Expression of EAAT2 as well as Sxc^- glutamate transporters is decreased in the context of alcoholism. Incidentally, total extracellular glutamate increases most likely due to an imbalance between glutamate uptake (EAAT2) and release (Sxc^-) [141, 169, 170]. Increase in the expression of EAAT2 by treatment with β-lactam antibiotic ceftriaxone decreased alcohol dependence [161, 179].

7.11 Epilepsy

In epilepsy, astrocytes undergo substantial pathological remodelling, which greatly affects their homeostatic capabilities and is linked to pathophysiology of this disease. In particular, the epileptic astroglial phenotype includes changes (mutations and/or expression levels) in ion channels, receptors and transporters [19, 196]. Abnormal electrophysiological characteristics have been observed in astrocytes isolated from patients with mesial temporal lobe epilepsy and associated sclerosis. These astrocytes, in addition, have severe impairment of intercellular coupling [19]. Astrocytes in sclerotic tissue up-regulated the expression of GFAP, suggesting thus their activation. Decrease in K^+ buffering seems to be the dominant feature of astroglial remodelling in epileptic brains, which results in an increase of extracellular K^+ concentration [124,

140]. Such an increase in extracellular K^+ can be sufficient to instigate seizures [206]. Abnormal astroglial K^+ buffering, at least in part, is linked to a significant down-regulation of inward rectifier $K_{ir}4.1$ channels. Here, decreases in $K_{ir}4.1$ current density and protein content have been found in astrocytes from the human sclerotic CA1 hippocampal area [24, 93, 94]. Genetic deletion of *KCNJ10* gene encoding $K_{ir}4.1$ channel specifically from astroglia resulted in impaired K^+ buffering, depolarisation of astrocytes, motor impairments and early death [59]. Other studies confirmed this finding by demonstrating that deletion of $K_{ir}4.1$ channels induces epileptiform symptoms in animals [196]. Mutations of *KCNJ10* gene in humans are associated with the development of SeSAME syndrome (also called EAST syndrome), an autosomal recessive disorder characterised by epilepsy, ataxia, sensorineural deafness, wasting renal tubulopathy, mental retardation and electrolyte imbalance [22, 182]. Whether the modifications of Na^+/K^+ ATPase (NKA), another critical component of astroglial K^+ buffering (NKA is primarily responsible for K^+ uptake, whereas $K_{ir}4.1$ channels for K^+ release and shuttling back to neurones [29, 114]) contribute to SeSAME, it remains to be explored. One of the forms of migraine, the familial hemiplegic migraine type 2, is however associated with loss-of-function mutation of astroglial specific α2 subunits of NKA [39]. Considering fundamental similarities of pathogenesis of migraine and epilepsy we may expect some abnormalities of astroglial NKA in the later pathology.

Epileptic astrocytes also demonstrate compromised glutamate uptake and homeostasis [51]. Deletion of the astroglial EAAT2 glutamate transporter results in an epileptiform phenotype with lethal spontaneous seizures, increased susceptibility to acute cortical injury and seizures after administration of sub-convulsive doses of pentylenetetrazole [202]. Similarly, seizures and epileptiform phenotype were triggered by pharmacological inhibition of EAAT by intracerebroventricular injections of DL-threo-beta-benzyloxyaspartate [58]. Down-regulation of glutamine synthetase was also linked to epilepsy through affecting inhibition in neuronal networks [151]. Animals subjected to long-lasting pharmacological blockade of glutamine synthetase demonstrated seizures [20, 224], whereas levels of glutamine synthetase were found to be significantly decreased in the human hippocampus and amygdala of patients with temporal lobe epilepsy [64]. Finally, loss-of-function mutations of glutamine synthetase induced severe seizures [87]. Astrocytes can also contribute to pathogenesis of epilepsy through anomalous adenosine homeostasis, resulting from modified expression of the astroglia-specific adenosine kinase, which is the key enzyme for adenosine turnover in the CNS [10, 23]. Expression of adenosine kinase is high in tissues from subjects with pharmacologically refractory temporal lobe epilepsy [10, 11, 128]. Increase in expression and activity of adenosine kinase diminishes the availability of adenosine, thus increasing neuronal network excitability and increasing probability of seizures [117, 118].

7.12 Epilogue

Since the inception of neurobiology, we have had a conceptual roller coaster ride in regards to the role of neuroglia in pathology of the brain. Two centuries ago our founding fathers of gliology had a clear vision on the active role of glia, i.e. glia is more than putty and has prominent roles in pathophysiology of the brain. Awkwardly, the twentieth century brought a different view where starring role has been solely played by neurones. This dominant neurono-centric approach has been challenged by the resurgence of neurogliopathology in the past 20 years. While we here presented the astrocyte-centric view of the brain pathology, we surely support the notion that it is the interaction between neurones and glia that underlies physiology and pathology of the brain. These two major cellular constituents interact, so that perturbing one will affect the other. Thus, only intellectually acceptable approaches to grapple with the management of the brain diseases will be those that have gestalt assets.

Acknowledgments VP's work is supported by a grant from the National Institute of General Medical Sciences of the National Institutes of Health (R01GM123971). VP is an Honorary Professor at University of Rijeka, Croatia.

References

1. Achucarro N (1910) Some pathological findings in the neuroglia and in the ganglion cells of the cortex in senile conditions. Bull Gov Hosp Insane 2:81–90
2. Alberdi E, Wyssenbach A, Alberdi M, Sanchez-Gomez MV, Cavaliere F, Rodriguez JJ, Verkhratsky A, Matute C (2013) Ca^{2+}-dependent endoplasmic reticulum stress correlates with astrogliosis in oligomeric amyloid beta-treated astrocytes and in a model of Alzheimer's disease. Aging Cell 12:292–302
3. Albrecht J, Zielinska M, Norenberg MD (2010) Glutamine as a mediator of ammonia neurotoxicity: A critical appraisal. Biochem Pharmacol 80:1303–1308
4. Alexander WS (1949) Progressive fibrinoid degeneration of fibrillary astrocytes associated with mental retardation in a hydrocephalic infant. Brain 72:373–381
5. Allen JW, Shanker G, Aschner M (2001) Methylmercury inhibits the in vitro uptake of the glutathione precursor, cystine, in astrocytes, but not in neurons. Brain Res 894:131–140
6. Alzheimer A (1910) Beiträge zur Kenntnis der pathologischen Neuroglia und ihrer Beziehungen zu den Abbauvorgängen im Nervengewebe. In: Nissl F, Alzheimer A (eds) Histologische und histopathologische Arbeiten über die Grosshirnrinde mit besonderer Berücksichtigung der pathologischen Anatomie der Geisteskrankheiten. Jena, Gustav Fischer, pp 401–562
7. Anderson MA, Burda JE, Ren Y, Ao Y, O'Shea TM, Kawaguchi R, Coppola G, Khakh BS, Deming TJ, Sofroniew MV (2016) Astrocyte scar formation aids central nervous system axon regeneration. Nature 532:195–200
8. Andriezen WL (1893) The neuroglia elements of the brain. Brit Med J 2:227–230
9. Armstrong V, Reichel CM, Doti JF, Crawford CA, McDougall SA (2004) Repeated amphetamine treatment causes a persistent elevation of glial fibrillary acidic protein in the caudate-putamen. Eur J Pharmacol 488:111–115
10. Aronica E, Sandau US, Iyer A, Boison D (2013) Glial adenosine kinase–a neuropathological marker of the epileptic brain. Neurochem Int 63:688–695

11. Aronica E, Zurolo E, Iyer A, de Groot M, Anink J, Carbonell C, van Vliet EA, Baayen JC, Boison D, Gorter JA (2011) Upregulation of adenosine kinase in astrocytes in experimental and human temporal lobe epilepsy. Epilepsia 52:1645–1655
12. Baker DA, Madayag A, Kristiansen LV, Meador-Woodruff JH, Haroutunian V, Raju I (2008) Contribution of cystine-glutamate antiporters to the psychotomimetic effects of phencyclidine. Neuropsychopharmacology 33:1760–1772
13. Banasr M, Duman RS (2008) Glial loss in the prefrontal cortex is sufficient to induce depressive-like behaviors. Biol Psychiatry 64:863–870
14. Barley K, Dracheva S, Byne W (2009) Subcortical oligodendrocyte- and astrocyte-associated gene expression in subjects with schizophrenia, major depression and bipolar disorder. Schizophr Res 112:54–64
15. Basler T, Meier-Hellmann A, Bredle D, Reinhart K (2002) Amino acid imbalance early in septic encephalopathy. Intensive Care Med 28:293–298
16. Bauer D, Gupta D, Harotunian V, Meador-Woodruff JH, McCullumsmith RE (2008) Abnormal expression of glutamate transporter and transporter interacting molecules in prefrontal cortex in elderly patients with schizophrenia. Schizophr Res 104:108–120
17. Bauer D, Haroutunian V, Meador-Woodruff JH, McCullumsmith RE (2010) Abnormal glycosylation of EAAT1 and EAAT2 in prefrontal cortex of elderly patients with schizophrenia. Schizophr Res 117:92–98
18. Bechtholt-Gompf AJ, Walther HV, Adams MA, Carlezon WA Jr, Ongur D, Cohen BM (2010) Blockade of astrocytic glutamate uptake in rats induces signs of anhedonia and impaired spatial memory. Neuropsychopharmacology 35:2049–2059
19. Bedner P, Dupper A, Huttmann K, Muller J, Herde MK, Dublin P, Deshpande T, Schramm J, Haussler U, Haas CA, Henneberger C, Theis M, Steinhauser C (2015) Astrocyte uncoupling as a cause of human temporal lobe epilepsy. Brain 138:1208–1222
20. Benedetti B, Matyash V, Kettenmann H (2011) Astrocytes control GABAergic inhibition of neurons in the mouse barrel cortex. J Physiol 589:1159–1172
21. Bernard R, Kerman IA, Thompson RC, Jones EG, Bunney WE, Barchas JD, Schatzberg AF, Myers RM, Akil H, Watson SJ (2011) Altered expression of glutamate signaling, growth factor, and glia genes in the locus coeruleus of patients with major depression. Mol Psychiatry 16:634–646
22. Bockenhauer D, Feather S, Stanescu HC, Bandulik S, Zdebik AA, Reichold M, Tobin J, Lieberer E, Sterner C, Landoure G, Arora R, Sirimanna T, Thompson D, Cross JH, van't Hoff W, Al Masri O, Tullus K, Yeung S, Anikster Y, Klootwijk E, Hubank M, Dillon MJ, Heitzmann D, Arcos-Burgos M, Knepper MA, Dobbie A, Gahl WA, Warth R, Sheridan E, Kleta R (2009) Epilepsy, ataxia, sensorineural deafness, tubulopathy, and KCNJ10 mutations. N Engl J Med 360, 1960–1970
23. Boison D, Aronica E (2015) Comorbidities in neurology: is adenosine the common link? Neuropharmacology 97:18–34
24. Bordey A, Sontheimer H (1998) Properties of human glial cells associated with epileptic seizure foci. Epilepsy Res 32:286–303
25. Bowers MS, Kalivas PW (2003) Forebrain astroglial plasticity is induced following withdrawal from repeated cocaine administration. Eur J Neurosci 17:1273–1278
26. Bowley MP, Drevets WC, Ongur D, Price JL (2002) Low glial numbers in the amygdala in major depressive disorder. Biol Psychiatry 52:404–412
27. Brambilla R, Bracchi-Ricard V, Hu WH, Frydel B, Bramwell A, Karmally S, Green EJ, Bethea JR (2005) Inhibition of astroglial nuclear factor κB reduces inflammation and improves functional recovery after spinal cord injury. J Exp Med 202:145–156
28. Braun K, Antemano R, Helmeke C, Buchner M, Poeggel G (2009) Juvenile separation stress induces rapid region- and layer-specific changes in S100ss- and glial fibrillary acidic protein-immunoreactivity in astrocytes of the rodent medial prefrontal cortex. Neuroscience 160:629–638
29. Breslin K, Wade JJ, Wong-Lin K, Harkin J, Flanagan B, Van Zalinge H, Hall S, Walker M, Verkhratsky A, McDaid L (2018) Potassium and sodium microdomains in thin astroglial processes: A computational model study. PLoS Comput Biol 14:e1006151

30. Bridges RJ, Natale NR, Patel SA (2012) System xc⁻ cystine/glutamate antiporter: an update on molecular pharmacology and roles within the CNS. Br J Pharmacol 165:20–34
31. Brusilow SW, Koehler RC, Traystman RJ, Cooper AJ (2010) Astrocyte glutamine synthetase: importance in hyperammonemic syndromes and potential target for therapy. Neurotherapeutics 7:452–470
32. Burda JE, Bernstein AM, Sofroniew MV (2016) Astrocyte roles in traumatic brain injury. Exp Neurol 275(Pt 3):305–315
33. Burda JE, Sofroniew MV (2014) Reactive gliosis and the multicellular response to CNS damage and disease. Neuron 81:229–248
34. Burnstock G, Fredholm BB, Verkhratsky A (2011) Adenosine and ATP receptors in the brain. Curr Top Med Chem 11:973–1011
35. Butterworth RF (2011) Hepatic encephalopathy: a central neuroinflammatory disorder? Hepatology 53:1372–1376
36. Büttner A, Weis S (2006) Neuropathological alterations in drug abusers. The involvement of neurons, glial and vascular systems. Forensic Sci Med Pathol 2:115–126
37. Campos de Carvalho AC, Roy C, Hertzberg EL, Tanowitz HB, Kessler JA, Weiss LM, Wittner M, Dermietzel R, Gao Y, Spray DC (1998) Gap junction disappearance in astrocytes and leptomeningeal cells as a consequence of protozoan infection. Brain Res 790:304–314
38. Candelario-Jalil E, Taheri S, Yang Y, Sood R, Grossetete M, Estrada EY, Fiebich BL, Rosenberg GA (2007) Cyclooxygenase inhibition limits blood-brain barrier disruption following intracerebral injection of tumor necrosis factor-alpha in the rat. J Pharmacol Exp Ther 323:488–498
39. Capuani C, Melone M, Tottene A, Bragina L, Crivellaro G, Santello M, Casari G, Conti F, Pietrobon D (2016) Defective glutamate and K^+ clearance by cortical astrocytes in familial hemiplegic migraine type 2. EMBO Mol Med 8:967–986
40. Carpentier PA, Duncan DS, Miller SD (2008) Glial toll-like receptor signaling in central nervous system infection and autoimmunity. Brain Behav Immun 22:140–147
41. Cekanaviciute E, Buckwalter MS (2016) Astrocytes: integrative regulators of neuroinflammation in stroke and other neurological diseases. Neurotherapeutics 13:685–701
42. Cellot G, Cherubini E (2014) GABAergic signaling as therapeutic target for autism spectrum disorders. Front Pediatr 2:70
43. Chauhan VS, Sterka DG Jr, Furr SR, Young AB, Marriott I (2009) NOD2 plays an important role in the inflammatory responses of microglia and astrocytes to bacterial CNS pathogens. Glia 57:414–423
44. Chen C, Jiang P, Xue H, Peterson SE, Tran HT, McCann AE, Parast MM, Li S, Pleasure DE, Laurent LC, Loring JF, Liu Y, Deng W (2014) Role of astroglia in Down's syndrome revealed by patient-derived human-induced pluripotent stem cells. Nat Commun 5:4430
45. Chepkova AN, Sergeeva OA, Gorg B, Haas HL, Klocker N, Haussinger D (2017) Impaired novelty acquisition and synaptic plasticity in congenital hyperammonemia caused by hepatic glutamine synthetase deficiency. Sci Rep 7:40190
46. Churchill MJ, Wesselingh SL, Cowley D, Pardo CA, McArthur JC, Brew BJ, Gorry PR (2009) Extensive astrocyte infection is prominent in human immunodeficiency virus-associated dementia. Ann Neurol 66:253–258
47. Combes V, Guillemin GJ, Chan-Ling T, Hunt NH, Grau GE (2012) The crossroads of neuroinflammation in infectious diseases: endothelial cells and astrocytes. Trends Parasitol 28:311–319
48. Costello JM (1977) A new syndrome: mental subnormality and nasal papillomata. Aust Paediatr J 13:114–118
49. Cotter D, Mackay D, Chana G, Beasley C, Landau S, Everall IP (2002) Reduced neuronal size and glial cell density in area 9 of the dorsolateral prefrontal cortex in subjects with major depressive disorder. Cereb Cortex 12:386–394
50. Cotter D, Mackay D, Landau S, Kerwin R, Everall I (2001) Reduced glial cell density and neuronal size in the anterior cingulate cortex in major depressive disorder. Arch Gen Psychiatry 58:545–553

51. Coulter DA, Eid T (2012) Astrocytic regulation of glutamate homeostasis in epilepsy. Glia 60:1215–1226
52. Cullen KM, Halliday GM (1994) Chronic alcoholics have substantial glial pathology in the forebrain and diencephalon. Alcohol Alcohol Suppl 2:253–257
53. Czeh B, Di Benedetto B (2013) Antidepressants act directly on astrocytes: evidences and functional consequences. Eur Neuropsychopharmacol 23:171–185
54. Czeh B, Nagy SA (2018) Clinical findings documenting cellular and molecular abnormalities of glia in depressive disorders. Front Mol Neurosci 11:56
55. Czeh B, Simon M, Schmelting B, Hiemke C, Fuchs E (2006) Astroglial plasticity in the hippocampus is affected by chronic psychosocial stress and concomitant fluoxetine treatment. Neuropsychopharmacology 31:1616–1626
56. Dantzer R, Kelley KW (2007) Twenty years of research on cytokine-induced sickness behavior. Brain Behav Immun 21:153–160
57. De Keyser J, Mostert JP, Koch MW (2008) Dysfunctional astrocytes as key players in the pathogenesis of central nervous system disorders. J Neurol Sci 267:3–16
58. Demarque M, Villeneuve N, Manent JB, Becq H, Represa A, Ben-Ari Y, Aniksztejn L (2004) Glutamate transporters prevent the generation of seizures in the developing rat neocortex. J Neurosci 24:3289–3294
59. Djukic B, Casper KB, Philpot BD, Chin LS, McCarthy KD (2007) Conditional knock-out of $K_{ir}4.1$ leads to glial membrane depolarization, inhibition of potassium and glutamate uptake, and enhanced short-term synaptic potentiation. J Neurosci 27:11354–11365
60. Dong L, Li B, Verkhratsky A, Peng L (2015) Cell type-specific in vivo expression of genes encoding signalling molecules in the brain in response to chronic mild stress and chronic treatment with fluoxetine. Psychopharmacology (Berl)
61. Dringen R, Gutterer JM, Hirrlinger J (2000) Glutathione metabolism in brain metabolic interaction between astrocytes and neurons in the defense against reactive oxygen species. Eur J Biochem 267:4912–4916
62. Dringen R, Hirrlinger J (2003) Glutathione pathways in the brain. Biol Chem 384:505–516
63. Dvoriantchikova G, Barakat D, Brambilla R, Agudelo C, Hernandez E, Bethea JR, Shestopalov VI, Ivanov D (2009) Inactivation of astroglial NF-kappa B promotes survival of retinal neurons following ischemic injury. Eur J Neurosci 30:175–185
64. Eid T, Tu N, Lee TS, Lai JC (2013) Regulation of astrocyte glutamine synthetase in epilepsy. Neurochem Int 63:670–681
65. El-Hage N, Podhaizer EM, Sturgill J, Hauser KF (2011) Toll-like receptor expression and activation in astroglia: differential regulation by HIV-1 Tat, gp120, and morphine. Immunol Invest 40:498–522
66. Esen N, Shuffield D, Syed MM, Kielian T (2007) Modulation of connexin expression and gap junction communication in astrocytes by the gram-positive bacterium S. aureus. Glia 55:104–117
67. Esen N, Tanga FY, DeLeo JA, Kielian T (2004) Toll-like receptor 2 (TLR2) mediates astrocyte activation in response to the Gram-positive bacterium Staphylococcus aureus. J Neurochem 88:746–758
68. Eugenin EA, Berman JW (2007) Gap junctions mediate human immunodeficiency virus-bystander killing in astrocytes. J Neurosci 27:12844–12850
69. Falkai P, Bogerts B (1986) Cell loss in the hippocampus of schizophrenics. Eur Arch Psychiatry Neurol Sci 236:154–161
70. Farina C, Aloisi F, Meinl E (2007) Astrocytes are active players in cerebral innate immunity. Trends Immunol 28:138–145
71. Fatemi SH, Folsom TD, Reutiman TJ, Lee S (2008) Expression of astrocytic markers aquaporin 4 and connexin 43 is altered in brains of subjects with autism. Synapse 62:501–507
72. Fattore L, Puddu MC, Picciau S, Cappai A, Fratta W, Serra GP, Spiga S (2002) Astroglial in vivo response to cocaine in mouse dentate gyrus: a quantitative and qualitative analysis by confocal microscopy. Neuroscience 110:1–6

73. Felipo V (2013) Hepatic encephalopathy: effects of liver failure on brain function. Nat Rev Neurosci 14:851–858
74. Ferrer I (2018) Astrogliopathy in tauopathies. Neuroglia 1:126–150
75. Franke H (1995) Influence of chronic alcohol treatment on the GFAP-immunoreactivity in astrocytes of the hippocampus in rats. Acta Histochem 97:263–271
76. Friend DM, Keefe KA (2013) Glial reactivity in resistance to methamphetamine-induced neurotoxicity. J Neurochem 125:566–574
77. Frommann C (1878) Untersuchungen über die Gewebsveränderungen bei der Multiplen Sklerose des Gehirns und Rückenmarks. Verlag von Gustav Fischer, Jena
78. Garcia AD, Petrova R, Eng L, Joyner AL (2010) Sonic hedgehog regulates discrete populations of astrocytes in the adult mouse forebrain. J Neurosci 30:13597–13608
79. Giaume C, Kirchhoff F, Matute C, Reichenbach A, Verkhratsky A (2007) Glia: the fulcrum of brain diseases. Cell Death Differ 14:1324–1335
80. Giovedi S, Corradi A, Fassio A, Benfenati F (2014) Involvement of synaptic genes in the pathogenesis of autism spectrum disorders: the case of synapsins. Front Pediatr 2:94
81. Gleichman AJ, Carmichael ST (2014) Astrocytic therapies for neuronal repair in stroke. Neurosci Lett 565:47–52
82. Gorg B, Morwinsky A, Keitel V, Qvartskhava N, Schror K, Haussinger D (2010) Ammonia triggers exocytotic release of L-glutamate from cultured rat astrocytes. Glia 58:691–705
83. Gray LR, Turville SG, Hitchen TL, Cheng WJ, Ellett AM, Salimi H, Roche MJ, Wesselingh SL, Gorry PR, Churchill MJ (2014) HIV-1 entry and trans-infection of astrocytes involves CD81 vesicles. PLoS ONE 9:e90620
84. Guilarte TR, Nihei MK, McGlothan JL, Howard AS (2003) Methamphetamine-induced deficits of brain monoaminergic neuronal markers: distal axotomy or neuronal plasticity. Neuroscience 122:499–513
85. Gunnarson E, Axehult G, Baturina G, Zelenin S, Zelenina M, Aperia A (2005) Lead induces increased water permeability in astrocytes expressing aquaporin 4. Neuroscience 136:105–114
86. Haack N, Dublin P, Rose CR (2014) Dysbalance of astrocyte calcium under hyperammonemic conditions. PLoS ONE 9:e105832
87. Haberle J, Shahbeck N, Ibrahim K, Hoffmann GF, Ben-Omran T (2011) Natural course of glutamine synthetase deficiency in a 3 year old patient. Mol Genet Metab 103:89–91
88. Harvey BK, Airavaara M, Hinzman J, Wires EM, Chiocco MJ, Howard DB, Shen H, Gerhardt G, Hoffer BJ, Wang Y (2011) Targeted over-expression of glutamate transporter 1 (GLT-1) reduces ischemic brain injury in a rat model of stroke. PLoS ONE 6:e22135
89. Heming N, Mazeraud A, Verdonk F, Bozza FA, Chretien F, Sharshar T (2017) Neuroanatomy of sepsis-associated encephalopathy. Crit Care 21:65
90. Heneka MT, Rodriguez JJ, Verkhratsky A (2010) Neuroglia in neurodegeneration. Brain Res Rev 63:189–211
91. Herman M, Ciancanelli M, Ou YH, Lorenzo L, Klaudel-Dreszler M, Pauwels E, Sancho-Shimizu V, Perez de Diego R, Abhyankar A, Israelsson E, Guo Y, Cardon A, Rozenberg F, Lebon P, Tardieu M, Heropolitanska-Pliszka E, Chaussabel D, White MA, Abel L, Zhang SY, Casanova JL (2012) Heterozygous TBK1 mutations impair TLR3 immunity and underlie herpes simplex encephalitis of childhood. J Exp Med 209:1567–1582
92. Herrmann JE, Imura T, Song B, Qi J, Ao Y, Nguyen TK, Korsak RA, Takeda K, Akira S, Sofroniew MV (2008) STAT3 is a critical regulator of astrogliosis and scar formation after spinal cord injury. J Neurosci 28:7231–7243
93. Heuser K, Nagelhus EA, Tauboll E, Indahl U, Berg PR, Lien S, Nakken S, Gjerstad L, Ottersen OP (2010) Variants of the genes encoding AQP4 and $K_{ir}4.1$ are associated with subgroups of patients with temporal lobe epilepsy. Epilepsy Res 88:55–64
94. Hinterkeuser S, Schroder W, Hager G, Seifert G, Blumcke I, Elger CE, Schramm J, Steinhauser C (2000) Astrocytes in the hippocampus of patients with temporal lobe epilepsy display changes in potassium conductances. Eur J Neurosci 12:2087–2096
95. Hol EM, Pekny M (2015) Glial fibrillary acidic protein (GFAP) and the astrocyte intermediate filament system in diseases of the central nervous system. Curr Opin Cell Biol 32:121–130

96. Iovino F, Orihuela CJ, Moorlag HE, Molema G, Bijlsma JJ (2013) Interactions between blood-borne Streptococcus pneumoniae and the blood-brain barrier preceding meningitis. PLoS ONE 8:e68408
97. Jacobs S, Doering LC (2010) Astrocytes prevent abnormal neuronal development in the fragile x mouse. J Neurosci 30:4508–4514
98. Jacobs S, Nathwani M, Doering LC (2010) Fragile X astrocytes induce developmental delays in dendrite maturation and synaptic protein expression. BMC Neurosci 11:132
99. Jassam YN, Izzy S, Whalen M, McGavern DB, El Khoury J (2017) Neuroimmunology of traumatic brain injury: time for a paradigm shift. Neuron 95:1246–1265
100. Kadoi Y, Saito S (1996) An alteration in the gamma-aminobutyric acid receptor system in experimentally induced septic shock in rats. Crit Care Med 24:298–305
101. Kanemaru K, Kubota J, Sekiya H, Hirose K, Okubo Y, Iino M (2013) Calcium-dependent N-cadherin up-regulation mediates reactive astrogliosis and neuroprotection after brain injury. Proc Natl Acad Sci U S A 110:11612–11617
102. Karlsen AS, Pakkenberg B (2011) Total numbers of neurons and glial cells in cortex and basal ganglia of aged brains with Down syndrome–a stereological study. Cereb Cortex 21:2519–2524
103. Karlsson RM, Tanaka K, Heilig M, Holmes A (2008) Loss of glial glutamate and aspartate transporter (excitatory amino acid transporter 1) causes locomotor hyperactivity and exaggerated responses to psychotomimetics: rescue by haloperidol and metabotropic glutamate 2/3 agonist. Biol Psychiatry 64:810–814
104. Karlsson RM, Tanaka K, Saksida LM, Bussey TJ, Heilig M, Holmes A (2009) Assessment of glutamate transporter GLAST (EAAT1)-deficient mice for phenotypes relevant to the negative and executive/cognitive symptoms of schizophrenia. Neuropsychopharmacology 34:1578–1589
105. Kelly T, Kafitz KW, Roderigo C, Rose CR (2009) Ammonium-evoked alterations in intracellular sodium and pH reduce glial glutamate transport activity. Glia 57:921–934
106. Kelly T, Rose CR (2010) Ammonium influx pathways into astrocytes and neurones of hippocampal slices. J Neurochem 115:1123–1136
107. Kidd SA, Lachiewicz A, Barbouth D, Blitz RK, Delahunty C, McBrien D, Visootsak J, Berry-Kravis E (2014) Fragile X syndrome: a review of associated medical problems. Pediatrics 134:995–1005
108. Kigerl KA, de Rivero Vaccari JP, Dietrich WD, Popovich PG, Keane RW (2014) Pattern recognition receptors and central nervous system repair. Exp Neurol 258:5–16
109. Kirischuk S, Parpura V, Verkhratsky A (2012) Sodium dynamics: another key to astroglial excitability? Trends Neurosci 35:497–506
110. Klein RS, Hunter CA (2017) Protective and pathological immunity during central nervous system infections. Immunity 46:891–909
111. Koss WA, Sadowski RN, Sherrill LK, Gulley JM, Juraska JM (2012) Effects of ethanol during adolescence on the number of neurons and glia in the medial prefrontal cortex and basolateral amygdala of adult male and female rats. Brain Res 1466:24–32
112. Krencik R, Hokanson KC, Narayan AR, Dvornik J, Rooney GE, Rauen KA, Weiss LA, Rowitch DH, Ullian EM (2015) Dysregulation of astrocyte extracellular signaling in Costello syndrome. Sci Transl Med 7:286ra266
113. Lanciotti A, Brignone MS, Bertini E, Petrucci TC, Aloisi F, Ambrosini E (2013) Astrocytes: emerging stars in leukodystrophy pathogenesis. Transl Neurosci 4
114. Larsen BR, Assentoft M, Cotrina ML, Hua SZ, Nedergaard M, Kaila K, Voipio J, MacAulay N (2014) Contributions of the Na^+/K^+-ATPase, NKCC1, and $K_{ir}4.1$ to hippocampal K^+ clearance and volume responses. Glia 62:608–622
115. Lauritzen M, Dreier JP, Fabricius M, Hartings JA, Graf R, Strong AJ (2011) Clinical relevance of cortical spreading depression in neurological disorders: migraine, malignant stroke, subarachnoid and intracranial hemorrhage, and traumatic brain injury. J Cereb Blood Flow Metab 31:17–35

116. Li L, Lundkvist A, Andersson D, Wilhelmsson U, Nagai N, Pardo AC, Nodin C, Stahlberg A, Aprico K, Larsson K, Yabe T, Moons L, Fotheringham A, Davies I, Carmeliet P, Schwartz JP, Pekna M, Kubista M, Blomstrand F, Maragakis N, Nilsson M, Pekny M (2008) Protective role of reactive astrocytes in brain ischemia. J Cereb Blood Flow Metab 28:468–481
117. Li T, Lan JQ, Boison D (2008) Uncoupling of astrogliosis from epileptogenesis in adenosine kinase (ADK) transgenic mice. Neuron Glia Biol 4:91–99
118. Li T, Quan Lan J, Fredholm BB, Simon RP, Boison D (2007) Adenosine dysfunction in astrogliosis: cause for seizure generation? Neuron Glia Biol 3:353–366
119. Liang C, Du T, Zhou J, Verkhratsky A, Peng L (2014) Ammonium increases Ca^{2+} signalling and up-regulates expression of TRPC1 gene in astrocytes in primary cultures and in the in vivo brain. Neurochem Res 39:2127–2135
120. Liddelow SA, Barres BA (2017) Reactive astrocytes: production, function, and therapeutic potential. Immunity 46:957–967
121. Liddelow SA, Guttenplan KA, Clarke LE, Bennett FC, Bohlen CJ, Schirmer L, Bennett ML, Munch AE, Chung WS, Peterson TC, Wilton DK, Frouin A, Napier BA, Panicker N, Kumar M, Buckwalter MS, Rowitch DH, Dawson VL, Dawson TM, Stevens B, Barres BA (2017) Neurotoxic reactive astrocytes are induced by activated microglia. Nature 541:481–487
122. Liu X, Chauhan VS, Young AB, Marriott I (2010) NOD2 mediates inflammatory responses of primary murine glia to Streptococcus pneumoniae. Glia 58:839–847
123. Liu Z, Song D, Yan E, Verkhratsky A, Peng L (2015) Chronic treatment with anti-bipolar drugs suppresses glutamate release from astroglial cultures. Amino Acids 47:1045–1051
124. Lothman EW, Somjen GG (1976) Functions of primary afferents and responses of extracellular K+ during spinal epileptiform seizures. Electroencephalogr Clin Neurophysiol 41:253–267
125. Makar TK, Nedergaard M, Preuss A, Gelbard AS, Perumal AS, Cooper AJ (1994) Vitamin E, ascorbate, glutathione, glutathione disulfide, and enzymes of glutathione metabolism in cultures of chick astrocytes and neurons: evidence that astrocytes play an important role in antioxidative processes in the brain. J Neurochem 62:45–53
126. Marcaggi P, Coles JA (2001) Ammonium in nervous tissue: transport across cell membranes, fluxes from neurons to glial cells, and role in signalling. Prog Neurobiol 64:157–183
127. Marin-Valencia I, Roe CR, Pascual JM (2010). Pyruvate carboxylase deficiency: mechanisms, mimics and anaplerosis. Mol Genet Metab 1:9–17
128. Masino SA, Li T, Theofilas P, Sandau US, Ruskin DN, Fredholm BB, Geiger JD, Aronica E, Boison D (2011) A ketogenic diet suppresses seizures in mice through adenosine A(1) receptors. J Clin Invest 121:2679–2683
129. Mc AD, Araki S (1958) Minamata disease: an unusual neurological disorder caused by contaminated fish. Lancet 2:629–631
130. McGinnis WR (2004) Oxidative stress in autism. Altern Ther Health Med 10:22–36; quiz 37, 92
131. Messing A, Brenner M, Feany MB, Nedergaard M, Goldman JE (2012) Alexander disease. J Neurosci 32:5017–5023
132. Miguel-Hidalgo J, Shoyama Y, Wanzo V (2009) Infusion of gliotoxins or a gap junction blocker in the prelimbic cortex increases alcohol preference in Wistar rats. J Psychopharmacol 23:550–557
133. Miguel-Hidalgo JJ (2005) Lower packing density of glial fibrillary acidic protein-immunoreactive astrocytes in the prelimbic cortex of alcohol-naive and alcohol-drinking alcohol-preferring rats as compared with alcohol-nonpreferring and Wistar rats. Alcohol Clin Exp Res 29:766–772
134. Miguel-Hidalgo JJ (2009) The role of glial cells in drug abuse. Curr Drug Abuse Rev 2:76–82
135. Miguel-Hidalgo JJ, Overholser JC, Meltzer HY, Stockmeier CA, Rajkowska G (2006) Reduced glial and neuronal packing density in the orbitofrontal cortex in alcohol dependence and its relationship with suicide and duration of alcohol dependence. Alcohol Clin Exp Res 30:1845–1855
136. Minami T, Okazaki J, Kawabata A, Kawaki H, Okazaki Y, Tohno Y (1998) Roles of nitric oxide and prostaglandins in the increased permeability of the blood-brain barrier caused by lipopolysaccharide. Environ Toxicol Pharmacol 5:35–41

137. Minkiewicz J, de Rivero Vaccari JP, Keane RW (2013) Human astrocytes express a novel NLRP2 inflammasome. Glia 61:1113–1121
138. Miyajima H (2003) Aceruloplasminemia, an iron metabolic disorder. Neuropathology 23:345–350
139. Montana V, Verkhratsky A, Parpura V (2014) Pathological role for exocytotic glutamate release from astrocytes in hepatic encephalopathy. Curr Neuropharmacol 12:324–333
140. Moody WJ, Futamachi KJ, Prince DA (1974) Extracellular potassium activity during epileptogenesis. Exp Neurol 42:248–263
141. Moussawi K, Riegel A, Nair S, Kalivas PW (2011) Extracellular glutamate: functional compartments operate in different concentration ranges. Front Syst Neurosci 5:94
142. Nedergaard M. (1996). Spreading depression as a contributor to ischemic brain damage. Adv Neurol 71:75–83; discussion 83–74
143. Norenberg MD (1987) The role of astrocytes in hepatic encephalopathy. Neurochem Pathol 6:13–33
144. Obara-Michlewska M, Ruszkiewicz J, Zielinska M, Verkhratsky A, Albrecht J (2014) Astroglial NMDA receptors inhibit expression of K4.1 channels in glutamate-overexposed astrocytes in vitro and in the brain of rats with acute liver failure. Neurochem Int
145. Oehmichen M, Meissner C, Reiter A, Birkholz M (1996) Neuropathology in non-human immunodeficiency virus-infected drug addicts: hypoxic brain damage after chronic intravenous drug abuse. Acta Neuropathol 91:642–646
146. Ohnuma T, Tessler S, Arai H, Faull RL, McKenna PJ, Emson PC (2000) Gene expression of metabotropic glutamate receptor 5 and excitatory amino acid transporter 2 in the schizophrenic hippocampus. Brain Res Mol Brain Res 85:24–31
147. Oide T, Yoshida K, Kaneko K, Ohta M, Arima K (2006) Iron overload and antioxidative role of perivascular astrocytes in aceruloplasminemia. Neuropathol Appl Neurobiol 32:170–176
148. Okada S, Nakamura M, Katoh H, Miyao T, Shimazaki T, Ishii K, Yamane J, Yoshimura A, Iwamoto Y, Toyama Y, Okano H (2006) Conditional ablation of Stat3 or Socs3 discloses a dual role for reactive astrocytes after spinal cord injury. Nat Med 12:829–834
149. Ongur D, Drevets WC, Price JL (1998) Glial reduction in the subgenual prefrontal cortex in mood disorders. Proc Natl Acad Sci U S A 95:13290–13295
150. Orellana JA, Saez JC, Bennett MV, Berman JW, Morgello S, Eugenin EA (2014) HIV increases the release of dickkopf-1 protein from human astrocytes by a Cx43 hemichannel-dependent mechanism. J Neurochem 128:752–763
151. Ortinski PI, Dong J, Mungenast A, Yue C, Takano H, Watson DJ, Haydon PG, Coulter DA (2010) Selective induction of astrocytic gliosis generates deficits in neuronal inhibition. Nat Neurosci 13:584–591
152. Parpura V, Heneka MT, Montana V, Oliet SH, Schousboe A, Haydon PG, Stout RF Jr, Spray DC, Reichenbach A, Pannicke T, Pekny M, Pekna M, Zorec R, Verkhratsky A (2012) Glial cells in (patho)physiology. J Neurochem 121:4–27
153. Patel SC, Suresh S, Kumar U, Hu CY, Cooney A, Blanchette-Mackie EJ, Neufeld EB, Patel RC, Brady RO, Patel YC, Pentchev PG, Ong WY (1999) Localization of Niemann-Pick C1 protein in astrocytes: implications for neuronal degeneration in Niemann-Pick type C disease. Proc Natl Acad Sci U S A 96:1657–1662
154. Pekny M, Johansson CB, Eliasson C, Stakeberg J, Wallen A, Perlmann T, Lendahl U, Betsholtz C, Berthold CH, Frisen J (1999) Abnormal reaction to central nervous system injury in mice lacking glial fibrillary acidic protein and vimentin. J Cell Biol 145:503–514
155. Pekny M, Pekna M (2014) Astrocyte reactivity and reactive astrogliosis: costs and benefits. Physiol Rev 94:1077–1098
156. Pekny M, Pekna M, Messing A, Steinhauser C, Lee JM, Parpura V, Hol EM, Sofroniew MV, Verkhratsky A (2016) Astrocytes: a central element in neurological diseases. Acta Neuropathol 131:323–345
157. Pekny M, Wilhelmsson U, Pekna M (2014) The dual role of astrocyte activation and reactive gliosis. Neurosci Lett 565:30–38

158. Piva S, McCreadie VA, Latronico N, Name J (2015) Neuroinflammation in sepsis: sepsis associated delirium. Cardiovasc Hematol Disord Drug Targ 1:10–18
159. Potokar M, Korva M, Jorgacevski J, Avsic-Zupanc T, Zorec R (2014) Tick-borne encephalitis virus infects rat astrocytes but does not affect their viability. PLoS ONE 9:e86219
160. Prust M, Wang J, Morizono H, Messing A, Brenner M, Gordon E, Hartka T, Sokohl A, Schiffmann R, Gordish-Dressman H, Albin R, Amartino H, Brockman K, Dinopoulos A, Dotti MT, Fain D, Fernandez R, Ferreira J, Fleming J, Gill D, Griebel M, Heilstedt H, Kaplan P, Lewis D, Nakagawa M, Pedersen R, Reddy A, Sawaishi Y, Schneider M, Sherr E, Takiyama Y, Wakabayashi K, Gorospe JR, Vanderver A (2011) GFAP mutations, age at onset, and clinical subtypes in Alexander disease. Neurology 77:1287–1294
161. Qrunfleh AM, Alazizi A, Sari Y (2013) Ceftriaxone, a beta-lactam antibiotic, attenuates relapse-like ethanol-drinking behavior in alcohol-preferring rats. J Psychopharmacol 27:541–549
162. Quaak I, Brouns MR, Van de Bor M (2013) The dynamics of autism spectrum disorders: how neurotoxic compounds and neurotransmitters interact. Int J Environ Res Public Health 10:3384–3408
163. Rajkowska G, Miguel-Hidalgo JJ, Makkos Z, Meltzer H, Overholser J, Stockmeier C (2002) Layer-specific reductions in GFAP-reactive astroglia in the dorsolateral prefrontal cortex in schizophrenia. Schizophr Res 57:127–138
164. Rajkowska G, Miguel-Hidalgo JJ, Wei J, Dilley G, Pittman SD, Meltzer HY, Overholser JC, Roth BL, Stockmeier CA (1999) Morphometric evidence for neuronal and glial prefrontal cell pathology in major depression. Biol Psychiatry 45:1085–1098
165. Rajkowska G, Stockmeier CA (2013) Astrocyte pathology in major depressive disorder: insights from human postmortem brain tissue. Curr Drug Targets 14:1225–1236
166. Rangroo Thrane V, Thrane AS, Wang F, Cotrina ML, Smith NA, Chen M, Xu Q, Kang N, Fujita T, Nagelhus EA, Nedergaard M (2013) Ammonia triggers neuronal disinhibition and seizures by impairing astrocyte potassium buffering. Nat Med 19:1643–1648
167. Rao VL, Dogan A, Todd KG, Bowen KK, Kim BT, Rothstein JD, Dempsey RJ (2001) Antisense knockdown of the glial glutamate transporter GLT-1, but not the neuronal glutamate transporter EAAC1, exacerbates transient focal cerebral ischemia-induced neuronal damage in rat brain. J Neurosci 21:1876–1883
168. Reinert LS, Harder L, Holm CK, Iversen MB, Horan KA, Dagnaes-Hansen F, Ulhoi BP, Holm TH, Mogensen TH, Owens T, Nyengaard JR, Thomsen AR, Paludan SR (2012) TLR3 deficiency renders astrocytes permissive to herpes simplex virus infection and facilitates establishment of CNS infection in mice. J Clin Invest 122:1368–1376
169. Reissner KJ, Kalivas PW (2010) Using glutamate homeostasis as a target for treating addictive disorders. Behav Pharmacol 21:514–522
170. Reissner KJ, Kalivas PW (2014) Emerging roles for glial pathology in addiction. In: Parpura V, Verkhratsky A (eds) Pathological potential of neuroglia: possible new targets for medical intervention. Sprinegr, New York, Heidelberg, Dordrecht, London, pp 397–418
171. Ren J, Song D, Bai Q, Verkhratsky A, Peng L (2015) Fluoxetine induces alkalinization of astroglial cytosol through stimulation of sodium-hydrogen exchanger 1: dissection of intracellular signaling pathways. Front Cell Neurosci 9:61
172. Rintala J, Jaatinen P, Kiianmaa K, Riikonen J, Kemppainen O, Sarviharju M, Hervonen A (2001) Dose-dependent decrease in glial fibrillary acidic protein-immunoreactivity in rat cerebellum after lifelong ethanol consumption. Alcohol 23:1–8
173. Rodriguez JJ, Butt AM, Gardenal E, Parpura V, Verkhratsky A (2016) Complex and differential glial responses in Alzheimer's disease and ageing. Curr Alzheimer Res 13:343–358
174. Rojas DC (2014) The role of glutamate and its receptors in autism and the use of glutamate receptor antagonists in treatment. J Neural Transm 121:891–905
175. Rose CF, Verkhratsky A, Parpura V (2013) Astrocyte glutamine synthetase: pivotal in health and disease. Biochem Soc Trans 41:1518–1524
176. Rosenbaum AI, Maxfield FR (2011) Niemann-Pick type C disease: molecular mechanisms and potential therapeutic approaches. J Neurochem 116:789–795

177. Rossi D, Brambilla L, Valori CF, Roncoroni C, Crugnola A, Yokota T, Bredesen DE, Volterra A (2008) Focal degeneration of astrocytes in amyotrophic lateral sclerosis. Cell Death Differ 15:1691–1700
178. Sanacora G, Banasr M (2013) From pathophysiology to novel antidepressant drugs: glial contributions to the pathology and treatment of mood disorders. Biol Psychiatry 73:1172–1179
179. Sari Y, Sreemantula SN, Lee MR, Choi DS (2013) Ceftriaxone treatment affects the levels of GLT1 and ENT1 as well as ethanol intake in alcohol-preferring rats. J Mol Neurosci 51:779–787
180. Sayre NL, Chen Y, Sifuentes M, Stoveken B, Lechleiter JD (2014) Purinergic receptor stimulation decreases ischemic brain damage by energizing astrocyte mitochondria. In: Parpura V, Verkhratsky A (eds) Glutamate and ATP at the interface of metabolism and signaling in the brain. Springer, Cham, Heidelberg, New York, Dordrecht, London, pp 121–150
181. Schmitt A, Steyskal C, Bernstein HG, Schneider-Axmann T, Parlapani E, Schaeffer EL, Gattaz WF, Bogerts B, Schmitz C, Falkai P (2009) Stereologic investigation of the posterior part of the hippocampus in schizophrenia. Acta Neuropathol 117:395–407
182. Scholl UI, Choi M, Liu T, Ramaekers VT, Hausler MG, Grimmer J, Tobe SW, Farhi A, Nelson-Williams C, Lifton RP (2009) Seizures, sensorineural deafness, ataxia, mental retardation, and electrolyte imbalance (SeSAME syndrome) caused by mutations in KCNJ10. Proc Natl Acad Sci U S A 106:5842–5847
183. Schwarcz R, Hunter CA (2007) Toxoplasma gondii and schizophrenia: linkage through astrocyte-derived kynurenic acid? Schizophr Bull 33:652–653
184. Sequeira A, Mamdani F, Ernst C, Vawter MP, Bunney WE, Lebel V, Rehal S, Klempan T, Gratton A, Benkelfat C, Rouleau GA, Mechawar N, Turecki G (2009) Global brain gene expression analysis links glutamatergic and GABAergic alterations to suicide and major depression. PLoS ONE 4:e6585
185. Shan D, Lucas EK, Drummond JB, Haroutunian V, Meador-Woodruff JH, McCullumsmith RE (2013) Abnormal expression of glutamate transporters in temporal lobe areas in elderly patients with schizophrenia. Schizophr Res 144:1–8
186. Sharshar T, Annane D, de la Grandmaison GL, Brouland JP, Hopkinson NS, Francoise G (2004) The neuropathology of septic shock. Brain Pathol 14:21–33
187. Shulyatnikova T, Verkhratsky A (2019) Astroglia in sepsis associated encephalopathy. Neurochem Res. E-pub ahead of print. https://doi.org/10.1007/s11064-019-02743-2
188. Skowronska M, Albrecht J (2012) Alterations of blood brain barrier function in hyperammonemia: an overview. Neurotox Res 21:236–244
189. Sofroniew MV (2005) Reactive astrocytes in neural repair and protection. Neuroscientist 11:400–407
190. Sofroniew MV (2009) Molecular dissection of reactive astrogliosis and glial scar formation. Trends Neurosci 32:638–647
191. Sofroniew MV (2014) Astrogliosis. Cold Spring Harb Perspect Biol 7:a020420
192. Sofroniew MV (2014) Multiple roles for astrocytes as effectors of cytokines and inflammatory mediators. Neuroscientist 20:160–172
193. Sofroniew MV (2015) Astrocyte barriers to neurotoxic inflammation. Nat Rev Neurosci 16:249–263
194. Song P, Zhao ZQ (2001) The involvement of glial cells in the development of morphine tolerance. Neurosci Res 39:281–286
195. Spodenkiewicz M, Diez-Fernandez C, Rufenacht V, Gemperle-Britschgi C, Haberle J (2016) Minireview on glutamine synthetase deficiency, an ultra-rare inborn error of amino acid biosynthesis. Biology (Basel) 5
196. Steinhauser C, Grunnet M, Carmignoto G (2015) Crucial role of astrocytes in temporal lobe epilepsy. Neuroscience
197. Stenzel W, Soltek S, Schluter D, Deckert M (2004) The intermediate filament GFAP is important for the control of experimental murine Staphylococcus aureus-induced brain abscess and Toxoplasma encephalitis. J Neuropathol Exp Neurol 63:631–640

198. Struys-Ponsar C, Guillard O, van den Bosch de Aguilar P (2000) Effects of aluminum exposure on glutamate metabolism: a possible explanation for its toxicity. Exp Neurol 163:157–164
199. Suarez-Fernandez MB, Soldado AB, Sanz-Medel A, Vega JA, Novelli A, Fernandez-Sanchez MT (1999) Aluminum-induced degeneration of astrocytes occurs via apoptosis and results in neuronal death. Brain Res 835:125–136
200. Suarez I, Bodega G, Ramos JA, Fernandez-Ruiz JJ, Fernandez B (2000) Neuronal and astroglial response to pre- and perinatal exposure to delta-9-tetra- hydrocannabinol in the rat substantia nigra. Dev Neurosci 22:253–263
201. Sun JD, Liu Y, Yuan YH, Li J, Chen NH (2012) Gap junction dysfunction in the prefrontal cortex induces depressive-like behaviors in rats. Neuropsychopharmacology 37:1305–1320
202. Tanaka K, Watase K, Manabe T, Yamada K, Watanabe M, Takahashi K, Iwama H, Nishikawa T, Ichihara N, Kikuchi T, Okuyama S, Kawashima N, Hori S, Takimoto M, Wada K (1997) Epilepsy and exacerbation of brain injury in mice lacking the glutamate transporter GLT-1. Science 276:1699–1702
203. Tang D, Kang R, Coyne CB, Zeh HJ, Lotze MT (2012) PAMPs and DAMPs: signal 0 s that spur autophagy and immunity. Immunol Rev 249:158–175
204. Tang G, Xu Z, Goldman JE (2006) Synergistic effects of the SAPK/JNK and the proteasome pathway on glial fibrillary acidic protein (GFAP) accumulation in Alexander disease. J Biol Chem 281:38634–38643
205. Tidyman WE, Rauen KA (2009) The RASopathies: developmental syndromes of Ras/MAPK pathway dysregulation. Curr Opin Genet Dev 19:230–236
206. Traynelis SF, Dingledine R (1988) Potassium-induced spontaneous electrographic seizures in the rat hippocampal slice. J Neurophysiol 59:259–276
207. Valori CF, Brambilla L, Martorana F, Rossi D (2014) The multifaceted role of glial cells in amyotrophic lateral sclerosis. Cell Mol Life Sci 71:287–297
208. van Gool WA, van de Beek D, Eikelenboom P (2010) Systemic infection and delirium: when cytokines and acetylcholine collide. Lancet 375:773–775
209. Varatharaj A, Galea I (2017) The blood-brain barrier in systemic inflammation. Brain Behav Immun 60:1–12
210. Verkhratsky A, Butt AM (2013) Glial physiology and pathophysiology. Wiley-Blackwell, Chichester
211. Verkhratsky A, Krishtal OA, Burnstock G (2009) Purinoceptors on neuroglia. Mol Neurobiol 39:190–208
212. Verkhratsky A, Marutle A, Rodriguez-Arellano JJ, Nordberg A (2015) Glial asthenia and functional paralysis: a new perspective on neurodegeneration and Alzheimer's disease. Neuroscientist 21:552–568
213. Verkhratsky A, Nedergaard M (2014) Astroglial cradle in the life of the synapse. Philos Trans R Soc Lond B Biol Sci 369:20130595
214. Verkhratsky A, Nedergaard M (2018) Physiology of astroglia. Physiol Rev 98:239–389
215. Verkhratsky A, Olabarria M, Noristani HN, Yeh CY, Rodriguez JJ (2010) Astrocytes in Alzheimer's disease. Neurotherapeutics 7:399–412
216. Verkhratsky A, Parpura V (2016) Astrogliopathology in neurological, neurodevelopmental and psychiatric disorders. Neurobiol Dis 85:254–261
217. Verkhratsky A, Rodriguez JJ, Parpura V (2013) Astroglia in neurological diseases. Future Neurol 8:149–158
218. Verkhratsky A, Rodriguez JJ, Steardo L (2014) Astrogliopathology: a central element of neuropsychiatric diseases? Neuroscientist 20:576–588
219. Verkhratsky A, Steardo L, Parpura V, Montana V (2016) Translational potential of astrocytes in brain disorders. Prog Neurobiol 144:188–205
220. Verkhratsky A, Zorec R, Parpura V (2017) Stratification of astrocytes in healthy and diseased brain. Brain Pathol 27:629–644
221. Verkhratsky A, Zorec R, Rodriguez JJ, Parpura V (2017) Neuroglia: functional paralysis and reactivity in Alzheimer's disease and other neurodegenerative pathologies. Adv Neurobiol 15:427–449

222. Virchow R (1858) Die Cellularpathologie in ihrer Begründung auf physiologische and pathologische Gewebelehre. Zwanzig Vorlesungen gehalten während der Monate Februar, März und April 1858 im pathologischen Institut zu Berlin. August Hirschwald, Berlin
223. Wang F, Du T, Liang C, Verkhratsky A, Peng L (2015) Ammonium increases Ca^{2+} signalling and upregulates expression of $Ca_V1.2$ gene in astrocytes in primary cultures and in the in vivo brain. Acta Physiol (Oxf) 214:261–274
224. Wang Y, Zaveri HP, Lee TS, Eid T (2009) The development of recurrent seizures after continuous intrahippocampal infusion of methionine sulfoximine in rats: a video-intracranial electroencephalographic study. Exp Neurol 220:293–302
225. Weber M, Scherf N, Kahl T, Braumann UD, Scheibe P, Kuska JP, Bayer R, Buttner A, Franke H (2013) Quantitative analysis of astrogliosis in drug-dependent humans. Brain Res 1500:72–87
226. Webster MJ, Knable MB, Johnston-Wilson N, Nagata K, Inagaki M, Yolken RH (2001) Immunohistochemical localization of phosphorylated glial fibrillary acidic protein in the prefrontal cortex and hippocampus from patients with schizophrenia, bipolar disorder, and depression. Brain Behav Immun 15:388–400
227. Weng YC, Kriz J (2007) Differential neuroprotective effects of a minocycline-based drug cocktail in transient and permanent focal cerebral ischemia. Exp Neurol 204:433–442
228. Wilson EH, Hunter CA (2004) The role of astrocytes in the immunopathogenesis of toxoplasmic encephalitis. Int J Parasitol 34:543–548
229. Wilson JX, Peters CE, Sitar SM, Daoust P, Gelb AW (2000) Glutamate stimulates ascorbate transport by astrocytes. Brain Res 858:61–66
230. Windrem MS, Osipovitch M, Liu Z, Bates J, Chandler-Militello D, Zou L, Munir J, Schanz S, McCoy K, Miller RH, Wang S, Nedergaard M, Findling RL, Tesar PJ, Goldman SA (2017) Human iPSC glial mouse chimeras reveal glial contributions to schizophrenia. Cell Stem Cell 21(195–208):e196
231. Xia M, Abazyan S, Jouroukhin Y, Pletnikov M (2014) Behavioral sequelae of astrocyte dysfunction: focus on animal models of schizophrenia. Schizophr Res
232. Yin Z, Milatovic D, Aschner JL, Syversen T, Rocha JB, Souza DO, Sidoryk M, Albrecht J, Aschner M (2007) Methylmercury induces oxidative injury, alterations in permeability and glutamine transport in cultured astrocytes. Brain Res 1131:1–10
233. Zamanian JL, Xu L, Foo LC, Nouri N, Zhou L, Giffard RG, Barres BA (2012) Genomic analysis of reactive astrogliosis. J Neurosci 32:6391–6410
234. Zeidan-Chulia F, Salmina AB, Malinovskaya NA, Noda M, Verkhratsky A, Moreira JC (2014) The glial perspective of autism spectrum disorders. Neurosci Biobehav Rev 38:160–172
235. Zhang L, Li L, Wang B, Qian DM, Song XM, Hu M (2013) Human cytomegalovirus infection modulates thrombospondins 1 and 2 in primary fetal astrocytes. NeuroReport 24:526–535
236. Zhang L, Li L, Wang B, Qian DM, Song XX, Hu M (2014) HCMV induces dysregulation of glutamate uptake and transporter expression in human fetal astrocytes. Neurochem Res 39:2407–2418
237. Zhang M, Strnatka D, Donohue C, Hallows JL, Vincent I, Erickson RP (2008) Astrocyte-only Npc1 reduces neuronal cholesterol and triples life span of Npc1-/- mice. J Neurosci Res 86:2848–2856
238. Zhang Y, Jin Y, Behr MJ, Feustel PJ, Morrison JP, Kimelberg HK (2005) Behavioral and histological neuroprotection by tamoxifen after reversible focal cerebral ischemia. Exp Neurol 196:41–46
239. Zhao Y, Rempe DA (2010) Targeting astrocytes for stroke therapy. Neurotherapeutics 7:439–451
240. Zorec R, Zupanc TA, Verkhratsky A (2019) Astrogliopathology in the infectious insults of the brain. Neurosci Lett 689:56–62

93. [illegible]
94. [illegible]
95. [illegible]
96. [illegible]
97. [illegible]
98. [illegible]
99. [illegible]
100. [illegible]

Chapter 8
Neuroglia in Ageing

Alexei Verkhratsky, Robert Zorec, Jose Julio Rodriguez-Arellano and Vladimir Parpura

Abstract Ageing reduces the functional capacity of all organs, so does that of the nervous system; the latter is evident in the reduction of cognitive abilities, learning and memory. While the exact mechanisms of ageing of the nervous system remain elusive, it is without doubt that morpho-functional changes in a variety of neuroglial cells contribute to this process. The age-dependent changes in neuroglia are characterised by a progressive loss of function. This reduces glial ability to homeostatically nurture, protect and regenerate the nervous tissue. Such neuroglial paralysis also facilitates neurodegenerative processes. Ageing of neuroglia is variable and can be affected by environmental factors and comorbidities.

Keywords Ageing · Astrocyte · Microglia · NG2 cells · Oligodendrocytes

---

A. Verkhratsky (✉)
Faculty of Biology, Medicine and Health, The University of Manchester, Manchester M13 9PT, UK
e-mail: Alexej.Verkhratsky@manchester.ac.uk

Faculty of Health and Medical Sciences, Center for Basic and Translational Neuroscience, University of Copenhagen, 2200 Copenhagen, Denmark

A. Verkhratsky · J. J. Rodriguez-Arellano
Achucarro Center for Neuroscience, IKERBASQUE, Basque Foundation for Science, 48011 Bilbao, Spain

R. Zorec
Laboratory of Neuroendocrinology-Molecular Cell Physiology, Faculty of Medicine, Institute of Pathophysiology, University of Ljubljana, Ljubljana, Slovenia

Celica BIOMEDICAL, Ljubljana, Slovenia

V. Parpura
Department of Neurobiology, The University of Alabama at Birmingham, Birmingham, AL, USA

A. Verkhratsky et al. (eds.), *Neuroglia in Neurodegenerative Diseases*, Advances in Experimental Medicine and Biology 1175,
https://doi.org/10.1007/978-981-13-9913-8_8

8.1 Mechanisms of Ageing

Ageing reduces the functional capacity of all organs and systems ultimately weakening the whole organism, reducing its adaptability, wearing its defensive systems and bringing it to death through age-dependent diseases. The nervous system similarly undergoes senescence that often impairs upon cognitive abilities, affects learning capabilities and enfeebles memory. Nonetheless, brain sustains ageing with surprising tenacity; the cognitive functions attain the summit in middle age and often sustain into old age, while the decline in other systems (including skeleto-muscular, cardiovascular or endocrine) progresses much faster. What are the mechanisms of ageing and which molecular and cellular processes underlie the ageing of the brain remains a matter of intense polemics. Indeed, there are in excess of 300 theories of ageing, which highlight various pathways, many of which do contribute to this process [52].

Probably, the very first coherent theory of ageing was proposed by August Weismann [99], who considered ageing as a very natural process favoured by natural selection to prevent competition of species with their own progeny. According to this view, mechanisms of ageing could be many and they can be different in different organisms. The only common future is guaranteed termination of individual life after a presumed fulfilment of the reproductive duties. One of the widely considered pathways is the activation of endoplasmic reticulum stress/unfolded protein responses that positively correlate with longevity and negatively with fertility [86]. Another relatively old group of theories regards ageing as a result of mechanochemical deterioration of molecules and basic systems such as organelles or cell membranes; this was initially proposed as 'hysteresis of colloids' [72, 73]. The mitochondrial or free radical theory of ageing links the damage of biological systems to reactive oxygen species and regards mitochondria as the age-defining clock [29, 30]. By reducing caloric intake this may resist the rate of ageing [48]. The gene regulation theories assume that changes in gene expression define ageing process [39], while the telomere theory postulates that it is the telomere DNA localised at the end of chromosomes which determines the life span: the shortening of telomeres eventually brings to the arrest of cell replication and death [28], The inflammatory theory of ageing became popular in recent years leading to the concept of 'inflammaging' [21]. Additionally, several theories look into the role of signalling systems both at organism (e.g., neuroendocrine or immune theories of ageing [20]) and cellular (e.g., calcium theory of ageing [40, 45, 89]) levels.

8.2 Ageing of the Brain

The maintenance of cognitive capacity of the brain over most of the human's lifespan results, most likely, from prominent neuroplasticity, remarkably long development and high degree of homeostatic and protective capabilities of neuroglia. The human brain is optimised for learning, with numerous mechanisms from adult neurogenesis (which supplies the hippocampus with new neurones [54]) and adult myelination (which lasts well into the fourth decade of human life [1, 98], while oligodendroglial progenitors are present throughout the brain across the whole lifespan and probably contribute to late-life regenerative myelination), to the highly sophisticated glymphatic system that purges the brain from toxic waste products [36], thus maintaining neural environment. Ageing affects cognition components in a rather distinct way. The age-dependent decline mainly affects the real-time processing and formation of new memories and behaviours, whereas the capacity to analyse semantic and long-term memories suffer much less [18, 35]. For example, a group of young adults were significantly better than the group of old people in recalling a list of words. However the ability of elders to use complex processing activities was indistinguishable from the youngsters [47]. This benign or physiological brain ageing is not granted to all, and age-dependent pathologies, most notably of neurodegenerative nature, affect a substantial part of population.

Age is the main risk factor for neurodegenerative diseases, which are often considered as a natural outcome of senescence process. However, there is a fundamental difference between physiological ageing and neurodegeneration. The latter reflects massive neuronal death and atrophy of the brain tissue, whereas the former is not associated with a substantial neuronal loss. The overall number of neurones is not significantly affected in physiological ageing in rodents, primates and humans [6, 12, 19, 100]. Likewise, the number and density of synapses are not significantly affected by ageing [23, 80], albeit synaptic size is reduced [56].

Factors which determine the fateful difference between physiological and pathological ageing are many. These are represented by genetic factors (the best example being familial Alzheimer's disease or Huntington disease), the environment and life style (including diet, education, mental or physical activity) and the associated pathology (such as vascular disorders and ischaemic lesions). Another fundamental factor that defines the degree of cognitive deficit of ageing and age-dependent neuropathologies is known as the cognitive reserve. The cognitive reserve is an intrinsic quality of an individual brain that determines the neurological deficit when a similar brain damage results in very different cognitive outcomes in different subjects [82, 103]. The cognitive reserve in turn is defined by (i) neuronal reserve, which is the status of neuronal networks acquired during the life span through learning and cognitive load and (ii) neuronal compensation that reflects the defensive, plastic and regenerative capacities of the individual brain. To a large extent, the neuronal compensation is defined by neuroglia, which is responsible for neuroprotection, regeneration and post-lesion remodelling of the neural circuitry. The role of neuroglia is therefore fundamental in defining physiological versus pathological senescence; the failure of glial cells to protect and sustain the neuronal networks, the neural tissue and the CNS as an organ facilitates the progression from physiological to pathological brain ageing [93].

8.3 Astroglia in Physiological Ageing

8.3.1 Morphology and Gene Profiling

Age-dependent changes in astroglial morphology and gene expression are complex and region specific. Total number of astrocytes in physiologically aged human brain does not seem to change significantly, even in centenarians [19, 59]. When it comes to astroglial morphological profiles and expression of glial fibrillary acidic protein (GFAP), which are indicative of astroglial reactivity, the data remain quite controversial. Both a decrease [7] and an increase [14] in the number of GFAP-positive astrocytes, in particular in hypothalamic areas [27], as well as astroglial atrophy and astroglial hypertrophy were observed. The volume of astroglial territorial domains has been found to almost double in 21-month-old mice when compared to 5-month-old animals [26]. Increase of GFAP expression and hypertrophy of GFAP-positive astrocytes have been described in the hippocampus of aged rodents [7, 34, 49] and humans [10, 55]. Ageing had a distinct effect on different subpopulation of astrocytes in a region-dependent manner (Fig. 8.1, [69]). The densities of GFAP-positive astrocytes in the CA1 region and dentate gyrus of the hippocampus of old (24-month-old) mice demonstrated prominent hypertrophy when compared to young (3-month-old) or adult (9-month-old) controls. To the contrary, GFAP-positive profiles of astrocytes in the entorhinal cortex of old animals were atrophic when compared to the young or adult mice. Ageing results in a substantial decrease in the number and complexity of processes of astrocytes in the entorhinal cortex. The astrocytes immunoreactive to

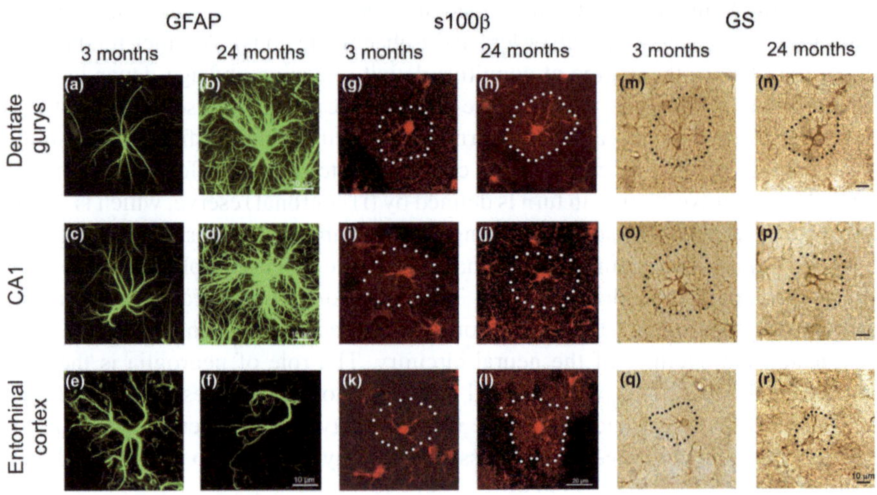

Fig. 8.1 Age-dependent remodelling of astroglial profiles in different brain areas. Confocal images showing glial fibrillary acidic protein—GFAP (a to f), s100β (g to l) and glutamine synthetase—GS (m to r) immunolabelled astrocytes in the dentate gyrus and CA1 hippocampal areas as well as in the entorhinal cortex of mice at 3 and 24 months. Modified from [69]

s100β protein were hypertrophic in the aged dentate gyrus but not in the CA1 region of the hippocampus as well as in the entorhinal cortex, whereas the profiles of a subpopulation of astrocytes labelled with glutamine synthetase were atrophic in the hippocampus with no changes in the entorhinal cortex [69]. Glutamine synthetase is a central enzyme necessary for operation of glutamine–glutamate/GABA shuttle, as well as for ammonium detoxification [70]. Suppression of expression of this enzyme may therefore affect neurotransmission and promote astroglial synthesis of GABA, an inhibitory neurotransmitter [22]. In parallel, hypertrophy of GFAP-positive astrocytes may be connected with environmental stimulation and plasticity representing the neural compensation. Exposure to the enriched environment is known for its positive effects on learning and memory, which occur in parallel with an increase in GFAP-positive astroglial profiles [68, 75].

The transcriptomic analysis of aged astrocytes similarly found a complex modification in genes expression. For example, astroglial cells from the cerebral cortex of aged mice demonstrated an increase in genes related to immune response with a decrease in expression of GFAP and genes related to neuroprotection and neuronal support [58]. Comparison of RNA-Seq from old and young astrocytes in the motor and visual cortices, hypothalamus and cerebellum revealed region specificity, with more significant changes in astroglia from the hippocampus and cerebellum [4], where astrocytes increased an expression of proinflammatory genes, genes encoding GFAP and Serpin3n, and genes linked to synaptic elimination such as complement component 3 and 4b [4]. Very similar results have been obtained in analysing the gene expression profiles of astrocytes from the hippocampus, cortex and striatum. Ageing affected hippocampal and striatal astrocytes the most with up-regulation of inflammatory genes and genes related to synaptic elimination [8]. Analysis of the gene expression profiles of different brain cells from ten brain regions of post-mortem tissue of humans, aged between 16 and 102 years, found that changes in astrocytes and oligodendrocytes were more prominent and complex compared to other cell types [81]; in particular, no age-dependent changes in neuronal gene expression pattern were identified. Again, these results indicate that functional preservation of neuroglia is critical for maintaining the ageing brain.

8.3.2 Astroglial Function

Although the data on physiology of aged astrocytes are rather limited, there are some hints for age-dependent remodelling of signalling and homeostatic profiles of astroglial cells. The resting membrane potential (around −80 mV) and membrane input resistance of astrocytes in cortical slices (from animals aged between 1 and 21 months) does not change much in ageing; if anything the input resistance is somewhat smaller in young adult mice (3–6-month-old—Fig. 8.2, [44]). Astrocytes from older mice express major types of receptors and are capable to generate ionotropic receptor-driven glial "postsynaptic" currents in response to neuronal activity [24, 44]. The density of ionotropic glutamate (AMPA and NMDA) receptors and P2X

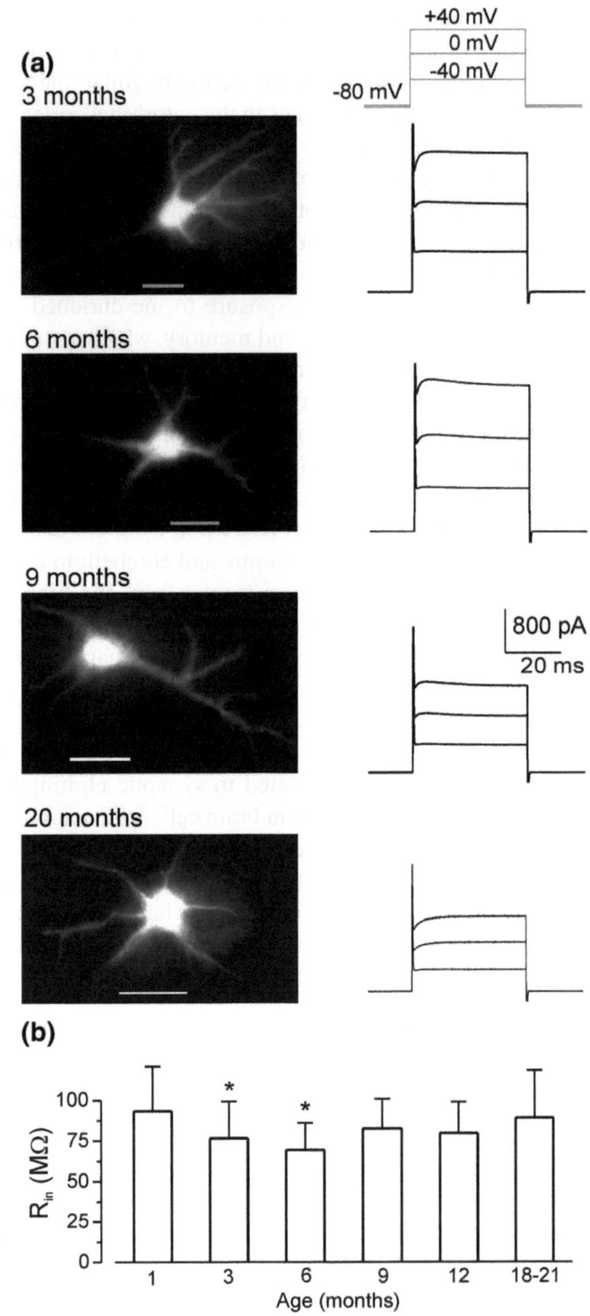

Fig. 8.2 Basic electrophysiological properties of cortical astrocytes in brain slices from mice of different ages. **a** Experiments were performed on mice expressing EGFP under control of human GFAP promoter. Representative EGFP images (*left column*) of cortical layer II astrocytes of different age groups and corresponding whole-cell currents (*right column*) evoked by depolarizing steps from the holding potential of −80 mV; the voltage protocol is shown on the top. **b** Age-related changes in the input resistance of cortical astrocytes. Data are presented as mean ± SD for 15 cells in each age group; decrease in the astrocyte R_{in} measured in 3 and 6 months in comparison to 1 month was statistically significant with $P < 0.05$ (one-way ANOVA). Reproduced with permission from [44]

purinoceptors, as well as the density of plasmalemmal glutamate transporter currents demonstrate bell-shaped age dependency (Fig. 8.3). Ionic currents generated by the above receptors and transporters are maximal in young adult (3- to 6-month-old) animals; at 9–21 months of age these currents are much smaller, although they are similar to currents recorded in 1-month-old animals [44].

Astrocytes are endowed with specific type of excitability, known as ionic excitability, which is associated with spatially and temporally organised fluctuations in the cytosolic concentration of several ions, including Ca^{2+}, Na^{+}, Cl^{-} and possibly K^{+} and H^{+} [95, 97]. Intracellular Ca^{2+} and Na^{+} signalling are of particular importance [95, 96] being involved in regulation of numerous astroglial physiological processes such as secretion [94] or homeostatic transport [41, 71]. Neurotransmitter or synaptically induced astroglial Ca^{2+} signals are age dependent. For example, Ca^{2+} signals are the largest in young adult mice and are relatively small in old and very young animals (Fig. 8.4; [44]). This dependence may be reflected in the functional expression of astrocytic receptors. Most likely an increase in the density of receptors, as well as in the density of plasmalemmal glutamate transporters and in the amplitude of Ca^{2+} signals, occur in the period of maximal environmental stimulation associated

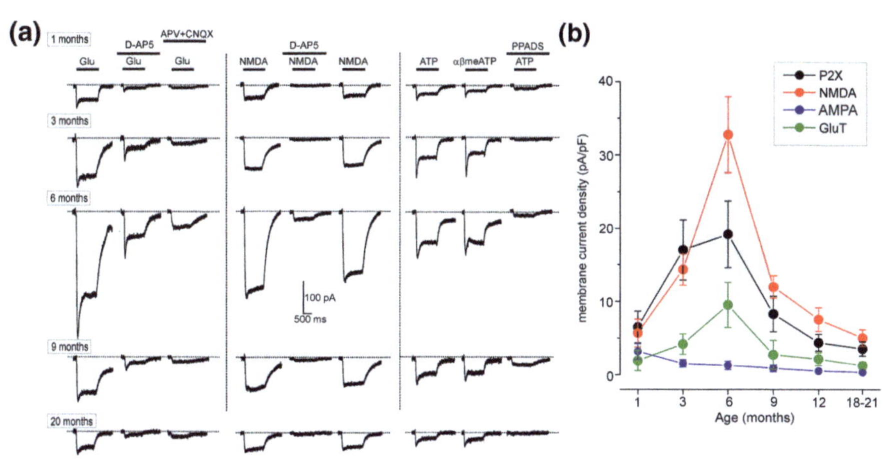

Fig. 8.3 Ageing affects the density of plasmalemmal glutamate transporters and ionotropic receptor-mediated currents in acutely isolated single cortical astrocytes. **a** Representative whole-cell currents elicited in the acutely isolated astrocytes by application of 100 μM glutamate (left column), 10 μM NMDA (middle column) and 10 μM ATP/αβmeATP (a potent and stable agonist at $P2X_1$, $P2X_3$, $P2X_{2/3}$, $P2X_{1/5}$ and $P2X_{4/6}$ receptors is also a weak partial agonist at human and mouse P2X4 receptors, but an antagonist at the rat $P2X_4$ receptor; it has little or no effect at other P2X and P2Y receptors), at holding potential of −80 mV. Glutamate- and NMDA-evoked currents were inhibited by 10 μM D-AP5, an NMDA antagonist and 30 μM CNQX, an AMPA receptor antagonist; ATP-evoked currents were inhibited by 10 μM PPADS, a selective purinergic P2X antagonist. **b** The density of currents mediated by P2X, NMDA and AMPA receptors and plasmalemmal glutamate transporters (GluT) in cortical astrocytes (mean ± SD for 9–12 cells for each age group); statistical significance of difference between average value for 1 month and corresponding values for 3 and 6 months $P < 0.02$ (ANOVA) for all types of currents

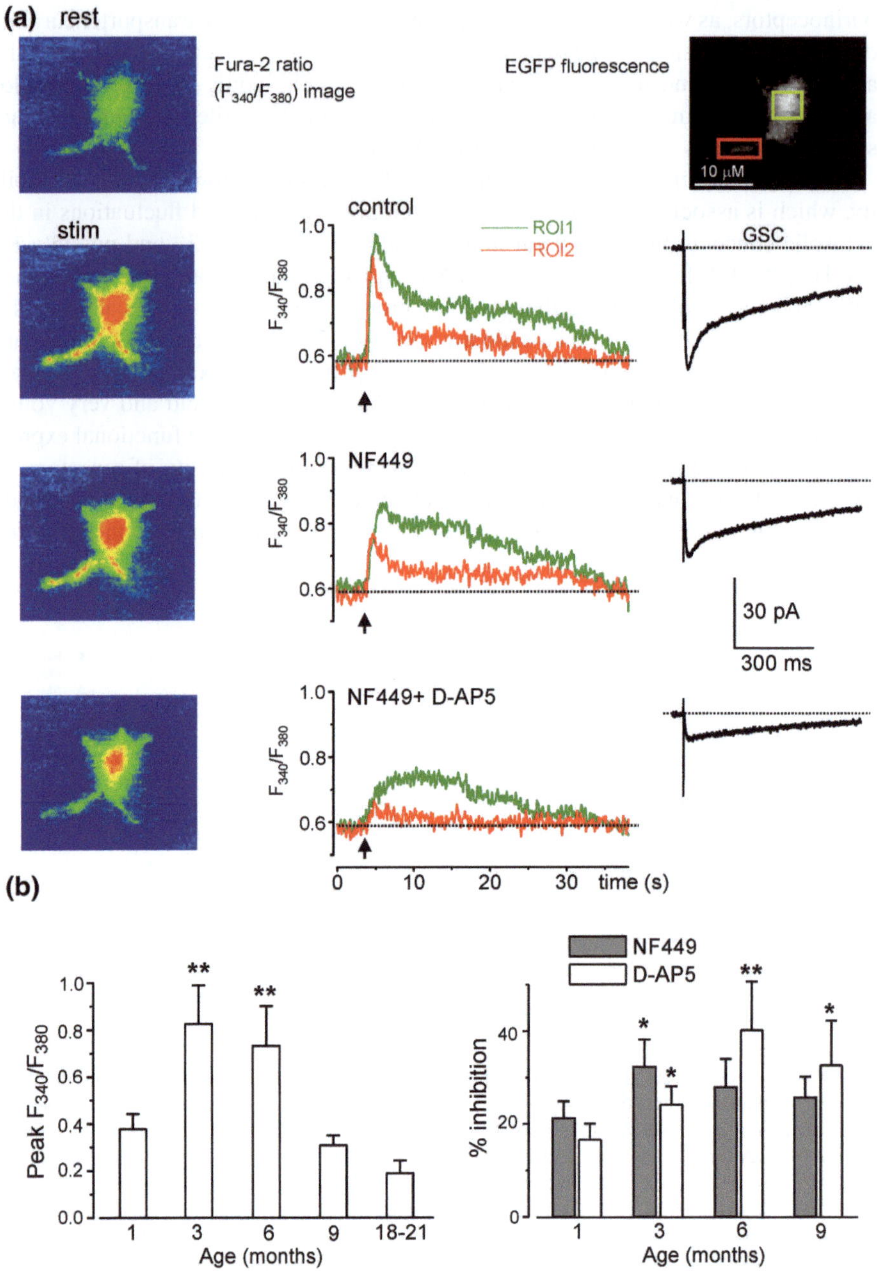
(a)
rest
Fura-2 ratio
(F340/F380) image
EGFP fluorescence
10 µM
stim
control
ROI1
ROI2
GSC
F340/F380
NF449
30 pA
300 ms
NF449+ D-AP5
time (s)
(b)
Peak F340/F380
Age (months)
NF449
D-AP5
% inhibition
Age (months)

◄**Fig. 8.4** Age-dependent changes in synaptically induced ionotropic Ca^{2+} signals in protoplasmic astrocytes in situ in cortical slices. **a** Cortical layer II astrocyte of 9-month-old mouse was loaded with the Ca^{2+} indicator Fura-2 in situ via patch pipette. Fluorescence images were recorded simultaneously with glial currents evoked by neuronal afferent stimulation in presence of a mixture of TBOA (GluT blocker) and CNQX in control, and after consecutive application of 10 nM NF-449 (selective antagonist of P2X receptors) and at that in a company of 30 μM D-AP5. Representative images (pseudo-colour, pipette image subtracted) and glial synaptic currents (GSC, right column) were recorded before (rest) and after stimulation as indicated. Ca^{2+} transients (middle column) are expressed as F_{340}/F_{380} ratio averaged over the corresponding regions of interest shown in the GFAP image of astrocyte (*top right*). **b** Age-related changes in the astrocytic Ca^{2+} signalling. *Left panel*, average peak amplitudes of $[Ca^{2+}]_i$ increases, induced by stimulation of neuronal afferents in cortical astrocytes of different ages. *Right panel*, average inhibitory effect of antagonists of P2X (NF449) and NMDA (D-AP5) receptors on the amplitudes of $[Ca^{2+}]_i$ increases responses in cortical astrocytes. Data are presented as mean ± SD for 3–4 cells for each age group; * $P < 0.05$, ** $P < 0.01$ one-way ANOVA compared to 1 month. Reproduced with permission from [44]

with intense learning; in younger and older ages synaptic activity is lower, which is reflected in a decrease in receptors expression.

A decrease in astroglial gap junctional coupling was found in old (20–27-month-old mice) neocortical astroglial syncytia [63]; there were no changes at earlier ages (up to 14-month-old [9]). Astrocytes in older brains down-regulate expression of aquaporin 4 (AQP4). A decrease in the density of these channels in the perivascular endfeet affects clearance of the brain parenchyma through the glymphatic pathway [42]. This decrease in AQP4 in the endfeet may be linked to the deficits in vesicular trafficking, which is the key pathway in delivery of numerous molecules to specific locations at the plasmalemma [64]. Ageing affects astroglial metabolic pathways, as an age dependent increase in oxidative metabolism was reported in older astrocytes, which may limit their ability to supply neurones with metabolic substrates [37]. There is also evidence of age-dependent alterations in astroglial ability to produce lactate and hence to operate lactate shuttle [31]. Similarly, ageing is associated with an increase in the ratio of glutamate to glutamine in the brain that indicates some aberrations in the operation of the glutamate/GABA-glutamine shuttle [16, 32]. Ageing is also associated with a decrease in the brain levels of glutathione, mainly produced in astrocytes; this limits the ability of astroglia to resist the oxidative damage to the neural tissue [17, 50].

8.4 Oligodendroglia in Physiological Ageing

The human brain has a disproportionally large white matter when compared to other mammals and even high primates [76], as indeed the white matter occupies >50% of the human brain. Additionally, the level of myelination is well developed in the grey matter [43], further demonstrating the importance of connectome to the cognition and intelligence. The anatomical prevalence of the white matter in the human brain is also associated with very long development: myelination attains its peak at ~45–47 years of age, with a subsequent slow and yet progressive age-dependent decline [1]. Normal ageing causes rather substantial shrinkage of the white matter which diminishes by ~11%; in comparison, the volume of the grey matter is decreased by only ~3% [33]. The highest degree of age-dependent alterations of the white matter is detected in the prefrontal cortex and associative tracts [66], which suffer early in Alzheimer's disease [15]. Incidentally, these brain regions emerge late in evolution and they are the slowest to develop, which instigated a 'last in, first out' hypothesis of the white matter ageing [66, 90]. Conceptually, changes in the white matter can be considered as a valuable marker of ageing [90], and moreover, accelerated degeneration of the white matter seems to indicate development of neurodegeneration and profound cognitive decline [65, 90].

Cells of the oligodendroglial lineage represented by oligodendrocytes and their precursors (also known as NG2 glia [13]) are, arguably, the most numerous glial cells in the human brain. Cells of the oligodendroglial lineage, in contrast to astrocytes, are highly vulnerable to excitotoxicity and to oxidative stress. The oligodendroglial precursors/NG2 cells, as well as more mature oligodendroglia, express several types of ionotropic glutamate receptors (including NMDA receptors) and P2X purinoceptors, which all can mediate excitotoxic Ca^{2+} overload and cause cell death [51, 53, 74, 92]. Furthermore, oligodendrocytes are highly vulnerable to oxidative damage, which is stipulated by a rather low content of antioxidants. In particular oligodendroglial cells contain two times less of glutathione compared to astrocytes, and yet they experience six times more of oxidative stress in physiological conditions [38, 87].

Ageing is associated with a significant decrease, by up to 30%, of the total number of oligodendrocytes [19, 59]. Rather surprisingly, in monkeys the number of oligodendroglial cells has been claimed to increase with age; for example, in the visual cortex of old monkeys the number of oligodendrocytes increased by 50% [62]. Notably, these oligodendrocytes also showed aberrant atrophic morphology and a deficiency in myelin production, which defined decreased CNS myelination in old primates [62]. The age-dependent myelin deficiencies are also associated with vasculature lesions in the white matter that add strain on oligodendrocytes and promote their degeneration [3, 101]. Ageing is also associated with a diminished capacity of remyelination supported by the NG2 glia. Notwithstanding the fact that the population of NG2-oligodendroglial precursors does not change numerically in the old brain, the capacity of NG2 cells to differentiate into mature oligodendrocytes is reduced. The NG2 cells in the old brain tend to retain their precursor status, so that the time of differentiation into mature myelinating phenotype is increased by almost two times [102]. All in all, age-dependent changes in the white matter are prominent and may be the leading cause of age-dependent cognitive decline.

8.5 Microglia in the Ageing Brain

Microglia in the ageing human brain undergoes rather idiosyncratic metamorphoses, which are not present in laboratory animals. Fundamentally, human microglia gradually degenerates, thus, reducing the defensive capabilities of the senescent nervous tissue.

In animals, the ageing process results in complex changes in microglial numbers and state. In old rats, microglial numbers decreased in the nigrostriatal system and cerebral cortex [77], and remained unchanged in the hippocampus [91]. In contrast, in old rhesus monkeys the densities of microglial cells increased, while these cells showed signs of increased phagocytosis [61]. In humans, ageing is associated with dystrophy and degeneration of microglia which resulted in deterioration of neuroprotective and defensive functions of these cells [84]. Morphological features of dystrophic aged microglia include deramification, spheroid formation, gnarling and fragmentation of processes [84]. The processes of aged microglial cells are shorter with less branching and reduced arborized area; the total number of microglia seems not to change with age [11]. Microglial dystrophy and a loss of function arguably increase the vulnerability of the old brain to neurodegeneration and may facilitate evolution of age-dependent cognitive disorders, including Alzheimer's disease [83]. The age-dependent microglial dystrophy can be associated with cytoskeleton abnormalities that underlie the cytorrhexis, rupturing of cells [88]. Microglial cells can accumulate tau [5] and the aged microglia (in marmosets) were reported to contain hyperphosphorylated tau [67]; this microgliatauopathy can be a factor that initiates microglial degeneration and dystrophy [67]. The prevalence of dystrophic microglia limits the neuroinflammatory capabilities of the old brain tissue, questioning the concept of inflammaging.

There is also evidence for age-dependent microglial activation in normal ageing, especially in rodents [57, 60] and in *Macaca nemestrina* monkeys [78]. There is an overall trend of hyperreactivity of microglia in aged mice [25, 46, 79], which is strikingly different to the dystrophy and a loss of function of human aged microglia, questioning the validity of rodents as an experimental models for brain ageing.

Aged human microglial cells are represented by two morphologically distinct classes identified as dystrophic or senescent microglia and dark microglia. The dystrophic microglial cells [85] are characterised by spherical swellings of processes, dilatation of the endoplasmic reticulum and abundance of lipofuscin deposits (that emerge from incomplete lysosomal degradation and endolysosomal stress and overload). The dystrophic microglial cells have been identified both in old brains and in high densities around senile plaques of Alzheimer's diseases patients [88]. Dystrophic microglial cells have fragmented processes and have a substantially diminished activation capacity [85, 88]. The dark microglia have been defined so because of the electron-dense cytoplasm and nucleoplasm, which in electron microscopy appear as dark as mitochondria [2]. The dark microglia are also characterised by ultrathin and highly ramified processes that frequently enwrap synaptic elements, axons and dendrites. This may indicate that dark microglial cells are involved in eliminating synapses [2]. In addition, dark microglia have altered expression of classical marker IBA1 and they do not express the $P2Y_{12}$ purinoceptor, which is considered as a marker for healthy surveillance microglia. Dark microglial cells cumulate with ageing and even more so in age-dependent pathologies [2].

8.6 Conclusions

All types of neuroglial cells undergo age-dependent remodelling which seems to be critical to define a physiological or pathological outcome of ageing process. In general, the age-dependent changes in neuroglial cells are characterised by a progressive loss of function which limits neuroprotection and regenerative potential of the neural tissue. This process of neuroglial senescence, however, is variable and most likely individually tailored by the lifestyle, environmental stress and comorbidities. Neuroglial paralysis facilitates emergence of neurodegeneration and cognitive decline, and hence a neuroglial state represents a potential therapeutic target for age-associated neurological disorders.

Acknowledgments VP's work is supported by a grant from the National Institute of General Medical Sciences of the National Institutes of Health (R01GM123971). VP is an Honorary Professor at University of Rijeka, Croatia.

References

1. Bartzokis G, Beckson M, Lu PH, Nuechterlein KH, Edwards N, Mintz J (2001) Age-related changes in frontal and temporal lobe volumes in men: a magnetic resonance imaging study. Arch Gen Psychiatry 58:461–465
2. Bisht K, Sharma KP, Lecours C, Sanchez MG, El Hajj H, Milior G, Olmos-Alonso A, Gomez-Nicola D, Luheshi G, Vallieres L, Branchi I, Maggi L, Limatola C, Butovsky O, Tremblay ME (2016) Dark microglia: a new phenotype predominantly associated with pathological states. Glia 64:826–839
3. Black S, Gao F, Bilbao J (2009) Understanding white matter disease: imaging-pathological correlations in vascular cognitive impairment. Stroke 40:S48–52
4. Boisvert MM, Erikson GA, Shokhirev MN, Allen NJ (2018) The aging astrocyte transcriptome from multiple regions of the mouse brain. Cell Rep 22:269–285
5. Bolos M, Llorens-Martin M, Jurado-Arjona J, Hernandez F, Rabano A, Avila J (2016) Direct evidence of internalization of tau by microglia in vitro and in vivo. J Alzheimer's Dis 50:77–87
6. Calhoun ME, Kurth D, Phinney AL, Long JM, Hengemihle J, Mouton PR, Ingram DK, Jucker M (1998) Hippocampal neuron and synaptophysin-positive bouton number in aging C57BL/6 mice. Neurobiol Aging 19:599–606
7. Cerbai F, Lana D, Nosi D, Petkova-Kirova P, Zecchi S, Brothers HM, Wenk GL, Giovannini MG (2012) The neuron-astrocyte-microglia triad in normal brain ageing and in a model of neuroinflammation in the rat hippocampus. PLoS ONE 7:e45250
8. Clarke LE, Liddelow SA, Chakraborty C, Munch AE, Heiman M, Barres BA (2018) Normal aging induces A1-like astrocyte reactivity. Proc Natl Acad Sci USA 115:E1896–E1905

9. Cruz NF, Ball KK, Dienel GA (2010) Astrocytic gap junctional communication is reduced in amyloid-beta-treated cultured astrocytes, but not in Alzheimer's disease transgenic mice. ASN Neuro 2:e00041
10. David JP, Ghozali F, Fallet-Bianco C, Wattez A, Delaine S, Boniface B, Di Menza C, Delacourte A (1997) Glial reaction in the hippocampal formation is highly correlated with aging in human brain. Neurosci Lett 235:53–56
11. Davies DS, Ma J, Jegathees T, Goldsbury C (2017) Microglia show altered morphology and reduced arborization in human brain during aging and Alzheimer's disease. Brain Pathol 27:795–808
12. Dickstein D, Kabaso D, Rocher A, Luebke J, Wearne S, Hof P (2006) Changes in the structural complexity of the aged brain. Aging Cell 6:275–284
13. Dimou L, Gallo V (2015) NG2-glia and their functions in the central nervous system. Glia 63:1429–1451
14. Diniz DG, Foro CA, Rego CM, Gloria DA, de Oliveira FR, Paes JM, de Sousa AA, Tokuhashi TP, Trindade LS, Turiel MC, Vasconcelos EG, Torres JB, Cunnigham C, Perry VH, Vasconcelos PF, Diniz CW (2010) Environmental impoverishment and aging alter object recognition, spatial learning, and dentate gyrus astrocytes. Eur J Neurosci 32:509–519
15. Douaud G, Groves AR, Tamnes CK, Westlye LT, Duff EP, Engvig A, Walhovd KB, James A, Gass A, Monsch AU, Matthews PM, Fjell AM, Smith SM, Johansen-Berg H (2014) A common brain network links development, aging, and vulnerability to disease. Proc Natl Acad Sci USA 111:17648–17653
16. Duarte JM, Do KQ, Gruetter R (2014) Longitudinal neurochemical modifications in the aging mouse brain measured in vivo by 1H magnetic resonance spectroscopy. Neurobiol Aging 35:1660–1668
17. Emir UE, Raatz S, McPherson S, Hodges JS, Torkelson C, Tawfik P, White T, Terpstra M (2011) Noninvasive quantification of ascorbate and glutathione concentration in the elderly human brain. NMR Biomed 24:888–894
18. Erickson CA, Barnes CA (2003) The neurobiology of memory changes in normal aging. Exp Gerontol 38:61–69
19. Fabricius K, Jacobsen JS, Pakkenberg B (2013) Effect of age on neocortical brain cells in 90+ year old human females—a cell counting study. Neurobiol Aging 34:91–99
20. Fabris N (1991) Neuroendocrine-immune interactions: a theoretical approach to aging. Arch Gerontol Geriatr 12:219–230
21. Franceschi C (2007) Inflammaging as a major characteristic of old people: can it be prevented or cured? Nutr Rev 65:S173–176
22. Garaschuk O, Verkhratsky A (2019) GABAergic astrocytes in Alzheimer's disease. Aging (Albany NY) 11:1602–1604
23. Geinisman Y, Ganeshina O, Yoshida R, Berry RW, Disterhoft JF, Gallagher M (2004) Aging, spatial learning, and total synapse number in the rat CA1 stratum radiatum. Neurobiol Aging 25:407–416
24. Gomez-Gonzalo M, Martin-Fernandez M, Martinez-Murillo R, Mederos S, Hernandez-Vivanco A, Jamison S, Fernandez AP, Serrano J, Calero P, Futch HS, Corpas R, Sanfeliu C, Perea G, Araque A (2017) Neuron-astrocyte signaling is preserved in the aging brain. Glia 65:569–580
25. Griffin R, Nally R, Nolan Y, McCartney Y, Linden J, Lynch MA (2006) The age-related attenuation in long-term potentiation is associated with microglial activation. J Neurochem 99:1263–1272
26. Grosche A, Grosche J, Tackenberg M, Scheller D, Gerstner G, Gumprecht A, Pannicke T, Hirrlinger PG, Wilhelmsson U, Huttmann K, Hartig W, Steinhauser C, Pekny M, Reichenbach A (2013) Versatile and simple approach to determine astrocyte territories in mouse neocortex and hippocampus. PLoS ONE 8:e69143
27. Hardy RN, Simsek ZD, Curry B, Core SL, Beltz T, Xue B, Johnson AK, Thunhorst RL, Curtis KS (2018) Aging affects isoproterenol-induced water drinking, astrocyte density, and central neuronal activation in female Brown Norway rats. Physiol Behav 192:90–97

28. Harley CB, Vaziri H, Counter CM, Allsopp RC (1992) The telomere hypothesis of cellular aging. Exp Gerontol 27:375–382
29. Harman D (1965) The free radical theory of aging: effect of age on serum copper levels. J Gerontol 20:151–153
30. Harman D (1972) The biologic clock: the mitochondria? J Am Geriatr Soc 20:145–147
31. Harris JL, Choi IY, Brooks WM (2015) Probing astrocyte metabolism in vivo: proton magnetic resonance spectroscopy in the injured and aging brain. Front Aging Neurosci 7:202
32. Harris JL, Yeh HW, Swerdlow RH, Choi IY, Lee P, Brooks WM (2014) High-field proton magnetic resonance spectroscopy reveals metabolic effects of normal brain aging. Neurobiol Aging 35:1686–1694
33. Haug H, Eggers R (1991) Morphometry of the human cortex cerebri and corpus striatum during aging. Neurobiol Aging 12:336–338; discussion 352–335
34. Hayakawa N, Kato H, Araki T (2007) Age-related changes of astorocytes, oligodendrocytes and microglia in the mouse hippocampal CA1 sector. Mech Ageing Dev 128:311–316
35. Hedden T, Gabrieli JD (2004) Insights into the ageing mind: a view from cognitive neuroscience. Nat Rev Neurosci 5:87–96
36. Jessen NA, Munk AS, Lundgaard I, Nedergaard M (2015) The glymphatic system: a beginner's guide. Neurochem Res 40:2583–2599
37. Jiang T, Cadenas E (2014) Astrocytic metabolic and inflammatory changes as a function of age. Aging Cell
38. Juurlink BH, Thorburne SK, Hertz L (1998) Peroxide-scavenging deficit underlies oligodendrocyte susceptibility to oxidative stress. Glia 22:371–378
39. Kanungo MS (1975) A model for ageing. J Theoret Biol 53:253–261
40. Khachaturian ZS (1987) Hypothesis on the regulation of cytosol calcium concentration and the aging brain. Neurobiol Aging 8:345–346
41. Kirischuk S, Parpura V, Verkhratsky A (2012) Sodium dynamics: another key to astroglial excitability? Trends Neurosci 35:497–506
42. Kress BT, Iliff JJ, Xia M, Wang M, Wei H, Zeppenfeld D, Xie L, Kang H, Xu Q, Liew J, Plog BA, Ding F, Deane R, Nedergaard M (2014) Impairment of paravascular clearance pathways in the aging brain. Ann Neurol
43. Krimer LS, Hyde TM, Herman MM, Saunders RC (1997) The entorhinal cortex: an examination of cyto- and myeloarchitectonic organization in humans. Cereb Cortex 7:722–731
44. Lalo U, Palygin O, North RA, Verkhratsky A, Pankratov Y (2011) Age-dependent remodelling of ionotropic signalling in cortical astroglia. Aging Cell 10:392–402
45. Landfield PW (1987) 'Increased calcium-current' hypothesis of brain aging. Neurobiol Aging 8:346–347
46. Lee DC, Ruiz CR, Lebson L, Selenica ML, Rizer J, Hunt JB Jr, Rojiani R, Reid P, Kammath S, Nash K, Dickey CA, Gordon M, Morgan D (2013) Aging enhances classical activation but mitigates alternative activation in the central nervous system. Neurobiol Aging 34:1610–1620
47. Logan JM, Sanders AL, Snyder AZ, Morris JC, Buckner RL (2002) Under-recruitment and nonselective recruitment: dissociable neural mechanisms associated with aging. Neuron 33:827–840
48. Lopez-Otin C, Galluzzi L, Freije JMP, Madeo F, Kroemer G (2016) Metabolic control of longevity. Cell 166:802–821
49. Lynch AM, Murphy KJ, Deighan BF, O'Reilly JA, Gun'ko YK, Cowley TR, Gonzalez-Reyes RE, Lynch MA (2010) The impact of glial activation in the aging brain. Aging Dis 1:262–278
50. Maher P (2005) The effects of stress and aging on glutathione metabolism. Ageing Res Rev 4:288–314
51. Matute C, Alberdi E, Domercq M, Sanchez-Gomez MV, Perez-Samartin A, Rodriguez-Antiguedad A, Perez-Cerda F (2007) Excitotoxic damage to white matter. J Anat 210:693–702
52. Medvedev ZA (1990) An attempt at a rational classification of theories of ageing. Biol Rev Camb Philos Soc 65:375–398
53. Micu I, Jiang Q, Coderre E, Ridsdale A, Zhang L, Woulfe J, Yin X, Trapp BD, McRory JE, Rehak R, Zamponi GW, Wang W, Stys PK (2006) NMDA receptors mediate calcium accumulation in myelin during chemical ischaemia. Nature 439:988–992

54. Ming GL, Song H (2011) Adult neurogenesis in the mammalian brain: significant answers and significant questions. Neuron 70:687–702
55. Nichols NR, Day JR, Laping NJ, Johnson SA, Finch CE (1993) GFAP mRNA increases with age in rat and human brain. Neurobiol Aging 14:421–429
56. Nicholson DA, Yoshida R, Berry RW, Gallagher M, Geinisman Y (2004) Reduction in size of perforated postsynaptic densities in hippocampal axospinous synapses and age-related spatial learning impairments. J Neurosci 24:7648–7653
57. Ogura K, Ogawa M, Yoshida M (1994) Effects of ageing on microglia in the normal rat brain: immunohistochemical observations. NeuroReport 5:1224–1226
58. Orre M, Kamphuis W, Osborn LM, Jansen AH, Kooijman L, Bossers K, Hol EM (2014) Isolation of glia from Alzheimer's mice reveals inflammation and dysfunction. Neurobiol Aging
59. Pelvig DP, Pakkenberg H, Stark AK, Pakkenberg B (2008) Neocortical glial cell numbers in human brains. Neurobiol Aging 29:1754–1762
60. Perry VH, Matyszak MK, Fearn S (1993) Altered antigen expression of microglia in the aged rodent CNS. Glia 7:60–67
61. Peters A, Josephson K, Vincent SL (1991) Effects of aging on the neuroglial cells and pericytes within area 17 of the rhesus monkey cerebral cortex. Anat Rec 229:384–398
62. Peters A, Sethares C (2004) Oligodendrocytes, their progenitors and other neuroglial cells in the aging primate cerebral cortex. Cereb Cortex 14:995–1007
63. Peters O, Schipke CG, Philipps A, Haas B, Pannasch U, Wang LP, Benedetti B, Kingston AE, Kettenmann H (2009) Astrocyte function is modified by Alzheimer's disease-like pathology in aged mice. J Alzheimers Dis 18:177–189
64. Potokar M, Stenovec M, Jorgacevski J, Holen T, Kreft M, Ottersen OP, Zorec R (2013) Regulation of AQP4 surface expression via vesicle mobility in astrocytes. Glia 61:917–928
65. Provenzano FA, Muraskin J, Tosto G, Narkhede A, Wasserman BT, Griffith EY, Guzman VA, Meier IB, Zimmerman ME, Brickman AM, Alzheimer's Disease Neuroimaging I (2013) White matter hyperintensities and cerebral amyloidosis: necessary and sufficient for clinical expression of Alzheimer disease? JAMA Neurol 70, 455–461
66. Raz N, Lindenberger U, Rodrigue KM, Kennedy KM, Head D, Williamson A, Dahle C, Gerstorf D, Acker JD (2005) Regional brain changes in aging healthy adults: general trends, individual differences and modifiers. Cereb Cortex 15:1676–1689
67. Rodriguez-Callejas JD, Fuchs E, Perez-Cruz C (2016) Evidence of Tau hyperphosphorylation and dystrophic microglia in the common marmoset. Front Aging Neurosci 8:315
68. Rodriguez JJ, Terzieva S, Olabarria M, Lanza RG, Verkhratsky A (2013) Enriched environment and physical activity reverse astrogliodegeneration in the hippocampus of AD transgenic mice. Cell Death Dis 4:e678
69. Rodriguez JJ, Yeh CY, Terzieva S, Olabarria M, Kulijewicz-Nawrot M, Verkhratsky A (2014) Complex and region-specific changes in astroglial markers in the aging brain. Neurobiol Aging 35:15–23
70. Rose CF, Verkhratsky A, Parpura V (2013) Astrocyte glutamine synthetase: pivotal in health and disease. Biochem Soc Trans 41:1518–1524
71. Rose CR, Verkhratsky A (2016) Principles of sodium homeostasis and sodium signalling in astroglia. Glia
72. Ruzicka V (1924) Beitrage zum Stadium der Protoplasmahysteretischen Vorgange (Zur Kausalitat der Alterns). Archiv fur mikroskopische Anatomie und Entwicklungsmechanik 101:459–482
73. Ruzicka V (1926) Altern und Verjungung won Standpunkt der allegemeinen Biologie. Praha, Praha
74. Salter MG, Fern R (2005) NMDA receptors are expressed in developing oligodendrocyte processes and mediate injury. Nature 438:1167–1171
75. Sampedro-Piquero P, De Bartolo P, Petrosini L, Zancada-Menendez C, Arias JL, Begega A (2014) Astrocytic plasticity as a possible mediator of the cognitive improvements after environmental enrichment in aged rats. Neurobiol Learn Mem 114:16–25

76. Schoenemann PT, Sheehan MJ, Glotzer LD (2005) Prefrontal white matter volume is disproportionately larger in humans than in other primates. Nat Neurosci 8:242–252
77. Sharaf A, Krieglstein K, Spittau B (2013) Distribution of microglia in the postnatal murine nigrostriatal system. Cell Tissue Res 351:373–382
78. Sheffield LG, Berman NE (1998) Microglial expression of MHC class II increases in normal aging of nonhuman primates. Neurobiol Aging 19:47–55
79. Sierra A, Gottfried-Blackmore AC, McEwen BS, Bulloch K (2007) Microglia derived from aging mice exhibit an altered inflammatory profile. Glia 55:412–424
80. Smith TD, Adams MM, Gallagher M, Morrison JH, Rapp PR (2000) Circuit-specific alterations in hippocampal synaptophysin immunoreactivity predict spatial learning impairment in aged rats. J Neurosci 20:6587–6593
81. Soreq L, Consortium UKBE, North American Brain Expression C, Rose J, Soreq E, Hardy J, Trabzuni D, Cookson MR, Smith C, Ryten M, Patani R, Ule J (2017) Major shifts in glial regional identity are a transcriptional hallmark of human brain aging. Cell Rep 18:557–570
82. Stern Y (2009) Cognitive reserve. Neuropsychologia 47:2015–2028
83. Streit WJ, Braak H, Xue QS, Bechmann I (2009) Dystrophic (senescent) rather than activated microglial cells are associated with tau pathology and likely precede neurodegeneration in Alzheimer's disease. Acta Neuropathol 118:475–485
84. Streit WJ, Sammons NW, Kuhns AJ, Sparks DL (2004) Dystrophic microglia in the aging human brain. Glia 45:208–212
85. Streit WJ, Xue QS, Tischer J, Bechmann I (2014) Microglial pathology. Acta Neuropathol Commun 2:142
86. Taylor RC (2016) Aging and the UPR(ER). Brain Res 1648:588–593
87. Thorburne SK, Juurlink BH (1996) Low glutathione and high iron govern the susceptibility of oligodendroglial precursors to oxidative stress. J Neurochem 67:1014–1022
88. Tischer J, Krueger M, Mueller W, Staszewski O, Prinz M, Streit WJ, Bechmann I (2016) Inhomogeneous distribution of Iba-1 characterizes microglial pathology in Alzheimer's disease. Glia 64:1562–1572
89. Toescu EC, Verkhratsky A (2007) The importance of being subtle: small changes in calcium homeostasis control cognitive decline in normal aging. Aging Cell 6:267–273
90. Tse KH, Herrup K (2017) DNA damage in the oligodendrocyte lineage and its role in brain aging. Mech Ageing Dev 161:37–50
91. VanGuilder HD, Bixler GV, Brucklacher RM, Farley JA, Yan H, Warrington JP, Sonntag WE, Freeman WM (2011) Concurrent hippocampal induction of MHC II pathway components and glial activation with advanced aging is not correlated with cognitive impairment. J Neuroinflammation 8:138
92. Verkhratsky A, Kirchhoff F (2007) NMDA Receptors in glia. Neuroscientist 13:28–37
93. Verkhratsky A, Marutle A, Rodriguez-Arellano JJ, Nordberg A (2015) Glial asthenia and functional paralysis: a new perspective on neurodegeneration and Alzheimer's disease. Neuroscientist 21:552–568
94. Verkhratsky A, Matteoli M, Parpura V, Mothet JP, Zorec R (2016) Astrocytes as secretory cells of the central nervous system: idiosyncrasies of vesicular secretion. EMBO J 35:239–257
95. Verkhratsky A, Nedergaard M (2018) Physiology of Astroglia. Physiol Rev 98:239–389
96. Verkhratsky A, Rodriguez JJ, Parpura V (2012) Calcium signalling in astroglia. Mol Cell Endocrinol 353:45–56
97. Verkhratsky A, Untiet V, Rose CR (2019) Ionic signalling in astroglia beyond calcium. J Physiol
98. Walhovd KB, Johansen-Berg H, Karadottir RT (2014) Unraveling the secrets of white matter—bridging the gap between cellular, animal and human imaging studies. Neuroscience 276C:2–13
99. Weismann A (1881) Ueber die Dauer des Lebens. Vortrag, in der 2. allgemeinen Sitzung d. 54. Versammlung Deutscher Naturforscher u. Aerzte in Salzburg, am 21, Sept 1881
100. West MJ (1993) Regionally specific loss of neurons in the aging human hippocampus. Neurobiol Aging 14:287–293

101. Young VG, Halliday GM, Kril JJ (2008) Neuropathologic correlates of white matter hyperintensities. Neurology 71:804–811
102. Zhu X, Hill RA, Dietrich D, Komitova M, Suzuki R, Nishiyama A (2011) Age-dependent fate and lineage restriction of single NG2 cells. Development 138:745–753
103. Zorec R, Parpura V, Verkhratsky A (2018) Preventing neurodegeneration by adrenergic astroglial excitation. FEBS J

Chapter 9
Astroglia in Leukodystrophies

M. S. Jorge and Marianna Bugiani

Abstract Leukodystrophies are genetically determined disorders affecting the white matter of the central nervous system. The combination of MRI pattern recognition and next-generation sequencing for the definition of novel disease entities has recently demonstrated that many leukodystrophies are due to the primary involvement and/or mutations in genes selectively expressed by cell types other than the oligodendrocytes, the myelin-forming cells in the brain. This has led to a new definition of leukodystrophies as genetic white matter disorders resulting from the involvement of any white matter structural component. As a result, the research has shifted its main focus from oligodendrocytes to other types of neuroglia. Astrocytes are the housekeeping cells of the nervous system, responsible for maintaining homeostasis and normal brain physiology and to orchestrate repair upon injury. Several lines of evidence show that astrocytic interactions with the other white matter cellular constituents play a primary pathophysiologic role in many leukodystrophies. These are thus now classified as astrocytopathies. This chapter addresses how the crosstalk between astrocytes, other glial cells, axons and non-neural cells are essential for the integrity and maintenance of the white matter in health. It also addresses the current knowledge of the cellular pathomechanisms of astrocytic leukodystrophies, and specifically Alexander disease, vanishing white matter, megalencephalic leukoencephalopathy with subcortical cysts and Aicardi–Goutière Syndrome.

Keywords Leukodystrophy · Astrocytopathy · Astrocytes

9.1 Introduction

The white matter (WM) of the brain consists of densely packed glial cells, including oligodendrocytes, oligodendrocyte progenitor cells (OPC), astrocytes and microglia, myelinated and unmyelinated axons and blood vessels with their cellular components [61, 76, 266]. The axonal tracts connect different grey matter areas to each other,

M. S. Jorge · M. Bugiani (✉)
Department of Pathology, Free University Medical Centre, Amsterdam, The Netherlands
e-mail: m.bugiani@amsterdamumc.nl

A. Verkhratsky et al. (eds.), *Neuroglia in Neurodegenerative Diseases*,
Advances in Experimental Medicine and Biology 1175,
https://doi.org/10.1007/978-981-13-9913-8_9

creating functional pathways and networks [77]. Astrocytes are a heterogeneous cell population with respect to their developmental origin, morphology, function, physiological properties and environment [162, 202, 218]. They derive from radial glia cells (RGC), outer radial glial cells (oRGs), OPC and precursors situated in the spinal cord [84, 202]. Radial glial cells are multipotent progenitor bipolar-shaped cells, which reside in the embryonic ventricular zone (VZ) and are derived from the neuroepithelial cells [14, 78, 104, 116]. oRGs are a type of subventricular zone (SVZ) progenitor cells, which are generated from radial glia asymmetrical divisions in the VZ [232]. OPCs are a cell population that also express the proteoglycan NG2 and are therefore also designated as NG2-glial [269]. These cells differentiate into oligodendrocytes and astrocytes during embryogenesis [59, 179]. However, OPCs survive in the adult brain and keep their potential to differentiate into oligodendrocytes under certain conditions, such as in the case of myelin damage. OPC proliferation and differentiation are also modulated by astrocyte-derived factors [59, 120, 191].

Astroglia exist in the white and grey matter, the former having a fibrous morphology and the latter a protoplasmic morphology [13, 71, 154]. Structurally, fibrous astrocytes are constituted by intermediate filaments such as vimentin, nestin and glial acidic fibrillary protein (GFAP) [99] and express the CD44 surface receptor [139, 209]. CD44 is the receptor for the extracellular component hyaluronan that negatively regulates OPC differentiation [8, 68]. $GFAP^{-/-}$ mice show abnormal white matter architecture and impaired long-term myelination, showing the importance of this protein on the global structure and function of the white matter [137]. Considering that astrocytes develop a specific morphology and molecular profile depending on their function and location in, and within, the white or grey matter, fibrous astrocytes conceivably adapt to their surroundings, showing a selective vulnerability upon certain conditions [71, 257]. This could explain why the white matter is not always homogeneously affected in leukodystrophies.

9.2 Role of Astrocytes in Maintaining the Integrity of the White Matter

Astrocytes promote white matter integrity and homeostasis through the cooperation with other glial cells, axons and non-neural cells. They control ion-water homeostasis [55], thereby modulating neuronal activity and myelin water content [211]; provide metabolic support to neurons and other cells in the CNS by regulating the glucose metabolism and, in part, brain lipid metabolism [102, 178]; and stimulate the cerebrospinal fluid (CSF) and the interstitial fluid (ISF) flow [107, 108, 212]. Furthermore, astrocytes are involved in the formation and maintenance of the blood–brain barrier (BBB) by the release of glial-derived neurotrophic factor (GDNF), angiopoietin-1, angiotensin II, bone morphogenetic protein (BMP) signalling and microglia recruitment, therefore inducing and maintaining BBB-related functions in endothelial cells and brain homeostasis [7, 225, 258]. The BBB is essential to protect the brain against the toxicity of many substances, the entrance of pathogens and influx of ions such as K^+ and Ca^{2+} [15]. This diffusion barrier is composed of endothelial cells glued together by tight junctions, pericytes and astrocyte endfeet [1, 9, 260]. These cellular elements along with neurons and the extracellular matrix form the 'neurovascular unit' [98]. Moreover, astrocytes contact with the blood vessels inducing alterations in their shape which are crucial for proper metabolic support to neurons [186]. Astrocytic intracellular Ca^{2+} increase is responsible to initiate this process that is triggered by increased neuronal activity. Nitric oxide (NO) is an essential compound dictating whether the blood vessels undergo contraction or dilatation [87]. Astrocytes also control local blood flow in the CNS in order to coincide with neuronal metabolic demands as a response to synaptic activity [106].

Astrocytes are responsible for many basilar regulatory functions that play a crucial role in maintaining the normal physiological conditions in the CNS [16, 111, 119]. They promote proper synaptic transmission through the uptake of K^+ and glutamate [167, 256] and modulate synaptic plasticity by the release of gliotransmitters, growth factors and via astrocytic Ca^{2+} signalling [120, 171]. Furthermore, they are involved in cognition, regulation of the circadian rhythm and formation and pruning of synapses [41, 222, 269], amongst others. Astrocytes present three different phenotypes, depending on whether the CNS is under normal physiological conditions or upon injury such as inflammation. In the healthy CNS, astrocytes are 'quiescent' performing all the functions described above [69]. However, upon injury, astrocytes show an active and a reactive phenotype [120], along a spectrum spanning between hypertrophic and scar-forming cells [118]. The active phenotype is reported as the intermediary form between quiescent and reactive phenotypes, meaning that astrocytes show a mild reaction to injury [136]. When the insult persists or pathogenesis begins, this reaction gradually increases into a severe response [160].

9.3 Astrogliosis

Astrocytes become reactive when there is a disruption in the normal function of the brain such as upon injuries, inflammatory processes or demyelinating conditions [264]. This process is called astrogliosis, which constitutes a pathological hallmark of the diseased tissue. This phenomenon is heterogeneous, not an all or none event but, on the contrary, it is a gradually changing process tailored on the type of CNS injury origin and its severity [220]. As a consequence, astrocytes change their morphology, molecular profile, signalling pathways, function and even their interactions with other cells, becoming active or reactive [217, 218]. We will not discuss further all of these alterations. Despite this, it is crucial to mention the GFAP overexpression, its intracytoplasmatic aggregates in astrocytes, IFN-α upregulation and KCa3.1 potassium channel [72, 153, 176], because they characterize some of the white matter disorders that will be addressed later on. GFAP, known to regulate many cellular pathways, is present in five isoforms (GFAP $\alpha, \beta, \gamma, \delta, \kappa$) [19, 161, 185, 195, 265] with different expressions and distributions in various pathologies, potentially shaping the outline and features of each disorder. Notwithstanding, the activity of each isoform is still not clearly defined and future research is needed to understand the functional consequences. Since the reactive astrocyte population is heterogeneous, and therefore difficult to classify in different stages, several outcomes may result from astrogliosis [78]. KCa3.1 channels play an important role in regulating ion-water homeostasis and astrocyte activation, which may be due to their upregulation in vacuolating leukodystrophies [23, 54, 197]. Astrocyte loss of function and astrogliosis can occur in different time scales with the former sometimes taking place before the latter [220]. Although they are not synonymous of each other, this is a characteristic consequence when an astrogliosis process is taking place. Loss of astrocyte coupling and inability to buffer glutamate and potassium are some of the after effects of astrogliosis [78], which impair neuronal activity and synaptic transmission by promoting an excitotoxic environment [196]. On the other hand, astrocytes along with microglia contribute to tissue repair and axon regeneration by participating in the formation of the glial scar [82, 210]. Glial scar is created when astrocytic processes surround an injured tissue, establishing a barrier to pathogens and inflammatory cells [33, 218, 220]. Hence, reactive astrocytes can play neuroprotective and detrimental roles depending on the pathological context [78].

9.4 Leukodystrophies

Two key works, carried out by Morell and Seitelberger in 1984, introduced the term leukodystrophy as a disease category providing a set of criteria for the classification. Both the authors have pinpointed the importance of oligodendrocytes, as the cells primarily affected, due to the fact that the pathogenesis of leukodystrophies was centred on myelin. Although at that time, a small number of pathologies were known and no genetic linkage had been established, their work provoked a great deal of research. This research was much sought after the introduction of more sensitive techniques such as magnetic resonance imaging (MRI), magnetic resonance spectroscopy (MRS) and whole exome sequencing (WES) that greatly improved clinical diagnosis [117]. Leukodystrophies were traditionally considered myelin disorders thus, for many years, the research on white matter diseases was focused on oligodendrocytes, being these the myelin-forming cells [85, 158, 208]. The few studies that have emphasized astrocytes as important contributors to the disease mechanisms of leukodystrophies have focused on their role in regulating myelin formation and function and have always approached interactions with other glial cells from the myelin perspective [144]. However, according to a new definition and classification of leukodystrophies [249], these encephalopathies are genetic and inherited disorders affecting the white matter of the CNS as a whole. This means that they are provoked by the dysfunction of cellular mechanisms, which in health are responsible for the integrity and maintenance of the white matter, and by the impairment of the crosstalk between all white matter structural components including oligodendrocytes, astrocytes, microglia, blood vessels and other non-neural cells [249]. In some of these diseases, astrocytes are known to acquire abnormal and cytotoxic functions due to mutations in astrocyte-specific gene products. Alexander Disease (AxD), Vanishing White Matter Disease (VWM), Megalencephalic Leukoencephalopathy with subcortical cysts (MLC), and Aicardi Goutières Syndrome (AGS) are examples of the leukodystrophies classified as astrocytopathies meaning that astrocytes are the primarily cells affected [24, 62, 177], although the pathomechanisms of the various disorders may differ [24, 243, 249]. The impaired cell-to-cell interactions and astrocytic dysfunction, contribute to cellular, molecular and morphological alterations in the brain and, thus, to the pathogenesis of leukodystrophies.

9.5 Alexander Disease

AxD is an autosomal dominant monogenic inherited pathology caused by mutations, most of them occurring de novo, in the coding region of *GFAP* [3, 24, 134, 138, 152, 254]. *GFAP* encodes for an intermediate filament protein called glial fibrillary acidic protein (GFAP), which is only expressed by astrocytes in the CNS and is responsible for their cytoskeletal architecture [182]. This disorder presents with psychomotor impairment, spasticity, intellectual disability and possibly seizures, accompanied by other symptoms that depend on the age at onset [86, 151, 199]. Furthermore, different phenotypes have been described, according to genotype, age of onset and pathological alterations in the brain. However, AxD is currently classified only into two types—type I and type II—the former with early-onset and showing developmental delay and macrocephaly, and the later with later-onset and presenting with bulbar signs, including dysphagia and dysphonia [151, 172, 181].

Neuropathological changes in AxD include widespread degeneration of the white matter, prevailing in the frontal white matter also with cystic degeneration, and variable loss of neurons. These can be assessed through MRI and macroscopically at autopsy [74, 242]. Cerebellar and brain stem atrophy, dilation of lateral ventricles and swelling of basal nuclei and thalami are some of the features seen in type I AxD. On the other hand, ventricular garlands and atrophy of medulla oblongata and cervical spinal cord are typical of type II AxD [74, 242, 246, 247].

Several studies have evaluated the occurrence of cellular and molecular changes in the AxD brain [91, 138, 151, 182, 228]. AxD astrocytes acquire a reactive phenotype characterized by hypertrophy of terminal cell processes with intracytoplasmatic aggregates named Rosenthal fibres (RF), which are the pathological hallmark of the disease. RF are comprised of mutant GFAP, ubiquitin, vimentin, nestin, plectin, p-JNK, p62, synemin, stress proteins—hsp-27 and αB-crystallin, and the 20S proteasome subunit [110, 177, 254]. αB-crystallin plays a crucial role in maintaining rearrangement of disruptive intermediate filaments networks, as RF [125, 228]. In addition, overexpression of GFAP mutant protein, possibly resulting from the dysregulation between synthesis and degradation, leads to aberrant morphology, polyploidy and astrocytic dysfunction and modulates various pathomechanisms of AxD [112, 228]. A previous work using transgenic mice with increased wild-type (WT) GFAP expression, showed that GFAP accumulation triggers cellular-stress and immune astrocytic responses [226, 227], by increasing the expression of complement components and macrophage-specific markers. As a result, this induces microglia and complement-dependent pathways activation [165, 166]. The complement cascade is part of the innate immune system and becomes activated when a set of soluble proteases binds to pathogens. It is initiated by a complement component, which can be either C1q (classical activation) or C3 (alternative activation). Other complement proteins become activated in a cascade, which ultimately results in the formation of a C3 protein, opsonizing subsets of synapses for elimination and leading to cellular responses such as pro-inflammatory signalling and receptor-mediated phagocytosis, among others [103, 130]. Additionally, the increase in expression of chemokines and cytokines like TNFα, IL1β and IL-6, along with the decreased expression of genes related with Ca^{2+} signalling, make astrocytes more vulnerable to the effects of stress and apoptosis [91, 182]. Additional data suggests that hyaluronan, a major constituent of the extracellular matrix known to prevent astrocyte and OPC differentiation is increased in AxD and other diseases where lack of myelin is observed. This glycosaminoglycan is mostly produced by astrocytes as its main receptor, CD44, is also expressed by these glial cells [139, 209].

9.5.1 Compromised Functions of Astrocytes in AxD

Accumulation of mutant GFAP inside astrocytic cell processes promotes the disturbance of its molecular profile, morphology, function and many cellular pathways that will alter their homeostasis and eventually normal brain physiology and interactions between cells [166].

Microglia are the resident immune cells of the central nervous system, which act as sentinels during pathological changes, responding to them by remodeling synaptic connections and their function in brain networks [126, 251]. Nonetheless, astrocytes also contribute to the regulation of the immune and anti-inflammatory responses in the CNS [42, 135, 219] through the expression of pattern-recognition receptors (PRRs). This PRR are also present in microglia and are responsible for sensing infectious agents entrance and endogenous danger signals [73]. Some studies suggest that these two glial cells are also coupled into gap junctions, this intercellular contact having the important function of modulating the activity of each other [94, 105, 219]. Specifically, astrocytes regulate migration, activation and proliferation of microglia via the release of anti- and pro-inflammatory factors [73]. On the other hand, microglia control astrocytic activation, proliferation and reactivity by the release of cytokines [174] and NO. NO induces astrocyte cell death by apoptosis, which was suggested as a mechanism that can control excessive reactivity [183]. Moreover, astrocytes and glial also dialogue in order to achieve other goals like the formation and maintenance of the BBB, [268] and the clearance of myelin debris in demyelination [216]. In AxD, astrocytes mediate microglial activation which, in return, induces stress and immune responses in astrocytes. Several lines of evidence [165, 166] corroborate that this cell-to-cell interaction may be the reason for perpetuated inflammation, affecting the function of other cells, including oligodendrocytes. Notably, damage to these cells can result in aberrant myelin formation. In addition, loss of astrocyte coupling due to decreased expression of connexins, which are the proteins responsible for the assembly of gap junctions, is observed. A previous study showed that the lack of connexins is related to demyelinating events [145].

Proper interaction between astrocytes and oligodendrocytes is essential for the support of the white matter. Astrocytes and oligodendrocytes are coupled through A/O gap junctions, allowing intercellular communication [169]. Along with A/A gap junctions between astrocytes, this population forms the panglial syncytium [159]. This functional structure allows the transportation of substances synthesized by astrocytes to oligodendrocytes in one-way direction and may act on K^+ buffering [155, 184, 234]. Astrocytes and oligodendrocytes require energy to pursue their functions, withdrawing glucose from blood vessels through the glucose transporter 1(GLUT1) [135]. However, when glucose is low in the blood or during increased neuronal activity, astrocytes act as the main energy suppliers to oligodendrocytes and neurons [40, 178]. Since they hold the capacity to store glycogen [28], they release glycogen-derived lactate and guide its transport to neurons through astrocyte–neuron lactate transfer shuttle (ANLTS) [510, 193]. Astrocytes synthesize various cytokines, chemokines and growth factors that regulate the migration, maturation, proliferation and differentiation of OPCs [44], ultimately modulating myelination [12, 44, 61, 120]. Ciliary neurotrophic factor (CNTF) [160], brain-derived neurotrophic factor (BDNF) [156, 262], leukemia inhibitory factor (LIF) [109] are examples of proteins that increase myelination by promoting oligodendrogenesis. TGF-alpha and BMP2/4 [12] are examples of factors that hamper myelination by preventing OPCs differentiation into myelin-forming cells [37, 44, 144, 205, 206].

In addition, the ability to buffer glutamate is impaired due to the defective expression of glutamate transporters—EAAT2/GLT—[231]. Moreover, it is observed a decrease in inwardly rectifying K^+ channels due to a deficient buffer of potassium [166]. Therefore, astrocytic dysfunction in buffering glutamate and potassium, in addition to lack of astrocyte coupling may lead to cytotoxicity, followed by loss of oligodendrocytes, selective neuronal death, possible axonal dysfunction and synaptic activity impairment. Together, these astrocytic dysfunctions contribute to neuronal excitotoxicity [11, 97] and prolonged neuronal depolarization, explaining one of the possible causes of neuronal death and the occurrence of epileptic seizures in this disease.

9.6 Vanishing White Matter

Vanishing white matter (VWM) is an autosomal recessive polygenic disorder, caused by mutations in any of the five genes (*EIF2B1, EIF2B2, EIF2B3, EIF2B4* and *EIF2B5*) that encode the subunits of the eukaryotic initiation translation factor 2B (eIF2B). eIF2B is responsible for the regulation of protein synthesis [29, 132, 239,

243, 246, 247], being specifically involved in the translation of messenger RNA (mRNA) into polypeptides. Adverse conditions like thermal, chemical and oxidative stress and physical trauma trigger a cellular-stress response, which leads to inhibition of protein synthesis to preserve cellular energy and prevent cellular death. This mechanism aims at decreasing accumulation of denatured and misfolded proteins, although it is compromised when elF2B shows to be dysfunctional and when mRNAs have peculiar ORFs that allow them to escape the inhibition of translation [43, 67, 200, 259]. This may be the possible reason behind the trigger of VWM, which is provoked by febrile infections, minor head trauma and acute fright [239, 240, 255].

Characterized, in every variant of the disease, by mild cognitive decline, occasional epileptic seizures and motor symptoms, like spasticity and cerebellar ataxia, this disease can also present other symptoms due to its broad phenotypic variation, which is influenced by genetics and environment [29, 80, 148, 180, 240, 244]. The onset of the most common variant is in childhood with an age range between 2 and 6 years [96, 201, 239]. However, the onset of the disease can vary depending on the variant of the disease. VWM is one of the most prevalent leukodystrophies in children and described as most frequent in Caucasian populations [241]. Nevertheless, one of the variants of the pathology seems to be more common among the Cree Indians with its onset in early childhood, with age at onset ranging between 3 and 9 months [80]. In other variants, the onset might be in adolescence or adulthood and is accompanied with the same symptoms described above in addition to psychiatric symptoms [148, 164, 180]. In female patients, the ovaries can also be affected [79]. Although VWM is chronic and progressive, it presents episodes of occasional seizures, increasing motor deterioration, hypotonia and loss of consciousness. Normally, the outcome is death after a few months, years or decades depending on onset age and disease severity, which are often inversely related. This means that the earlier the onset of the disease, the greater the severity thus, the younger the patient, the shorter the average life expectancy will be. The most severe variant of this disorder is the one that has an antenatal onset, in which the development of the foetus is completely compromised [246, 247].

Cellular, molecular and morphological changes are described through different stages of VWM including the pre-symptomatic phase. Some of these changes, assessed by MRI [245] consist in rarefaction and cavitation or cystic degeneration of the cerebral white matter, particularly that in the fronto-parietal lobe. Additional pathological findings include abnormally thin myelin sheaths, sometimes showing vacuolization; a certain degree of loss of myelin and axons; increased numbers of oligodendrocytes and OPC and abnormal morphology of astrocytes [194, 239, 240, 245, 246, 247]. The cause of myelin vacuolization is still not fully understood. However, a study suggests that myelin vacuolization is the result of astrocyte-specific proteins deficiency and lack of connexins expressed in those cells [55].

Cerebellum and brainstem show mild white matter signal abnormalities, the spinal cord is usually spared and the cerebral cortex is always spared. In this disorder, the brain undergoes gradual modifications, which are the result of the phenotypic variation but also of the stage of the disease. In an advanced phase, a large part of cerebral white matter ‘disappears’ by substitution with fluid with features and signal properties similar to cerebrospinal fluid (CSF) [239, 240, 245]. The assessment of the brain metabolic changes of patients with VWM is often performed with the aid of MRS, which shows the presence of lactate and glucose, compatible with the features described above in the later stage of the disease [20, 246, 247].

9.6.1 Compromised Functions of Astrocytes in VWM

Astrocytes and oligodendrocytes are especially affected in this disorder, although it has been proven that astroglia is the determinant in VWM [62]. Like all cells in the body, both of them express eIF2B, which makes it difficult to directly assign just one of them to the pathogenesis of VWM. Nonetheless, several studies have been addressing astrocytic dysfunction as the major contributory pathomechanism for the disorder [30, 62]. Previous work has shown that VWM astrocytes present abnormal molecular profile, overexpressing the GFAP delta isoform and heat shock protein αB-crystallin [30], which may justify their abnormal morphology, dysfunction and metabolic stress. Furthermore, the proportion between GFAP delta and alpha isoform is disturbed due to the lacking of upregulation of the GFAP alpha isoform, which may be responsible for the impaired crosstalk with other cell types, including oligodendrocytes and the inability to build up a scar in cavitated lesions in the white matter [30]. Several lines of evidence suggest that scarce and disproportionate gliosis and consequent lacking of scar formation are not related with a deficient proliferation index of VWM astrocytes but, on the contrary, to astrocytic immaturity. A few studies showed increased proliferation of astrocytes in vivo and in vitro, after mechanical stress, [30, 31, 58], revealing this was not the cause for meager gliosis. In contrast, the under-expression of protein S100β, which is crucial for astrocytic differentiation [8, 31], was detected in VWM astrocytes, unveiling their immature phenotype in this disease. The lack of astrocytic maturation contributes to impaired reactive gliosis, which may redound in the development of cavitated white matter.

In addition to all of the changes in the astrocytes, the crosstalk with oligodendrocytes and OPC within this pathology is impaired, explaining the lack of myelin, a typical feature of VWM. In mice, it was demonstrated that VWM astrocytes inhibit wild-type (WT) OPC differentiation into oligodendrocytes, impeding the formation of myelin-forming cells. On the other hand, VWM OPCs are able to go through a normal differentiation process if they are co-cultured with WT astrocytes [30, 62]. This suggests that astrocytic dysfunction is primary and that oligodendrocytes are the cells secondarily affected in this disorder. One possible reason for this occurrence is the increase in expression of hyaluronan, which follows the same pattern of the activity described above for AxD thus, inhibiting OPC maturation and (re)myelination. Accumulation of a particular form of this glycosaminoglycan, the high-molecular-weight (HMW) hyaluronan, was found in the frontal white matter of patients [31], which is consistent with the white matter abnormalities in fronto-parietal lobe showed in MRI. By contrast, less HMW hyaluronan is expressed in the cerebellum, which is also compatible with it being less affected at MRI. This suggests a link between astroglia, hyaluronan accumulation, OPC maturation and, in general, the severity of affected areas [31]. Studies using mice expressing VWM mutations showed that increased level of hyaluronan correlate with the course and severity of the pathology [62].

Astrocytes interact with neurons through the provision of trophic support [121, 221] and protection against oxidative stress [70, 147, 204] and excitotoxicity [11, 97]. Some axons are enwrapped with myelin, which is a modified plasma membrane that acts as an electrical insulator, increasing the velocity of nerve impulse propagation and decreasing energy expenditure [4]. In return, they promote the myelin formation by regulating astrocyte-derived pro-myelinating factors release and OPC differentiation. Nonetheless, a study showed that unmyelinated axons also contribute to the regulation of myelination through the vesicular release of glutamate in the white matter [267].

Another evidence that shows astrocytes as the primary cells affected in this disease is their effect on axonal pathology. Reduced axonal thickness correlates with tissue damage severity, suggesting impairment in the velocity of nerve impulse propagation and in consequence, in synaptic activity [76]. In addition to abnormalities in axonal diameter, myelin thickness and axon–myelin ratio, a recent work found an increase in the number of unmyelinated axons in the corpus callosum of VWM mice compared to WT animals [122]. The consequent increase in axonal density was showed, in the same study, to be promoted by VWM astrocytes.

In conclusion, the astrocyte dysfunction and the aberrant interaction with other glial cells and axons greatly influence the course of VWM pathogenesis

The phenotypic variation of the leukodystrophies may be influenced by astrocytes heterogeneity in terms of morphology, function, origin and even their responses to insults. Corroborating this hypothesis, evidence suggests that astrocyte location affects the abundance of GFAP [195]. Overall, it is reasonable to state that the heterogeneity of astrocytic function and its activity pattern drive the course (selective vulnerability, disease severity and repair potential) of astrocytopathies.

9.7 Megalencephalic Leukoencephalopathy with Subcortical Cysts

Megalencephalic leukoencephalopathy with subcortical cysts (MLC) [236, 237] is an autosomal polygenic inherited disorder, caused by recessive mutations in any of two genes—*MLC1* [132] or *GLIALCAM* [142]—or dominant mutations in *GLIALCAM* [95]. The former encodes a highly hydrophobic membrane protein that shows low homology to ion channels [21, 132, 229] such as the potassium channel Kv1.1, and transporters; while the latter encodes an adhesion-like molecule [22, 113, 143] that acts as an MLC1-chaperone, as a secondary subunit of the ClC-2 chloride channel [113] and additionally as a guiding molecule to direct MLC1 to cell junctions [143].

Mutations in the MLC1 affect the majority of the population of patients with MLC with a prevalence of 80% [22, 132, 133, 157].

This leukodystrophy is marked by mild cognitive decline, epileptic seizures and motor symptoms, like spasticity and ataxia, with its onset in infancy [95, 213, 233, 236, 237]. Morphological alterations in the brain such as diffuse MRI signal changes, swelling and, eventually, atrophy of the cerebral white matter intramyelinic edema, presence of subcortical cysts mainly in the anterior temporal, frontal and parietal regions [236, 237] and, pathologically, vacuoles in myelin [175, 238] and astrocytic endfeet [32, 63, 66] are some of the main features of MLC. These were discovered through MRI, MRS [56] and brain biopsy [248].

9.7.1 Compromised Functions of Astrocytes in MLC

In the CNS, MLC1 is expressed at the A/A junctions, in astrocytic endfeet contacting the blood– and CSF–brain barriers, and in Bergmann glia of the cerebellum [21, 63, 203, 229, 230]. GlialCAM is co-localized with MLC1 at astrocytic endfeet [35, 100, 142], but is also present in oligodendrocytes processes [55, 75, 113], unlike MLC1 [6, 66, 203]. Moreover, *GlialCAM* is important for the correct targeting of MLC1 [142, 143], GlialCAM [36, 100] and the voltage-gated osmosensitive Cl-channel (CLC-2) to astrocyte junctions [113]. A previous study, using GlialCAM-null mice, showed absent expression of MLC1 and under-expression of CLC-2 revealing a direct interaction between GlialCAM and these two proteins [32, 113]. Mutations in *GlialCAM* lead to defects in trafficking and, therefore, to altered sites, which have been suggested to affect the course of the disease [32, 66, 100, 214]. Furthermore, mutations in *GlialCAM* also cause dysfunction of CLC-2, because the ability to change functional properties like rectification is impaired [114]. The functional interaction between MLC1 and ion transporters such as the sodium/potassium-ATPase pump (Na, K-ATPase) [25, 26], ion and water channels, including KCa3.1, VRAC, the transient receptor potential cation channel subfamily V, member 4 (TRPV4) [66, 128, 129], inward rectifier potassium channel 4.1 (Kir4.1), the water channel aquaporin-4 (AQP4) [128] and cell signalling pathways, among which the EGFR/ERK [129] and PLCγ1 pathways [253], are involved in several cellular processes such as proliferation, maturation, astrocyte activation and apoptosis. The dysfunctional interaction between MLC1, GlialCAM and all the proteins described above contributes to dysfunction of the white matter and, thus, to the alteration of cerebral homeostasis [65].

MLC1 is implicated in the regulation of volume alterations in astrocytes and, therefore, in the promotion of their osmotic balance [155]. In response to the disturbance of ion-water homeostasis or oxidative stress, a cascade of events follows such as astrocyte swelling and activation [34, 81]. The return to osmotic balance is achieved through the activation of volume regulated anion channel (VRAC) and calcium-activated KCa3.1 potassium channel, and EGFR/ERK and PLCγ1 pathways [253], which will modulate the activation of the regulatory volume decrease (RVD) [173, 224]. VRAC activity is focused on the control of astrocytes concentration gradient, by regulating their water flux and organic osmolytes transport, such as chloride. However, in the context of MLC, MLC1 is under-expressed in astrocytes [64] together with reduced expression of GlialCAM [192], result in defective chloride currents, hindering VRAC activity and, therefore, the adjustment of changes in the intracellular and extracellular environment [66, 155, 192]. Thus, RVD is impaired, which contributes to the long-term maintenance of swollen astrocytes. A previous study proposes a possible causal relation between disturbed fluid homeostasis and vacuolization in astrocyte end-feet [35]. Using *Clcn2*$^{-}$ mice, it has been demonstrated that ClC-2 is involved in an astrocytic vacuolation phenotype similar to MLC [18]. Findings from an earlier study, in which expression of MLC1 protein was downregulated in order to mimic the expression levels of this protein in the diseased brain, show the appearance of vacuoles in astrocytes, a phenotype not presented prior to depletion of MLC1 [63]. Furthermore, an increase in both MLC1 and GlialCAM expression is able to revert this phenotype [35]. These results suggest a correlation between MLC1 and vacuolization in astrocytes, being the former a possible cause for the latter. However, it is not known which mechanism might be involved.

The KCa3.1 are potassium channels with Ca^+ entry-induced opening, located together with MLC1, in astrocytes processes that are part of the neurovascular unit [141]. The increase in intracellular Ca^+ concentration is mediated by TRPV4, which demonstrate a crosstalk between different ion channels and MLC1 protein [128]. Thus, previous work suggests that MLC1 and KCa3.1 may cooperate to regulate BBB permeability, Na^+ influx, and ion-water homeostasis [38, 129, 149]. These astrocytic functions are compromised, leading to morphological, cellular and molecular changes observed in MLC but also in many other leukodystrophies.

Astroglia clearance of K^+ by uptake and spatial buffer is impaired in several leukodystrophies [124, 146, 256], among them MLC, due to dysfunctional interaction of several proteins [27], such as MLC1, GlialCAM, connexins [163, 261], Kir4.1 [60, 93, 207] and AQP4 [17, 92, 223]. Astrocyte function impairment leads to hampered regulation of extracellular potassium concentration ($[K^+]_o$) resulting in astrocyte swelling and, consequently, in the development of cerebral white matter edema [129, 248]. Furthermore, the increase in $[K^+]_o$, due to slow K^+ kinetics, and defective glutamate uptake also plays a role in sustaining and spreading depolarization and, thus, in enhancing excitability followed by epileptic seizures [45, 65, 115]. Although mutations in *CLCN2* have not been found in MLC, several lines of evidence show that activation of CLC-2 is dependent on interaction with MLC1 and GlialCAM and depolarization. The formation of this tertiary complex may be important in delaying exacerbated brain activity, counteracting membrane voltage to more negative values through chloride influx [215]. The impairment of all of the astrocytic functions described above drives the course of the disease, which suggests that the cells that play the central role in MLC are the astrocytes [88].

Regarding cellular processes, WT MLC1 interferes in astrocyte proliferation by down-regulating the activation of various signalling pathways, such as ERK1/2 and PLCγ1, blocking KCa3.1 channel and promoting epidermal growth factor receptor (EGFR) degradation, all known to promote astrocyte growth. On the contrary, mutations in MLC1 favour astrocyte proliferation. This is corroborated by evidence showing that MLC1 expression increases when cell proliferation is inhibited [63].

Studies using MLC mouse models, such as *GlialCAM*-null mice and *Mlc1*-null mice [32, 66, 75, 100] are important to better understand the cellular and molecular pathomechanisms of the disease, especially in early stages, which is difficult to assess and study in MLC patients, unveiling the pathophysiological consequences of the reduced expression in both genes, in MLC [32, 66].

9.8 Aicardi–Goutières Syndrome

Aicardi–Goutières syndrome (AGS) [2, 50] is an autosomal polygenic inherited neurodegenerative leukodystrophy, caused by mutations, most of which are recessive [187], in several genes (TREX1 [48], RNASEH2A, RNASEH2B, RNASEH2C [49], SAMHD1 [186], ADAR1 [189] and IFIH1/MDA5 [190]). For the purpose of this chapter, we will only discuss three-prime repair exonuclease 1 (TREX1) mutations for two reasons: they are the only ones in which a relationship with astrocytes has been shown and are also the most common and lethal mutations found in patients with AGS. In addition, further details regarding innate immune signalling pathways and its activation have been addressed in AxD and, for this reason, will not be explained again in this section.

Characterized by severe cognitive decline, irritability, seizures and motor deterioration symptoms, such as spasticity and dystonia, this early-onset disease presents other symptoms due to its multisystem involvement [51, 123]. Cellular and morphological changes identified by neuroimaging and other assessment tools show encephalopathy, increased levels of IFN-α in the CSF [57, 101, 131], cerebral and brain stem atrophy, absence of myelin, calcification of basal ganglia and frontotemporal swelling with presence of cysts, although not always, [47, 89, 90, 127, 140, 168, 170, 188, 235, 245] in cerebral areas that coincide with the location of cysts observed in AxD, VWM and MLC [252].

9.8.1 Compromised Functions in AGS

TREX1 prevents the cytosolic accumulation of DNA and RNA, in physiological conditions. Mutations in this gene cause impairment of exonuclease [39, 150] and exoribonuclease function [263], which leads to the incorrect recognition of DNA and RNA as antigen [46]. This triggers the activation and/or deregulation of several innate immune signalling pathways, such as the cGAS–STING pathway [83], and increased production of monocytes, cytokines, among which interferon type 1 (IFN-α,) the biomarker for the disease, GFAP and other proteins [52, 53, 198, 250]. Previous studies show that in CNS, overproduction and release of IFN signature may be attributed to astrocytes and microglia. It was demonstrated that IFN-α co-localizes with GFAP and disrupts astrocyte-specific protein function, in addition to triggering astrogliosis [198, 250]. Regarding the mutations in the other genes, it is still unclear the contribution of astrocytes to AGS.

9.9 Conclusions

Astrocytes are a heterogeneous cell population in terms of origin, morphology, function and even their responses to brain insults. They play a role in modulating cellular and molecular mechanisms, activity and morphology of the brain. In this review, we describe how astrocytic function and factors secreted by astrocytes in addition to interaction with axons, glial cells and non-neural cells contribute to integrity and maintenance of white matter in health. We show that, in disease, astrocytic dysfunction and dysregulation of these mechanisms trigger abnormal responses in the brain, possibly also leading to leukodystrophies. In addition, we suggest that astrocytes heterogeneity influence the phenotypic variation observed in astrocytopathies. AxD and AGS are leukodystrophies, characterized by cytokines upregulation, GFAP and IFN-α, respectively. VWM and MLC are disorders also driven by an astrocytic defect in ion-water homeostasis.

Together, these conclusions suggest that astrocytes play an essential role in the pathophysiology of selected leukodystrophies, the astrocytopathies. Therefore, comprehending astrocytes physiology, function and interactions with other cells may provide new potential therapeutic targets.

References

1. Abbott NJ, Rönnbäck L, Hansson E (2006) Astrocyte–endothelial interactions at the blood––brain barrier. Nat Rev Neurosci 7(1):41
2. Aicardi J, Goutières F (1984) A progressive familial encephalopathy in infancy with calcifications of the basal ganglia and chronic cerebrospinal fluid lymphocytosis. Ann Neurol 15:49–54
3. Alexander WS (1949) Progressive fibrinoid degeneration of fibrillary astrocytes associated with mental retardation in a hydrocephalic infant. Brain 72(3):373–381
4. Alizadeh A, Dyck SM, Karimi-Abdolrezaee S (2015) Myelin damage and repair in pathologic CNS: challenges and prospects. Front Mol Neurosci 8:35
5. Allaman I, Belanger M, Magistretti PJ (2011) Astrocyte-neuron metabolic relationships: for better and for worse. Trends Neurosci 34:76–87
6. Ambrosini E, Serafini B, Lanciotti A, Tosini F, Scialpi F, Psaila R, Di Girolamo F, Petrucci TC, Aloisi F (2008) Biochemical characterization of MLC1 protein in astrocytes and its association with the dystrophin-glycoprotein complex. Mol Cell Neurosci 37:480–493
7. Araya R, Kudo M, Kawano M, Ishii K, Hashikawa T, Iwasato T, Yamada M (2008) BMP signaling through BMPRIA in astrocytes is essential for proper cerebral angiogenesis and formation of the blood-brain-barrier. Mol Cell Neurosci 38(3):417–430
8. Back SA, Tuohy TM, Chen H, Wallingford N, Craig A, Struve J, Luo NL, Banine F, Liu Y, Chang A, Trapp BD, Bebo BF Jr, Rao MS, Sherman LS (2005) Hyaluronan accumu-

lates in demyelinated lesions and inhibits oligodendrocyte progenitor maturation. Nat Med 11:966–72)
9. Ballabh P, Braun A, Nedergaard M (2004) The blood–brain barrier: an overview: structure, regulation, and clinical implications. Neurobiol Dis 16(1):1–13
10. Baltan S (2015) Can lactate serve as an energy substrate for axons in good times and in bad, in sickness and in health? Metab Brain Dis 30(1):25–30
11. Bambrick L, Kristian T, Fiskum G (2004) Astrocyte mitochondrial mechanisms of ischemic brain injury and neuroprotection. Neurochem Res 29(3):601–608
12. Barnett SC, Linington C (2013) Myelination: do astrocytes play a role? Neuroscientist 19:442–450
13. Barres BA (2008) The mystery and magic of glia: a perspective on their roles in health and disease. Neuron 60(3):430–440
14. Barry DS, Pakan JM, O'keeffe GW, McDermott KW (2013) The spatial and temporal arrangement of the radial glial scaffold suggests a role in axon tract formation in the developing spinal cord. J Anat 222(2):203–213
15. Bechmann I, Galea I, Perry VH (2007) What is the blood-brain barrier (not)? Trends Immunol 28(1):5–11
16. Bélanger M, Magistretti PJ (2009) The role of astroglia in neuroprotection. Dialogues Clin Neurosci 11(3):281–295
17. Binder DK, Yao X, Zador Z, Sick TJ, Verkman AS, Manley GT (2006) Increased seizure duration and slowed potassiumkinetics in mice lacking aquaporin-4 water channels. Glia 53:631–636
18. Blanz J, Schweizer M, Auberson M, Maier H, Muenscher A, Hubner CA, Jentsch TJ (2007) Leukoencephalopathy upon disruption of the chloride channel ClC-2. J Neurosci 27:6581–6589
19. Blechingberg J, Holm IE, Nielsen KB, Jensen TH, Jorgensen AL, Nielsen AL (2007) Identification and characterization of GFAPkappa, a novel glial fibrillary acidic protein isoform. Glia 55:497–507
20. Bluml S, Philippart M, Schiff mann R, Seymour K, Ross BD (2003) Membrane phospholipids and high-energy metabolites in childhood ataxia with CNS hypomyelination. Neurology 61:648–54
21. Boor PK, de Groot K, Waisfisz Q, Kamphorst W, Oudejans CB, Powers JM et al (2005) MLC1: a novel protein in distal astroglial processes. J Neuropathol Exp Neurol 64:412–419
22. Boor IPK, de Groot K, Mejaski-Bosnjak V, Brenner C, van der Knaap MS, Scheper GC, Pronk JC (2006) Megalencephalic leukoencephalopathy with subcortical cysts: an update and extended mutation analysis of MLC1. Hum Mutat 27:505–512
23. Bouhy D, Ghasemlou N, Lively S, Redensek A, Rathore KI, Schlichter LC, David S (2011) Inhibition of the Ca^{2+} dependent K^{+} channel, KCNN4/KCa3.1, improves tissue protection and locomotor recovery after spinal cord injury. J Neurosci 31:16298–16308
24. Brenner M, Johnson AB, Boespflug-Tanguy O, Rodriguez D, Goldman JE, Messing A (2001) Mutations in GFAP, encoding glial fibrillary acidic protein, are associated with Alexander disease. Nat Genet 27(1):117
25. Brignone MS, Lanciotti A, Macioce P, Macchia G, Gaetani M, Aloisi F, Petrucci TC, Ambrosini E (2011) The beta1 subunit of the Na, K-ATPase pump interacts with megalencephalic leucoencephalopathy with subcortical cysts protein 1 (MLC1) in brain astrocytes: new insights into MLC pathogenesis. Hum Mol Genet 20:90–103
26. Brignone MS, Lanciotti A, Visentin S et al (2014) Megalencephalic leukoencephalopathy with subcortical cysts protein-1 modulates endosomal pH and protein trafficking in astrocytes: relevance to MLC disease pathogenesis. Neurobiol Disease 66(100):1–18
27. Brignone MS, Lanciotti A, Camerini S et al (2015) MLC1 protein: a likely link between leukodystrophies and brain channelopathies. Front Cell Neurosci 9:66
28. Brown AM, Tekkök SB, Ransom BR (2003) Glycogen regulation and functional role in mouse white matter. J Physiol 549(Pt 2):501–512

29. Bugiani M, Boor I, Powers JM, Scheper GC, van der Knaap MS (2010) Leukoencephalopathy with vanishing white matter: a review. J Neuropathol Exp Neurol 69:987–996
30. Bugiani M, Boor I, van Kollenburg B, Postma N, Polder E, van Berkel C, Goldman SA (2011) Defective glial maturation in vanishing white matter disease. J Neuropathol Exp Neurol 70(1):69–82
31. Bugiani M, Postma N, Polder E, Dieleman N, Scheffer PG, Sim FJ, Boor I (2013) Hyaluronan accumulation and arrested oligodendrocyte progenitor maturation in vanishing white matter disease. Brain 136(1):209–222
32. Bugiani M, Dubey M, Breur M, Postma NL, Dekker MP, Ter Braak T et al (2017) Megalencephalic leukoencephalopathy with cysts: the Glialcam-null mouse model. Ann Clin Transl Neurol 4:450–465
33. Burda JE, Sofroniew MV (2014) Reactive gliosis and the multicellular response to CNS damage and disease. Neuron 81(2):229–248
34. Cai L, Du T, Song D, Li B, Hertz L, Peng L (2011) Astrocyte ERK phosphorylation precedes K(+)-induced swelling but follows hypotonicity-induced swelling. Neuropathology 31:250–264
35. Capdevila-Nortes X, Lopez-Hernandez T, Apaja PM et al (2013) Insights into MLC pathogenesis: GlialCAM is an MLC1 chaperone required for proper activation of volume-regulated anion currents. Hum Mol Genet 22:4405–4416
36. Capdevila-Nortes X, Jeworutzki E, Elorza-Vidal X, Barrallo-Gimeno A, Pusch M, Estévez R (2015) Structural determinants of interaction, trafficking and function in the ClC-2/MLC1 subunit GlialCAM involved in leukodystrophy. J Physiol 593(Pt 18):4165–4180
37. Chang MY, Son H, Lee YS, Lee SH (2003) Neurons and astrocytes secrete factors that cause stem cells to differentiate into neurons and astrocytes, respectively. Mol Cell Neurosci 23:414–426
38. Chen YJ, Wallace BK, Yuen N, Jenkins DP, Wulff H, O'Donnell ME (2015) Blood-brain barrier KCa3.1 channels: evidence for a role in brain Na uptake and edema in ischemic stroke. Stroke 46:237–244
39. Chen Q, Sun L, Chen ZJ (2016) Regulation and function of the cGASSTING pathway of cytosolic DNA sensing. Nat Immunol 17(10):1142–1149
40. Chih CP, Lipton P, Roberts EL Jr (2001) Do active cerebral neurons really use lactate rather than glucose? Trends Neurosci 24(10):573–578
41. Clarke LE, Barres BA (2013) Emerging roles of astrocytes in neural circuit development. Nat Rev Neurosci 14(5):311
42. Claycomb KI, Johnson KM, Winokur PN, Sacino AV, Crocker SJ (2013) Astrocyte regulation of CNS inflammation and remyelination. Brain Sci 3(3):1109–1127
43. Clemens MJ (2001) Initiation factor eIF2 alpha phosphorylation in stress responses and apoptosis. Prog Mol Subcell Biol 27:57–89
44. Clemente D, Ortega MC, Melero-Jerez C, De Castro F (2013) The effect of glia-glia interactions on oligodendrocyte precursor cell biology during development and in demyelinating diseases. Front Cell Neurosci 7:268
45. Coulter DA, Steinhauser C (2015) Role of astrocytes in epilepsy. Cold Spring Harb Perspect Med 5:a022434
46. Crow YJ, Jackson AP, Roberts E, van Beusekom E, Barth P, Corry P et al (2000) Aicardi-Goutieres syndrome displays genetic heterogeneity with one locus (AGS1) on chromosome 3p21. Am J Hum Genet 67:213–221
47. Crow YJ, Massey RF, Innes JR et al (2004) Congenital glaucoma and brain stem atrophy as features of Aicardi–Goutières syndrome. Am J Med Genet Part A 129A(3):303–307
48. Crow YJ, Hayward BE, Parmar R et al (2006a) Mutations in the gene encoding the 3′-5′ DNA exonuclease TREX1 cause Aicardi-Goutieres syndrome at the AGS1 locus. Nat Genet 38(8):917–920
49. Crow YJ, Leitch A, Hayward BE et al (2006b) Mutations in genes encoding ribonuclease H2 subunits cause Aicardi-Goutieres syndrome and mimic congenital viral brain infection. Nat Genet 38(8):910–916

50. Crow YJ (2013) Aicardi–Goutières syndrome. Handbook of Clinical Neurology, vol 113. pp 1629–1635
51. Crow YJ, Manel N (2015) Aicardi-Goutieres syndrome and the type I interferonopathies. Nat Rev Immunol 15:429–440
52. Cuadrado E, Jansen MH, Anink J, De Filippis L, Vescovi AL, Watts C et al (2013) Chronic exposure of astrocytes to interferon-alpha reveals molecular changes related to Aicardi-Goutieres syndrome. Brain 136(245–258):19
53. Cuadrado E, Michailidou I, van Bodegraven EJ, Jansen MH, Sluijs JA, Geerts D et al (2015) Phenotypic variation in Aicardi-Goutieres syndrome explained by cell-specific IFN-stimulated gene response and cytokine release. J Immunol 194:3623–3633
54. D'Alessandro G, Catalano M, Sciaccaluga M, Chece G, Cipriani R, Rosito M, Grimaldi A, Lauro C, Cantore G, Santoro A (2013) KCa3.1 channels are involved in the infiltrative behavior of glioblastoma in vivo. Cell Death Disease 4:e773
55. Depienne C, Bugiani M, Dupuits C, Galanaud D, Touitou V, Postma N, van Berkel C, Polder E, Tollard E, Darios F et al (2013) Brain white matter oedema due to ClC-2 chloride channel deficiency: an observational analytical study. Lancet Neurol 12:659–668
56. De Stefano N, Balestri P, Dotti MT, Grosso S, Mortilla M, Morgese G et al (2001) Severe metabolic abnormalities in the white matter of patients with vacuolating megalencephalic leukoencephalopathy with subcortical cysts. A proton MR spectroscopic imaging study. J Neurol 248:403–409
57. Diamond J (2004) Autosomal dominant IFIH1 gain-of-function mutations cause Aicardi-Goutières syndrome. Clin Genet 86:473–474
58. Dietrich J, Lacagnina M, Gass D, Richfield E, Mayer-Proschel M, Noble M, Torres C, Proschel C (2005) EIF2B5 mutations compromise GFAP $^{+}$ astrocyte generation in vanishing white matter leukodystrophy. Nat Med 11:277–283
59. Dimou L, Gallo V (2015) NG2-glia and their functions in the central nervous system. Glia 63(8):1429–1451
60. Djukic B, Casper KB, Philpot BD, Chin LS, McCarthy KD (2007) Conditional knock-out of Kir4.1 leads to glial membrane depolarization, inhibition of potassium and glutamate uptake, and enhanced short-term synaptic potentiation. J Neurosci 27:11354–11365
61. Domingues HS, Portugal CC, Socodato R, Relvas JB (2016) Oligodendrocyte, astrocyte, and microglia crosstalk in myelin development, damage, and repair. Front Cell Dev Biol 4:71
62. Dooves S, Bugiani M, Postma NL, Polder E, Land N, Horan ST, van Deijk AL, van de Kreeke A, Jacobs G, Vuong C et al (2016) Astrocytes are central in the pathomechanisms of vanishing white matter. J Clin Investig 126:1512–1524
63. Duarri A, Lopez de Heredia M, Capdevila-Nortes X, Ridder MC, Montolio M, Lopez-Hernandez T, Boor I, Lien CF, Hagemann T, Messing A et al (2011) Knockdown of MLC1 in primary astrocytes causes cell vacuolation: a MLC disease cell model. Neurobiol Disease 43:228–238
64. Duarri A, Teijido O, Lopez-Hernandez T et al (2008) Molecular pathogenesis of megalencephalic leukoencephalopathy with subcortical cysts: mutations in MLC1 cause folding defects. Hum Mol Genet 17:3728–3739
65. Dubey M, Brouwers E, Hamilton EMC, Stiedl O, Bugiani M, Koch H, Min R (2018) Seizures and disturbed brain potassium dynamics in the leukodystrophy megalencephalic leukoencephalopathy with subcortical cysts. Annu Neurol 83(3):636–649. https://doi.org/10.1002/ana.25190
66. Dubey M, Bugiani M, Ridder MC, Postma NL, Brouwers E, Polder E, Jacobs JG, Baayen JC, Klooster J, Kamermans M et al (2015) Mice with megalencephalic leukoencephalopathy with cysts: a developmental angle. Annu Neurol 77:114–131
67. Duncan RF, Hershey JW (1989) Protein synthesis and protein phosphorylation during heat stress, recovery, and adaptation. J Cell Biol 109:1467–1481
68. Dzwonek J, Wilczynski GM (2015) CD44: molecular interactions, signaling and functions in the nervous system. Front Cell Neurosci 9:175

69. Eddleston M, Mucke L (1993) Molecular profile of reactive astrocytes—implications for their role in neurologic disease. Neuroscience 54(1):15–36
70. Emerit J, Edeas M, Bricaire F (2004) Neurodegenerative diseases and oxidative stress. Biomed Pharmacother 58(1):39–46
71. Emsley JG, Macklis JD (2006) Astroglial heterogeneity closely reflects the neuronal-defined anatomy of the adult murine CNS. Neuron Glia Biol 2(3):175–186
72. Eng LF, Ghirnikar RS (1994) GFAP and astrogliosis. Brain Pathol 4(3):229–237
73. Farina C, Aloisi F, Meinl E (2007) Astrocytes are active players in cerebral innate immunity. Trends Immunol 28(3):138–145
74. Farina L, Pareyson D, Minati L, Ceccherini I, Chiapparini L, Romano S, Savoiardo M (2008) Can MR imaging diagnose adult-onset Alexander disease? Am J Neuroradiol 29(6):1190–1196
75. Favre-Kontula L, Rolland A, Bernasconi L, Karmirantzou M, Power C, Antonsson B, Boschert U (2008) GlialCAM, an immunoglobulin-like cell adhesion molecule is expressed in glial cells of the central nervous system. Glia 56:633–645
76. Fields RD (2008) White matter in learning, cognition and psychiatric disorders. Trends Neurosci 31(7):361–370
77. Filley CM, Fields RD (2016) White matter and cognition: making the connection. J Neurophysiol 116(5):2093–2104
78. Filous AR, Silver J (2016) Targeting astrocytes in CNS injury and disease: a translational research approach. Prog Neurobiol 144:173–187
79. Fogli A, Rodriguez D, Eymard-Pierre E, Bouhour F, Labauge P, Meaney BF, Zeesman S, Kaneski CR, Schiffmann R, Boespflug-Tanguy O (2003) Ovarian failure related to eukaryotic initiation factor 2B mutations. Am J Hum Genet 72:1544–1550
80. Fogli A, Wong K, Eymard-Pierre E, Wenger J, Bouffard JP, Goldin E, Black DN, Boespflug-Tanguy O, Schiffmann R (2002) Cree leukoencephalopathy and CACH/VWM disease are allelic at the EIF2B5 locus. Annu Neurol 52:506–510
81. Franco R, Lezama R, Ordaz B, Pasantes-Morales H (2004) Epidermal growth factor receptor is activated by hyposmolarity and is an early signal modulating osmolyte efflux pathways in Swiss 3T3 fibroblasts. Pflugers Arch 447:830–839
82. Gao Z, Zhu Q, Zhang Y, Zhao Y, Cai L, Shields CB, Cai J (2013) Reciprocal modulation between microglia and astrocyte in reactive gliosis following the CNS injury. Mol Neurobiol 48(3):690–701
83. Gao D, Li T, Li XD, Chen X, Li QZ, Wight-Carter M, Chen ZJ (2015) Activation of cyclic GMP-AMP synthase by self-DNA causes autoimmune diseases. Proc Natl Acad Sci USA 112:E5699–E5705
84. Ge WP, Jia JM (2016) Local production of astrocytes in the cerebral cortex. Neuroscience 323:3–9
85. Gordon HB, Letsou A, Bonkowsky JL (2014) The Leukodystrophies. Semin Neurol 34(03):312–320
86. Gordon N (2003) Alexander disease. Eur J Paediatr Neurol 7(6):395–399
87. Gordon GR, Mulligan SJ, MacVicar BA (2007) Astrocyte control of the cerebrovasculature. Glia 55(12):1214–1221
88. Gorospe JR, Maletkovic J (2006) Alexander disease and megalencephalic leukoencephalopathy with subcortical cysts: leukodystrophies arising from astrocyte dysfunction. Ment Retard Dev Disabil Res Rev 12:113–122
89. Goutières F, Aicardi J, Barth PG, Lebon P (1998) Aicardi-Goutières syndrome: an update and results of interferon-alpha studies. Ann Neurol 44(6):900–907
90. Goutieres F (2005) Aicardi–Goutieres syndrome. Brain Dev 27(3):201–206
91. Hagemann TL, Gaeta SA, Smith MA, Johnson DA, Johnson JA, Messing A (2005) Gene expression analysis in mice with elevated glial fibrillary acidic protein and Rosenthal fibers reveals a stress response followed by glial activation and neuronal dysfunction. Hum Mol Genet 14(16):2443–2458

92. Haj-Yasein NN, Bugge CE, Jensen V, Østby I, Ottersen OP, Hvalby Ø, Nagelhus EA (2015) Deletion of aquaporin-4 increases extracellular K^+ concentration during synaptic stimulation in mouse hippocampus. Brain Struct Funct 220(4):2469–2474
93. Haj-Yasein NN, Jensen V, Vindedal GF, Gundersen GA, Klungland A, Ottersen OP, Hvalby O, Nagelhus EA (2011) Evidence that compromised K^+ spatial buffering contributes to the epileptogenic effect of mutations in the human kir4.1 gene (KCNJ10). Glia 59:1635–1642
94. Hamby ME, Coppola G, Ao Y, Geschwind DH, Khakh BS, Sofroniew MV (2012) Inflammatory mediators alter the astrocyte transcriptome and calcium signaling elicited by multiple G-protein-coupled receptors. J Neurosci Off J Soc Neurosci 32(42):14489–14510
95. Hamilton EMC, Tekturk P, Cialdella F, van Rappard DF, Wolf NI, Yalcinkaya C, van der Knaap MS (2018) Megalencephalic leukoencephalopathy with subcortical cysts: characterization of disease variants. Neurology 90(16):e1395–e1403
96. Hanefeld F, Holzbach U, Kruse B, Wilichowski E, Christen HJ, Frahm J (1993) Diffuse white matter disease in three children: an encephalopathy with unique features on magnetic resonance imaging and proton magnetic resonance spectroscopy. Neuropediatrics 24:244–248
97. Hansson E, Rönnbäck L (2003) Glial neuronal signaling in the central nervous system. FASEB J 17(3):341–348
98. Hawkins BT, Davis TP (2005) The blood-brain barrier/neurovascular unit in health and disease. Pharmacol Rev 57(2):173–185
99. Hirako Y, Yamakawa H, Tsujimura Y, Nishizawa Y, Okumura M, Usukura J, Matsumoto H, Jackson KW, Owaribe K, Ohara O (2003) Characterization of mammalian synemin, an intermediate filament protein present in all four classes of muscle cells and some neuroglial cells: co-localization and interaction with type III intermediate filament proteins and keratins. Cell Tissue Res 313:195–207
100. Hoegg-Beiler MB, Sirisi S, Orozco IJ, Ferrer I, Hohensee S, Auberson M, Gödde K, Vilches C, de Heredia ML, Nunes V et al (2014) Disrupting MLC1 and GlialCAM and ClC-2 interactions in leukodystrophy entails glial chloride channel dysfunction. Nat Commun 5:3475
101. Hofer MJ, Campbell IL (2013) Type I interferon in neurological disease-the devil from within. Cytokine Growth Factor Rev 24:257–267
102. Hofmann K, Rodriguez-Rodriguez R, Gaebler A, Casals N, Scheller A, Kuerschner L (2017) Astrocytes and oligodendrocytes in grey and white matter regions of the brain metabolize fatty acids. Nature 7:10779
103. Hong S, Beja-Glasser VF, Nfonoyim BM, Frouin A, Li S, Ramakrishnan S, Lemere CA (2016) Complement and microglia mediate early synapse loss in Alzheimer mouse models. Science 352(6286):712–716
104. Howard BM, Mo Z, Filipovic R, Moore AR, Antic SD, Zecevic N (2008) Radial glia cells in the developing human brain. Neuroscientist 14(5):459–473
105. Hu X, Yuan Y, Wang D, Su Z (2016) Heterogeneous astrocytes: active players in CNS. Brain Res Bull 125:1–18
106. Iadecola C, Nedergaard M (2007) Glial regulation of the cerebral microvasculature. Nat Neurosci 10(11):1369
107. Iliff JJ, Wang M, Liao Y, Plogg BA, Peng W, Gundersen GA, Benveniste H, Vates GE, Deane R, Goldman SA, Nagelhus EA, Nedergaard, M (2012) A paravascular pathway facilitates CSF flow through the brain parenchyma and the clearance of interstitial solutes, including amyloid β. Sci Transl Med 4(147):147ra111
108. Iliff JJ, Nedergaard M (2013) Is there a cerebral lymphatic system? Stroke J Cereb Circ 44(6 Suppl 1):S93–S95
109. Ishibashi T, Dakin KA, Stevens B, Lee PR, Kozlov SV, Stewart CL, Fields RD (2006) Astrocytes promote myelination in response to electrical impulses. Neuron 49(6):823–832
110. Iwaki T, Iwaki A, Tateishi J, Sakaki Y, Goldman JE (1993) Alpha B-crystallin and 27-kd heat shock protein are regulated by stress conditions in the central nervous system and accumulate in Rosenthal fibers. Am J Pathol 143(2):487
111. Jäkel S, Dimou L (2017) Glial cells and their function in the adult brain: a journey through the history of their ablation. Front Cell Neurosci 11:24. https://doi.org/10.3389/fncel.2017.00024

112. Jany PL, Hagemann TL, Messing A (2013) GFAP expression as an indicator of disease severity in mouse models of Alexander disease. ASN Neuro 5(1):e00109
113. Jeworutzki E, Lopez-Hernandez T, Capdevila-Nortes X, Sirisi S, Bengtsson L, Montolio M, Zifarelli G, Arnedo T, Muller CS, Schulte U et al (2012) GlialCAM, a protein defective in a leukodystrophy, serves as a ClC-2 Cl(-) channel auxiliary subunit. Neuron 73:951–961
114. Jeworutzki E, Lagostena L, Elorza-Vidal X, López-Hernández T, Estévez R, Pusch M (2014) GlialCAM, a CLC-2 Cl(-) channel subunit, activates the slow gate of CLC chloride channels. Biophys J 107:1105–1116
115. Kager H, Wadman WJ, Somjen GG (2000) Simulated seizures and spreading depression in a neuron model incorporating interstitial space and ion concentrations. J Neurophysiol 84:495–512
116. Kessaris N, Pringle N, Richardson WD (2008) Specification of CNS glia from neural stem cells in the embryonic neuroepithelium. Philos Trans R Soc B Biol Sci 363(1489):71–85
117. Kevelam SH, Steenweg ME, Srivastava S, Helman G, Naidu S, Schiffmann R, Blaser S, Vanderver A, Wolf NI, van der Knaap MS (2016) Update on Leukodystrophies: a historical perspective and adapted definition. Neuropediatrics 47(6):349–354
118. Khakh BS, Sofroniew MV (2015) Diversity of astrocyte functions and phenotypes in neural circuits. Nat Neurosci 18(7):942
119. Kimelberg HK, Nedergaard M (2010) Functions of astrocytes and their potential as therapeutic targets. Neurotherapeutics 7:338–353
120. Kıray H, Lindsay SL, Hosseinzadeh S, Barnett SC (2016) The multifaceted role of astrocytes in regulating myelination. Exp Neurol 283:541–549
121. Kirchhoff F, Dringen R, Giaume C (2001) Pathways of neuron-astrocyte interactions and their possible role in neuroprotection. Eur Arch Psychiatr Clin Neurosci 251(4):159–169
122. Klok MD, Bugiani M, de Vries SI, Gerritsen W, Breur M, Sluis S, Heine VM, Kole MH, Baron W, Knaap MS (2018) Axonal abnormalities in vanishing white matter. Ann Clin Transl Neurol
123. Klok MD, Bakels HS, Postma NL, van Spaendonk RML, van der Knaap MS, Bugiani M (2015) Interferon-α and the calcifying microangiopathy in Aicardi-Goutières syndrome. Ann Clin Transl Neurol 2(7):774–779
124. Kofuji P, Newman EA (2004) Potassium buffering in the central nervous system. Neuroscience 129:1043–1054
125. Koyama Y, Goldman JE (1999) Formation of GFAP cytoplasmic inclusions in astrocytes and their disaggregation by αB-crystallin. Am J Pathol 154(5):1563–1572
126. Kraft AD, Harry GJ (2011) Features of microglia and neuroinflammation relevant to environmental exposure and neurotoxicity. Int J Environ Res Public Health 8(7):2980–3018
127. Kumar D, Rittey C, Cameron AH et al (1998) Recognizable inherited syndrome of progressive central nervous system degeneration and generalized intracranial calcification with overlapping phenotype of the syndrome of Aicardi and Goutières. Am J Med Genet Part A 75:508–515
128. Lanciotti A, Brignone MS, Molinari P et al (2012) Megalencephalic leukoencephalopathy with subcortical cysts protein 1 functionally cooperates with the TRPV4 cation channel to activate the response of astrocytes to osmotic stress: dysregulation by pathological mutations. Hum Mol Genet 21:2166–2180
129. Lanciotti A, Brignone MS, Visentin S et al (2016) Megalencephalic leukoencephalopathy with subcortical cysts protein-1 regulates epidermal growth factor receptor signaling in astrocytes. Hum Mol Genet 25:1543–1558
130. Larkin PB, Muchowski PJ (2012) Genetic deficiency of complement component 3 does not alter disease progression in a mouse model of Huntington's disease. J Huntington's Dis 1(1):107–118
131. Lebon P, Badoual J, Ponsot G, Goutieres F, Hemeury-Cukier F, Aicardi J (1988) Intrathecal synthesis of interferon-alpha in infants with progressive familial encephalopathy. J Neurol Sci 84:201–208

132. Leegwater PA, Yuan BQ, van der Steen J, Mulders J, Konst AA, Boor PK, Mejaski-Bosnjak V, van der Maarel SM, Frants RR, Oudejans CB et al (2001) Mutations of MLC1 (KIAA0027), encoding a putative membrane protein, cause megalencephalic leukoencephalopathy with subcortical cysts. Am J Hum Genet 68:831–838
133. Leegwater PA, Boor PK, Yuan BQ, van der Steen J, Visser A, Konst AA, Oudejans CB, Schutgens RB, Pronk JC, van der Knaap MS (2002) Identification of novel mutations in MLC1 responsible for megalencephalic leukoencephalopathy with subcortical cysts. Hum Genet 110:279–283
134. Li R, Johnson AB, Salomons G, Goldman JE, Naidu S, Quinlan R, Siebert JR (2005) Glial fibrillary acidic protein mutations in infantile, juvenile, and adult forms of Alexander disease. Ann Neurol 57(3):310–326
135. Li J, Zhang L, Chu Y, Namaka M, Deng B, Kong J, Bi X (2016) Astrocytes in oligodendrocyte lineage development and white matter pathology. Front Cell Neurosci 10:119
136. Liberto CM, Albrecht PJ, Herx LM, Yong VW, Levison SW (2004) Pro-regenerative properties of cytokine-activated astrocytes. J Neurochem 89(5):1092–1100
137. Liedtke W, Edelmann W, Bieri PL, Chiu FC, Cowan NJ, Kucherlapati R, Raine CS (1996) GFAP is necessary for the integrity of CNS white matter architecture and long-term maintenance of myelination. Neuron 17(4):607–615
138. Liem RK, Messing A (2009) Dysfunctions of neuronal and glial intermediate filaments in disease. J Clin Investig 119(7):1814–1824
139. Liu Y, Han SS, Wu Y, Tuohy TM, Xue H, Cai J, Back SA, Sherman LS, Fischer I, Rao MS (2004) CD44 expression identifies astrocyte-restricted precursor cells. Dev Biol 276(1):31–46
140. Livingston JH, Stivaros S, van der Knaap MS, Crow YJ (2013) Recognizable phenotypes associated with intracranial calcification. Dev Med Child Neurol 55:46–57
141. Longden TA, Dunn KM, Draheim HJ, Nelson MT, Weston AH, Edwards G (2011) Intermediate-conductance calcium-activated potassium channels participate in neurovascular coupling. Br J Pharmacol 164:922–933
142. Lopez-Hernandez T, Ridder MC, Montolio M, Capdevila-Nortes X, Polder E, Sirisi S, Duarri A, Schulte U, Fakler B, Nunes V et al (2011) Mutant GlialCAM causes megalencephalic leukoencephalopathy with subcortical cysts, benign familial macrocephaly, and macrocephaly with retardation and autism. Am J Human Genet 88:422–432
143. Lopez-Hernandez T, Sirisi S, Capdevila-Nortes X et al (2011) Molecular mechanisms of MLC1 and GLIALCAM mutations in megalencephalic leukoencephalopathy with subcortical cysts. Hum Mol Genet 20:3266–3277
144. Lundgaard I, Osório MJ, Kress BT, Sanggaard S, Nedergaard M (2014) White matter astrocytes in health and disease. Neuroscience 276:161–173
145. Lutz SE, Zhao Y, Gulinello M, Lee SC, Raine CS, Brosnan CF (2009) Deletion of astrocyte connexins 43 and 30 leads to a dysmyelinating phenotype and hippocampal CA1 vacuolation. J Neurosci 29:7743–7752
146. MacAulay N, Zeuthen T (2012) Glial K^+ clearance and cell swelling: key roles for cotransporters and pumps. Neurochem Res 37:2299–2309
147. Maragakis NJ, Rothstein JD (2006) Mechanisms of disease: astrocytes in neurodegenerative disease. Nat Clin Practice Neurol 2(12):679–689
148. Matsui M, Mizutani K, Miki Y, Mezaki T, Takahashi Y, Shibasaki H (2003) Adult-onset leukoencephalopathy with vanishing white matter. Eur J Radiol 46:90–92
149. Mauler F, Hinz V, Horváth E, Schuhmacher J, Hofmann HA, Wirtz S, Hahn MG, Urbahns K (2004) Selective intermediate-/small-conductance calcium-activated potassium channel (KCNN4) blockers are potent and effective therapeutics in experimental brain oedema and traumatic brain injury caused by acute subdural haematoma. Eur J Neurosci 20:1761–1768
150. Mazur DJ, Perrino FW (1999) Identification and expression of the TREX1 and TREX2 cDNA sequences encoding mammalian 3′–>5′ exonucleases. J Biol Chem 274:19655–19660
151. Messing A, Brenner M, Feany MB, Nedergaard M, Goldman JE (2012) Alexander disease. J Neurosci 32(15):5017–5023

152. Messing A, Head MW, Galles K, Galbreath EJ, Goldman JE, Brenner M (1998) Fatal encephalopathy with astrocyte inclusions in GFAP transgenic mice. Am J Pathol 152(2):391–398
153. Middeldorp J, Hol EM (2011) GFAP in health and disease. Prog Neurobiol 93(3):421–443
154. Miller RH, Raff MC (1984) Fibrous and protoplasmic astrocytes are biochemically and developmentally distinct. J Neurosci 4(2):585–592
155. Min R, van der Knaap MS (2018) Genetic defects disrupting glial ion and water homeostasis in the brain. Brain Pathol 28:372–387
156. Miyamoto N, Maki T, Shindo A, Liang AC, Maeda M, Egawa N, Arai K (2015) Astrocytes promote oligodendrogenesis after white matter damage via brain-derived neurotrophic factor. J Neurosci 35(41):14002–14008
157. Montagna G, Teijido O, Eymard-Pierre E, Muraki K, Cohen B, Loizzo A, Grosso P, Tedeschi G, Palacín M, Boespflug-Tanguy O, Bertini E, Santorelli FM, Estévez R (2016) Vacuolating megalencephalic leukoencephalopathy with subcortical cysts: functional studies of novel variants in MLC1. Hum Mutat 27:292
158. Morell P, Wiesmann U (1984) A correlative synopsis of the leukodystrophies. Neuropediatrics 15:62–65
159. Nagy JI, Rash JE (2000) Connexins and gap junctions of astrocytes and oligodendrocytes in the CNS. Brain Res Rev 32(1):29–44
160. Nash B, Ioannidou K, Barnett SC (2011) Astrocyte phenotypes and their relationship to myelination. J Anat 219(1):44–52
161. Nielsen AL, Holm IE, Johansen M, Bonven B, Jorgensen P, Jorgensen AL (2002) A new splice variant of glial fibrillary acidic protein, GFAP epsilon, interacts with the presenilin proteins. J Biol Chem 277:29983–29991
162. Oberheim NA, Goldman SA, Nedergaard M (2012) Heterogeneity of astrocytic form and function. Methods Mol Biol (Clifton, NJ) 814:23–45
163. Odermatt B, Wellershaus K, Wallraff A, Seifert G, Degen J, Euwens C, Fuss B, Büssow H, Steinhäuser C, Willecke K (2003) Connexin 47 (Cx47)-deficient mice with enhancend green fluorescent protein reporter gene reveal predominant oligodendrocytic expression of Cx47 and display vacuolized myelin in the CNS. J Neurosci 23:4549–4559
164. Ohtake H, Shimohata T, Terajima K et al (2004) Adult-onset leukoencephalopathy with vanishing white matter with a missense mutation in EIF2B5. Neurology 62:1601–1603
165. Olabarria M, Putilina M, Riemer EC et al (2015) Acta Neuropathol 130:469
166. Olabarria M, Goldman J (2017) Disorders of astrocytes: alexander disease as a model. Annu Rev Pathol 12(1):131–152
167. Oliet SH, Piet R, Poulain DA (2001) Control of glutamate clearance and synaptic efficacy by glial coverage of neurons. Science 292:923–926
168. Orcesi S, La Piana R, Fazzi E (2009) Aicardi-Goutieres syndrome. Br Med Bull 89:183–201
169. Orthmann-Murphy JL, Abrams CK, Scherer SS (2008) Gap junctions couple astrocytes and oligodendrocytes. J Mol Neurosci 35(1):101–116
170. Ostergaard JR, Christensen T (2004) Aicardi-Goutieres syndrome: neuroradiological findings after nine years of follow-up. Eur J Paediatr Neurol 8(5):243–246
171. Ota Y, Zanetti AT, Hallock RM (2013) The role of astrocytes in the regulation of synaptic plasticity and memory formation. Neural Plast 2013
172. Pareyson D, Fancellu R, Mariotti C, Romano S, Salmaggi A, Carella F, Ceccherini I (2008) Adult-onset Alexander disease: a series of eleven unrelated cases with review of the literature. Brain 131(9):2321–2331
173. Pasantes-Morales H, Lezama RA, Ramos-Mandujano G, Tuz KL (2006) Mechanisms of cell volume regulation in hypo-osmolality. Am J Med 119:S4–S11
174. Pascual O, Achour SB, Rostaing P, Triller A, Bessis A (2012) Microglia activation triggers astrocyte-mediated modulation of excitatory neurotransmission. Proc Natl Acad Sci 109(4):E197–E205
175. Pascual-Castroviejo I, Van der Knaap MS, Pronk JC, Garcia-Segura JM, Gutierez-Molina M, Pascual-Pascual SI (2005) Vacuolating megalencephalic leukoencephalopathy: 24 year follow-up of two siblings. Neurologia 20:33–40

176. Pekny M, Nilsson M (2005) Astrocyte activation and reactive gliosis. Glia 50(4):427–434
177. Pekny M, Pekna M, Messing A, Steinhäuser C, Lee JM, Parpura V, Verkhratsky A (2016) Astrocytes: a central element in neurological diseases. Acta Neuropathol 131(3):323–345
178. Pellerin L, Bouzier-Sore AK, Aubert A, Serres S, Merle M, Costalat R, Magistretti PJ (2007) Activity-dependent regulation of energy metabolism by astrocytes: an update. Glia 55(12):1251–1262
179. Polito A, Reynolds R (2005) NG2-expressing cells as oligodendrocyte progenitors in the normal and demyelinated adult central nervous system. J Anat 207(6):707–716
180. Prass K, Bruck W, Schroder NW, Bender A, Prass M, Wolf T, Van der Knaap MS, Zschenderlein R (2001) Adult-onset leukoencephalopathy with vanishing white matter presenting with dementia. Annu Neurol 50:665–668
181. Prust M, Wang J, Morizono H, Messing A, Brenner M, Gordon E, Albin R (2011) GFAP mutations, age at onset, and clinical subtypes in Alexander disease. Neurology 77(13):1287–1294
182. Quinlan RA, Brenner M, Goldman JE, Messing A (2007) GFAP and its role in Alexander disease. Exp Cell Res 313(10):2077–2087
183. Quintas C, Pinho D, Pereira C, Saraiva L, Gonçalves J, Queiroz G (2014) Microglia P2Y 6 receptors mediate nitric oxide release and astrocyte apoptosis. J Neuroinflammation 11(1):141
184. Rash JE (2010) Molecular disruptions of the panglial syncytium block potassium siphoning and axonal saltatory conduction: pertinence to neuromyelitis optica and other demyelinating diseases of the central nervous system. Neuroscience 168:982–1008
185. Reeves SA, Helman LJ, Allison A, Israel MA (1989) Molecular cloning and primary structure of human glial fibrillary acidic protein. Proc Natl Acad Sci USA 86(13):5178–5182
186. Ricci G, Volpi L, Pasquali L, Petrozzi L, Siciliano G (2009) Astrocyte–neuron interactions in neurological disorders. J Biol Phys 35(4):317–336
187. Rice G, Newman WG, Dean J, Patrick T, Parmar R, Flintoff K et al (2007a) Heterozygous mutations in TREX1 cause familial chilblain lupus and dominant Aicardi-Goutieres syndrome. Am J Human Genet 80:811–815
188. Rice G, Patrick T, Parmar R, Taylor CF, Aeby A, Aicardi J et al (2007b) Clinical and molecular phenotype of Aicardi-Goutieres syndrome. Am J Human Genet 81:713–725
189. Rice GI, Kasher PR, Forte GM et al (2012) Mutations in ADAR1 cause Aicardi-Goutieres syndrome associated with a type I interferon signature. Nat Genet 44(11):1243–1248
190. Rice GI, Del Toro Duany Y, Jenkinson EM et al (2014) Gain-of-function mutations in IFIH1 cause a spectrum of human disease phenotypes associated with upregulated type I interferon signaling. Nat Genet 46(5):503–509
191. Richardson WD, Young KM, Tripathi RB, McKenzie I (2011) NG2-glia as multipotent neural stem cells: fact or fantasy? Neuron 70(4):661–673
192. Ridder MC, Boor I, Lodder JC, Postma NL, Capdevila-Nortes X, Duarri A, Brussaard AB, Estévez R, Scheper GC, Mansvelder HD, van der Knaap MS (2011) Megalencephalic leucoencephalopathy with cysts: defect in chloride currents and cell volume regulation. Brain 134:3342–3354
193. Rinholm JE, Hamilton NB, Kessaris N, Richardson WD, Bergersen LH, Attwell D (2011) Regulation of oligodendrocyte development and myelination by glucose and lactate. J Neurosci Off J Soc Neurosci 31(2):538–548
194. Rodriguez D, Gelot A, della Gaspera B, Robain O, Ponsot G, Sarlieve LL, Ghandour S, Pompidou A, Dautigny A, Aubourg P et al (1999) Increased density of oligodendrocytes in childhood ataxia with diffuse central hypomyelination (CACH) syndrome: neuropathological and biochemical study of two cases. Acta Neuropathol 97:469–480
195. Roelofs RF, Fischer DF, Houtman SH, Sluijs JA, Van HW, Van Leeuwen FW, Hol EM (2005) Adult human subventricular, subgranular, and subpial zones contain astrocytes with a specialized intermediate filament cytoskeleton. Glia 52:289–300
196. Rothstein JD, Dykes-Hoberg M, Pardo CA, Bristol LA, Jin L, Kuncl RW, Kanai Y, Hediger MA, Wang Y, Schielke JP, Welty DF (1996) Knockout of glutamate transporters reveals a major role for astroglial transport in excitotoxicity and clearance of glutamate. Neuron 16(3):675–686

197. Ruggieri P, Mangino G, Fioretti B, Catacuzzeno L, Puca R, Ponti D, Miscusi M, Franciolini F, Ragona G, Calogero A (2012) The inhibition of KCa3.1 channels activity reduces cell motility in glioblastoma derived cancer stem cells. PLoS One 7:e47825
198. Sase S, Takanohashi A, Vanderver A, Almad A (2018) Astrocytes, an active player in Aicardi-Goutières syndrome. Brain Pathol 28:399–407
199. Sawaishi Y (2009) Review of Alexander disease: beyond the classical concept of leukodystrophy. Brain Develop 31(7):493–498
200. Scheper GC, Mulder J, Kleijn M, Voorma HO, Thomas A, van Wijk R (1997) Inactivation of eIF2B and phosphorylation of PHAS-I in heat-shocked rat hepatoma cells. J Biol Chem 272:26850–26856
201. Schiffmann R, Moller JR, Trapp BD, Shih HH, Farrer RG, Katz DA, Alger JR, Parker CC, Hauer PE, Kaneski CR et al (1994) Childhood ataxia with diffuse central nervous system hypomyelination. Ann Neurol 35(3):331–340
202. Schitine C, Nogaroli L, Costa MR, Hedin-Pereira C (2015) Astrocyte heterogeneity in the brain: from development to disease. Front Cell Neurosci 9:76
203. Schmitt A, Gofferje V, Weber M, Meyer J, Mössner R, Lesch KP (2003) The brain-specific protein MLC1 implicated in megalencephalic leukoencephalopathy with subcortical cysts is expressed in glial cells in the murine brain. Glia 44:283–295
204. Schulz JB, Lindenau J, Seyfried J, Dichgans J. Glutathione, oxidative stress and neurodegeneration. Eur J Biochem 267(16):4904–11
205. See J, Zhang X, Eraydin N, Mun SB, Mamontov P, Golden JA et al (2004) Oligodendrocyte maturation is inhibited by bone morphogenetic protein. Mol Cell Neurosci 26:481–492
206. See JM, Grinspan JB (2009) Sending mixed signals: bone morphogenetic protein in myelination and demyelination. J Neuropatholoiy Exp Neurol 68:595–604
207. Seifert G, Huttmann K, Binder DK, Hartmann C, Wyczynski A, Neusch C, Steinhäuser C (2009) Analysis of astroglial K channel expression in the developing hippocampus reveals a predominant role of the Kir4.1 subunit. J Neurosci 29:7474–7488
208. Seitelberger F (1984) Structural manifestations of leukodystrophies. Neuropediatrics 15:53–61
209. Sherman LS, Back SA (2008) A 'GAG' reflex prevents repair of the damaged CNS. Trends Neurosci 31:44–52
210. Silver J, Miller JH (2004) Regeneration beyond the glial scar. Nat Rev Neurosci 5(2):146–156
211. Simard M, Nedergaard M (2004) The neurobiology of glia in the context of water and ion homeostasis. Neuroscience 129(4):877–896
212. Simon MJ, Iliff JJ (2016) Regulation of cerebrospinal fluid (CSF) flow in neurodegenerative, neurovascular and neuroinflammatory disease. Biochem Biophys Acta 1862(3):442–451
213. Singhal BS, Gorospe JR, Naidu S (2003) Megalencephalic leukoencephalopathy with subcortical cysts. J Child Neurol 18:646–652
214. Sirisi S, Folgueira M, Lopez-Hernandez T et al (2004) Megalencephalic leukoencephalopathy with subcortical cysts protein 1 regulates glial surface localization of GLIALCAM from fish to humans. Hum Mol Genet 23:5069–5086
215. Sirisi S, Elorza-Vidal X, Arnedo T, Armand-Ugón M, Callejo G, Estévez R (2017) Depolarization causes the formation of a ternary complex between GlialCAM, MLC1 and ClC-2 in astrocytes: implications in megalencephalic leukoencephalopathy. Hum Mol Genet 26(13):2436–2450
216. Skripuletz T, Bauer K, Kucman V, Gudi V, Pul R, Stangel M (2013) Astrocytes are essential for regeneration of oligodendrocytes and myelin after cuprizone-induced demyelination. Mult Scler J 19(11):52–53
217. Sofroniew MV (2009) Molecular dissection of reactive astrogliosis and glial scar formation. Trends Neurosci 32(12):638–647
218. Sofroniew MV, Vinters HV (2010) Astrocytes: biology and pathology. Acta Neuropathol 119(1):7–35
219. Sofroniew MV (2014) Multiple roles for astrocytes as effectors of cytokines and inflammatory mediators. Neurosci 20(2):160–172

220. Sofroniew MV (2015) Astrogliosis. Cold Spring Harb Perspect Biol 7(2)
221. Somjen GG (1988) Nervenkitt: notes on the history of the concept of neuroglia. Glia 1(1):2–9
222. Stevens B, Allen NJ, Vazquez LE, Howell GR, Christopherson KS, Nouri N, Micheva KD, Mehalow AK, Huberman AD, Stafford B, Sher A, Litke AM, Lambris JD, Smith SJ, John SW, Barres BA (2007) The classical complement cascade mediates CNS synapse elimination. Cell 131:1164–1178
223. Strohschein S, Huttmann K, Gabriel S, Binder DK, Heinemann U, Steinhauser C (2011) Impact of aquaporin-4 channels on K^+ buffering and gap junction coupling in the hippocampus. Glia 59:973–980
224. Su G, Kintner DB, Sun D (2002) Contribution of Na(+)-K(+)-Cl(−) cotransporter to high-[K(+)](o)- induced swelling and EAA release in astrocytes. Am J Physiol-Cell Physiol 282:C1136–C1146
225. Takeshita Y, Ransohoff RM (2015) Blood–brain barrier and neurological diseases. J Clin Exp Neuroimmunol 6(4):351–361
226. Tang G, Xu Z, Goldman JE (2006) Synergistic effects of the SAPK/JNK and the proteasome pathway on glial fibrillary acidic protein (GFAP) accumulation in Alexander disease. J Biol Chem 281(50):38634–38643
227. Tang G, Yue Z, Talloczy Z, Goldman JE (2008) Adaptive autophagy in Alexander disease-affected astrocytes. Autophagy 4(5):701–703
228. Tang G, Perng MD, Wilk S, Quinlan R, Goldman JE (2010) Oligomers of mutant glial fibrillary acidic protein (GFAP) inhibit the proteasome system in Alexander disease astrocytes, and the small heat shock protein αB-crystallin reverses the inhibition. J Biol Chem 285(14):10527–10537
229. Teijido O, Martinez A, Pusch M, Zorzano A, Soriano E, Del Rio JA, Palacin M, Estevez R (2004) Localization and functional analyses of the MLC1 protein involved in megalencephalic leukoencephalopathy with subcortical cysts. Hum Mol Genet 13:2581–2594
230. Teijido O, Casaroli-Marano R, Kharkovets T, Aguado F, Zorzano A, Palacín M, Soriano E, Martínez A, Estévez R (2007) Expression patterns of MLC1 protein in the central and peripheral nervous systems. Neurobiol Dis 26:532–545
231. Tian R, Wu X, Hagemann TL, Sosunov AA, Messing A, McKhann GM, Goldman JE (2010) Alexander disease mutant glial fibrillary acidic protein compromises glutamate transport in astrocytes. J Neuropathol Exp Neurol 69(4):335–345
232. Toda T, Shinmyo Y, Duong TAD, Masuda K, Kawasaki H (2016) An essential role of SVZ progenitors in cortical folding in gyrencephalic mammals. Scientific reports 6:29578
233. Topku M, Saatci I, Topcuoglu MA (1998) Megalencephaly and leukodystrophy with mild clinical course: a report on 12 new cases. Brain Dev 20:142–153
234. Tress O, Maglione M, May D, Pivneva T, Richter N, Seyfarth J, Binder S, Zlomuzica A, Seifert G, Theis M, Dere E, Kettenmann H, Willecke K (2012) Panglial gap junctional communication is essential for maintenance of myelin in the CNS. J Neurosci Off J Soc Neurosci 32:7499–7518
235. Uggetti C, La Piana R, Orcesi S, Egitto MG, Crow YJ, Fazzi E (2009) Aicardi-Goutieres syndrome: neuroradiologic findings and follow-up. AJNR Am J Neuroradiol 30(10):1971–1976
236. Van der Knaap MS, Barth PG, Stroink H, van Nieuwenhuizen O, Arts WF, Hoogenraad F, Valk J (1995a) Leukoencephalopathy with swelling and a discrepantly mild clinical course in eight children. Annu Neurol 37:324–334
237. Van der Knaap MS, Valk J, Barth PG, Smit LM, van Engelen BG, Tortori Donati P (1995b) Leukoencephalopathy with swelling in children and adolescents: MRI patterns and differential diagnosis. Neuroradiology 37:679–686
238. Van der Knaap MS, Barth PG, Vrensen GF, Valk J (1996) Histopathology of an infantile-onset spongiform leukoencephalopathy with a discrepantly mild clinical course. Acta Neuropathol 92:206–212
239. Van der Knaap MS, Barth PG, Gabreels FJ, Franzoni E, Begeer JH, Stroink H, Rotteveel JJ, Valk J (1997) A new leukoencephalopathy with vanishing white matter. Neurology 48:845–855

240. Van der Knaap MS, Kamphorst W, Barth PG, Kraaijeveld CL, Gut E, Valk J (1998) Phenotypic variation in leukoencephalopathy with vanishing white matter. Neurology 51:540–547
241. Van der Knaap MS, Breiter SN, Naidu S, Hart AA, Valk J (1999) Defining and categorizing leukoencephalopathies of unknown origin: MR imaging approach. Radiology 213:121–133
242. Van Der Knaap MS, Naidu S, Breiter SN, Blaser S, Stroink H, Springer S,Begeer JC, van Coster R, Barth PG, Thomas NH, Valk J, Powers JM (2001). Alexander disease: diagnosis with MR imaging. Am J Neuroradiol 22(3):541–552
243. Van der Knaap MS, Leegwater PA, Konst AA, Visser A, Naidu S, Oudejans CB, Schutgens RB, Pronk JC (2002) Mutations in each of the five subunits of translation initiation factor eIF2B can cause leukoencephalopathy with vanishing white matter. Annu Neurol 51:264–270
244. Van der Knaap MS, van Berkel CG, Herms J, van Coster R, Baethmann M, Naidu S, Boltshauser E, Willemsen MA, Plecko B, Hoffmann GF et al (2003) eIF2B-related disorders: antenatal onset and involvement of multiple organs. Am J Hum Genet 73:1199–1207
245. Van der Knaap MS, Valk J (2005) GM1 gangliosidosis. Magnetic resonance of myelination and myelin disorders. Springer, Berlin Heidelberg New York, pp 96–102
246. Van der Knaap MS, Pronk JC, Scheper GC (2006) Vanishing white matter disease. Lancet Neurol 5:413–423
247. Van der Knaap MS, Ramesh V, Schiffmann R, Blaser S, Kyllerman M, Gholkar A et al (2006) Alexander disease ventricular garlands and abnormalities of the medulla and spinal cord. Neurology 66(4):494–498
248. Van der Knaap MS, Boor I, Estevez R (2012) Megalencephalic leukoencephalopathy with subcortical cysts: chronic white matter oedema due to a defect in brain ion and water homoeostasis. Lancet Neurol 11:973–985
249. Van der Knaap MS, Bugiani M (2017) Leukodystrophies: a proposed classification system based on pathological changes and pathogenetic mechanisms. Acta Neuropathol 134(3):351–382
250. Van Heteren JT, Rozenberg F, Aronica E et al (2008) Astrocytes produce interferon-alpha and CXCL10, but not IL-6 or CXCL8, in Aicardi-Goutières syndrome. Glia 56:568–578
251. Van Rossum D, Hanisch UK (2004) Microglia. Metab Brain Dis 19(3–4):393–411
252. Vanderver A, Prust M, Kadom N, Demarest S, Crow YJ, Helman G, Orcesi S, La Piana R, Uggetti C, Wang J, Gordisch-Dressman H, van der Knaap MS, Livingston JH (2015) Early onset Aicardi Goutières syndrome: MRI pattern recognition. J Child Neurol 30(10):1343–1348
253. Varela D, Simon F, Olivero P, Armisén R, Leiva-Salcedo E, Jørgensen F, Sala F, Stutzin A (2007) Activation of H_2O_2-induced VSOR Cl-currents in HTC cells require phospholipase C gamma1 phosphorylation and Ca^{2+} mobilisation. Cell Physiol Biochem 20:773–780
254. Verkhratsky A, Parpura V (2014) Neurological and psychiatric disorders as a neuroglial failure. Period Biol 116(2):115–124
255. Vermeulen G, Seidl R, Mercimek-Mahmutoglu S, Rotteveel MD, Scheper GC, Van der Knaap MS (2005) Fright is a provoking factor in vanishing white matter disease. Annu Neurol 57:560–563
256. Walz W (2000) Role of astrocytes in the clearance of excess extracellular potassium. Neurochem Int 36:291–300
257. Wang DD, Bordey A (2008) The astrocyte odyssey. Prog Neurobiol 86(4):342–367. https://doi.org/10.1016/j.pneurobio.2008.09.015
258. Weiss N, Miller F, Cazaubon S, Couraud PO (2009) The blood-brain barrier in brain homeostasis and neurological diseases. Biochem Biophys Acta 1788(4):842–857
259. Welch WJ (1992) Mammalian stress response: cell physiology, structure/ function of stress proteins, and implications for medicine and disease. Physiol Rev 72:1063–1081
260. Wolburg H, Noell S, Mack A, Wolburg-Buchholz K, Fallier-Becker P (2009) Brain endothelial cells and the glio-vascular complex. Cell Tissue Res 335:75
261. Wu M, Moh MC, Schwarz H (2016) HepaCAM associates with connexin 43 and enhances its localization in cellular junctions. Sci Rep 6:36218

262. Xiao J, Wong AW, Willingham MM, van den Buuse M, Kilpatrick TJ, Murray SS (2010) Brain-derived neurotrophic factor promotes central nervous system myelination via a direct effect upon oligodendrocytes. Neurosignals 18(3):186–202
263. Yuan F, Dutta T, Wang L, Song L, Gu L, Qian L et al (2015) Human DNA exonuclease TREX1 is also an exoribonuclease that acts on single-stranded RNA. J Biol Chem 290:13344–13353
264. Zamanian J, Xu L, Foo L, Nouri N, Zhou L, Giffard R, Barres B (2012) Genomic analysis of reactive astrogliosis. J Neurosci Off J Soc Neurosci 32(18):6391–6410
265. Zelenika D, Grima B, Brenner M, Pessac B (1995) A novel glial fibrillary acidic protein mRNA lacking exon 1. Mol Brain Res 30:251–258
266. Zhang K, Sejnowski TJ (2000) A universal scaling law between gray matter and white matter of cerebral cortex. Proc Natl Acad Sci USA 97(10):5621–5626
267. Ziskin JL, Nishiyama A, Rubio M, Fukaya M, Bergles DE (2007) Vesicular release of glutamate from unmyelinated axons in white matter. Nat Neurosci 10(3):321
268. Zlokovic BV (2008) The blood-brain barrier in health and chronic neurodegenerative disorders. Neuron 57(2):178–201
269. Zuchero JB, Barres BA (2015) Glia in mammalian development and disease. Dev (Camb, Engl) 142(22):3805–3809

Chapter 10
Astrocytes in Motor Neuron Diseases

Chiara F. Valori, Giulia Guidotti, Liliana Brambilla and Daniela Rossi

Abstract Motor neuron disorders are highly debilitating and mostly fatal conditions for which only limited therapeutic options are available. To overcome this limitation and develop more effective therapeutic strategies, it is critical to discover the pathogenic mechanisms that trigger and sustain motor neuron degeneration with the greatest accuracy and detail. In the case of Amyotrophic Lateral Sclerosis (ALS), several genes have been associated with familial forms of the disease, whilst the vast majority of cases develop sporadically and no defined cause can be held responsible. On the contrary, the huge majority of Spinal Muscular Atrophy (SMA) occurrences are caused by loss-of-function mutations in a single gene, *SMN1*. Although the typical hallmark of both diseases is the loss of motor neurons, there is increasing awareness that pathological lesions are also present in the neighbouring glia, whose dysfunction clearly contributes to generating a toxic environment in the central nervous system. Here, ALS and SMA are sequentially presented, each disease section having a brief introduction, followed by a focussed discussion on the role of the astrocytes in the disease pathogenesis. Such a dissertation is substantiated by the findings that built awareness on the glial involvement and how the glial–neuronal interplay is perturbed, along with the appraisal of this new cellular site for possible therapeutic intervention.

Keywords Astrocytes · Motor neuron · Amyotrophic lateral sclerosis · Spinal muscular atrophy · Transgenic animal models

---

C. F. Valori
Department of Neuropathology, German Centre for Neurodegenerative Diseases (DZNE), 72076 Tübingen, Germany

G. Guidotti · L. Brambilla · D. Rossi (✉)
Laboratory for Research on Neurodegenerative Disorders, IRCCS Istituti Clinici Scientifici Maugeri, Via Maugeri 4, 27100 Pavia, Italy
e-mail: daniela.rossi@icsmaugeri.it

A. Verkhratsky et al. (eds.), *Neuroglia in Neurodegenerative Diseases*,
Advances in Experimental Medicine and Biology 1175,
https://doi.org/10.1007/978-981-13-9913-8_10

10.1 Amyotrophic Lateral Sclerosis

10.1.1 A Brief Introduction

Amyotrophic Lateral Sclerosis (ALS) is an adult-onset progressive and fatal neurodegenerative condition, caused by the demise of both cortical and spinal cord motor neurons. This extensive degeneration causes the patients to suffer from a wide array of symptoms, including dysphagia (difficulty in swallowing) and dysarthria (problems with the muscles that help produce speech) in the case of the so-called bulbar-onset ALS as well as fasciculations (muscle twitching), tremors and muscular weakness in the case of the so-called spinal-onset ALS. Diagnosis is formulated upon meeting El Escorial and Airlie House criteria, which evaluate the distribution and the progression of muscular weakness, and combine a neurological evaluation with neuroimaging and electrophysiological assessments (reviewed in [106, 291]). On the basis of these criteria, epidemiologic surveys have estimated that ALS has a prevalence of 4–6 individuals per 100,000 individuals per year, making it the most prevalent form of adult-onset motor neuron disease. The disease predominantly presents itself when the patient is 40–60 years of age, often with a rapid progression. The unfolding of the disease is monitored using different rating scales (reviewed in [106]) to examine the progression of the patient disability. Death usually occurs by a respiratory failure within 3–5 years from the symptom onset. In about 50% of patients, the decline in the quality of life is further accelerated by the onset of behavioural and linguistic abnormalities, which are monitored using an ad hoc scale, the Edinburgh Cognitive and Behavioural ALS Screen [1]. The degree of cognitive impairment might become so severe to call for an additional diagnosis of Frontotemporal Lobe Dementia in about 15% of occurrences [197, 282].

The therapeutic approach to ALS is multidisciplinary and integrates symptomatic therapies to ease muscle spasticity, excessive drooling and depression, with non-invasive assisted ventilation and gastrostomy, the latter to ensure an adequate nutritional intake. Although this strategy helps to prolong the lifespan for a few months [169], it is not aiming at slowing down the neurodegenerative process. To this end, there are only two drugs approved by the U.S. Food and Drug Administration (FDA) for the treatment of ALS: riluzole, which prolongs survival of ALS patients in late stages of the disease [65], and edaravone, a more recently approved agent, which appears to be effective in a subset of patients at early stages of the disease [97]. Additional studies are; however, required to evaluate edaravone's full potential and to assess its long-term effects [3].

The development of effective disease-halting therapies is therefore urgently needed. To successfully reach this goal, the scientific community has to move forward in the process of pinpointing molecular and cellular mechanisms driving the neurodegenerative cascade. In the vast majority of cases, the disease appears sporadically (sALS) and several epidemiological studies have identified risk factors both non-modifiable (e.g., ageing; genetic polymorphism) and modifiable (e.g., environmental factors, such as exposure to heavy metals or pesticides; viruses); the latter including also personal behaviour (e.g., smoking and vigorous physical activity). However, approximately 5–10% of patients have a familial form of ALS (fALS), where the disease is transmitted, mainly through an autosomal dominant pattern of inheritance. In the past few years, genetic studies have been able to link the disease with mutations in several genes, thus providing an explanation for almost all affected families. Interestingly, those genes encode proteins involved in specific cellular functions, such as redox homeostasis (*Superoxide dismutase 1* (*SOD1*) [240]); RNA/DNA metabolism (*TAR DNA binding protein 43* (*TARDBP*) [123, 268], *Fused in sarcoma* (*FUS*) [144, 292], *Matrin 3* (*MATR3*) [121], *T cell-restricted intracellular antigen-1* (*TIA1*) [163], *angiogenin* (*ANG*) [95], *hnRNPA2B1* and *hnRNPA1* [133]); vesicle trafficking/autophagy (*Alsin* (*ALS2*) [314], *Optineurin* (*OPTN*) [173], *TANK-binding kinase 1* (*TBK1*) [80], *Annexin 11A* (*ANXA11*) (Smith et al., 2017)); proteostasis (*valosin-containing protein* (*VCP/p97*) [120], *Ubiquilin 2* (*UBQLN2*) [48], *Sequestosome 1* (*SQSTM1*/p62) [67], *Ubiquilin 4* (*UBQLN4*) [61]); and cytoskeletal dynamics (*Profilin 1* [309], *Kif5A* [205]).

However, the most prevalent causes of fALS are mutations in *C9orf72* [47, 232], where the genetic lesion consists of the aberrant expansion of a hexanucleotide repeat (GGGGCC) sequence localized in the intronic/promoter region of the gene. It is presently highly debated as to whether toxicity arises from haploinsufficient protein expression or from the gain of a toxic function of RNA harbouring an expanded tract (formation of foci, accumulation of dipeptide proteins due to RNA translation or sequestration of RNA/DNA binding proteins; reviewed in [220]).

Neuropathological assessment of the central nervous system (CNS) from ALS patients reveals sclerosis of the pyramidal tract as well as motor cortex atrophy and severe loss of motor neurons associated with gliosis. Notably, the hallmark of surviving motor neurons is the presence of ubiquitin-positive proteinaceous inclusions. Although the composition of these aggregates has not been entirely detailed, in 97% of occurrences inclusions are enriched in the RNA/DNA binding protein TDP-43 [201]. The mechanisms driving protein accumulation into insoluble aggregates is still under investigation, though significant attention has been given to stress granules, i.e. membrane-less cytoplasmic organelles that transiently assemble upon different types of cellular insults. Stress granules are enriched in RNA-binding proteins (including many ALS-associated proteins, such as TIA1, TDP-43 and FUS) and trap most mRNAs in a translationally silent status until stress withdrawal (reviewed in [256]). Accumulating evidence suggests that, in ALS, those organelles cannot resolve, but they aberrantly evolve into pathological aggregates (reviewed in [275]).

These observations and the growing number of ALS-linked genes coding for DNA/RNA-binding proteins set the ground for the hypothesis that aberrant RNA metabolism is playing a role in motor neuron demise along with oxidative stress, excitotoxicity, DNA damage, impaired axonal transport and mitochondrial dysfunction (reviewed in [83, 106]).

In addition to these events, occurring cell-autonomously within motor neurons, the articulated and pivotal contribution of glial cells to motor neuron dysfunction and demise has been increasingly acknowledged. In particular, neuroinflammation-related events, oligodendrocyte cell death and phenotypic changes of the astrocytes have been reported (reviewed in [106]). Since each glial cell subpopulation is physiologically empowered to perform specialized tasks to ensure an optimal environment for neuronal survival and activity (reviewed in [301]), glia are likely to offer a distinct and specific contribution to neuronal loss under pathological circumstances. In the next sections of this chapter, we will focus on the impact of the astrocytes, first reporting histological findings describing astrocyte pathology in human specimens; then gathering genetic experiments supporting the hypothesis of their active involvement in ALS pathogenesis. Lastly, we will tackle mechanistic studies aiming at elucidating key molecular players while discussing the potential of astrocytes as a therapeutic target. With regard to microglia and oligodendrocytes, we refer the reader to other reviews addressing the role of these glial cell types in ALS [37, 145, 206].

10.1.2 Role of the Astrocytes in ALS Pathogenesis

10.1.2.1 Evidence from ALS Patients

For decades, astrocytosis has been recognized as a histological finding in ALS. In particular, neuropathologists have provided accurate morphological descriptions of this phenomenon, owing to a thorough investigation of different brain areas [125, 143, 196, 199]. Intriguingly, the characterization of astrocytosis in the spinal cord of ALS patients [253] allowed the early formulation of two key hypotheses. First, the morphology of these cells was suggesting an ongoing active process of remodelling, rather than a purely passive reaction. Second, because astrocytosis and the prevalence of dystrophic neurites were exacerbated where the corticospinal tract entered the grey matter, it was suggested a 'dying back' mechanism in the pathogenesis of ALS [253]. Moreover, the physiologically close association between astrocytes and motor neuronal cell bodies was partially lost in sALS spinal cord specimens, where a more loose interaction was reported [208]. Finally, isolated degenerating astrocytes were identified in the motor neuronal microenvironment in sALS spinal cords [172]. Although of outstanding importance, these findings depict an end-stage scenario in ALS, while it is of primary relevance to gain information about earlier events. The development of the glial [^{11}C]-PBR28 [6, 7, 321] or, even better, of the astrocyte-specific [11C](L)-deprenyl-D2 PET [119] radioligands, coupled to recent advances in neuroimaging techniques, paved the way to detect the development and localization of astrocytosis in vivo. These tools are indeed expected to enable the correlation between astrocytosis and neuronal loss during disease progression in patients. Furthermore, they hold the potential to monitor the efficacy of glial-targeted therapies.

Abnormal morphological changes are not the only neuropathological evidence of astrocyte involvement in ALS, but molecular and functional abnormalities were also reported. For example, hyaline [127] and ubiquitinated protein inclusions were shown not to be an exclusive hallmark of motor neurons, but were observed also in the astrocytes [31, 172, 177, 217]. In fALS-*SOD1* and sALS cases such inclusions were described to be enriched in misfolded SOD1 protein [76], despite early appraisals reporting reduced or absent SOD1 [208] and increased SOD2 immunoreactivity [26]. Taken together, this amount of evidence suggests that astrocytes themselves are subject to stress. Among others, oxidative stress conditions were proposed to strike the astrocytes, as indicated by their increased expression of several proteins induced by reactive oxygen species (ROS). These proteins include the calcium-binding protein S100beta [183]; cyclooxygenase 2 (COX-2) [166]; iNOS and nitrotyrosine [251]; and nNOS [8] in sALS patients, together with different advanced glycation and lipoxidation end products in fALS cases with SOD1 inclusions [127, 257–259]. However, spinal cord astrocytes have been described to counteract these insults by mounting a protective response via the upregulation of the c-Jun N-terminal kinases (JNKs) and the nuclear translocation of the nuclear factor kB (NF-kB) [182].

The early observation that ALS patients show abnormally elevated concentrations of the excitatory amino acids glutamate, aspartate and their metabolites in their cerebrospinal fluid (CSF) [244] led to the core discovery that astrocytes are functionally impaired, as they display reduced expression of the high-affinity plasmalemmal glutamate transporter EAAT2 [31, 243]. This protein is critical to quench glutamatergic synaptic signalling, thereby preventing excitotoxicity and modulating the energy metabolism in response to neuronal activity (reviewed in [236]). Given these critical roles in preserving brain homeostasis, it is not surprising that EAAT2 dysfunction could contribute to the pathogenesis of a growing number of CNS disorders (reviewed in [273]). In addition, glutamatergic signalling in astrocytes was reported to be perturbed as a consequence of the upregulation of several metabotropic glutamate receptors (mGluR1alpha, mGluR5, mGluR2/3) in the spinal cord of ALS patients [12, 172].

10.1.2.2 Generation and Characterization of Transgenic Animal Models

The evidence gathered from human post-mortem material collectively supports the hypothesis that the ALS-causing mechanisms are hitting also the astrocytes. However, neuropathological assessments of autoptic tissues allow to describe only the end-stage situation and do not permit to unveil whether astrocytic lesions can be considered just as 'collateral damage events' or whether astrocyte malfunctioning directly and critically contributes to neuronal demise. To shed light on these issues and to reproduce the complexity of the CNS, one needs to model the disorder in intact organisms, such as transgenic animals. These were first made possible by the milestone discovery of mutations in the *SOD1* gene in a subset of fALS [240]. The earliest and most extensively characterized model of ALS is a transgenic mouse that ubiquitously expresses high amounts of the mutant human SOD1 protein (hSOD1) carrying a single amino acid substitution at position 93, where glycine is substituted by alanine (hSOD1^{G93A}; [102]). These mice develop an early onset and quickly progressing phenotype consisting of a rapid escalation of tremors, muscular weakness, motor impairment and, finally, premature death. Histological analysis has revealed substantial motor neuron loss, presence of ubiquitin-positive inclusions in the surviving motor neurons as well as in glial cells [177, 217, 242, 270], astrocytosis and microgliosis [105], thus mimicking the human condition. A remarkably similar phenotype could also be described in mice overexpressing hSOD1^{G37R} [308] or hSOD1^{G85R} [31], where glycine at position 37 or 85, respectively, was replaced by arginine. Intriguingly, the latter strain was reported to develop astrocytic SOD1- and ubiquitin-positive inclusions as the earliest indication of the disease. Such observation has two implications: first, it suggests that astrocytes are directly damaged by mutant human SOD1 expression; second, it supports the hypothesis that motor neuron demise may be a consequence of glial dysfunction [31]. To test this theory, the astrocyte-specific GFAP promoter was used to drive the expression of mouse SOD1^{G86R}, the murine orthologue of hSOD1^{G85R}. Although these mice displayed astrocytosis, they failed to develop motor impairment. Furthermore, histological assessments did not reveal motor neuron loss or microgliosis, thus suggesting that the restricted expression of mutant SOD1 in the astrocytes is not sufficient to cause neurodegeneration [93]. Subsequent genetic experiments, however, argued against this early conclusion. In particular, thorough histological analyses of chimeric mice revealed that wild-type motor neurons display pathological ubiquitinated protein inclusions when surrounded by mutant SOD1-expressing non-neuronal cells. Complementary, the survival of mutant SOD1-expressing neurons was prolonged in the presence of wild-type non-neuronal cells [41]. In keeping with this evidence, selective ablation of different mutant SOD1s from astrocytes slowed down disease progression and extended survival in ALS mouse models [303, 310]. Transplantation studies provided further evidence in support of the hypotheses that ALS astrocytes are intrinsically neurotoxic and that healthy cells retain their neuroprotective phenotype in an ALS environment. Briefly, transplanting hSOD1^{G93A}-expressing astrocyte precursors in the spinal cord was reported to induce an ALS-like phenotype with motor neuron degeneration in wild-type rats [213], while introducing wild-type astrocyte precursors extended the lifespan in hSOD1^{G93A} rats [153]. More recently, the tremendous advancements in the field of stem-cell technology enabled researchers to perform analogous transplantation experiments with human cells. In particular, induced pluripotent stem cell (iPSC)-derived glial precursors from a SOD1^{D90A} fALS patient [38] or a sALS patient [228] transplanted into the spinal cord of wild-type mice predominantly differentiated into astrocytes and induced both motor impairment and histological signs of motor neuron degeneration. Complementary to this evidence, iPSC- or human embryonic stem-cell-derived healthy donor glial precursors transplanted in the spinal cord of hSOD1^{G93A}-overexpressing mice at the early symptomatic stage ameliorated the mouse phenotype [116, 137].

Although these studies provide indisputable evidence that astrocytes play a role in the pathogenesis of ALS, they almost exclusively focus on the ALS-*SOD1* subtype. Therefore, it is reasonable to postulate that their relevance might be restricted to this specific form of the disease. The continuous development of newer models aiming at recapitulating other familial forms, such as ALS-*TDP-43*, is meant to allow additional investigations on astrocyte dysfunction in order to determine whether they represent core mechanisms of ALS pathogenesis. An important contribution to the comprehension of this issue came from the generation of transgenic rats with astrocyte-specific inducible expression of human TDP-43 harbouring the substitution of methionine to valine at position 337 ($hTDP43^{M337V}$) [279]. Upon transgene induction, the animals underwent a rapid deterioration of their motor functions with motor neuron loss, accumulation of ubiquitinated inclusions in the astrocytes, astrocytosis and microgliosis [279]. Since it has been proposed that TDP-43 mutations lead to a loss of its function, it is particularly relevant that mice with TDP-43 knockdown predominantly occurring in the astrocytes developed motor dysfunction, electromyographical abnormalities, paralysis and spinal cord motor neuron loss [311]. Furthermore, neuromuscular junction abnormalities were present in fruit flies with expression of different ALS-associated TDP-43 variants in various glial populations, including the astrocytes [64].

From this large amount of evidence, it clearly emerges that fALS-associated proteins can trigger astrocyte dysfunction, and their aberrant phenotype is critical to induce motor neuron demise. Furthermore, healthy astrocytes retain their protective abilities even in a disease microenvironment, an observation that has tremendous implications for the development of new cell therapies. One may hypothesize that implanting healthy cells should halt the disease progression in ALS patients. To explore this therapeutic option in the clinical practice, a phase I/IIa clinical trial (ClinicalTrials.gov Identifier: NCT03482050) has recently started (April 2018) recruiting patients to assess not only safety and tolerability but also disease progression rates upon intrathecal administration of human stem-cell-derived astrocytes.

10.1.2.3 Intrinsic Dysfunctions of the Astrocytes

To fully exploit the potential of astrocytes as targets for ALS therapy, it would be extremely valuable to understand not only the molecular mechanisms underlying astrocyte abnormalities but also how their communication with motor neurons is impaired. Several lines of research are trying to give an answer to the following key questions: are astrocytes directly suffering from fALS-driven toxicity? are they releasing toxic substances or, when stressed, do they become unable to provide trophic and metabolic support to neurons?

As previously mentioned, ubiquitinated protein inclusions were identified in the astrocytes of both human post-mortem material and transgenic animal models of ALS [31, 127, 136, 172, 177, 217]. The biological relevance of such inclusions has remained elusive until our group demonstrated that cells harbouring ubiquinated protein aggregates are degenerating astrocytes expressing also activated caspase-3, a marker of apoptosis. Damaged astrocytes appear at the pre-symptomatic stage, prior to the loss of motor neuronal cell bodies, and their number ramps up during disease progression [172, 242]. Remarkably, it has been recently demonstrated that the accumulation of such apoptotic astrocytes is also shortening the lifespan of $hSOD1^{G93A}$ mice [136], thus consolidating the hypothesis that neuroglia degeneration is an active driver of disease progression. Furthermore, in $hSOD1^{G93A}$ mice at the end stage, it has been shown that activated astrocytes overexpress some BH3-only members of the Bcl-2 family proteins, which are normally channelling detrimental signals from mitochondria into an apoptotic pathway of cell death [58, 140]. However, those cells do not simultaneously express further markers of apoptosis nor display aberrant morphologies, thus suggesting that such proteins might possess other functions than apoptosis induction [58, 140], or they might be marking cells which are committed to die, but have not yet fully executed the cell death program. Finally, evidence of astrocyte degeneration came from the characterization of human astrocytes differentiated from $TDP43^{M337V}$ [255] and VCP^{R191Q} (arginine to glutamine substitution in position 191) or VCP^{R155C} (arginine to cysteine substitution in position 155)—expressing inducible pluripotent stem cells (iPSCs) [104].

From a mechanistic point of view, the observation that apoptotic astrocytes are located in the immediate neighbourhood of glutamatergic terminals suggested to investigate whether $hSOD1^{G93A}$ astrocytes were susceptible to glutamate toxicity. In vitro experiments demonstrated that the viability of ALS astrocytes is unaffected under standard culturing conditions. On the contrary, glutamate triggers astrocyte apoptosis by activating its metabotropic receptor 5 (mGluR5) followed by downstream aberrant calcium (Ca^{2+}) release from the intracellular stores [172, 242] (Fig. 10.1), a phenotype recently linked to the downregulation of protein kinase C isoform epsilon [297]. Interestingly, in the past few years, the evidence of abnormal Ca^{2+} signalling in ALS astrocytes has been confirmed and extended to other experimental settings by several independent groups. In particular, in $hSOD1^{G93A}$ expressing astrocytes, it has been discovered a chain of events triggered by the excessive activity of the store-operated Ca^{2+} entry mechanism, which leads to Ca^{2+} overfilling of the endoplasmic reticulum (ER), thereby producing an augmented Ca^{2+} release from the ER in response to ATP stimulation [5, 128] (Fig. 10.1). The overexpression of Connexin-43 [5], a molecule responsible for the formation of gap junctions and hemichannels within the astrocyte network, has been also implicated in the elevation of intracellular Ca^{2+} levels. Moreover, sustained Ca^{2+} increases deriving from the extracellular environment have been observed in the astrocytes upon treatment with recombinant mutant SOD1 [187].

Curiously, degeneration is not the only consequence of the overexpression of ALS-associated proteins in the astrocytes. Another subpopulation of astrocytes with high proliferation rate could be identified in $hSOD1^{G93A}$ transgenic rats [52] and mice [39]. It is yet to mention that in vivo ablation of proliferating astrocytes did not ameliorate the ALS phenotype in different mouse models [152], thus suggesting that the contribution of such glial cell population to disease pathogenesis is likely limited.

In the wake of recent discoveries pointing at stress granules as precursors of pathological lesions in ALS, dynamics of these membrane-less organelles has been investigated in relevant neuronal populations as well as in the astrocytes, thus uncovering their distinctive behaviour. In particular, it has been recently demonstrated that astrocytes resolve stress granules with faster kinetics than neurons [131].

Finally, to provide a comprehensive and unbiased picture of the repercussions of ALS-associated proteins on the astrocyte homeostasis, different groups have set out to investigate transcriptional, translation and metabolic changes either in cell

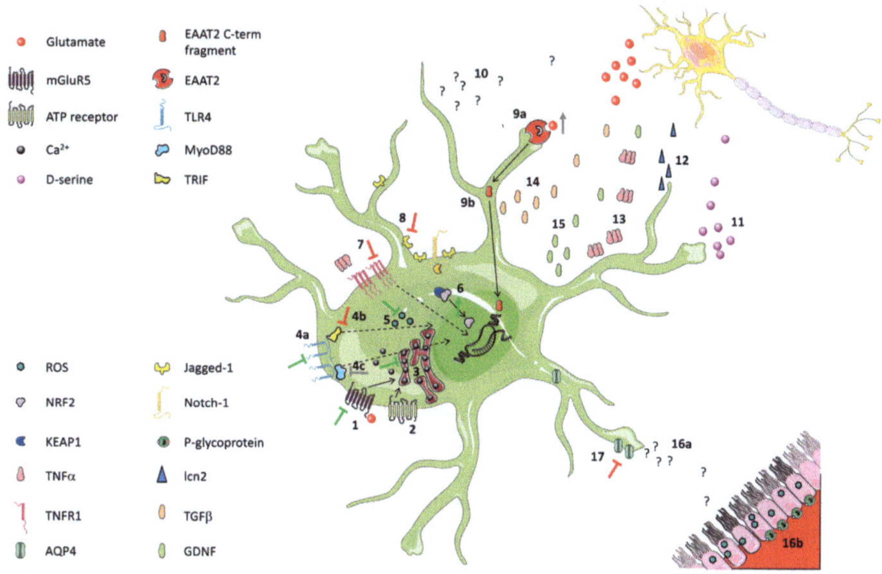

Fig. 10.1 Aberrant pathways in ALS astrocytes: a window of therapeutic opportunity. In ALS astrocytes, several signalling pathways are intrinsically aberrant. Activation of mGluR5 (1) or ATP receptors (2) leads to abnormal calcium signalling (3) and apoptosis. Moreover, upregulated TLR4 (4a) signals through its adaptor TRIF (4b) to counteract cell demise or through MyoD88 (4c). Accumulation of reactive oxygen species (ROS; 5) and subsequent activation of the transcription factor NRF2 (6), which is released from its KEAP1 inhibitor, are also hallmarks of ALS astrocytes. Upregulation of TNFR1 (7) and the Notch ligand Jagged-1 (8) have also been described. In terms of cross-talk with neurons, ALS astrocytes display reduced expression of the glutamate transporter EAAT2 (9a), which become abnormally SUMOylated and cleaved (9b). The excitotoxic milieu is then worsened by the release of an unidentified proteinaceous factor (10) and D-serine (11). Toxicity triggered by factors including lipocalin (lcn2; 12), TNFα (13) and TGFβ (14) is opposed by the release of the neurotrophic GDNF (15). Astrocytes also contribute to the impairment of the BBB by releasing unidentified neuroinflammatory factors (16a) that lead to the upregulation of the transporter P-glycoprotein (16b). Finally, upregulation and mislocalization of the water channel aquaporin-4 (AQP4; 17) is also contributing to BBB leakage. Therapeutic intervention (inhibition, blunt arrow; activation, pointed arrow) on these pathways proved to be beneficial (green), detrimental (red) or ineffective (grey) on disease progression. Details and references in the text

culture [113, 165] or in animal models along the disease progression [14, 69, 271]. In cell culture, among the differentially expressed genes, there is an enrichment in those encoding for secreted proteins. More specifically, in response to pathogenetic TDP-43 expression, neuroprotective mediators are downregulated whereas neurotoxic factors, such as lipocalin-2 (*lcn2*) and chitinase-3-like protein 1 (*Chi3l1*), are upregulated [113]. Furthermore, several metabolic pathways become dysregulated in response to glutamatergic stimuli or in co-culture with motor neurons, particularly the cellular shuttling of lactate between astrocytes and motoneurons [165]. Astrocytes captured from $hSOD1^{G93A}$ mice display an analogous metabolic dysregulation as well as a shift from a neuroprotective towards neurotoxic phenotype already at the pre-symptomatic stage [69]. Furthermore, their phenotype shows further changes along disease progression, when ALS astrocytes display faulty cholesterol homeostasis due to defective nuclear translocation of the sterol regulatory binding protein 2 (SREBP2), along with a pro-inflammatory phenotype characterized by enhanced phagocytosis and lysosomal activity [14]. Importantly, upregulation of the inflammatory process and downregulation of metabolism has been observed also in hSOD1^{G37R} astrocytes [271]. The peroxisome proliferator-activated receptors (PPARs) and the liver X receptors (LXRs) have been proposed as master regulators of the observed transcriptional change [271], although nuclear translocation of PPARs does not seem to enhance in spinal cord astrocytes from another mutant SOD1 mouse model of ALS (i.e. hSOD1^{G93A}) during disease progression [18].

These latest 'omics' studies not only describe how astrocytes adapt to the expression of noxious proteins but shed some light on how their interactions with neighbouring cells may be affected, in terms of both (i) receiving and transducing signals coming from the extracellular environment and (ii) providing an adequate response. These abilities are particularly critical in the context of astrocyte-motor neuron crosstalk. Thus, in the next paragraph, we will discuss how they might have lost their capacity to maintain a supportive microenvironment for motor neurons while gaining neurotoxic properties.

10.1.2.4 Astrocyte-Motor Neuron Communication Is Impaired in ALS

The evidence outlined so far suggests that the expression of ALS proteins in astrocytes affects their bidirectional communication with motor neurons. Many independent groups have demonstrated that ALS astrocytes are toxic to the motor cells in mice [25, 50, 69, 132, 198, 225, 280] and, more importantly, in human co-culture systems [49, 103, 167, 180, 231, 267]. Moreover, such toxicity appears to be mediated by one or more soluble factors, as conditioned medium from ALS-*SOD1* (hSOD1^{G93A} or hSOD1^{G86R}) or ALS-*TDP-43* (hTDP43^{A315T}) astrocytes is sufficient to induce motor neuron hyperexcitability, mitochondrial damage with enhanced oxidative stress and activation of a deadly c-Abl signalling cascade [81, 238, 239]. The toxicity of conditioned medium from hSOD1^{G93A}-expressing astrocytes has been confirmed also in vivo through its chronic administration in rat lumbar spinal cords, where it rapidly induces motor dysfunction and motor neuron loss [230]. These findings have been further expanded by exploring the toxicity of the astrocyte medium from either mouse cells expressing hFUS^{R521G} (position 521: arginine to glycine substitution) [132] or from iPSC-derived astrocytes harbouring a repeat expansion in the *C9orf72* gene [164].

These observations globally prompted the hunt for neurotoxic factor(s), and several molecules have been subsequently implicated in the harmful cascade, fuelling an extensive scientific debate along the last years. Astrocytes are well known to release a wide variety of bioactive molecules via the activation of an array of mechanisms (reviewed in [300]). The most obvious candidate to investigate in the context of ALS was initially the excitatory amino acid **glutamate**, a molecule whose implication is supported by several findings. First, it is well known that excessive glutamatergic stimulation triggers neuronal demise through a distinctive mechanism of cell death named 'excitotoxicity'. Second, in ALS patients, astrocytes display reduced expression and function of the astrocyte-specific glutamate plasmalemmal transporter EAAT2 [31, 243], a condition that leads to CSF accumulation of glutamate [244]. Importantly, deficient expression of EAAT2 (GLT-1 in rodents) has been consistently reported in several animal models of both ALS-*SOD1* [31, 100, 112, 215, 243] and ALS-*TDP-43* [51, 279] (Fig. 10.1). As yet, there is no common consensus about the pathway that mediates such depletion, and a large number of potential mechanisms have been implicated. Early evidence excluded that EAAT2 downregulation had a genetic origin, as ALS patients do not segregate *EAAT2* mutations [9, 117]. Recently, it has been discovered that the transcription factor Yin Yang 1 (YY1) is a negative regulator of *EAAT2* expression [126]. In hSOD1^{G93A} mouse astrocytes, a detrimental signalling cascade is set in motion by the overexpression of astrocytic elevated gene 1, leading to nuclear accumulation of YY1 and subsequently reduced transcription of *EAAT2* [315]. Furthermore, other groups have scrutinized all the steps affecting the maturation and the stability of *EAAT2* primary transcript, starting from the occurrence of alternative splicing events. Thus, Lin et al. demonstrated the accumulation of aberrant mRNA species that limits the availability of a translationally competent transcript, either by its enhanced degradation or through a dominant-negative effect [160]. However, subsequent studies revealed that those mRNA species are not specific for ALS cases, as they can be detected also in cohorts of controls and patients affected by other CNS disorders [99, 111, 181]. Another important post-transcriptional modification of the mRNA may be RNA editing. Flomen et al. detected an aberrant editing event taking place in *EAAT2* primary transcripts, specifically in affected areas of the CNS of ALS patients. This modification generates an alternative polyadenylation site that shortens the mRNA half-life [72]. Finally, *EAAT2* mRNA could be regulated also by signalling from neurons. In particular, glutamate released upon neuronal activation induces the transcription of *EAAT2* [313] and modulates its mobility into the membrane [4]. Intriguingly, transcriptional activation is mediated by the expression of the nuclear protein Kappa-B Motif-Binding Phosphoprotein (KBBP, [313]) the murine homolog of the heterogeneous nuclear RNA/DNA-binding protein K (hnRNP K), a protein recruited by RNA foci in ALS patients with *C9orf72* hexanucleotide expansion [43] and showing a complex regulatory interplay with TDP-43 [194, 195].

However, glutamate might not be the only molecule released by neurons and acting on the astrocytes to modulate EAAT2 activity. MicroRNA 124a (miR-124a) is an additional interesting candidate. It has been reported that this noncoding RNA is enclosed in exosomes released by neurons and internalized by the astrocytes, where it positively regulates EAAT2 expression [193]. In the context of ALS, miR-124a is downregulated [193, 320], an event that can possibly cause the loss of *EAAT2*. MiR-124a downregulation might have even a broader impact on the astrocyte phenotype, as it also regulates the expression of the transcription factors *Sox2* and *Sox9*, which govern astrocyte differentiation [320]. Importantly, astrocytes themselves are competent for the release of exosomes (reviewed in [300]) and, although an early report did not unravel any abnormalities in the profile of miRNAs released from hSOD1^{G93A} astrocytes [122], further studies are needed to investigate this topic thoroughly.

The studies discussed so far mainly focused on the possible modulation of EAAT2 expression and function by acting at various levels of its transcript metabolism. However, its activity can be limited also at the protein level by post-translational modifications. For example, EAAT2 was reported to be cleaved by caspase-3 [28] and subsequently SUMOylated on the resulting C-terminal fragment [89] (Fig. 10.1). In hSOD1^{G93A} astrocytes and spinal cord, caspase-3 cleaves EAAT2 after an aspartic acid in the position 504, thus inactivating the transporter [28]. More recently, hSOD1^{G93A} mice have been crossed with knock-in animals, where the endogenous $EAAT2^{WT/WT}$ has been replaced with the $EAAT2^{D504N/D504N}$ variant (position 504: aspartic acid to asparagine substitution), which destroys the caspase-3 consensus sequence and makes the protein resistant to cleavage without altering its properties as a transporter. The resulting mice display delayed disease progression and increased lifespan, in the absence of significant neuroprotection, thus demonstrating that EAAT2 inactivation drives the late phase of the disease [241]. Furthermore, in cell cultures and hSOD1^{G93A} spinal cords, the SUMOylated C-terminal fragment redistributes to the nucleus of astrocyte, where it triggers glial- and neurotoxicity [75, 89]. More recently, it has been demonstrated that also a fraction of the full-length EAAT2 protein is SUMOylated, and this latter modulates its insertion into the cell membrane. Interestingly, the extent of this post-translational modification is unaffected by the expression of hSOD1^{G93A} in astrocytes and does not change during the course of ALS in mice [74].

From a therapeutic standpoint, EAAT2 dysfunction is a particularly attractive target and several approaches have been attempted to rescue its downregulation, hoping to significantly halt disease progression. A validation of this hypothesis came from early studies in transgenic mice. In particular, genetic overexpression of EAAT2 itself [100] or of its modulator Peroxisome Proliferator-Activated Receptor γ Coactivator 1α (PGC1α; [156]) in hSOD1^{G93A} mice improved their motor functions. More recently, EAAT2 overexpression was achieved through intraspinal injection of a viral vector driving its expression specifically in the astrocytes at the symptomatic stage of hSOD1^{G93A}. Unfortunately, despite successful and sustained astrocyte transduction, no prolonged survival was recorded [154]. In parallel, attempts to identify small drugs able to pharmacologically enhance EAAT2 activity gained momentum from the seminal discovery that β-lactam antibiotics increase EAAT2 expression. In particular, the cephalosporin ceftriaxone gave extremely promising results in terms of lifespan extension when chronically administered to hSOD1^{G93A} mice [245]. Since antibiotics have been long used in clinical practice, a phase I/II clinical trial was promptly started and demonstrated that ceftriaxone has a very good safety profile and can reach the CNS in therapeutic amounts in ALS patients [23]. It was therefore quite disappointing when ceftriaxone failed to provide any therapeutic benefit in a phase III clinical trial aiming at assessing its efficacy [45]. Despite this shortcoming, other drugs have been investigated in preclinical testing such as harmine [155], LDN/OSU-021320 [42, 139] and, more recently, MC1568 [146]. The latter drug is an inhibitor of class II histone deacetylases (HDACs), enzymes regulating gene transcription by shaping the epigenetic landscape and involved in a growing number of neurodegenerative conditions (reviewed in [276]). In primary cultures of mouse glia, MC1568 administration increased both EAAT2 expression and its SUMOylation, thereby failing to boost transport activity. Nevertheless, upon chronic administration in hSOD1^{G93A} mice, this drug sustained EAAT2 expression and activity, thus inducing a transient improvement of the motor function. Unfortunately, this beneficial effect was only temporary, as the treatment was unsuccessful in extending the lifespan [32, 146]. Taken together, these studies suggest that post-translational modifications play a crucial role in tuning EAAT2 function. Therefore, it is tempting to speculate that drugs that promote glutamate transport should be associated with a SUMOylation inhibitor to achieve a successful therapeutic outcome in ALS. Intriguingly, glutamate accumulation is also a feature of hTDP-43^{WT} overexpressing mice, even in the absence of detectable EAAT2 loss. In this specific model, the amino acid accumulation is attributed to astrocyte malfunction and it is coupled with reduced glutamate metabolism [109].

The evidence of a role for excitotoxic stress in the pathogenesis of ALS was further reinforced by the discovery that ALS astrocyte release D-serine, an amino acid necessary for the full activation of neuronal N-methyl-D-aspartate (NMDA) glutamatergic receptors [249, 250]. Intriguingly, genetic analysis of an ALS family could identify a single loss-of-function mutation (position 199: aspartate to tryptophane) in the gene coding for the enzyme D-aminoacid oxidase (DAO), deputed to the catabolism of D-serine [190]. Furthermore, co-culture experiments demonstrated that the expression of this hDAO^{D199W} mutant in astrocytes is sufficient to trigger impaired autophagy, ubiquitinylated protein accumulation and, ultimately, motor neuron cell death [190, 221]. In vivo, hDAO^{D199W} overexpressing mice developed motor dysfunction and age-dependent motor neuron loss [138]. Curiously, crossing these animals with hSOD1^{G93A} overexpressing mice had only a modest worsening of the phenotype in females, while the severity of motor dysfunction was unaffected in males [138]. In addition, two different viral-mediated gene therapy approaches, boosting DAO expression, proved to be neuroprotective and led to extended lifespan in hSOD1^{G93A} mice [161, 304]. Finally, an intriguing link between D-serine overproduction and EAAT2 dysfunction has been proposed in hSOD1^{G93A} expressing astrocytes by showing aberrant expression of the protein interacting with C kinase 1 (PICK1), an interacting partner of both EAAT2 and serine racemase (i.e. the enzyme converting L-serine into D-serine) [73]. Further functional studies are, however, necessary to clarify the mechanistic relevance of such early observation in the context of ALS pathogenesis.

Astrocytes can modulate excitotoxicity also by regulating the composition of AMPA receptors on motor neurons, thus tuning their vulnerability to glutamatergic stimuli. In particular, they control the expression of the GluA2 subunit, the presence of which renders AMPA-type glutamate receptors impermeable to calcium influx from the extracellular environment under physiological conditions and, therefore, it increases the resistance to the glutamatergic insult [266]. Remarkably, hSOD1^{G93A} [290] and hFUS^{R521G} [132] astrocytes lose their ability to induce upregulation of GluA2 in neurons which, in turn, become more vulnerable to excitotoxic stimuli.

Increased production of **ROS** has been extensively investigated as another possible culprit of astrocyte-mediated neuronal toxicity, and various species have been detected in different familial subtypes of ALS [35, 81, 167, 189, 238, 239, 294] (Fig. 10.1). Consistently, increasing astrocytic antioxidant activity through pharmacological or genetic approaches showed neuroprotective potential [35, 167, 189, 223, 294].

A particularly appealing target was identified in the transcription factor nuclear erythroid 2-related factor 2 (NRF2), a master regulator of the oxidative stress response in the astrocytes. The regulation of this molecule and its target genes has been described in detail elsewhere [158]. Briefly, under physiological conditions, NRF2 is sequestered in the cytoplasm by its inhibitor Kelch-like ECH associated protein 1 (KEAP1) (Fig. 10.1). Upon chemical or oxidative insult, KEAP1 is rapidly degraded and NRF2 can translocate to the nucleus where it competes with Broad-complex, tramtrack, bric-à-brac (BTB) and CNC homology 1 (BACH1) to interact with small Maf proteins. Such interaction leads to the formation of heterodimers that can bind to the antioxidant response element (ARE) in the promoter region of its target genes, ultimately boosting their expression. Several studies of human genetics [21] as well as investigations performed in both cell and animal models [56, 94, 135, 188, 194, 224, 229, 278, 302] and in human post-mortem material [15, 248] converge to suggest that the NRF2 signalling is affected in ALS and could be exploited as a therapeutic target. Several compounds apt at sustaining NRF2 activity were identified in vitro and, subsequently, tested in hSOD1^{G93A}-expressing mice. Regrettably, the outcome of these treatments revealed only a modest improvement in motor performance and, in some cases, in survival [68, 175, 202]. In keeping with this, NRF2 overexpression in neurons through genetic manipulation [296] or viral-mediated gene therapy [200] could not extend lifespan in ALS mice. Furthermore, direct NRF2 ablation in hSOD1^{G93A} mice did not accelerate disease progression [101]. However, selective NRF2 activation in astrocytes had a clear neuroprotective effect in co-cultures [53, 293, 295] and in vivo, extending the lifespan of hSOD1^{G93A} mice [294]. Taken together, this amount of evidence has profound implications for the design of disease-modifying therapies based on tackling oxidative stress, namely, the therapeutic agent should specifically target the astrocytes. Furthermore, one should consider that not every target giving promising results in astrocyte cell cultures is confirmed in animal models [223]. When oxidative stress escapes the control systems, it can exert various detrimental effects by an autocrine/paracrine mechanism. For example, it was reported to cause defective glutamate homeostasis by reducing EAAT2 activity, an event that has been demonstrated in cells expressing different ALS-associated mutant hSOD1 [284–286]. These early findings set the ground for a novel gene therapy approach that was recently developed. Specifically, lentiviral vectors were designed to deliver *EAAT2*, *Glutamic dehydrogenase2 (GDH2)* and *NRF2*. In hSOD1^{G93A} astrocytes co-cultured with the motor neuron-like NSC-34 cells, only the simultaneous administration of these three vectors to glial cells conferred protection against a glutamatergic insult. Moreover, the cocktail ameliorated motor dysfunction, delayed onset and prolonged survival in vivo [20].

Since activation of astrocytes is part of the inflammatory response in the CNS, a number of **neuroinflammatory molecules** have also been investigated as putative neurotoxic agents in the astrocyte-driven neuronal demise. First, attention was given to the eicosanoid class and, in particular, to the arachidonic acid-derived metabolites prostaglandin (PG) D_2 and E_2, whose receptors are upregulated in hSOD1^{G93A} astrocytes and neurons [49, 141, 157]. Notably, delayed disease onset and prolonged survival were obtained by crossing ALS mice with animals devoid of the PGE_2 receptor EP2 [157]. Mechanistically, it has been speculated that pre-stimulation with low doses of PGE_2 can induce its EP2 receptor, thereby exposing neurons to the toxic action of a subsequent prostaglandin challenge [141]. More recently, the expression of the catabolic enzyme 15-hydroxyprostaglandin dehydrogenase (15-PGDH) was reported to increase with the disease progression in the spinal cord of hSOD1^{G93A} mice, and its immunoreactivity was associated with astrocytes in the white matter [191]. This suggests an unbalance between the synthesis and degradation of PGE_2 in the spinal cord. Remarkably, arachidonic acid serves as a precursor also for another class of inflammatory mediators, namely, leukotrienes. In this biochemical metabolic cascade, it is processed by different lipoxygenases to generate intermediates known as 12- and 15-hydroxyeicosatetraenoic acid (12-HETE; 15-HETE), which were reported to accumulate in the brain of hSOD1^{G93A} mice with the disease progression. Importantly, pharmacological treatments with nitrosylated fatty acids, such as nitro-oleic acid (NO_2-OA), prevented 12-HETE accumulation and improved motor function of ALS mice [283]. Remarkably, this compound is activating NRF2 signalling in hSOD1^{G93A} astrocytes, thereby blocking their toxicity towards neurons in cell cultures [53] and in vivo [283]. These findings suggest that drugs simultaneously tackling different aspects of astrocyte malfunction might grant a better rescue of their toxic phenotype.

A particularly interesting neuroinflammatory and neurotoxic factor is the secreted lcn2 protein, which has been implicated in a wide variety of cellular processes within the CNS, both under physiological and pathological conditions (reviewed in [118]). In the context of ALS pathogenesis, lcn2 is strongly upregulated in the astrocytes in rodent models of different subtypes of fALS, namely, ALS-*SOD1*, ALS-*TDP-43* and ALS-*FUS* [24, 113, 279] (Fig. 10.1). In vitro studies allowed depicting a mechanistic loop where neuronal expression of ALS-associated TDP-43 and FUS mutants not only induce the astrocytic release of lcn2, but also sensitize neurons to lcn2 toxicity [24]. Further investigations are certainly needed to identify the missing pieces of this puzzle, though the reported observations seem to suggest that testing an lcn2 receptor antagonist could be a possible strategy to halt ALS progression.

Another class of inflammatory molecules that could mediate neurotoxicity is that of cytokines. hSOD1^{G93A} astrocytes were in fact reported to release interferon γ (IFNγ), which triggers a detrimental signalling cascade on motor neurons expressing the lymphotoxin-β receptor (LT-βR) and the adapter Light [2]. Interestingly, disrupting this axis by genetic ablation of *Light* [2] or by infusing IFNγ neutralizing antibodies in the CNS ameliorates the phenotype of hSOD1^{G93A} mice [210]. More complex is the role played by the cytokine Tumor Necrosis Factor α (TNFα; Fig. 10.1) which, in addition to its roles as immunomodulatory factor (reviewed in [124]), is also critically modulating neurotransmission (reviewed in [246]). Regarding its specific involvement in ALS, astrocytes expressing $hFUS^{R521G}$ were recently described to display enhanced release of TNFα, coupled to perturbed AMPA receptor trafficking and neurotoxicity in co-cultures, thus revealing a deleterious paracrine action of astrocytic TNFα in ALS-*FUS* [132]. Yet, in both transgenic $hSOD1^{G93A}$ co-cultures and mice, this pleitropic cytokine was reported to trigger opposite effects, depending on the TNF receptor (TNFR) engaged. Thus, activation of TNFR2 was shown to trigger motor neuron loss [280], whereas we demonstrated that stimulation of TNFR1 in the astrocytes can be beneficial for motor neurons by prompting the expression and secretion of the glial cell line derived neurotrophic factor (GDNF, [30]), a powerful neuroprotective agent in ALS and several other CNS disorders (reviewed in [114]) (Fig. 10.1). More specifically, we reported that this neuroprotective response is completely abolished by genetic ablation of TNFR1 in vitro and in vivo. In $hSOD1^{G93A}$ mice and in sALS patients, we showed progressive upregulation of the entire TNF-TNFR1-GDNF axis. We, therefore, speculated that disrupting this protective cascade by transferring $hSOD1^{G93A}$ mice on a TNFR1 knockout background would prevent astrocytes from mounting a neuroprotective reaction, thus exacerbating neuronal loss and shortening lifespan. Consistent with this hypothesis, disease onset was not anticipated in $hSOD1^{G93A}$ mice lacking TNFR1, whilst disease progression was accelerated and neuronal loss became more severe [30]. The relevance of sustained release of glial-derived GDNF has been recently corroborated by a study describing the positive impact on the disease course of the implantation of human progenitor cells overexpressing GDNF and differentiating into astrocytes [277]. Thus, there is great hope that a phase I/IIa clinical trial (ClinicalTrials.gov Identifier NCT02943850), aimed at assessing the safety of such cells in ALS patients, will give positive results and will be quickly transferred to the clinic to determine the efficacy of this approach.

Growth factors are another category of mediators that was surprisingly reported to play a role in the toxic interplay between astrocytes and motor neurons in ALS. The earliest evidence of their involvement came from the investigation of the transcriptome of astrocytes extracted through laser capture microdissection from presymptomatic $hSOD1^{G93A}$ mouse spinal cord sections. In particular, upregulation of nerve growth factor β (β-NGF) was observed, and extensive in vitro characterization demonstrated that ALS astrocytes release toxic amounts of the immature form of this neurotrophin. This latter activates an apoptotic cell death program in motor neurons upon interacting with its receptor p75 ($p75^{NTR}$) [69]. Intriguingly, in symptomatic $hSOD1^{G93A}$ mice, motor neurons display strong immunoreactivity for the fibroblast growth factor 1 (FGF-1), which activates its receptor FGFR1 on astrocytes to mediate the release of NGF [34]. A possible explanation for this toxic effect of NGF came from a very recent paper showing that the neurotrophin undergoes nitration and glycation in the spinal cord of $hSOD1^{G93A}$ mice. These post-translational modifications grant NGF aberrant contacts with the receptor for advanced glycation end products (RAGE), which interacts with $p75^{NTR}$ to activate a toxic cascade [134]. However, boosting the NRF2-mediated antioxidant response in astrocytes prevented NGF-induced motor neuron death [224]. An unbiased assessment of the transcriptome of $hSOD1^{G93A}$ astrocytes identified a core role also for Transforming Growth

Factor β (TGFβ) signalling in ALS [225] (Fig. 10.1). In hSOD1^{G93A} mice, TGFβ1 was shown to be upregulated in the astrocytes and indirectly drive disease progression and motor neuron cell death by reducing the protective component of microglia and T-cells [63]. Moreover, astrocyte-derived TGFβ1 was reported to increase cell stress directly in human motor neurons by inhibiting autophagy, which in turn causes pathological protein inclusions [281].

Besides releasing a wide array of neurotoxic molecules, ALS astrocytes can be deleterious to neurons also by failing their duties as supportive cells. As already mentioned while discussing the dysregulation of glutamatergic homeostasis, ALS astrocytes lose their ability to sustain GluA2 expression on motor neurons, thus favouring the influx of calcium ions through AMPA receptors [132, 165, 290]. Furthermore, hSOD1^{G93A}-expressing astrocytes were shown to reduce metabolic support from lactate release [69]. More recently, evidence started to accumulate that ALS pathogenesis is also characterized by impaired processing of 'help-me' signals that neurons launch to astrocytes, either because the message itself is impaired or because astrocytes fail to react adequately. For example, the fALS-associated protein angiogenin (Ang) is a motor neuron-secreted RNAse that is internalized by the astrocytes [264]. Within astrocytes, Ang cleaves RNA to trigger in return a neuroprotective response characterized by the release of several trophic proteins [263]. Intriguingly, the ALS-associated mutant Ang^{K40I} (position 40: lysine to isoleucine substitution) fails to enroll this neuroprotective response from astrocytes because it is devoid of RNA cleavage activity, despite being correctly released and internalized [264]. Furthermore, Tyzack et al. have shown that ephrin type B receptor 1 (EphB1) is upregulated in injured motor neurons, which activate a potent ephrin-B1-mediated neuroprotective response in the astrocytes. Yet, the EphB1-ephrin-B1 pathway was shown to be disrupted in hSOD1^{G93A} mice and in iPSC-derived astrocytes from a SOD1^{D90A} patient, thus implicating that astrocytes fail to support motor neurons in ALS [287]. Finally, a supporting environment can be maintained upon the concomitant expression of major histocompatibility complex class I (MHCI) molecules on motor neurons and their receptor killer cell immunoglobulin-like receptor KIR3DL2 on the astrocytes. However, this axis seems to be disrupted in ALS, as motor neurons from hSOD1^{G93A} mice and from both sporadic and familial cases display reduced MHCI expression because of the action of a soluble factor released by ALS astrocytes [267].

In the previous paragraphs, we have reported the current evidence on the molecular mediators implicated in astrocyte-induced motor neuron toxicity as well as on the loss-of-function astrocytes can experience in ALS. Those studies have been complemented by others, aiming at elucidating the cellular pathways that become aberrant within astroglial cells, thereby forcing astrocytes to acquire a toxic phenotype. For example, abnormal activation of Jagged-1/Notch signalling was recently reported in reactive astrocytes in the spinal cord of hSOD1^{G93A} mice and sALS patients. Astrocyte-specific inactivation of Jagged-1 intensified the activation of Notch signalling and accelerated the disease progression in mice, thus suggesting that Notch overactivation contributes to ALS pathogenesis and it is mitigated by the upregulation of Jagged-1 in reactive astrocytes (Fig. 10.1) [207]. The toll-like receptor 4 (TLR4) signalling cascade is another pathway that captured particular attention, as this receptor controls the expression of various proteins through the activation of its downstream effector NF-kB (reviewed in [88]). There is evidence indicating that TLR4 and its endogenous ligand, the damage-associated molecular pattern molecule High Mobility Group Box-1 (HMGB1), are upregulated in the astrocytes and microglia during disease progression in hSOD1^{G93A} mice [149] (Fig. 10.1). TLR4 signalling is transduced along two different pathways depending on two diverse interaction partners, the myeloid differentiation factor 88 (MyoD88) and the TIR domain-containing adaptor inducing interferon-β (TRIF) (Fig. 10.1). Intriguingly, MyoD88 ablation in hSOD1^{G93A} mice does not modify the disease course, while hSOD1^{G93A} animals lacking TRIF have a shorter lifespan and display a higher number of aberrant astrocytes. This suggests that TRIF plays an important role in protecting the microenvironment surrounding motor neurons [136]. In keeping with this, we recently demonstrated that activation of TLR4 (and RAGE) by HMGB1 can trigger the production of the trophic factors GDNF and brain-derived neurotrophic factor (BDNF) by the astrocytes, which are likely to exert neuroprotective activities [29]. Furthermore, different groups have investigated the potential role of the TLR4 and RAGE downstream transcription factor NF-kB in ALS astrocytes. This protein complex is not only at the crossroad of many inflammatory pathways, but it is also directly activated by different proteins implicated in fALS, including TDP-43 [272], Ubiquilin 2 [227], hnRNPA1 [107] and FUS [288]. Human astrocytes derived from sALS and fALS patients were shown to differentially express an array of cytokines, whose expression is driven by NF-kB, thus suggesting the involvement of this transcription factor in the disease [103]. Furthermore, preserved motor function was observed in TDP-43 overexpressing mice [272] and prolonged survival was obtained in hSOD1^{G93A} and hSOD1^{G37R} mice, upon pharmacological treatment with the NF-kB inhibitor Withaferin A [218], also able to block the TLR4-TNFα axis in the astrocytes [171]. Although these papers suggest that targeting NF-kB might be a useful therapeutic tool in ALS, early evidence demonstrated that this factor should be specifically targeted in cells others than the astrocytes. In particular, crossing hSOD1^{G93A} mice with animals that express the dominant-negative IkBα, under the control of the GFAP promoter, did not affect disease onset or progression [44]. These results were then replicated by Frakes et al., who demonstrated that overexpression of another repressor of the NF-kB signalling, through gene therapy or genetic manipulation in the astrocytes, did not rescue motor impairment nor extended survival in hSOD1^{G93A} mice [79]. However, a very recent investigation shed light on the role of astrocytic NF-kB activation in vivo, in the hSOD1^{G93A} mouse model of ALS. Taking advantage of astrocyte-restricted conditional expression of constitutively active NF-kB, Ouali Alami and collaborators revealed a multifaceted and stage-specific response of the transcription factor [211]. While prolonged NF-kB activation in the astrocytes was shown to accelerate disease progression, its activation in the pre-symptomatic phase induced neuroprotective effects on motor neurons [211]. The most straightforward conclusion arising from these observations is that, when planning NF-kB modulation for therapeutic purposes, one should carefully consider the stage of the disease.

10.1.2.5 Astrocyte Faulty Regulation of the Blood–Brain Barrier

The astrocyte social network does not exclusively include neurons and other glial cells, but comprises additional cell types and structures, notably constituting the blood–brain barrier (BBB). This is a highly specialized formation consisting of microvascular endothelial cells, pericytes and perivascular astrocytes. Serving as a highly guarded border, the BBB carefully regulates the exchange of substances between the bloodstream and the brain parenchyma through different mechanisms (reviewed in [159]). As demonstrated in several studies (reviewed in [209]), astrocytes are regulators of different BBB functions owing to the localization of their endfeet around the capillaries. In the context of ALS, evidence of BBB leakage has been provided both in hSOD1^{G93A} mice and in sALS [84–86, 186, 269, 306, 318], and it has been associated with the overexpression of transporter proteins, such as the P-glycoprotein [36, 84]. Of note, this phenomenon seems to be secondary to aberrant signalling from astrocytes. In particular, both mouse hSOD1^{G93A} and human SOD1^{A4V} astrocytes were reported to trigger NF-kB translocation and P-glycoprotein upregulation through enhanced oxidative stress, while in astroglial cells from a patient expressing FUS^{H517Q} mutation, the mechanism was described to involve the release of pro-inflammatory factors [229]. Clarifying in detail these pathways would be of outstanding importance to identify new potential targets for therapeutic intervention. Interestingly, riluzole was found to be a substrate of the P-glycoprotein and to modulate the activity of other transporters [184, 185]. Thus, a formulation of riluzole and verapamil, a blocker of voltage-dependent calcium channels [62] that can reduce P-glycoprotein activity, was placed under development in order to block the efflux of riluzole from the CNS parenchyma. This cocktail of drugs gave promising results in cell culture experiments [312].

Astrocytes are likely to contribute to BBB dysfunction in ALS also via the upregulation of aquaporin-4 (AQP4; [16, 46, 203, 305]), an important regulator of both the brain water homeostasis (reviewed in [82]) and the glymphatic system, i.e. a CNS cleansing system that interacts with and complements the BBB (reviewed in [299]). Interestingly, AQP4 upregulation is also associated with its mislocalization in hSOD1^{G93A} mouse and ALS patient spinal cord, but not in other models of gliosis, thereby suggesting that this regulation is not part of a generic inflammatory response, but it is caused by some yet unidentified ALS-specific events [46, 305] (Fig. 10.1). In support of this hypothesis, TDP-43 depletion in astrocytes was shown to cause AQP4 overexpression, though it reduced surface levels [131]. While restoring the BBB function appears an interesting therapeutic approach in ALS, an early attempt to pursue this strategy proved disappointing. Thus, the generation of hSOD1^{G93A} transgenic mice lacking AQP4 led to the rescue of BBB leakage, but also to an earlier onset of the disease and shortened lifespan [305]. This result suggests that the extent and the timing of AQP4 depletion are critical to ensure a therapeutic effect, as AQP4 overexpression might also play a beneficial role clearing up toxins that would otherwise accumulate into the CNS parenchyma.

10.2 Spinal Muscular Atrophy

10.2.1 A Brief Introduction

Spinal muscular atrophy (SMA) is a progressive neurodegenerative disease caused by the loss of spinal motor neurons and affecting approximately 1 in 6000 to 1 in 10000 live births [298]. In the vast majority of occurrences (~96%), the disease is caused by deletions or gene conversion events in the *Survival of Motor Neuron 1 (SMN1)* gene on chromosome 5q13 [151], causing a reduction in the levels of its translated product, the SMN protein. This genetic defect is transmitted as an autosomal recessive trait and leads to an extremely variable clinical presentation (reviewed in [274]). SMA patients are subclassified into 5 categories on the basis of the age of onset and symptom severity. Briefly, in the most aggressive form of the disease, i.e. SMA type 0, muscular weakness is already evident at birth, when respiratory difficulties can be observed and lead to death within few weeks. SMA type I, also known as Werdnig-Hoffmann disease, appears within 6 months of age. Affected children never reach the ability to sit unaided and die within 3 years of age. SMA type II (or Dubowitz syndrome) has an age of onset between 6 and 36 months of age and a milder severity. Patients live to adulthood with substantial motor disabilities, but their life expectancy is reduced due to respiratory complications. SMA type III (Kugelberg Welander disease) appears at around 3 years of age and does not shorten life expectancy, although the quality of life is reduced by significant muscular weakness. SMA type IV is an adulthood onset disorder and it is the mildest form of the disease, leading to some degree of motor disability without affecting life expectancy. Since a unique genetic defect causes such dramatically different clinical phenotypes, it was soon speculated that other genes might act as disease modifiers. The architecture of the 5q13 locus supports this hypothesis. This is an unstable region of the human genome, characterized by duplications and inversions generating a variable number of copies of a more centromeric gene, called *SMN2*. This evolutionary backup gene is almost identical to *SMN1*, differing principally in a single nucleotide at the beginning of exon 7. However, this point variant has a deep impact on SMN expression as it leads to exon 7 skipping. The resulting transcript is translated in a shorter protein with an alternative C-terminus, conventionally referred to as SMNΔ7, which is unstable and rapidly degraded. Only 10% of *SMN2* primary transcript is processed with exon 7 inclusion and can be translated into the full-length stable SMN protein (reviewed in [260]). Consequently, the number of *SMN2* copies that SMA patients carry critically tunes residual SMN protein expression, thereby modulating disease severity. Importantly, many different animal models could be generated by reducing the expression of the endogenous *SMN* gene (reviewed in [60]).

On the basis of this evidence, it has clearly emerged that even a small increase in SMN expression could lead to substantial therapeutic benefit. Thus, several approaches have been developed to achieve this goal [96]. The first strategy aims at delivering the whole SMN cDNA through viral-mediated gene therapy. Since the initial proof-of-principle study in transgenic mice [13], a substantial advancement into the development of an effective therapy came from the discovery of the adeno-associated virus serotype 9 (AAV9) as vector apt to cross the BBB [77]. This approach has been successfully exploited in transgenic mice [11, 19, 55, 78, 90, 91, 179, 235, 289] and larger animal models of the disease [57, 115, 179], obtaining a tremendous impact in both preserving animal motor functions and extending lifespan. These promising preclinical findings paved the way to a phase I clinical trial whose outcome suggests that a single injection of therapeutic virus leads to improved motor abilities and extends life expectancy in SMA type I patients [176]. Thus, the restoration of SMN levels by means of AAV9-based gene therapy (Zolgensma®) has been recently approved by the U.S. FDA for the treatment of pediatric patients, less than 2 years of age. Complementary, the existence of the *SMN2* gene in humans was found to offer an exceptional opportunity for an alternative approach to raise SMN expression, taking advantage of the development of antisense oligonucleotide (ASO) technology as a toolbox to modulate protein expression (reviewed in [233]). Extensive studies on the molecular mechanisms regulating *SMN2* primary transcript splicing [260] led to the development of Nusinersen (Spinraza™), an ASO able to prevent exon 7 skipping, thereby allowing full-length SMN expression. In two recent clinical trials, this drug was administered to children with type II and type III SMA (Phase I study; NCT01494701; NCT01780246; [40]) and, later, in patients with infantile-onset SMA (phase II study; NCT01839656; [70]). Promising signs of efficacy in motor function were observed, which have encouraged the design of sham-controlled, phase III clinical studies for in infantile- and late-onset SMA (NCT02193074; NCT02292537; [71, 178]). The overall findings of these studies have supported the recent approval of Nusinersen by the U.S. FDA for treatment of 5q SMA, followed by the European Medicine Agency (EMA) and by other national drug management authorities worldwide (Canada, Japan, Brazil, Italy and Switzerland).

Despite these advancements, many questions still deserve to be addressed in order to be able to develop even more effective therapies to cure SMA patients. First, are there other tools to modulate SMN expression, for instance, by regulating its stability? Second, what are the molecular mechanisms linking SMN deficiency to motor neuron demise? Evidence from genetic studies and animal models of the disease have highlighted the possibility that other genes may play a critical role in modulating SMA severity in either SMN-dependent (i.e. *Uba1*) or SMN-independent ways (i.e. *Plastin3*; reviewed in [307]). It is, therefore, crucial to understand the molecular mechanisms driving SMA pathogenesis. Many hypotheses can be formulated on the basis of the growing number of cellular functions in which SMN has been involved so far (reviewed in [261]). The SMN protein is composed of 294 amino acids that are arranged in functionally distinct domains. Each of them specifically mediates the interaction with nucleic acids or with a plethora of other proteins, either in the nucleus or in the cytoplasm. Based on this, the functions proposed for SMN include, but are not limited to, trafficking and remodelling of small nuclear ribonucleoproteins (snRNPs), modulation of RNA/DNA metabolism, signal transduction pathways affecting actin-cytoskeletal remodelling, and endocytosis and autophagy. Among the different cellular formations, the neuromuscular junction emerges as a trans-cellular structure that is particularly vulnerable to reduced expression of SMN (for a specific review on the topic see [27]).

Finally, 'non-5q' SMA is an umbrella term to group a minority of cases (~ 4%) who develop the disease in the absence of mutations in *SMN1*. These patients often display symptoms other than proximal muscular atrophy and show a diverse pattern of inheritance. Only recent advances in whole-genome sequencing allowed to identify a growing number of causative genes (reviewed in [222]).

10.2.2 The Role of Astrocytes in the Pathogenesis of 5q SMA

Although reactive gliosis has been reported upon neuropathological assessment of 5q [10, 87, 142] and some forms of non-5q SMA [54, 108], the actual role played by astrocytes in the pathogenesis of these disorders has not been specifically addressed for a long time. Only in recent years, evidence has been accumulated that suggests a causal role for astrocyte dysfunction in the pathogenesis of 5q SMA, whilst its implication in non-5q SMA remains neglected.

The possibility that SMA may be a non-cell autonomous disease arose from the observation that astrocytosis precedes the loss of motor neuron cell bodies in vitro and in vivo [174, 219] as well as from the phenotypic analysis of several transgenic mice where SMN expression was modulated in motor neurons. In one instance, reduced SMN expression in motor neurons (and oligodendrocytes) was achieved by crossing Olig2-Cre expressing mice with others harbouring a *floxed* version of SMN exon 7 [216]. The progeny of this breeding displayed a mild SMA-like phenotype with no reduction in lifespan, in sharp contrast with the aggressive and fatal phenotype exhibited by mice where SMN expression was systemically reduced [216]. Complementary, several studies in different mouse models of SMA investigated the benefit of selective SMN reintroduction in motor neurons. These investigations failed to show a complete rescue of the detrimental phenotype [92, 148, 162, 170, 212]. Taken together, this early evidence strongly argues in favour of the theory that toxicity triggered by SMN deficiency has a non-cell autonomous component. To directly address this hypothesis, SMN expression was selectively reintroduced in the astrocytes, in a transgenic mouse model of the disease, by viral-mediated gene therapy [234]. In this study, Rindt et al. designed a vector where the SMN coding sequence was under the control of the GFAP promoter, and was further modified to harbour the consensus sequence of a miRNA selectively expressed by motor neurons, in order to prevent any SMN expression in those cells. Importantly, mice injected with such vector displayed enhanced lifespan, reduced neuromuscular junction pathology and diminished gliosis [234], thus demonstrating that targeting astrocyte dysfunction can be beneficial to control SMA. Yet, the exact molecular mechanisms mediating astrocyte involvement remain to be fully elucidated.

The earliest evidence that SMN deficiency might be directly perturbing astrocyte physiological functions came from the discovery that, both in a human astrocytoma cell line treated with siRNA against *SMN1* as well as in astrocytes from the spinal cord of SMA mice, there is an upregulation of Jagged-1, a Notch ligand. Furthermore, in the surrounding motor neurons, Notch receptors are correspondingly upregulated with evidence of activation of the downstream signalling cascade [33]. Other studies took advantage of the possibility of culturing astrocytes from mouse models of SMA and, more recently, from the differentiation of SMA patient-derived iPSCs into astrocytes. Using these models, abnormal calcium signalling was reported in both human [174] and mouse cells [319]. Specifically, in human SMA astrocytes, basal calcium concentrations were increased, while the cell response to ATP stimulation was reduced in comparison with non-SMA controls [174]. In striking contrast, mouse SMA astrocytes did not display any difference upon resting conditions, whereas their response to stimuli was enhanced [319]. Taken together, this amount of evidence suggests that SMN deficiency is intrinsically detrimental for the astrocyte homeostasis. But what is the impact on the cross-talk with neurons? Early experiments aiming at assessing the endurance of iPSC-derived motor neurons revealed reduced survival of cells differentiated from patient tissues [59, 247], thus suggesting that SMN deficiency can lead to cell-autonomous cell death. However, those early cultures contained significant contamination of astrocytes, which might have contributed to cell demise. Indeed, later refinement of the differentiation protocol allowed to obtain homogenous motor neuron populations, and these revealed comparable survival rates between cells from SMA patients and healthy controls [262].

This evidence supports the hypothesis that SMA astrocytes might trigger motor neuron sufferance either by losing their supporting functions or by gaining toxic activities. Consistent with these latter hypotheses, human SMA astrocytes were reported to display reduced release of the trophic factor GDNF [174]. Furthermore, co-culture experiments provided evidence that SMA astrocytes can induce neurotoxicity by releasing miR-146a, although follow-up studies are still necessary to elucidate the exact mechanism triggered by this molecule [262]. Conversely, mouse SMA astrocytes did not show reduced support to motor neurons in a model of contact co-cultures, although neurons exhibited a lower density of synaptic contacts [319]. Since this detrimental effect on motor neuronal function was not replicated in non-contact co-cultures, it was postulated that it may be caused by an antigen exposed on the cell membrane. In keeping with this view, SMA astrocytes were shown to exhibit reduced surface expression of Ephrin B2 [319]. Finally, another group provided evidence that also SMA astrocyte conditioned medium can lead to neuronal sufferance in terms of reduced neurite outgrowth by decreasing the release of the Monocyte Chemoactive Protein 1 (MCP1) [168].

10.3 Discussion/Perspective

As extensively documented in the previous sections, astrocytes appear to play complex and critical roles in both ALS and SMA, thereby emerging as key elements in the pathogenesis and progression of motor neuron diseases. Many aspects driving their detrimental effects appear to be intermingled. First, their own homeostasis seems to be directly disrupted by respective disease proteins. Secondly, they sense and react to the distressed status of their neighbouring neurons, becoming reactive. Multiple evidence suggests that 'reactive astrocytosis' can lead to both loss of support to motor neurons and gain of new aberrant toxic functions. However, a comprehensive consensus on the molecular mechanisms disrupting the astrocyte–neuron crosstalk in motor neuron degeneration is still missing and needs to be pursued. To achieve this goal, several lines of investigation are currently underway.

To start with, the majority of findings concerning the role of astrocytes in the pathogenesis of ALS have been inferred using mutant SOD1 overexpressing models of the disease. It is only very recently that this horizon has been widened including evidence from other molecular subtypes of the disease. The same need applies, even with stronger urgency, to SMA research. In this context, the investigations on the astrocytes are still an underappreciated topic, limited to the 5q-linked form of the disease, while no mechanistic studies have been conducted in non-5q SMA. Furthermore, in ALS and SMA, the field is currently progressing along parallel, unconnected paths and no side-by-side experiments have been so far presented. Yet, some analogies and differences start emerging (summarized in Table 10.1). The more refined this type of analysis will be, the earlier it will be possible to distinguish shared from distinctive pathways and envision suitable therapeutic strategies. For instance, we have discussed findings describing the dysregulation of intracellular calcium homeostasis in the context of both ALS and SMA. This is a particularly intriguing evidence, considering that calcium concentrations regulate several signalling cascades coupled with correct neuronal interplay (reviewed in [98]). An interesting difference between the two diseases relates to the neuroprotective factor GDNF. We have demonstrated that, in ALS, its production is not only retained, but even boosted, during disease progression. By contrast, human SMA astrocytes displayed reduced GDNF expression and release, thus suggesting that this event is not specifically caused by gliosis, but it is rather due to a disease-specific dysregulation. Further studies are needed to prove whether restoring GDNF expression in astrocytes in animal models of SMA would be sufficient to sustain neuroprotection.

Another point for discussion stems from the collective interpretation of the evidence gathered so far, namely, the assumption that astrocytes located within the same CNS region are *bona fide* a homogenous population. However, recent studies (reviewed by [17]) challenged this view. It is now believed that, although astrocytes express an array of core transcripts that provide them with the ability to fulfil their housekeeping duties towards neurons, they express also distinct subsets of transcripts making them apt to tune the activity of specific neuronal populations even within individual brain regions. Intriguingly, single-cell transcriptome analysis within the

Table 10.1 Analogies and differences of the astrocyte features in ALS and SMA

	ALS		SMA	
Correcting the pathogenic genetic alteration in astrocytes ameliorates the phenotype in animal models of the disease	Selective ablation of mutant SOD1 from astrocytes slows down disease progression and prolongs survival	Yamanaka et al. [310], Wang et al. [303]	Selective restoration of SMN expression in astrocytes by viral-mediated gene therapy prolongs survival of SMA mice	Rindt et al. [234]
	In chimeric mice, motor neurons expressing mutant SOD1 resist the pathogenic insult when surrounded by non-transgenic glial cells	Clement et al. [41]		
	Transplanting non-transgenic astrocyte precursors ameliorates the phenotype of $hSOD1^{G93A}$ rats	Lepore et al. [153]		
	Transplanting human iPSC- or embryonic stem cell-derived astrocyte precursors ameliorates the phenotype of $hSOD1^{G93A}$ mice	Kondo et al. [137], Izrael et al. [116]		
Disease-astrocytes are toxic to motor neuron	In chimeric mice, non-transgenic motor neurons display signs of cellular stress when surrounded by mutant SOD1 expressing glia	Clement et al. [41]		

(continued)

Table 10.1 (continued)

	ALS		SMA	
	Transplanting hSOD1^{G93A} astrocyte precursors triggers motor neuron loss in non-transgenic rats	Papadeas et al. [213]		
	Transplanting human iPSC-derived ALS astrocyte precursors triggers motor impairment in non-transgenic mice	Chen et al. [38], Qian et al. [228]		
	Mouse ALS astrocytes are toxic to motor neurons in co-cultures	Ferraiuolo et al. [69], Di Giorgio et al. [50], Nagai et al. [198], Bilsland et al. [25], Phatnani et al. [310], Tortarolo et al. [280], Kia et al. [132]	Mouse SMA astrocytes do not trigger motor neuron demise in co-cultures, but the number of synaptic contact is reduced	Zhou et al. [319]
	Human ALS astrocytes are toxic to motor neurons in contact co-cultures	Di Giorgio et al. [49], Marchetto et al. [167], Haidet-Phillips et al. [103], Meyer et al. [180], Re et al. [231], Song et al. [267]	Human SMA astrocytes are toxic to motor neuron in contact co-cultures	Sison et al. [262]
	Conditioned medium from ALS-astrocytes is sufficient to trigger motor neuron toxicity in vitro and in vivo	Kia et al. [132], Fritz et al. [81], Rojas et al. [238, 239], Ramirez-Jarquin et al. [230], Madill et al. [164]	Conditioned medium from SMA astrocytes triggers reduced neurite outgrowth in motor neurons	Martin et al. [168]

(continued)

Table 10.1 (continued)

	ALS		SMA	
Astrocytes display an abnormal intracellular calcium homeostasis	Abnormal calcium accumulation and release from intracellular stores upon different stimuli	Martorana et al. [172], Vergouts et al. [297], Kawamata et al. [128], Almad et al. [5], Milosevic et al. [187]	In human iPSC-derived SMA astrocytes, the intracellular calcium concentration is increased, upon standard culturing conditions, and the response to ATP stimulation is reduced	McGivern et al. [174]
			Mouse SMA astrocyte do not display any difference in the resting calcium concentration, but their response to ATP stimulation is enhanced	Zhou et al. [319]
GDNF supply to motor neurons	During disease progression, enhanced stimulation of the TNF-TNFR1 axis prompts augmented GDNF release from ALS astrocytes	Brambilla et al. [30]	Human iPSC-derived SMA astrocytes display reduced release of GDNF	McGivern et al. [174]
Activation of the Notch signalling pathway	Notch ligand Jagged-1 is abnormally overexpressed in $hSOD1^{G93A}$ expressing mouse astrocytes. Preventing this event has detrimental consequences on the phenotype	Nonneman et al. [207]	SMN depleted astrocytes display enhanced expression of Jagged-1 in vitro and in SMA mice	Caraballo-Miralles et al. [310]

mouse cortex and the hippocampus has revealed the presence of two subpopulations of astrocytes (defined as type 1 and type 2; see [317]). In the ventral spinal cord, adult astrocytes display a regionally distinct phenotype, enabling their classification into three populations (VA1, VA2 and VA3), which develop from progenitors allocated in defined domains [110]. Functionally, the expression of domain-specific proteins is pivotal to ensure correct motor neuron circuitry [192] and electrophysiological properties [130]. It is likely that such diversity is the result of an intricate interplay between the unfolding of an intrinsic patterning program and the microenvironment, which is modulated by inputs from surrounding neurons and other glial cell populations (reviewed in [66]). This landscape becomes even more complex with ageing or when a neuroinflammatory insult strikes. Astrocytes then mount a phenotypical switch toward a reactive status, which limits or exacerbates neuronal loss, depending on the experimental context [316]. It is, therefore, tempting to speculate that implementing these findings to address the role of astrocytes in the pathogenesis of ALS and SMA might have a tremendous impact in several ways. For instance, a different transcriptional profile might explain (i) why some astrocytes undergo cell sufferance and death in the context of ALS and others 'simply' develop a reactive phenotype; or (ii) which cells mount a protective response and which ones exacerbate neurodegeneration. The challenge is now to link a specific astrocyte subpopulation defined by distinct molecular features under physiological conditions with their fate in the context of disease development, an issue that would probably need computational models of gliosis to be addressed. Also, since we have previously discussed that astrocytes are a potential therapeutic target, the pre-clinical investigation should be refined including the possibility that only specific subpopulations of astrocytes should be manipulated. In keeping with this, the efficiency of transplant studies might also be perfected by infusing only the most beneficial subtype of cells. Finally, an aspect that should be carefully considered is the impact of physical activity on astroglial cells. Several lines of evidence indicate that enriched environment and physical exercise importantly contribute to reduce reactive astrocytosis [129, 147, 237, 252, 254] and to alleviate the neuroinflammatory response in various animal models of injury and disease [22, 150, 204, 214, 226]. This supports the view that rehabilitative training can be used to favour the morphological remodelling and to improve the functional performance of the astrocytes. What remains to be clarified is whether this is a general issue or whether it is effective on specific astrocytic subpopulations.

References

1. Abrahams S, Newton J, Niven E, Foley J, Bak TH (2014) Screening for cognition and behaviour changes in ALS. Amyotrophic lateral sclerosis and frontotemporal degeneration 15:9–14
2. Aebischer J, Cassina P, Otsmane B, Moumen A, Seilhean D, Meininger V, Barbeito L, Pettmann B, Raoul C (2011) IFNgamma triggers a LIGHT-dependent selective death of motoneurons contributing to the non-cell-autonomous effects of mutant SOD1. Cell Death Differ 18:754–768

3. Al-Chalabi A et al (2017) July 2017 ENCALS statement on edaravone. Amyotrophic lateral sclerosis and frontotemporal degeneration 18:471–474
4. Al Awabdh S, Gupta-Agarwal S, Sheehan DF, Muir J, Norkett R, Twelvetrees AE, Griffin LD, Kittler JT (2016) Neuronal activity mediated regulation of glutamate transporter GLT-1 surface diffusion in rat astrocytes in dissociated and slice cultures. Glia 64:1252–1264
5. Almad AA, Doreswamy A, Gross SK, Richard JP, Huo Y, Haughey N, Maragakis NJ (2016) Connexin 43 in astrocytes contributes to motor neuron toxicity in amyotrophic lateral sclerosis. Glia 64:1154–1169
6. Alshikho MJ, Zurcher NR, Loggia ML, Cernasov P, Chonde DB, Izquierdo Garcia D, Yasek JE, Akeju O, Catana C, Rosen BR, Cudkowicz ME, Hooker JM, Atassi N (2016) Glial activation colocalizes with structural abnormalities in amyotrophic lateral sclerosis. Neurology 87:2554–2561
7. Alshikho MJ, Zurcher NR, Loggia ML, Cernasov P, Reynolds B, Pijanowski O, Chonde DB, Izquierdo Garcia D, Mainero C, Catana C, Chan J, Babu S, Paganoni S, Hooker JM, Atassi N (2018) Integrated MRI and [(11) C]-PBR28 PET Imaging in Amyotrophic Lateral sclerosis. Ann Neurol
8. Anneser JM, Cookson MR, Ince PG, Shaw PJ, Borasio GD (2001) Glial cells of the spinal cord and subcortical white matter up-regulate neuronal nitric oxide synthase in sporadic amyotrophic lateral sclerosis. Exp Neurol 171:418–421
9. Aoki M, Lin CL, Rothstein JD, Geller BA, Hosler BA, Munsat TL, Horvitz HR, Brown RH Jr (1998) Mutations in the glutamate transporter EAAT2 gene do not cause abnormal EAAT2 transcripts in amyotrophic lateral sclerosis. Ann Neurol 43:645–653
10. Araki S, Hayashi M, Tamagawa K, Saito M, Kato S, Komori T, Sakakihara Y, Mizutani T, Oda M (2003) Neuropathological analysis in spinal muscular atrophy type II. Acta Neuropathol 106:441–448
11. Armbruster N, Lattanzi A, Jeavons M, Van Wittenberghe L, Gjata B, Marais T, Martin S, Vignaud A, Voit T, Mavilio F, Barkats M, Buj-Bello A (2016) Efficacy and biodistribution analysis of intracerebroventricular administration of an optimized scAAV9-SMN1 vector in a mouse model of spinal muscular atrophy. Mol Ther Methods Clin Dev 3:16060
12. Aronica E, Catania MV, Geurts J, Yankaya B, Troost D (2001) Immunohistochemical localization of group I and II metabotropic glutamate receptors in control and amyotrophic lateral sclerosis human spinal cord: upregulation in reactive astrocytes. Neuroscience 105:509–520
13. Azzouz M, Le T, Ralph GS, Walmsley L, Monani UR, Lee DC, Wilkes F, Mitrophanous KA, Kingsman SM, Burghes AH, Mazarakis ND (2004) Lentivector-mediated SMN replacement in a mouse model of spinal muscular atrophy. J Clin Investig 114:1726–1731
14. Baker DJ, Blackburn DJ, Keatinge M, Sokhi D, Viskaitis P, Heath PR, Ferraiuolo L, Kirby J, Shaw PJ (2015) Lysosomal and phagocytic activity is increased in astrocytes during disease progression in the SOD1 (G93A) mouse model of amyotrophic lateral sclerosis. Front Cell Neurosci 9:410
15. Bakkar N, Kousari A, Kovalik T, Li Y, Bowser R (2015) RBM45 Modulates the Antioxidant Response in Amyotrophic Lateral Sclerosis through Interactions with KEAP1. Mol Cell Biol 35:2385–2399
16. Bataveljic D, Nikolic L, Milosevic M, Todorovic N, Andjus PR (2012) Changes in the astrocytic aquaporin-4 and inwardly rectifying potassium channel expression in the brain of the amyotrophic lateral sclerosis SOD1(G93A) rat model. Glia 60:1991–2003
17. Ben Haim L, Rowitch DH (2017) Functional diversity of astrocytes in neural circuit regulation. Nat Rev Neurosci 18:31–41
18. Benedusi V, Martorana F, Brambilla L, Maggi A, Rossi D (2012) The peroxisome proliferator-activated receptor gamma (PPARgamma) controls natural protective mechanisms against lipid peroxidation in amyotrophic lateral sclerosis. J Biol Chem 287:35899–35911
19. Benkhelifa-Ziyyat S, Besse A, Roda M, Duque S, Astord S, Carcenac R, Marais T, Barkats M (2013) Intramuscular scAAV9-SMN injection mediates widespread gene delivery to the spinal cord and decreases disease severity in SMA mice. Mol Ther: J Am Soc Gene Ther 21:282–290

20. Benkler C, Barhum Y, Ben-Zur T, Offen D (2016) Multifactorial gene therapy enhancing the glutamate uptake system and reducing oxidative stress delays symptom onset and prolongs survival in the SOD1-G93A ALS mouse model. J Mol Neurosci: MN 58:46–58
21. Bergstrom P, von Otter M, Nilsson S, Nilsson AC, Nilsson M, Andersen PM, Hammarsten O, Zetterberg H (2014) Association of NFE2L2 and KEAP1 haplotypes with amyotrophic lateral sclerosis. Amyotroph Lateral Scler Front Degener 15:130–137
22. Bernardes D, Oliveira-Lima OC, Silva TV, Faraco CC, Leite HR, Juliano MA, Santos DM, Bethea JR, Brambilla R, Orian JM, Arantes RM, Carvalho-Tavares J (2013) Differential brain and spinal cord cytokine and BDNF levels in experimental autoimmune encephalomyelitis are modulated by prior and regular exercise. J Neuroimmunol 264:24–34
23. Berry JD, Shefner JM, Conwit R, Schoenfeld D, Keroack M, Felsenstein D, Krivickas L, David WS, Vriesendorp F, Pestronk A, Caress JB, Katz J, Simpson E, Rosenfeld J, Pascuzzi R, Glass J, Rezania K, Rothstein JD, Greenblatt DJ, Cudkowicz ME (2013) Design and initial results of a multi-phase randomized trial of ceftriaxone in amyotrophic lateral sclerosis. PLoS ONE 8:e61177
24. Bi F, Huang C, Tong J, Qiu G, Huang B, Wu Q, Li F, Xu Z, Bowser R, Xia XG, Zhou H (2013) Reactive astrocytes secrete lcn2 to promote neuron death. Proc Natl Acad Sci USA 110:4069–4074
25. Bilsland LG, Nirmalananthan N, Yip J, Greensmith L, Duchen MR (2008) Expression of mutant SOD1 in astrocytes induces functional deficits in motoneuron mitochondria. J Neurochem 107:1271–1283
26. Blaauwgeers HG, Vianney de Jong JM, Verspaget HW, van den Berg FM, Troost D (1996) Enhanced superoxide dismutase-2 immunoreactivity of astrocytes and occasional neurons in amyotrophic lateral sclerosis. J Neurol Sci 140:21–29
27. Boido M, Vercelli A (2016) Neuromuscular Junctions as key contributors and therapeutic targets in spinal muscular atrophy. Front Neuroanat 10:6
28. Boston-Howes W, Gibb SL, Williams EO, Pasinelli P, Brown RH Jr, Trotti D (2006) Caspase-3 cleaves and inactivates the glutamate transporter EAAT2. J Biol Chem 281:14076–14084
29. Brambilla L, Martorana F, Guidotti G, Rossi D (2018) Dysregulation of astrocytic HMGB1 signaling in amyotrophic lateral sclerosis. Front Neurosci 12
30. Brambilla L, Guidotti G, Martorana F, Iyer AM, Aronica E, Valori CF, Rossi D (2016) Disruption of the astrocytic TNFR1-GDNF axis accelerates motor neuron degeneration and disease progression in amyotrophic lateral sclerosis. Hum Mol Genet 25:3080–3095
31. Bruijn LI, Becher MW, Lee MK, Anderson KL, Jenkins NA, Copeland NG, Sisodia SS, Rothstein JD, Borchelt DR, Price DL, Cleveland DW (1997) ALS-linked SOD1 mutant G85R mediates damage to astrocytes and promotes rapidly progressive disease with SOD1-containing inclusions. Neuron 18:327–338
32. Buonvicino D, Felici R, Ranieri G, Caramelli R, Lapucci A, Cavone L, Muzzi M, Di Pietro L, Bernardini C, Zwergel C, Valente S, Mai A, Chiarugi A (2018) Effects of class II-selective histone deacetylase inhibitor on neuromuscular function and disease progression in SOD1-ALS Mice. Neuroscience 379:228–238
33. Caraballo-Miralles V, Cardona-Rossinyol A, Garcera A, Torres-Benito L, Soler RM, Tabares L, Llado J, Olmos G (2013) Notch signaling pathway is activated in motoneurons of spinal muscular atrophy. Int J Mol Sci 14:11424–11437
34. Cassina P, Pehar M, Vargas MR, Castellanos R, Barbeito AG, Estevez AG, Thompson JA, Beckman JS, Barbeito L (2005) Astrocyte activation by fibroblast growth factor-1 and motor neuron apoptosis: implications for amyotrophic lateral sclerosis. J Neurochem 93:38–46
35. Cassina P, Cassina A, Pehar M, Castellanos R, Gandelman M, de Leon A, Robinson KM, Mason RP, Beckman JS, Barbeito L, Radi R (2008) Mitochondrial dysfunction in SOD1G93A-bearing astrocytes promotes motor neuron degeneration: prevention by mitochondrial-targeted antioxidants. J Neurosci: Off J Soc Neurosci 28:4115–4122
36. Chan GN, Evans RA, Banks DB, Mesev EV, Miller DS, Cannon RE (2017) Selective induction of P-glycoprotein at the CNS barriers during symptomatic stage of an ALS animal model. Neurosci Lett 639:103–113

37. Chen H, Kankel MW, Su SC, Han SWS, Ofengeim D (2018) Exploring the genetics and non-cell autonomous mechanisms underlying ALS/FTLD. Cell Death Differ 25:646–660
38. Chen H, Qian K, Chen W, Hu B, Blackbourn LWt, Du Z, Ma L, Liu H, Knobel KM, Ayala M, Zhang SC (2015) Human-derived neural progenitors functionally replace astrocytes in adult mice. J Clin Investig 125:1033–1042
39. Chen Y, Guan Y, Liu H, Wu X, Yu L, Wang S, Zhao C, Du H, Wang X (2012) Activation of the Wnt/beta-catenin signaling pathway is associated with glial proliferation in the adult spinal cord of ALS transgenic mice. Biochem Biophys Res Commun 420:397–403
40. Chiriboga CA, Swoboda KJ, Darras BT, Iannaccone ST, Montes J, De Vivo DC, Norris DA, Bennett CF, Bishop KM (2016) Results from a phase 1 study of nusinersen (ISIS-SMN(Rx)) in children with spinal muscular atrophy. Neurology 86:890–897
41. Clement AM, Nguyen MD, Roberts EA, Garcia ML, Boillee S, Rule M, McMahon AP, Doucette W, Siwek D, Ferrante RJ, Brown RH Jr, Julien JP, Goldstein LS, Cleveland DW (2003) Wild-type nonneuronal cells extend survival of SOD1 mutant motor neurons in ALS mice. Science 302:113–117
42. Colton CK, Kong Q, Lai L, Zhu MX, Seyb KI, Cuny GD, Xian J, Glicksman MA, Lin CL (2010) Identification of translational activators of glial glutamate transporter EAAT2 through cell-based high-throughput screening: an approach to prevent excitotoxicity. J Biomol Screen 15:653–662
43. Cooper-Knock J, Higginbottom A, Stopford MJ, Highley JR, Ince PG, Wharton SB, Pickering-Brown S, Kirby J, Hautbergue GM, Shaw PJ (2015) Antisense RNA foci in the motor neurons of C9ORF72-ALS patients are associated with TDP-43 proteinopathy. Acta Neuropathol 130:63–75
44. Crosio C, Valle C, Casciati A, Iaccarino C, Carri MT (2011) Astroglial inhibition of NF-kappaB does not ameliorate disease onset and progression in a mouse model for amyotrophic lateral sclerosis (ALS). PLoS ONE 6:e17187
45. Cudkowicz ME et al (2014) Safety and efficacy of ceftriaxone for amyotrophic lateral sclerosis: a multi-stage, randomised, double-blind, placebo-controlled trial. Lancet Neurol 13:1083–1091
46. Dai J, Lin W, Zheng M, Liu Q, He B, Luo C, Lu X, Pei Z, Su H, Yao X (2017) Alterations in AQP4 expression and polarization in the course of motor neuron degeneration in SOD1G93A mice. Mol Med Rep 16:1739–1746
47. DeJesus-Hernandez M et al (2011) Expanded GGGGCC hexanucleotide repeat in noncoding region of C9ORF72 causes chromosome 9p-linked FTD and ALS. Neuron 72:245–256
48. Deng HX et al (2011) Mutations in UBQLN2 cause dominant X-linked juvenile and adult-onset ALS and ALS/dementia. Nature 477:211–215
49. Di Giorgio FP, Boulting GL, Bobrowicz S, Eggan KC (2008) Human embryonic stem cell-derived motor neurons are sensitive to the toxic effect of glial cells carrying an ALS-causing mutation. Cell Stem Cell 3:637–648
50. Di Giorgio FP, Carrasco MA, Siao MC, Maniatis T, Eggan K (2007) Non-cell autonomous effect of glia on motor neurons in an embryonic stem cell-based ALS model. Nat Neurosci 10:608–614
51. Diaper DC, Adachi Y, Lazarou L, Greenstein M, Simoes FA, Di Domenico A, Solomon DA, Lowe S, Alsubaie R, Cheng D, Buckley S, Humphrey DM, Shaw CE, Hirth F (2013) Drosophila TDP-43 dysfunction in glia and muscle cells cause cytological and behavioural phenotypes that characterize ALS and FTLD. Hum Mol Genet
52. Diaz-Amarilla P, Olivera-Bravo S, Trias E, Cragnolini A, Martinez-Palma L, Cassina P, Beckman J, Barbeito L (2011) Phenotypically aberrant astrocytes that promote motoneuron damage in a model of inherited amyotrophic lateral sclerosis. Proc Natl Acad Sci USA 108:18126–18131
53. Diaz-Amarilla P, Miquel E, Trostchansky A, Trias E, Ferreira AM, Freeman BA, Cassina P, Barbeito L, Vargas MR, Rubbo H (2016) Electrophilic nitro-fatty acids prevent astrocyte-mediated toxicity to motor neurons in a cell model of familial amyotrophic lateral sclerosis via nuclear factor erythroid 2-related factor activation. Free Radic Biol Med 95:112–120

54. Dlamini N, Josifova DJ, Paine SM, Wraige E, Pitt M, Murphy AJ, King A, Buk S, Smith F, Abbs S, Sewry C, Jacques TS, Jungbluth H (2013) Clinical and neuropathological features of X-linked spinal muscular atrophy (SMAX2) associated with a novel mutation in the UBA1 gene. Neuromuscul Disord: NMD 23:391–398
55. Dominguez E, Marais T, Chatauret N, Benkhelifa-Ziyyat S, Duque S, Ravassard P, Carcenac R, Astord S, Pereira de Moura A, Voit T, Barkats M (2011) Intravenous scAAV9 delivery of a codon-optimized SMN1 sequence rescues SMA mice. Hum Mol Genet 20:681–693
56. Duan W, Li X, Shi J, Guo Y, Li Z, Li C (2010) Mutant TAR DNA-binding protein-43 induces oxidative injury in motor neuron-like cell. Neuroscience 169:1621–1629
57. Duque SI, Arnold WD, Odermatt P, Li X, Porensky PN, Schmelzer L, Meyer K, Kolb SJ, Schumperli D, Kaspar BK, Burghes AH (2015) A large animal model of spinal muscular atrophy and correction of phenotype. Ann Neurol 77:399–414
58. Duval N, Sumner WA, Andrianakos AG, Gray JJ, Bouchard RJ, Wilkins HM, Linseman DA (2018) The Bcl-2 Homology-3 Domain (BH3)-only proteins, bid, DP5/Hrk, and BNip3L, are upregulated in reactive astrocytes of end-stage mutant SOD1 mouse spinal cord. Front Cell Neurosci 12:15
59. Ebert AD, Yu J, Rose FF Jr, Mattis VB, Lorson CL, Thomson JA, Svendsen CN (2009) Induced pluripotent stem cells from a spinal muscular atrophy patient. Nature 457:277–280
60. Edens BM, Ajroud-Driss S, Ma L, Ma YC (2015) Molecular mechanisms and animal models of spinal muscular atrophy. Biochem Biophys Acta 1852:685–692
61. Edens BM, Yan J, Miller N, Deng HX, Siddique T, Ma YC (2017) A novel ALS-associated variant in UBQLN4 regulates motor axon morphogenesis. eLife 6
62. Elliott WJ, Ram CV (2011) Calcium channel blockers. J Clin Hypertens (Greenwich) 13:687–689
63. Endo F, Komine O, Fujimori-Tonou N, Katsuno M, Jin S, Watanabe S, Sobue G, Dezawa M, Wyss-Coray T, Yamanaka K (2015) Astrocyte-derived TGF-beta1 accelerates disease progression in ALS mice by interfering with the neuroprotective functions of microglia and T cells. Cell Rep 11:592–604
64. Estes PS, Daniel SG, McCallum AP, Boehringer AV, Sukhina AS, Zwick RA, Zarnescu DC (2013) Motor neurons and glia exhibit specific, individualized responses to TDP-43 expression in a Drosophila model of ALS. Dis Model Mech
65. Fang T, Al Khleifat A, Meurgey JH, Jones A, Leigh PN, Bensimon G, Al-Chalabi A (2018) Stage at which riluzole treatment prolongs survival in patients with amyotrophic lateral sclerosis: a retrospective analysis of data from a dose-ranging study. Lancet Neurol 17:416–422
66. Farmer WT, Murai K (2017) Resolving Astrocyte Heterogeneity in the CNS. Frontiers in cellular neuroscience 11:300
67. Fecto F, Yan J, Vemula SP, Liu E, Yang Y, Chen W, Zheng JG, Shi Y, Siddique N, Arrat H, Donkervoort S, Ajroud-Driss S, Sufit RL, Heller SL, Deng HX, Siddique T (2011) SQSTM1 mutations in familial and sporadic amyotrophic lateral sclerosis. Arch Neurol 68:1440–1446
68. Feng X, Peng Y, Liu M, Cui L (2012) DL-3-n-butylphthalide extends survival by attenuating glial activation in a mouse model of amyotrophic lateral sclerosis. Neuropharmacology 62:1004–1010
69. Ferraiuolo L, Higginbottom A, Heath PR, Barber S, Greenald D, Kirby J, Shaw PJ (2011) Dysregulation of astrocyte-motoneuron cross-talk in mutant superoxide dismutase 1-related amyotrophic lateral sclerosis. Brain: J Neurol 134:2627–2641
70. Finkel RS, Chiriboga CA, Vajsar J, Day JW, Montes J, De Vivo DC, Yamashita M, Rigo F, Hung G, Schneider E, Norris DA, Xia S, Bennett CF, Bishop KM (2016) Treatment of infantile-onset spinal muscular atrophy with nusinersen: a phase 2, open-label, dose-escalation study. Lancet 388:3017–3026
71. Finkel RS et al (2017) Nusinersen versus sham control in infantile-onset spinal muscular atrophy. N Engl J Med 377:1723–1732
72. Flomen R, Makoff A (2011) Increased RNA editing in EAAT2 pre-mRNA from amyotrophic lateral sclerosis patients: involvement of a cryptic polyadenylation site. Neurosci Lett 497:139–143

73. Focant MC, Goursaud S, Boucherie C, Dumont AO, Hermans E (2013) PICK1 expression in reactive astrocytes within the spinal cord of amyotrophic lateral sclerosis (ALS) rats. Neuropathol Appl Neurobiol 39:231–242
74. Foran E, Rosenblum L, Bogush A, Pasinelli P, Trotti D (2014) Sumoylation of the astroglial glutamate transporter EAAT2 governs its intracellular compartmentalization. Glia 62:1241–1253
75. Foran E, Bogush A, Goffredo M, Roncaglia P, Gustincich S, Pasinelli P, Trotti D (2011) Motor neuron impairment mediated by a sumoylated fragment of the glial glutamate transporter EAAT2. Glia 59:1719–1731
76. Forsberg K, Andersen PM, Marklund SL, Brannstrom T (2011) Glial nuclear aggregates of superoxide dismutase-1 are regularly present in patients with amyotrophic lateral sclerosis. Acta Neuropathol 121:623–634
77. Foust KD, Nurre E, Montgomery CL, Hernandez A, Chan CM, Kaspar BK (2009) Intravascular AAV9 preferentially targets neonatal neurons and adult astrocytes. Nat Biotechnol 27:59–65
78. Foust KD, Wang X, McGovern VL, Braun L, Bevan AK, Haidet AM, Le TT, Morales PR, Rich MM, Burghes AH, Kaspar BK (2010) Rescue of the spinal muscular atrophy phenotype in a mouse model by early postnatal delivery of SMN. Nat Biotechnol 28:271–274
79. Frakes AE, Ferraiuolo L, Haidet-Phillips AM, Schmelzer L, Braun L, Miranda CJ, Ladner KJ, Bevan AK, Foust KD, Godbout JP, Popovich PG, Guttridge DC, Kaspar BK (2014) Microglia induce motor neuron death via the classical NF-kappaB pathway in amyotrophic lateral sclerosis. Neuron 81:1009–1023
80. Freischmidt A et al (2015) Haploinsufficiency of TBK1 causes familial ALS and fronto-temporal dementia. Nat Neurosci 18:631–636
81. Fritz E, Izaurieta P, Weiss A, Mir FR, Rojas P, Gonzalez D, Rojas F, Brown RH, Madrid R, van Zundert B (2013) Mutant SOD1-expressing astrocytes release toxic factors that trigger motor neuron death by inducing hyper-excitability. J Neurophysiol
82. Fukuda AM, Badaut J (2012) Aquaporin 4: a player in cerebral edema and neuroinflammation. J Neuroinf 9:279
83. Gao FB, Almeida S, Lopez-Gonzalez R (2017) Dysregulated molecular pathways in amyotrophic lateral sclerosis-frontotemporal dementia spectrum disorder. EMBO J 36:2931–2950
84. Garbuzova-Davis S, Haller E, Saporta S, Kolomey I, Nicosia SV, Sanberg PR (2007) Ultrastructure of blood-brain barrier and blood-spinal cord barrier in SOD1 mice modeling ALS. Brain Res 1157:126–137
85. Garbuzova-Davis S, Saporta S, Haller E, Kolomey I, Bennett SP, Potter H, Sanberg PR (2007) Evidence of compromised blood-spinal cord barrier in early and late symptomatic SOD1 mice modeling ALS. PLoS ONE 2:e1205
86. Garbuzova-Davis S, Hernandez-Ontiveros DG, Rodrigues MC, Haller E, Frisina-Deyo A, Mirtyl S, Sallot S, Saporta S, Borlongan CV, Sanberg PR (2012) Impaired blood-brain/spinal cord barrier in ALS patients. Brain Res 1469:114–128
87. Garcia-Cabezas MA, Garcia-Alix A, Martin Y, Gutierrez M, Hernandez C, Rodriguez JI, Morales C (2004) Neonatal spinal muscular atrophy with multiple contractures, bone fractures, respiratory insufficiency and 5q13 deletion. Acta Neuropathol 107:475–478
88. Garcia Bueno B, Caso JR, Madrigal JL, Leza JC (2016) Innate immune receptor Toll-like receptor 4 signalling in neuropsychiatric diseases. Neurosci Biobehav Rev 64:134–147
89. Gibb SL, Boston-Howes W, Lavina ZS, Gustincich S, Brown RH Jr, Pasinelli P, Trotti D (2007) A caspase-3-cleaved fragment of the glial glutamate transporter EAAT2 is sumoylated and targeted to promyelocytic leukemia nuclear bodies in mutant SOD1-linked amyotrophic lateral sclerosis. J Biol Chem 282:32480–32490
90. Glascock JJ, Shababi M, Wetz MJ, Krogman MM, Lorson CL (2012) Direct central nervous system delivery provides enhanced protection following vector mediated gene replacement in a severe model of spinal muscular atrophy. Biochem Biophys Res Commun 417:376–381
91. Glascock JJ, Osman EY, Wetz MJ, Krogman MM, Shababi M, Lorson CL (2012) Decreasing disease severity in symptomatic, Smn(-/-);SMN2(+/+), spinal muscular atrophy mice following scAAV9-SMN delivery. Hum Gene Ther 23:330–335

92. Gogliotti RG, Quinlan KA, Barlow CB, Heier CR, Heckman CJ, Didonato CJ (2012) Motor neuron rescue in spinal muscular atrophy mice demonstrates that sensory-motor defects are a consequence, not a cause, of motor neuron dysfunction. J Neurosci: Off J Soc Neurosci 32:3818–3829
93. Gong YH, Parsadanian AS, Andreeva A, Snider WD, Elliott JL (2000) Restricted expression of G86R Cu/Zn superoxide dismutase in astrocytes results in astrocytosis but does not cause motoneuron degeneration. J Neurosci: Off J Soc Neurosci 20:660–665
94. Goode A, Rea S, Sultana M, Shaw B, Searle MS, Layfield R (2016) ALS-FTLD associated mutations of SQSTM1 impact on Keap1-Nrf2 signalling. Mol Cell Neurosci 76:52–58
95. Greenway MJ, Andersen PM, Russ C, Ennis S, Cashman S, Donaghy C, Patterson V, Swingler R, Kieran D, Prehn J, Morrison KE, Green A, Acharya KR, Brown RH Jr, Hardiman O (2006) ANG mutations segregate with familial and 'sporadic' amyotrophic lateral sclerosis. Nat Genet 38:411–413
96. Groen EJN, Talbot K, Gillingwater TH (2018) Advances in therapy for spinal muscular atrophy: promises and challenges. Nat Rev Neurol 14:214–224
97. Group W, Group EM-AS (2017) Safety and efficacy of edaravone in well defined patients with amyotrophic lateral sclerosis: a randomised, double-blind, placebo-controlled trial. Lancet Neurol 16:505–512
98. Guerra-Gomes S, Sousa N, Pinto L, Oliveira JF (2017) Functional roles of astrocyte calcium elevations: from synapses to behavior. Front Cell Neurosci 11:427
99. Guo H, Lai L, Butchbach ME, Lin CL (2002) Human glioma cells and undifferentiated primary astrocytes that express aberrant EAAT2 mRNA inhibit normal EAAT2 protein expression and prevent cell death. Mol Cell Neurosci 21:546–560
100. Guo H, Lai L, Butchbach ME, Stockinger MP, Shan X, Bishop GA, Lin CL (2003) Increased expression of the glial glutamate transporter EAAT2 modulates excitotoxicity and delays the onset but not the outcome of ALS in mice. Hum Mol Genet 12:2519–2532
101. Guo Y, Zhang Y, Wen D, Duan W, An T, Shi P, Wang J, Li Z, Chen X, Li C (2013) The modest impact of transcription factor Nrf2 on the course of disease in an ALS animal model. Lab Investig; J Tech Methods Pathol 93:825–833
102. Gurney ME, Pu H, Chiu AY, Dal Canto MC, Polchow CY, Alexander DD, Caliendo J, Hentati A, Kwon YW, Deng HX et al (1994) Motor neuron degeneration in mice that express a human Cu, Zn superoxide dismutase mutation. Science 264:1772–1775
103. Haidet-Phillips AM, Hester ME, Miranda CJ, Meyer K, Braun L, Frakes A, Song S, Likhite S, Murtha MJ, Foust KD, Rao M, Eagle A, Kammesheidt A, Christensen A, Mendell JR, Burghes AH, Kaspar BK (2011) Astrocytes from familial and sporadic ALS patients are toxic to motor neurons. Nat Biotechnol 29:824–828
104. Hall CE et al (2017) Progressive Motor Neuron Pathology and the Role of Astrocytes in a Human Stem Cell Model of VCP-Related ALS. Cell reports 19:1739–1749
105. Hall ED, Oostveen JA, Gurney ME (1998) Relationship of microglial and astrocytic activation to disease onset and progression in a transgenic model of familial ALS. Glia 23:249–256
106. Hardiman O, Al-Chalabi A, Chio A, Corr EM, Logroscino G, Robberecht W, Shaw PJ, Simmons Z, van den Berg LH (2017) Amyotrophic lateral sclerosis. Nat Rev Dis Prim 3:17085
107. Hay DC, Kemp GD, Dargemont C, Hay RT (2001) Interaction between hnRNPA1 and IkappaBalpha is required for maximal activation of NF-kappaB-dependent transcription. Mol Cell Biol 21:3482–3490
108. Herva R, Conradi NG, Kalimo H, Leisti J, Sourander P (1988) A syndrome of multiple congenital contractures: neuropathological analysis on five fetal cases. Am J Med Genet 29:67–76
109. Heyburn L, Hebron ML, Smith J, Winston C, Bechara J, Li Z, Lonskaya I, Burns MP, Harris BT, Moussa CE (2016) Tyrosine kinase inhibition reverses TDP-43 effects on synaptic protein expression, astrocytic function and amino acid dis-homeostasis. J Neurochem 139:610–623
110. Hochstim C, Deneen B, Lukaszewicz A, Zhou Q, Anderson DJ (2008) Identification of positionally distinct astrocyte subtypes whose identities are specified by a homeodomain code. Cell 133:510–522

111. Honig LS, Chambliss DD, Bigio EH, Carroll SL, Elliott JL (2000) Glutamate transporter EAAT2 splice variants occur not only in ALS, but also in AD and controls. Neurology 55:1082–1088
112. Howland DS, Liu J, She Y, Goad B, Maragakis NJ, Kim B, Erickson J, Kulik J, DeVito L, Psaltis G, DeGennaro LJ, Cleveland DW, Rothstein JD (2002) Focal loss of the glutamate transporter EAAT2 in a transgenic rat model of SOD1 mutant-mediated amyotrophic lateral sclerosis (ALS). Proc Natl Acad Sci USA 99:1604–1609
113. Huang C, Huang B, Bi F, Yan LH, Tong J, Huang J, Xia XG, Zhou H (2014) Profiling the genes affected by pathogenic TDP-43 in astrocytes. J Neurochem 129:932–939
114. Ibáñez CF, Andressoo J-O (2017) Biology of GDNF and its receptors — Relevance for disorders of the central nervous system. Neurobiol Dis 97:80–89
115. Iyer CC, Wang X, Renusch SR, Duque SI, Wehr AM, Mo XM, McGovern VL, Arnold WD, Burghes AH, Kolb SJ (2017) SMN blood levels in a porcine model of spinal muscular atrophy. J Neuromuscul Dis 4:59–66
116. Izrael M, Slutsky SG, Admoni T, Cohen L, Granit A, Hasson A, Itskovitz-Eldor J, Krush Paker L, Kuperstein G, Lavon N, Yehezkel Ionescu S, Solmesky LJ, Zaguri R, Zhuravlev A, Volman E, Chebath J, Revel M (2018) Safety and efficacy of human embryonic stem cell-derived astrocytes following intrathecal transplantation in SOD1(G93A) and NSG animal models. Stem Cell Res Ther 9:152
117. Jackson M, Steers G, Leigh PN, Morrison KE (1999) Polymorphisms in the glutamate transporter gene EAAT2 in European ALS patients. J Neurol 246:1140–1144
118. Jha MK, Lee S, Park DH, Kook H, Park KG, Lee IK, Suk K (2015) Diverse functional roles of lipocalin-2 in the central nervous system. Neurosci Biobehav Rev 49:135–156
119. Johansson A, Engler H, Blomquist G, Scott B, Wall A, Aquilonius SM, Langstrom B, Askmark H (2007) Evidence for astrocytosis in ALS demonstrated by [11C](L)-deprenyl-D2 PET. J Neurol Sci 255:17–22
120. Johnson JO et al (2010) Exome sequencing reveals VCP mutations as a cause of familial ALS. Neuron 68:857–864
121. Johnson JO et al (2014) Mutations in the Matrin 3 gene cause familial amyotrophic lateral sclerosis. Nat Neurosci 17:664–666
122. Jovicic A, Gitler AD (2017) Distinct repertoires of microRNAs present in mouse astrocytes compared to astrocyte-secreted exosomes. PLoS ONE 12:e0171418
123. Kabashi E, Valdmanis PN, Dion P, Spiegelman D, McConkey BJ, Vande Velde C, Bouchard JP, Lacomblez L, Pochigaeva K, Salachas F, Pradat PF, Camu W, Meininger V, Dupre N, Rouleau GA (2008) TARDBP mutations in individuals with sporadic and familial amyotrophic lateral sclerosis. Nat Genet 40:572–574
124. Kalliolias GD, Ivashkiv LB (2016) TNF biology, pathogenic mechanisms and emerging therapeutic strategies. Nat Rev Rheumatol 12:49–62
125. Kamo H, Haebara H, Akiguchi I, Kameyama M, Kimura H, McGeer PL (1987) A distinctive distribution of reactive astroglia in the precentral cortex in amyotrophic lateral sclerosis. Acta Neuropathol 74:33–38
126. Karki P, Webb A, Smith K, Johnson J Jr, Lee K, Son DS, Aschner M, Lee E (2014) Yin Yang 1 is a repressor of glutamate transporter EAAT2, and it mediates manganese-induced decrease of EAAT2 expression in astrocytes. Mol Cell Biol 34:1280–1289
127. Kato S, Saito M, Hirano A, Ohama E (1999) Recent advances in research on neuropathological aspects of familial amyotrophic lateral sclerosis with superoxide dismutase 1 gene mutations: neuronal Lewy body-like hyaline inclusions and astrocytic hyaline inclusions. Histol Histopathol 14:973–989
128. Kawamata H, Ng SK, Diaz N, Burstein S, Morel L, Osgood A, Sider B, Higashimori H, Haydon PG, Manfredi G, Yang Y (2014) Abnormal intracellular calcium signaling and SNARE-dependent exocytosis contributes to SOD1G93A astrocyte-mediated toxicity in amyotrophic lateral sclerosis. J Neurosci: Off J Soc Neurosci 34:2331–2348
129. Keiner S, Wurm F, Kunze A, Witte OW, Redecker C (2008) Rehabilitative therapies differentially alter proliferation and survival of glial cell populations in the perilesional zone of cortical infarcts. Glia 56:516–527

130. Kelley KW, Ben Haim L, Schirmer L, Tyzack GE, Tolman M, Miller JG, Tsai HH, Chang SM, Molofsky AV, Yang Y, Patani R, Lakatos A, Ullian EM, Rowitch DH (2018) Kir4.1-dependent astrocyte-fast motor neuron interactions are required for peak strength. Neuron 98:306–319 e307
131. Khalfallah Y, Kuta R, Grasmuck C, Prat A, Durham HD, Vande Velde C (2018) TDP-43 regulation of stress granule dynamics in neurodegenerative disease-relevant cell types. Sci Rep 8:7551
132. Kia A, McAvoy K, Krishnamurthy K, Trotti D, Pasinelli P (2018) Astrocytes expressing ALS-linked mutant FUS induce motor neuron death through release of tumor necrosis factor-alpha. Glia 66:1016–1033
133. Kim HJ et al (2013) Mutations in prion-like domains in hnRNPA2B1 and hnRNPA1 cause multisystem proteinopathy and ALS. Nature 495:467–473
134. Kim MJ, Vargas MR, Harlan BA, Killoy KM, Ball LE, Comte-Walters S, Gooz M, Yamamoto Y, Beckman JS, Barbeito L, Pehar M (2018) Nitration and glycation turn mature NGF into a toxic factor for motor neurons: a role for p75(NTR) and RAGE signaling in ALS. Antioxid Redox Signal 28:1587–1602
135. Kirby J, Halligan E, Baptista MJ, Allen S, Heath PR, Holden H, Barber SC, Loynes CA, Wood-Allum CA, Lunec J, Shaw PJ (2005) Mutant SOD1 alters the motor neuronal transcriptome: implications for familial ALS. Brain: J Neurol 128:1686–1706
136. Komine O, Yamashita H, Fujimori-Tonou N, Koike M, Jin S, Moriwaki Y, Endo F, Watanabe S, Uematsu S, Akira S, Uchiyama Y, Takahashi R, Misawa H, Yamanaka K (2018) Innate immune adaptor TRIF deficiency accelerates disease progression of ALS mice with accumulation of aberrantly activated astrocytes. Cell Death Differ
137. Kondo T, Funayama M, Tsukita K, Hotta A, Yasuda A, Nori S, Kaneko S, Nakamura M, Takahashi R, Okano H, Yamanaka S, Inoue H (2014) Focal transplantation of human iPSC-derived glial-rich neural progenitors improves lifespan of ALS mice. Stem Cell Rep 3:242–249
138. Kondori NR, Paul P, Robbins JP, Liu K, Hildyard JCW, Wells DJ, de Belleroche JS (2017) Characterisation of the pathogenic effects of the in vivo expression of an ALS-linked mutation in D-amino acid oxidase: Phenotype and loss of spinal cord motor neurons. PLoS ONE 12:e0188912
139. Kong Q, Chang LC, Takahashi K, Liu Q, Schulte DA, Lai L, Ibabao B, Lin Y, Stouffer N, Das Mukhopadhyay C, Xing X, Seyb KI, Cuny GD, Glicksman MA, Lin CL (2014) Small-molecule activator of glutamate transporter EAAT2 translation provides neuroprotection. J Clin Investig 124:1255–1267
140. Konig HG, Coughlan KS, Kinsella S, Breen BA, Prehn JH (2014) The BCL-2 family protein Bid is critical for pro-inflammatory signaling in astrocytes. Neurobiol Dis 70:99–107
141. Kosuge Y, Miyagishi H, Yoneoka Y, Yoneda K, Nango H, Ishige K, Ito Y (2017) Pathophysiological role of prostaglandin E2-induced up-regulation of the EP2 receptor in motor neuron-like NSC-34 cells and lumbar motor neurons in ALS model mice. Neurochem Int
142. Kuru S, Sakai M, Konagaya M, Yoshida M, Hashizume Y, Saito K (2009) An autopsy case of spinal muscular atrophy type III (Kugelberg-Welander disease). Neuropathol: Off J Jpn Soc Neuropathol 29:63–67
143. Kushner PD, Stephenson DT, Wright S (1991) Reactive astrogliosis is widespread in the subcortical white matter of amyotrophic lateral sclerosis brain. J Neuropathol Exp Neurol 50:263–277
144. Kwiatkowski TJ Jr et al (2009) Mutations in the FUS/TLS gene on chromosome 16 cause familial amyotrophic lateral sclerosis. Science 323:1205–1208
145. Lall D, Baloh RH (2017) Microglia and C9orf72 in neuroinflammation and ALS and frontotemporal dementia. J Clin Investig 127:3250–3258
146. Lapucci A, Cavone L, Buonvicino D, Felici R, Gerace E, Zwergel C, Valente S, Mai A, Chiarugi A (2017) Effect of Class II HDAC inhibition on glutamate transporter expression and survival in SOD1-ALS mice. Neurosci Lett 656:120–125
147. Latimer CS, Searcy JL, Bridges MT, Brewer LD, Popovic J, Blalock EM, Landfield PW, Thibault O, Porter NM (2011) Reversal of glial and neurovascular markers of unhealthy brain aging by exercise in middle-aged female mice. PLoS ONE 6:e26812

148. Lee AJ, Awano T, Park GH, Monani UR (2012) Limited phenotypic effects of selectively augmenting the SMN protein in the neurons of a mouse model of severe spinal muscular atrophy. PLoS ONE 7:e46353
149. Lee JY, Lee JD, Phipps S, Noakes PG, Woodruff TM (2015) Absence of toll-like receptor 4 (TLR4) extends survival in the hSOD1 G93A mouse model of amyotrophic lateral sclerosis. Journal of neuroinflammation 12:90
150. Leem YH, Lee YI, Son HJ, Lee SH (2011) Chronic exercise ameliorates the neuroinflammation in mice carrying NSE/htau23. Biochem Biophys Res Commun 406:359–365
151. Lefebvre S, Burglen L, Reboullet S, Clermont O, Burlet P, Viollet L, Benichou B, Cruaud C, Millasseau P, Zeviani M et al (1995) Identification and characterization of a spinal muscular atrophy-determining gene. Cell 80:155–165
152. Lepore AC, Dejea C, Carmen J, Rauck B, Kerr DA, Sofroniew MV, Maragakis NJ (2008) Selective ablation of proliferating astrocytes does not affect disease outcome in either acute or chronic models of motor neuron degeneration. Exp Neurol 211:423–432
153. Lepore AC, Rauck B, Dejea C, Pardo AC, Rao MS, Rothstein JD, Maragakis NJ (2008) Focal transplantation-based astrocyte replacement is neuroprotective in a model of motor neuron disease. Nat Neurosci 11:1294–1301
154. Li K, Hala TJ, Seetharam S, Poulsen DJ, Wright MC, Lepore AC (2015) GLT1 overexpression in SOD1(G93A) mouse cervical spinal cord does not preserve diaphragm function or extend disease. Neurobiol Dis 78:12–23
155. Li Y, Sattler R, Yang EJ, Nunes A, Ayukawa Y, Akhtar S, Ji G, Zhang PW, Rothstein JD (2011) Harmine, a natural beta-carboline alkaloid, upregulates astroglial glutamate transporter expression. Neuropharmacology 60:1168–1175
156. Liang H, Ward WF, Jang YC, Bhattacharya A, Bokov AF, Li Y, Jernigan A, Richardson A, Van Remmen H (2011) PGC-1alpha protects neurons and alters disease progression in an amyotrophic lateral sclerosis mouse model. Muscle Nerve 44:947–956
157. Liang X, Wang Q, Shi J, Lokteva L, Breyer RM, Montine TJ, Andreasson K (2008) The prostaglandin E2 EP2 receptor accelerates disease progression and inflammation in a model of amyotrophic lateral sclerosis. Ann Neurol 64:304–314
158. Liddell JR (2017) Are astrocytes the predominant cell type for activation of Nrf2 in aging and neurodegeneration? Antioxidants (Basel) 6
159. Liebner S, Dijkhuizen RM, Reiss Y, Plate KH, Agalliu D, Constantin G (2018) Functional morphology of the blood-brain barrier in health and disease. Acta Neuropathol 135:311–336
160. Lin CL, Bristol LA, Jin L, Dykes-Hoberg M, Crawford T, Clawson L, Rothstein JD (1998) Aberrant RNA processing in a neurodegenerative disease: the cause for absent EAAT2, a glutamate transporter, in amyotrophic lateral sclerosis. Neuron 20:589–602
161. Lin H, Hu H, Duan W, Liu Y, Tan G, Li Z, Deng B, Song X, Wang W, Wen D, Wang Y, Li C (2018) Intramuscular delivery of scAAV9-hIGF1 prolongs survival in the hSOD1(G93A) ALS mouse model via upregulation of D-amino acid oxidase. Mol Neurobiol 55:682–695
162. Lutz CM, Kariya S, Patruni S, Osborne MA, Liu D, Henderson CE, Li DK, Pellizzoni L, Rojas J, Valenzuela DM, Murphy AJ, Winberg ML, Monani UR (2011) Postsymptomatic restoration of SMN rescues the disease phenotype in a mouse model of severe spinal muscular atrophy. J Clin Investig 121:3029–3041
163. Mackenzie IR et al (2017) TIA1 mutations in amyotrophic lateral sclerosis and frontotemporal dementia promote phase separation and alter stress granule dynamics. Neuron 95(808–816):e809
164. Madill M, McDonagh K, Ma J, Vajda A, McLoughlin P, O'Brien T, Hardiman O, Shen S (2017) Amyotrophic lateral sclerosis patient iPSC-derived astrocytes impair autophagy via non-cell autonomous mechanisms. Molecular brain 10:22
165. Madji Hounoum B, Mavel S, Coque E, Patin F, Vourc'h P, Marouillat S, Nadal-Desbarats L, Emond P, Corcia P, Andres CR, Raoul C, Blasco H (2017) Wildtype motoneurons, ALS-Linked SOD1 mutation and glutamate profoundly modify astrocyte metabolism and lactate shuttling. Glia 65:592–605

166. Maihofner C, Probst-Cousin S, Bergmann M, Neuhuber W, Neundorfer B, Heuss D (2003) Expression and localization of cyclooxygenase-1 and -2 in human sporadic amyotrophic lateral sclerosis. Eur J Neurosci 18:1527–1534
167. Marchetto MC, Muotri AR, Mu Y, Smith AM, Cezar GG, Gage FH (2008) Non-cell-autonomous effect of human SOD1 G37R astrocytes on motor neurons derived from human embryonic stem cells. Cell Stem Cell 3:649–657
168. Martin JE, Nguyen TT, Grunseich C, Nofziger JH, Lee PR, Fields D, Fischbeck KH, Foran E (2017) Decreased motor neuron support by SMA astrocytes due to diminished MCP1 secretion. J Neurosci: Off J Soc Neurosci 37:5309–5318
169. Martin S, Trevor-Jones E, Khan S, Shaw K, Marchment D, Kulka A, Ellis CE, Burman R, Turner MR, Carroll L, Mursaleen L, Leigh PN, Shaw CE, Pearce N, Stahl D, Al-Chalabi A (2017) The benefit of evolving multidisciplinary care in ALS: a diagnostic cohort survival comparison. Amyotroph Lateral Scler Front Degener 18:569–575
170. Martinez TL, Kong L, Wang X, Osborne MA, Crowder ME, Van Meerbeke JP, Xu X, Davis C, Wooley J, Goldhamer DJ, Lutz CM, Rich MM, Sumner CJ (2012) Survival motor neuron protein in motor neurons determines synaptic integrity in spinal muscular atrophy. J Neurosci: Off J Soc Neurosci 32:8703–8715
171. Martorana F, Guidotti G, Brambilla L, Rossi D (2015) Withaferin a inhibits nuclear factor-kappaB-Dependent Pro-inflammatory and stress response pathways in the astrocytes. Neural Plast 2015:381964
172. Martorana F, Brambilla L, Valori CF, Bergamaschi C, Roncoroni C, Aronica E, Volterra A, Bezzi P, Rossi D (2012) The BH4 domain of Bcl-X(L) rescues astrocyte degeneration in amyotrophic lateral sclerosis by modulating intracellular calcium signals. Hum Mol Genet 21:826–840
173. Maruyama H et al (2010) Mutations of optineurin in amyotrophic lateral sclerosis. Nature 465:223–226
174. McGivern JV, Patitucci TN, Nord JA, Barabas MA, Stucky CL, Ebert AD (2013) Spinal muscular atrophy astrocytes exhibit abnormal calcium regulation and reduced growth factor production. Glia 61:1418–1428
175. Mead RJ, Higginbottom A, Allen SP, Kirby J, Bennett E, Barber SC, Heath PR, Coluccia A, Patel N, Gardner I, Brancale A, Grierson AJ, Shaw PJ (2013) S[+] Apomorphine is a CNS penetrating activator of the Nrf2-ARE pathway with activity in mouse and patient fibroblast models of amyotrophic lateral sclerosis. Free Radic Biol Med 61C:438–452
176. Mendell JR et al (2017) Single-dose gene-replacement therapy for spinal muscular atrophy. N Engl J Med 377:1713–1722
177. Mendonca DM, Chimelli L, Martinez AM (2006) Expression of ubiquitin and proteasome in motorneurons and astrocytes of spinal cords from patients with amyotrophic lateral sclerosis. Neurosci Lett 404:315–319
178. Mercuri E et al (2018) Nusinersen versus sham control in later-onset spinal muscular atrophy. N Engl J Med 378:625–635
179. Meyer K, Ferraiuolo L, Schmelzer L, Braun L, McGovern V, Likhite S, Michels O, Govoni A, Fitzgerald J, Morales P, Foust KD, Mendell JR, Burghes AH, Kaspar BK (2015) Improving single injection CSF delivery of AAV9-mediated gene therapy for SMA: a dose-response study in mice and nonhuman primates. Mol Ther: J Am Soc Gene Ther 23:477–487
180. Meyer K, Ferraiuolo L, Miranda CJ, Likhite S, McElroy S, Renusch S, Ditsworth D, Lagier-Tourenne C, Smith RA, Ravits J, Burghes AH, Shaw PJ, Cleveland DW, Kolb SJ, Kaspar BK (2014) Direct conversion of patient fibroblasts demonstrates non-cell autonomous toxicity of astrocytes to motor neurons in familial and sporadic ALS. Proc Natl Acad Sci USA 111:829–832
181. Meyer T, Fromm A, Munch C, Schwalenstocker B, Fray AE, Ince PG, Stamm S, Gron G, Ludolph AC, Shaw PJ (1999) The RNA of the glutamate transporter EAAT2 is variably spliced in amyotrophic lateral sclerosis and normal individuals. J Neurol Sci 170:45–50
182. Migheli A, Piva R, Atzori C, Troost D, Schiffer D (1997) c-Jun, JNK/SAPK kinases and transcription factor NF-kappa B are selectively activated in astrocytes, but not motor neurons, in amyotrophic lateral sclerosis. J Neuropathol Exp Neurol 56:1314–1322

183. Migheli A, Cordera S, Bendotti C, Atzori C, Piva R, Schiffer D (1999) S-100beta protein is upregulated in astrocytes and motor neurons in the spinal cord of patients with amyotrophic lateral sclerosis. Neurosci Lett 261:25–28
184. Milane A, Fernandez C, Vautier S, Bensimon G, Meininger V, Farinotti R (2007) Minocycline and riluzole brain disposition: interactions with p-glycoprotein at the blood-brain barrier. J Neurochem 103:164–173
185. Milane A, Vautier S, Chacun H, Meininger V, Bensimon G, Farinotti R, Fernandez C (2009) Interactions between riluzole and ABCG2/BCRP transporter. Neurosci Lett 452:12–16
186. Milane A, Fernandez C, Dupuis L, Buyse M, Loeffler JP, Farinotti R, Meininger V, Bensimon G (2010) P-glycoprotein expression and function are increased in an animal model of amyotrophic lateral sclerosis. Neurosci Lett 472:166–170
187. Milosevic M, Bataveljic D, Nikolic L, Bijelic D, Andjus P (2016) The effect of amyotrophic lateral sclerosis-linked exogenous SOD1-G93A on electrophysiological properties and intracellular calcium in cultured rat astrocytes. Amyotroph Lateral Scler Front Degener 17:443–451
188. Mimoto T, Miyazaki K, Morimoto N, Kurata T, Satoh K, Ikeda Y, Abe K (2012) Impaired antioxydative Keap1/Nrf2 system and the downstream stress protein responses in the motor neuron of ALS model mice. Brain Res 1446:109–118
189. Miquel E, Cassina A, Martinez-Palma L, Bolatto C, Trias E, Gandelman M, Radi R, Barbeito L, Cassina P (2012) Modulation of astrocytic mitochondrial function by dichloroacetate improves survival and motor performance in inherited amyotrophic lateral sclerosis. PLoS ONE 7:e34776
190. Mitchell J, Paul P, Chen HJ, Morris A, Payling M, Falchi M, Habgood J, Panoutsou S, Winkler S, Tisato V, Hajitou A, Smith B, Vance C, Shaw C, Mazarakis ND, de Belleroche J (2010) Familial amyotrophic lateral sclerosis is associated with a mutation in D-amino acid oxidase. Proc Natl Acad Sci USA 107:7556–7561
191. Miyagishi H, Kosuge Y, Takano A, Endo M, Nango H, Yamagata-Murayama S, Hirose D, Kano R, Tanaka Y, Ishige K, Ito Y (2017) Increased expression of 15-hydroxyprostaglandin dehydrogenase in spinal astrocytes during disease progression in a model of amyotrophic lateral sclerosis. Cell Mol Neurobiol 37:445–452
192. Molofsky AV, Kelley KW, Tsai HH, Redmond SA, Chang SM, Madireddy L, Chan JR, Baranzini SE, Ullian EM, Rowitch DH (2014) Astrocyte-encoded positional cues maintain sensorimotor circuit integrity. Nature 509:189–194
193. Morel L, Regan M, Higashimori H, Ng SK, Esau C, Vidensky S, Rothstein J, Yang Y (2013) Neuronal exosomal miRNA-dependent translational regulation of astroglial glutamate transporter GLT1. J Biol Chem 288:7105–7116
194. Moujalled D, Grubman A, Acevedo K, Yang S, Ke YD, Moujalled DM, Duncan C, Caragounis A, Perera ND, Turner BJ, Prudencio M, Petrucelli L, Blair I, Ittner LM, Crouch PJ, Liddell JR, White AR (2017) TDP-43 mutations causing amyotrophic lateral sclerosis are associated with altered expression of RNA-binding protein hnRNP K and affect the Nrf2 antioxidant pathway. Hum Mol Genet 26:1732–1746
195. Moujalled D, James JL, Yang S, Zhang K, Duncan C, Moujalled DM, Parker SJ, Caragounis A, Lidgerwood G, Turner BJ, Atkin JD, Grubman A, Liddell JR, Proepper C, Boeckers TM, Kanninen KM, Blair I, Crouch PJ, White AR (2015) Phosphorylation of hnRNP K by cyclin-dependent kinase 2 controls cytosolic accumulation of TDP-43. Hum Mol Genet 24:1655–1669
196. Murayama S, Inoue K, Kawakami H, Bouldin TW, Suzuki K (1991) A unique pattern of astrocytosis in the primary motor area in amyotrophic lateral sclerosis. Acta Neuropathol 82:456–461
197. Murphy J et al (2016) Cognitive-behavioral screening reveals prevalent impairment in a large multicenter ALS cohort. Neurology 86:813–820
198. Nagai M, Re DB, Nagata T, Chalazonitis A, Jessell TM, Wichterle H, Przedborski S (2007) Astrocytes expressing ALS-linked mutated SOD1 release factors selectively toxic to motor neurons. Nat Neurosci 10:615–622

199. Nagy D, Kato T, Kushner PD (1994) Reactive astrocytes are widespread in the cortical gray matter of amyotrophic lateral sclerosis. J Neurosci Res 38:336–347
200. Nanou A, Higginbottom A, Valori CF, Wyles M, Ning K, Shaw P, Azzouz M (2013) Viral delivery of antioxidant genes as a therapeutic strategy in experimental models of amyotrophic lateral sclerosis. Mol Ther: J Am Soc Gene Ther
201. Neumann M, Sampathu DM, Kwong LK, Truax AC, Micsenyi MC, Chou TT, Bruce J, Schuck T, Grossman M, Clark CM, McCluskey LF, Miller BL, Masliah E, Mackenzie IR, Feldman H, Feiden W, Kretzschmar HA, Trojanowski JQ, Lee VM (2006) Ubiquitinated TDP-43 in frontotemporal lobar degeneration and amyotrophic lateral sclerosis. Science 314:130–133
202. Neymotin A, Calingasan NY, Wille E, Naseri N, Petri S, Damiano M, Liby KT, Risingsong R, Sporn M, Beal MF, Kiaei M (2011) Neuroprotective effect of Nrf2/ARE activators, CDDO ethylamide and CDDO trifluoroethylamide, in a mouse model of amyotrophic lateral sclerosis. Free Radic Biol Med 51:88–96
203. Nicaise C, Soyfoo MS, Authelet M, De Decker R, Bataveljic D, Delporte C, Pochet R (2009) Aquaporin-4 overexpression in rat ALS model. Anat Rec (Hoboken) 292:207–213
204. Nichol KE, Poon WW, Parachikova AI, Cribbs DH, Glabe CG, Cotman CW (2008) Exercise alters the immune profile in Tg2576 Alzheimer mice toward a response coincident with improved cognitive performance and decreased amyloid. J Neuroinflammation 5:13
205. Nicolas A et al (2018) Genome-wide analyses identify KIF5A as a novel ALS gene. Neuron 97(1268–1283):e1266
206. Nonneman A, Robberecht W, Van Den Bosch L (2014) The role of oligodendroglial dysfunction in amyotrophic lateral sclerosis. Neurodegener Dis Manag 4:223–239
207. Nonneman A, Criem N, Lewandowski SA, Nuyts R, Thal DR, Pfrieger FW, Ravits J, Van Damme P, Zwijsen A, Van Den Bosch L, Robberecht W (2018) Astrocyte-derived Jagged-1 mitigates deleterious Notch signaling in amyotrophic lateral sclerosis. Neurobiol Dis 119:26–40
208. O'Reilly SA, Roedica J, Nagy D, Hallewell RA, Alderson K, Marklund SL, Kuby J, Kushner PD (1995) Motor neuron-astrocyte interactions and levels of Cu, Zn superoxide dismutase in sporadic amyotrophic lateral sclerosis. Exp Neurol 131:203–210
209. Osipova ED, Semyachkina-Glushkovskaya OV, Morgun AV, Pisareva NV, Malinovskaya NA, Boitsova EB, Pozhilenkova EA, Belova OA, Salmin VV, Taranushenko TE, Noda M, Salmina AB (2018) Gliotransmitters and cytokines in the control of blood-brain barrier permeability. Rev Neurosci
210. Otsmane B, Aebischer J, Moumen A, Raoul C (2014) Cerebrospinal fluid-targeted delivery of neutralizing anti-IFNgamma antibody delays motor decline in an ALS mouse model. NeuroReport 25:49–54
211. Ouali Alami N et al (2018) NF-kappaB activation in astrocytes drives a stage-specific beneficial neuroimmunological response in ALS. EMBO J 37
212. Paez-Colasante X, Seaberg B, Martinez TL, Kong L, Sumner CJ, Rimer M (2013) Improvement of neuromuscular synaptic phenotypes without enhanced survival and motor function in severe spinal muscular atrophy mice selectively rescued in motor neurons. PLoS ONE 8:e75866
213. Papadeas ST, Kraig SE, O'Banion C, Lepore AC, Maragakis NJ (2011) Astrocytes carrying the superoxide dismutase 1 (SOD1G93A) mutation induce wild-type motor neuron degeneration in vivo. Proc Natl Acad Sci USA 108:17803–17808
214. Parachikova A, Nichol KE, Cotman CW (2008) Short-term exercise in aged Tg2576 mice alters neuroinflammation and improves cognition. Neurobiol Dis 30:121–129
215. Pardo AC, Wong V, Benson LM, Dykes M, Tanaka K, Rothstein JD, Maragakis NJ (2006) Loss of the astrocyte glutamate transporter GLT1 modifies disease in SOD1(G93A) mice. Exp Neurol 201:120–130
216. Park GH, Maeno-Hikichi Y, Awano T, Landmesser LT, Monani UR (2010) Reduced survival of motor neuron (SMN) protein in motor neuronal progenitors functions cell autonomously to cause spinal muscular atrophy in model mice expressing the human centromeric (SMN2) gene. J Neurosci: Off J Soc Neurosci 30:12005–12019

217. Pasinelli P, Houseweart MK, Brown RH Jr, Cleveland DW (2000) Caspase-1 and -3 are sequentially activated in motor neuron death in Cu, Zn superoxide dismutase-mediated familial amyotrophic lateral sclerosis. Proc Natl Acad Sci USA 97:13901–13906
218. Patel P, Julien JP, Kriz J (2015) Early-stage treatment with Withaferin A reduces levels of misfolded superoxide dismutase 1 and extends lifespan in a mouse model of amyotrophic lateral sclerosis. Neurother: J Am Soc Exp NeuroTherapeutics 12:217–233
219. Patitucci TN, Ebert AD (2016) SMN deficiency does not induce oxidative stress in SMA iPSC-derived astrocytes or motor neurons. Hum Mol Genet 25:514–523
220. Pattamatta A, Cleary JD, Ranum LPW (2018) All in the Family: Repeats and ALS/FTD. Trends Neurosci 41:247–250
221. Paul P, Murphy T, Oseni Z, Sivalokanathan S, de Belleroche JS (2014) Pathogenic effects of amyotrophic lateral sclerosis-linked mutation in D-amino acid oxidase are mediated by D-serine. Neurobiol Aging 35:876–885
222. Peeters K, Chamova T, Jordanova A (2014) Clinical and genetic diversity of SMN1-negative proximal spinal muscular atrophies. Brain: J Neurol 137:2879–2896
223. Pehar M, Beeson G, Beeson CC, Johnson JA, Vargas MR (2014) Mitochondria-targeted catalase reverts the neurotoxicity of hSOD1G(9)(3)A astrocytes without extending the survival of ALS-linked mutant hSOD1 mice. PLoS ONE 9:e103438
224. Pehar M, Vargas MR, Robinson KM, Cassina P, Diaz-Amarilla PJ, Hagen TM, Radi R, Barbeito L, Beckman JS (2007) Mitochondrial superoxide production and nuclear factor erythroid 2-related factor 2 activation in p75 neurotrophin receptor-induced motor neuron apoptosis. J Neurosci: Off J Soc Neurosci 27:7777–7785
225. Phatnani HP, Guarnieri P, Friedman BA, Carrasco MA, Muratet M, O'Keeffe S, Nwakeze C, Pauli-Behn F, Newberry KM, Meadows SK, Tapia JC, Myers RM, Maniatis T (2013) Intricate interplay between astrocytes and motor neurons in ALS. Proc Natl Acad Sci USA 110:E756–765
226. Piao CS, Stoica BA, Wu J, Sabirzhanov B, Zhao Z, Cabatbat R, Loane DJ, Faden AI (2013) Late exercise reduces neuroinflammation and cognitive dysfunction after traumatic brain injury. Neurobiol Dis 54:252–263
227. Picher-Martel V, Dutta K, Phaneuf D, Sobue G, Julien JP (2015) Ubiquilin-2 drives NF-kappaB activity and cytosolic TDP-43 aggregation in neuronal cells. Mol Brain 8:71
228. Qian K, Huang H, Peterson A, Hu B, Maragakis NJ, Ming GL, Chen H, Zhang SC (2017) Sporadic ALS astrocytes induce neuronal degeneration in vivo. Stem Cell Rep 8:843–855
229. Qosa H, Lichter J, Sarlo M, Markandaiah SS, McAvoy K, Richard JP, Jablonski MR, Maragakis NJ, Pasinelli P, Trotti D (2016) Astrocytes drive upregulation of the multidrug resistance transporter ABCB1 (P-Glycoprotein) in endothelial cells of the blood-brain barrier in mutant superoxide dismutase 1-linked amyotrophic lateral sclerosis. Glia 64:1298–1313
230. Ramirez-Jarquin UN, Rojas F, van Zundert B, Tapia R (2017) Chronic infusion of SOD1(G93A) astrocyte-secreted factors induces spinal motoneuron degeneration and neuromuscular dysfunction in healthy rats. J Cell Physiol 232:2610–2615
231. Re DB, Le Verche V, Yu C, Amoroso MW, Politi KA, Phani S, Ikiz B, Hoffmann L, Koolen M, Nagata T, Papadimitriou D, Nagy P, Mitsumoto H, Kariya S, Wichterle H, Henderson CE, Przedborski S (2014) Necroptosis drives motor neuron death in models of both sporadic and familial ALS. Neuron 81:1001–1008
232. Renton AE et al (2011) A hexanucleotide repeat expansion in C9ORF72 is the cause of chromosome 9p21-linked ALS-FTD. Neuron 72:257–268
233. Rinaldi C, Wood MJA (2018) Antisense oligonucleotides: the next frontier for treatment of neurological disorders. Nat Rev Neurol 14:9–21
234. Rindt H, Feng Z, Mazzasette C, Glascock JJ, Valdivia D, Pyles N, Crawford TO, Swoboda KJ, Patitucci TN, Ebert AD, Sumner CJ, Ko CP, Lorson CL (2015) Astrocytes influence the severity of spinal muscular atrophy. Hum Mol Genet 24:4094–4102
235. Robbins KL, Glascock JJ, Osman EY, Miller MR, Lorson CL (2014) Defining the therapeutic window in a severe animal model of spinal muscular atrophy. Hum Mol Genet 23:4559–4568

236. Robinson MB, Jackson JG (2016) Astroglial glutamate transporters coordinate excitatory signaling and brain energetics. Neurochem Int 98:56–71
237. Rodriguez JJ, Terzieva S, Olabarria M, Lanza RG, Verkhratsky A (2013) Enriched environment and physical activity reverse astrogliodegeneration in the hippocampus of AD transgenic mice. Cell Death Dis 4:e678
238. Rojas F, Cortes N, Abarzua S, Dyrda A, van Zundert B (2014) Astrocytes expressing mutant SOD1 and TDP43 trigger motoneuron death that is mediated via sodium channels and nitroxidative stress. Front Cell Neurosci 8:24
239. Rojas F, Gonzalez D, Cortes N, Ampuero E, Hernandez DE, Fritz E, Abarzua S, Martinez A, Elorza AA, Alvarez A, Court F, van Zundert B (2015) Reactive oxygen species trigger motoneuron death in non-cell-autonomous models of ALS through activation of c-Abl signaling. Front Cell Neurosci 9:203
240. Rosen DR, Siddique T, Patterson D, Figlewicz DA, Sapp P, Hentati A, Donaldson D, Goto J, O'Regan JP, Deng HX et al (1993) Mutations in Cu/Zn superoxide dismutase gene are associated with familial amyotrophic lateral sclerosis. Nature 362:59–62
241. Rosenblum LT, Shamamandri-Markandaiah S, Ghosh B, Foran E, Lepore AC, Pasinelli P, Trotti D (2017) Mutation of the caspase-3 cleavage site in the astroglial glutamate transporter EAAT2 delays disease progression and extends lifespan in the SOD1-G93A mouse model of ALS. Exp Neurol 292:145–153
242. Rossi D, Brambilla L, Valori CF, Roncoroni C, Crugnola A, Yokota T, Bredesen DE, Volterra A (2008) Focal degeneration of astrocytes in amyotrophic lateral sclerosis. Cell Death Differ 15:1691–1700
243. Rothstein JD, Van Kammen M, Levey AI, Martin LJ, Kuncl RW (1995) Selective loss of glial glutamate transporter GLT-1 in amyotrophic lateral sclerosis. Ann Neurol 38:73–84
244. Rothstein JD, Tsai G, Kuncl RW, Clawson L, Cornblath DR, Drachman DB, Pestronk A, Stauch BL, Coyle JT (1990) Abnormal excitatory amino acid metabolism in amyotrophic lateral sclerosis. Ann Neurol 28:18–25
245. Rothstein JD, Patel S, Regan MR, Haenggeli C, Huang YH, Bergles DE, Jin L, Dykes Hoberg M, Vidensky S, Chung DS, Toan SV, Bruijn LI, Su ZZ, Gupta P, Fisher PB (2005) Beta-lactam antibiotics offer neuroprotection by increasing glutamate transporter expression. Nature 433:73–77
246. Santello M, Volterra A (2012) TNFalpha in synaptic function: switching gears. Trends Neurosci 35:638–647
247. Sareen D, Ebert AD, Heins BM, McGivern JV, Ornelas L, Svendsen CN (2012) Inhibition of apoptosis blocks human motor neuron cell death in a stem cell model of spinal muscular atrophy. PLoS ONE 7:e39113
248. Sarlette A, Krampfl K, Grothe C, Neuhoff N, Dengler R, Petri S (2008) Nuclear erythroid 2-related factor 2-antioxidative response element signaling pathway in motor cortex and spinal cord in amyotrophic lateral sclerosis. J Neuropathol Exp Neurol 67:1055–1062
249. Sasabe J, Chiba T, Yamada M, Okamoto K, Nishimoto I, Matsuoka M, Aiso S (2007) D-serine is a key determinant of glutamate toxicity in amyotrophic lateral sclerosis. The EMBO journal 26:4149–4159
250. Sasabe J, Miyoshi Y, Suzuki M, Mita M, Konno R, Matsuoka M, Hamase K, Aiso S (2012) D-amino acid oxidase controls motoneuron degeneration through D-serine. Proc Natl Acad Sci USA 109:627–632
251. Sasaki S, Shibata N, Komori T, Iwata M (2000) iNOS and nitrotyrosine immunoreactivity in amyotrophic lateral sclerosis. Neurosci Lett 291:44–48
252. Saur L, Baptista PP, de Senna PN, Paim MF, do Nascimento P, Ilha J, Bagatini PB, Achaval M, Xavier LL (2014) Physical exercise increases GFAP expression and induces morphological changes in hippocampal astrocytes. Brain structure and function 219:293–302
253. Schiffer D, Cordera S, Cavalla P, Migheli A (1996) Reactive astrogliosis of the spinal cord in amyotrophic lateral sclerosis. J Neurol Sci 139(Suppl):27–33
254. Seo TB, Kim BK, Ko IG, Kim DH, Shin MS, Kim CJ, Yoon JH, Kim H (2010) Effect of treadmill exercise on Purkinje cell loss and astrocytic reaction in the cerebellum after traumatic brain injury. Neurosci Lett 481:178–182

255. Serio A, Bilican B, Barmada SJ, Ando DM, Zhao C, Siller R, Burr K, Haghi G, Story D, Nishimura AL, Carrasco MA, Phatnani HP, Shum C, Wilmut I, Maniatis T, Shaw CE, Finkbeiner S, Chandran S (2013) Astrocyte pathology and the absence of non-cell autonomy in an induced pluripotent stem cell model of TDP-43 proteinopathy. Proc Natl Acad Sci USA 110:4697–4702
256. Sheinberger J, Shav-Tal Y (2017) mRNPs meet stress granules. FEBS Lett 591:2534–2542
257. Shibata N, Nagai R, Uchida K, Horiuchi S, Yamada S, Hirano A, Kawaguchi M, Yamamoto T, Sasaki S, Kobayashi M (2001) Morphological evidence for lipid peroxidation and protein glycoxidation in spinal cords from sporadic amyotrophic lateral sclerosis patients. Brain Res 917:97–104
258. Shibata N, Yamada S, Uchida K, Hirano A, Sakoda S, Fujimura H, Sasaki S, Iwata M, Toi S, Kawaguchi M, Yamamoto T, Kobayashi M (2004) Accumulation of protein-bound 4-hydroxy-2-hexenal in spinal cords from patients with sporadic amyotrophic lateral sclerosis. Brain Res 1019:170–177
259. Shibata N et al (2007) Protein-bound crotonaldehyde accumulates in the spinal cord of superoxide dismutase-1 mutation-associated familial amyotrophic lateral sclerosis and its transgenic mouse model. Neuropathol: Off J Jpn Soc Neuropathol 27:49–61
260. Singh NN, Howell MD, Androphy EJ, Singh RN (2017) How the discovery of ISS-N1 led to the first medical therapy for spinal muscular atrophy. Gene Ther 24:520–526
261. Singh RN, Howell MD, Ottesen EW, Singh NN (2017) Diverse role of survival motor neuron protein. Biochem Biophys Acta 1860:299–315
262. Sison SL, Patitucci TN, Seminary ER, Villalon E, Lorson CL, Ebert AD (2017) Astrocyte-produced miR-146a as a mediator of motor neuron loss in spinal muscular atrophy. Hum Mol Genet 26:3409–3420
263. Skorupa A, Urbach S, Vigy O, King MA, Chaumont-Dubel S, Prehn JH, Marin P (2013) Angiogenin induces modifications in the astrocyte secretome: relevance to amyotrophic lateral sclerosis. J Proteomics 91:274–285
264. Skorupa A, King MA, Aparicio IM, Dussmann H, Coughlan K, Breen B, Kieran D, Concannon CG, Marin P, Prehn JH (2012) Motoneurons secrete angiogenin to induce RNA cleavage in astroglia. J Neurosci: Off J Soc Neurosci 32:5024–5038
265. Smith BN et al (2017) Mutations in the vesicular trafficking protein annexin A11 are associated with amyotrophic lateral sclerosis. Sci Transl Med 9
266. Sommer B, Kohler M, Sprengel R, Seeburg PH (1991) RNA editing in brain controls a determinant of ion flow in glutamate-gated channels. Cell 67:11–19
267. Song S, Miranda CJ, Braun L, Meyer K, Frakes AE, Ferraiuolo L, Likhite S, Bevan AK, Foust KD, McConnell MJ, Walker CM, Kaspar BK (2016) Major histocompatibility complex class I molecules protect motor neurons from astrocyte-induced toxicity in amyotrophic lateral sclerosis. Nat Med 22:397–403
268. Sreedharan J, Blair IP, Tripathi VB, Hu X, Vance C, Rogelj B, Ackerley S, Durnall JC, Williams KL, Buratti E, Baralle F, de Belleroche J, Mitchell JD, Leigh PN, Al-Chalabi A, Miller CC, Nicholson G, Shaw CE (2008) TDP-43 mutations in familial and sporadic amyotrophic lateral sclerosis. Science 319:1668–1672
269. Stamenkovic S, Pavicevic A, Mojovic M, Popovic-Bijelic A, Selakovic V, Andjus P, Bacic G (2017) In vivo EPR pharmacokinetic evaluation of the redox status and the blood brain barrier permeability in the SOD1(G93A) ALS rat model. Free Radic Biol Med 108:258–269
270. Stieber A, Gonatas JO, Gonatas NK (2000) Aggregates of mutant protein appear progressively in dendrites, in periaxonal processes of oligodendrocytes, and in neuronal and astrocytic perikarya of mice expressing the SOD1(G93A) mutation of familial amyotrophic lateral sclerosis. J Neurol Sci 177:114–123
271. Sun S, Sun Y, Ling SC, Ferraiuolo L, McAlonis-Downes M, Zou Y, Drenner K, Wang Y, Ditsworth D, Tokunaga S, Kopelevich A, Kaspar BK, Lagier-Tourenne C, Cleveland DW (2015) Translational profiling identifies a cascade of damage initiated in motor neurons and spreading to glia in mutant SOD1-mediated ALS. Proc Natl Acad Sci USA 112:E6993–7002

272. Swarup V, Phaneuf D, Dupre N, Petri S, Strong M, Kriz J, Julien JP (2011) Deregulation of TDP-43 in amyotrophic lateral sclerosis triggers nuclear factor kappaB-mediated pathogenic pathways. J Exp Med 208:2429–2447
273. Takahashi K, Foster JB, Lin CL (2015) Glutamate transporter EAAT2: regulation, function, and potential as a therapeutic target for neurological and psychiatric disease. Cell Mol Life Sci: CMLS 72:3489–3506
274. Talbot K, Tizzano EF (2017) The clinical landscape for SMA in a new therapeutic era. Gene Ther 24:529–533
275. Taylor JP, Brown RH Jr, Cleveland DW (2016) Decoding ALS: from genes to mechanism. Nature 539:197–206
276. Thomas EA, D'Mello SR (2018) Complex neuroprotective and neurotoxic effects of histone deacetylases. J Neurochem 145:96–110
277. Thomsen GM, Avalos P, Ma AA, Alkaslasi M, Cho N, Wyss L, Vit JP, Godoy M, Suezaki P, Shelest O, Bankiewicz KS, Svendsen CN (2018) Transplantation of neural progenitor cells expressing glial cell line-derived neurotrophic factor into the motor cortex as a strategy to treat amyotrophic lateral sclerosis. Stem Cells
278. Tian YP, Che FY, Su QP, Lu YC, You CP, Huang LM, Wang SG, Wang L, Yu JX (2017) Effects of mutant TDP-43 on the Nrf2/ARE pathway and protein expression of MafK and JDP2 in NSC-34 cells. Genet Mol Res: GMR 16
279. Tong J, Huang C, Bi F, Wu Q, Huang B, Liu X, Li F, Zhou H, Xia XG (2013) Expression of ALS-linked TDP-43 mutant in astrocytes causes non-cell-autonomous motor neuron death in rats. EMBO J
280. Tortarolo M, Vallarola A, Lidonnici D, Battaglia E, Gensano F, Spaltro G, Fiordaliso F, Corbelli A, Garetto S, Martini E, Pasetto L, Kallikourdis M, Bonetto V, Bendotti C (2015) Lack of TNF-alpha receptor type 2 protects motor neurons in a cellular model of amyotrophic lateral sclerosis and in mutant SOD1 mice but does not affect disease progression. J Neurochem 135:109–124
281. Tripathi P, Rodriguez-Muela N, Klim JR, de Boer AS, Agrawal S, Sandoe J, Lopes CS, Ogliari KS, Williams LA, Shear M, Rubin LL, Eggan K, Zhou Q (2017) Reactive astrocytes promote ALS-like degeneration and intracellular protein aggregation in human motor neurons by disrupting autophagy through TGF-beta1. Stem Cell Rep 9:667–680
282. Trojsi F et al (2017) Comorbidity of dementia with amyotrophic lateral sclerosis (ALS): insights from a large multicenter Italian cohort. J Neurol 264:2224–2231
283. Trostchansky A, Mastrogiovanni M, Miquel E, Rodriguez-Bottero S, Martinez-Palma L, Cassina P, Rubbo H (2018) Profile of arachidonic acid-derived inflammatory markers and its modulation by nitro-oleic acid in an inherited model of amyotrophic lateral sclerosis. Front Mol Neurosci 11:131
284. Trotti D, Nussberger S, Volterra A, Hediger MA (1997) Differential modulation of the uptake currents by redox interconversion of cysteine residues in the human neuronal glutamate transporter EAAC1. Eur J Neurosci 9:2207–2212
285. Trotti D, Rolfs A, Danbolt NC, Brown RH Jr, Hediger MA (1999) SOD1 mutants linked to amyotrophic lateral sclerosis selectively inactivate a glial glutamate transporter. Nat Neurosci 2:427–433
286. Trotti D, Rossi D, Gjesdal O, Levy LM, Racagni G, Danbolt NC, Volterra A (1996) Peroxynitrite inhibits glutamate transporter subtypes. J Biol Chem 271:5976–5979
287. Tyzack GE, Hall CE, Sibley CR, Cymes T, Forostyak S, Carlino G, Meyer IF, Schiavo G, Zhang SC, Gibbons GM, Newcombe J, Patani R, Lakatos A (2017) A neuroprotective astrocyte state is induced by neuronal signal EphB1 but fails in ALS models. Nat Commun 8:1164
288. Uranishi H, Tetsuka T, Yamashita M, Asamitsu K, Shimizu M, Itoh M, Okamoto T (2001) Involvement of the pro-oncoprotein TLS (translocated in liposarcoma) in nuclear factor-kappa B p65-mediated transcription as a coactivator. J Biol Chem 276:13395–13401
289. Valori CF, Ning K, Wyles M, Mead RJ, Grierson AJ, Shaw PJ, Azzouz M (2010) Systemic delivery of scAAV9 expressing SMN prolongs survival in a model of spinal muscular atrophy. Sci Transl Med 2:35ra42

290. Van Damme P, Bogaert E, Dewil M, Hersmus N, Kiraly D, Scheveneels W, Bockx I, Braeken D, Verpoorten N, Verhoeven K, Timmerman V, Herijgers P, Callewaert G, Carmeliet P, Van Den Bosch L, Robberecht W (2007) Astrocytes regulate GluR2 expression in motor neurons and their vulnerability to excitotoxicity. Proc Natl Acad Sci USA 104:14825–14830
291. van Es MA, Hardiman O, Chio A, Al-Chalabi A, Pasterkamp RJ, Veldink JH, van den Berg LH (2017) Amyotrophic lateral sclerosis. Lancet 390:2084–2098
292. Vance C et al (2009) Mutations in FUS, an RNA processing protein, cause familial amyotrophic lateral sclerosis type 6. Science 323:1208–1211
293. Vargas MR, Pehar M, Cassina P, Beckman JS, Barbeito L (2006) Increased glutathione biosynthesis by Nrf2 activation in astrocytes prevents p75NTR-dependent motor neuron apoptosis. J Neurochem 97:687–696
294. Vargas MR, Johnson DA, Sirkis DW, Messing A, Johnson JA (2008) Nrf2 activation in astrocytes protects against neurodegeneration in mouse models of familial amyotrophic lateral sclerosis. J Neurosci: Off J Soc Neurosci 28:13574–13581
295. Vargas MR, Pehar M, Cassina P, Martinez-Palma L, Thompson JA, Beckman JS, Barbeito L (2005) Fibroblast growth factor-1 induces heme oxygenase-1 via nuclear factor erythroid 2-related factor 2 (Nrf2) in spinal cord astrocytes: consequences for motor neuron survival. J Biol Chem 280:25571–25579
296. Vargas MR, Burton NC, Gan L, Johnson DA, Schafer M, Werner S, Johnson JA (2013) Absence of Nrf2 or Its selective overexpression in neurons and muscle does not affect survival in ALS-linked mutant hSOD1 mouse models. PLoS ONE 8:e56625
297. Vergouts M, Doyen PJ, Peeters M, Opsomer R, Hermans E (2018) Constitutive downregulation protein kinase C epsilon in hSOD1(G93A) astrocytes influences mGluR5 signaling and the regulation of glutamate uptake. Glia 66:749–761
298. Verhaart IEC, Robertson A, Wilson IJ, Aartsma-Rus A, Cameron S, Jones CC, Cook SF, Lochmuller H (2017) Prevalence, incidence and carrier frequency of 5q-linked spinal muscular atrophy - a literature review. Orphanet J Rare Dis 12:124
299. Verheggen ICM, Van Boxtel MPJ, Verhey FRJ, Jansen JFA, Backes WH (2018) Interaction between blood-brain barrier and glymphatic system in solute clearance. Neurosci Biobehav Rev 90:26–33
300. Verkhratsky A, Matteoli M, Parpura V, Mothet JP, Zorec R (2016) Astrocytes as secretory cells of the central nervous system: idiosyncrasies of vesicular secretion. EMBO J 35:239–257
301. von Bernhardi R, Eugenin-von Bernhardi J, Flores B, Eugenin Leon J (2016) Glial Cells and Integrity of the Nervous System. Adv Exp Med Biol 949:1–24
302. Wang F, Lu Y, Qi F, Su Q, Wang L, You C, Che F, Yu J (2014) Effect of the human SOD1-G93A gene on the Nrf2/ARE signaling pathway in NSC-34 cells. Mol Med Rep 9:2453–2458
303. Wang L, Gutmann DH, Roos RP (2011) Astrocyte loss of mutant SOD1 delays ALS disease onset and progression in G85R transgenic mice. Hum Mol Genet 20:286–293
304. Wang W, Duan W, Wang Y, Wen D, Liu Y, Li Z, Hu H, Cui H, Cui C, Lin H, Li C (2017) Intrathecal delivery of ssAAV9-DAO extends survival in SOD1(G93A) ALS mice. Neurochem Res 42:986–996
305. Watanabe-Matsumoto S, Moriwaki Y, Okuda T, Ohara S, Yamanaka K, Abe Y, Yasui M, Misawa H (2017) Dissociation of blood-brain barrier disruption and disease manifestation in an aquaporin-4-deficient mouse model of amyotrophic lateral sclerosis. Neurosci Res
306. Winkler EA, Sengillo JD, Sullivan JS, Henkel JS, Appel SH, Zlokovic BV (2013) Blood-spinal cord barrier breakdown and pericyte reductions in amyotrophic lateral sclerosis. Acta Neuropathol 125:111–120
307. Wirth B, Mendoza Ferreira N, Torres-Benito L (2017) Spinal Muscular Atrophy Disease Modifiers
308. Wong PC, Borchelt DR (1995) Motor neuron disease caused by mutations in superoxide dismutase 1. Curr Opin Neurol 8:294–301
309. Wu CH et al (2012) Mutations in the profilin 1 gene cause familial amyotrophic lateral sclerosis. Nature 488:499–503

310. Yamanaka K, Chun SJ, Boillee S, Fujimori-Tonou N, Yamashita H, Gutmann DH, Takahashi R, Misawa H, Cleveland DW (2008) Astrocytes as determinants of disease progression in inherited amyotrophic lateral sclerosis. Nat Neurosci 11:251–253
311. Yang C, Wang H, Qiao T, Yang B, Aliaga L, Qiu L, Tan W, Salameh J, McKenna-Yasek DM, Smith T, Peng L, Moore MJ, Brown RH Jr, Cai H, Xu Z (2014) Partial loss of TDP-43 function causes phenotypes of amyotrophic lateral sclerosis. Proc Natl Acad Sci USA 111:E1121–1129
312. Yang T, Ferrill L, Gallant L, McGillicuddy S, Fernandes T, Schields N, Bai S (2018) Verapamil and riluzole cocktail liposomes overcome pharmacoresistance by inhibiting P-glycoprotein in brain endothelial and astrocyte cells: A potent approach to treat amyotrophic lateral sclerosis. Eur J Pharm Sci: Off J Eur Fed Pharm Sci 120:30–39
313. Yang Y, Gozen O, Watkins A, Lorenzini I, Lepore A, Gao Y, Vidensky S, Brennan J, Poulsen D, Won Park J, Li Jeon N, Robinson MB, Rothstein JD (2009) Presynaptic regulation of astroglial excitatory neurotransmitter transporter GLT1. Neuron 61:880–894
314. Yang Y, Hentati A, Deng HX, Dabbagh O, Sasaki T, Hirano M, Hung WY, Ouahchi K, Yan J, Azim AC, Cole N, Gascon G, Yagmour A, Ben-Hamida M, Pericak-Vance M, Hentati F, Siddique T (2001) The gene encoding alsin, a protein with three guanine-nucleotide exchange factor domains, is mutated in a form of recessive amyotrophic lateral sclerosis. Nat Genet 29:160–165
315. Yin X, Wang S, Qi Y, Wang X, Jiang H, Wang T, Yang Y, Wang Y, Zhang C, Feng H (2018) Astrocyte elevated gene-1 is a novel regulator of astrogliosis and excitatory amino acid transporter-2 via interplaying with nuclear factor-kappaB signaling in astrocytes from amyotrophic lateral sclerosis mouse model with hSOD1(G93A) mutation. Mol Cell Neurosci 90:1–11
316. Zamanian JL, Xu L, Foo LC, Nouri N, Zhou L, Giffard RG, Barres BA (2012) Genomic analysis of reactive astrogliosis. J Neurosci: Off J Soc Neurosci 32:6391–6410
317. Zeisel A, Munoz-Manchado AB, Codeluppi S, Lonnerberg P, La Manno G, Jureus A, Marques S, Munguba H, He L, Betsholtz C, Rolny C, Castelo-Branco G, Hjerling-Leffler J, Linnarsson S (2015) Brain structure. Cell types in the mouse cortex and hippocampus revealed by single-cell RNA-seq. Science 347:1138–1142
318. Zhong Z, Deane R, Ali Z, Parisi M, Shapovalov Y, O'Banion MK, Stojanovic K, Sagare A, Boillee S, Cleveland DW, Zlokovic BV (2008) ALS-causing SOD1 mutants generate vascular changes prior to motor neuron degeneration. Nat Neurosci 11:420–422
319. Zhou C, Feng Z, Ko CP (2016) Defects in Motoneuron-Astrocyte Interactions in Spinal Muscular Atrophy. The Journal of neuroscience: the official journal of the Society for Neuroscience 36:2543–2553
320. Zhou F, Zhang C, Guan Y, Chen Y, Lu Q, Jie L, Gao H, Du H, Zhang H, Liu Y, Wang X (2018) Screening the expression characteristics of several miRNAs in G93A-SOD1 transgenic mouse: altered expression of miRNA-124 is associated with astrocyte differentiation by targeting Sox2 and Sox9. J Neurochem 145:51–67
321. Zurcher NR, Loggia ML, Lawson R, Chonde DB, Izquierdo-Garcia D, Yasek JE, Akeju O, Catana C, Rosen BR, Cudkowicz ME, Hooker JM, Atassi N (2015) Increased in vivo glial activation in patients with amyotrophic lateral sclerosis: assessed with [(11)C]-PBR28. NeuroImage Clinical 7:409–414

Chapter 11
Astroglia in Alzheimer's Disease

Alexei Verkhratsky, Vladimir Parpura, Jose Julio Rodriguez-Arellano and Robert Zorec

Abstract Alzheimer's disease is the most common cause of dementia. Cellular changes in the brains of the patients suffering from Alzheimer's disease occur well in advance of the clinical symptoms. At the cellular level, the most dramatic is a demise of neurones. As astroglial cells carry out homeostatic functions of the brain, it is certain that these cells are at least in part a cause of Alzheimer's disease. Historically, Alois Alzheimer himself has recognised this at the dawn of the disease description. However, the role of astroglia in this disease has been understudied. In this chapter, we summarise the various aspects of glial contribution to this disease and outline the potential of using these cells in prevention (exercise and environmental enrichment) and intervention of this devastating disease.

A. Verkhratsky (✉)
Faculty of Biology, Medicine and Health, The University of Manchester, Manchester M13 9PT, UK
e-mail: Alexej.Verkhratsky@manchester.ac.uk

Faculty of Health and Medical Sciences, Center for Basic and Translational Neuroscience, University of Copenhagen, 2200 Copenhagen, Denmark

Achucarro Center for Neuroscience, IKERBASQUE, Basque Foundation for Science, 48011 Bilbao, Spain

V. Parpura
Department of Neurobiology, The University of Alabama at Birmingham, Birmingham, AL, USA

University of Rijeka, Rijeka, Croatia

J. J. Rodriguez-Arellano
BioCruces Health Research Institute, IKERBASQUE, Basque Foundation for Science, Bilbao, Spain

Department of Neuroscience, The University of the Basque Country UPV/EHU, Plaza de Cruces 12, 48903 Barakaldo, Bizkaia, Spain

R. Zorec
Laboratory of Neuroendocrinology-Molecular Cell Physiology, Faculty of Medicine, Institute of Pathophysiology, University of Ljubljana, Ljubljana, Slovenia

Celica BIOMEDICAL, Ljubljana, Slovenia

A. Verkhratsky et al. (eds.), *Neuroglia in Neurodegenerative Diseases*,
Advances in Experimental Medicine and Biology 1175,
https://doi.org/10.1007/978-981-13-9913-8_11

Keywords Pathological ageing · Astrocytes · Alzheimer's disease · Astroglial atrophy · Neurodegeneration · Calcium signalling · Stem cells

11.1 Senile Dementia—The Outcome of Pathological Ageing

Dementia as a medical term was introduced in the first century AD by Aulus Cornelius Celsus in his fundamental discourse *De Medicina* [50] to characterise major cognitive impairments of the mankind. The term dementia originates from the prefix "de" (meaning "out of"), the stem "ment" ('mind') and the suffix "ia" (diseased condition). Historically, this term was used in a very broad sense to indicate chronic cognitive impairments associated with psychotic symptoms such as delusions or hallucinations. However, dementia was not associated with age-dependent cognitive decline, although from the very dawn of medical observations, these impairments were considered as an essential and inevitable part of ageing process. Already in the seventh century BC, Pythagoras defined the advanced ages of human life as "senium" when "the system returns to the imbecility of the first epoch of infancy" [26]. From Aristotle to Lucretius and Galen, the ageing was considered to be associated with mental decline, impossible to arrest or recuperate [26, 38, 121, 302]. This gloomy outlook was not, however, shared by Cicero who believed in selective development of a senescence-dependent cognitive decline: "senile imbecility does not occur in all old men, but only in those of feeble mind" [54].

Over centuries the mental weakness of old people was defined as senility, idiocy, morosis, dotage, etc.; in 1794, the term dementia was formalised by Philippe Pinel and this term was officially recognised by the French Law [291]. At the end of the nineteenth century, the definition of senile dementia became widespread and underlying histopathology became to be scrutinised. The specific lesions, the plaques (then known as *miliary foci*), were discovered by Block and Marinesco in the brain of old epileptic patient [36], and subsequently these plaques were observed in the post-mortem tissues of patients suffering from senile dementia by Redlich and by Otto Fischer [77, 225]. In 1903, Max Bielschowsky developed an improved version of the Golgi silver stain that allowed visualisation of neurofibrilles [30]. Using this new technique, Alois Alzheimer (in 1906) was able to visualise neurofibrillary tangles in the post-mortem brain of Mrs. Augusta D, whom he first seen in 1901 in Frankfurt with the symptoms of confusion, delusions and dementia. These tangles displayed extraordinary thickness and often merged into dense bundles reaching the surface of a neurone [7, 26]. The brain of Augusta D also contained senile plaques, and thus the case of early dementia associated with appearance of senile plaques and pathological neurofibrillary tangles has been reported in 1907 [7]. This was a rather unique description, which differed from the widespread dementia of the early twentieth century associated predominantly with neurosyphilis or vascular ischemic brain damage. The disease was named "Alzheimersche krankheit" by Emil Kraepelin in the 8th edition of his immensely influential textbook on Psychiatry (*Psychiatrie:*

Ein Lehrbuch fuer Studierende und Arzte). Kraepelin defined this new disease as a rapidly progressive, early-onset dementia distinct from the senile dementia [136]; for the history, people involved, histological details and controversies see [27, 109, 172].

Alzheimer's disease (AD) was rarely diagnosed in the first half of the twentieth century and was generally regarded as a rare pathology that affected relatively young persons. Only in 1960, the AD histopathology was related to the sporadic cases of age-dependent (i.e. senile) dementia and the notion of AD as senescent-associated pathology had been developed [244, 245, 290]. It seems also that the pandemic of the senile dementia observed in recent years has evolved over the last century. Detailed physiological investigation of organs and systems of an extended population (826 subjects) of elderly (80–100 years of age) inhabitants of the UK performed in 1889 revealed surprisingly little changes in their cognitive status. Furthermore "... the brain in many held out as well or better than other organs - which may be regarded one of the bright rays, if not the brightest, in the centenarian landscape" [111]. Indeed, in this study dementia was observed only in 2 out of 74 centenarians. This contrasts remarkably with our times, when >50% of people older than 85 demonstrate signs of severe cognitive impairment [343]. Of course, evolution of epidemiological changes may be interpreted from many angles, and yet it is impossible not to speculate that an increased environmental toxicity, changes in diet and mounting social pressure have contributed to a rise of sporadic AD in the modern population.

Increased prevalence of senile dementia at advanced age parallels increase in life expectancy. Modern definition regards AD as a severe neurodegenerative disorder associated with specific histopathological markers represented by (i) focal extracellular deposits of fibrillar β-amyloid generally known as neuritic or senile plaques in the brain parenchyma and the walls of blood vessels, and (ii) intraneuronal accumulation of neurofibrillary tangles composed of abnormal hyperphosphorylated tau filaments [39, 70]. AD affects specific brain regions associated with learning and memory, including the basal forebrain, the hippocampus and the neocortex. Clinical symptoms of AD are manifested by a progressive decline of cognitive functions including short- and long-term memory [185]. At advanced stages of the disease, clinical presentation of AD is complicated by a variety of behavioural disturbances including agitation, irritability, anxiety, delusions and depression [165].

Conceptually, two forms of AD are defined: (i) early-onset or familial Alzheimer's disease (FAD) and (ii) late-onset or sporadic AD, or SAD [35]. Epidemiologically, the late-onset SAD accounts for the absolute majority (95–99%) of AD cases in people above 65 years of age. The familial variant of AD is associated with mutations in three genes encoding for amyloid precursor protein (APP), presenilin-1 (PS-1) and presenilin-2 (PS-2), which are inherited in an autosomal dominant manner [23, 257, 294]. In contrast to SAD, which is linked to an old age, the familial AD occurs in much younger group of patients between 40 and 50 years old; the FAD is characterised by a rapid progression and idiosyncratic clinical manifestation. Anatomical and histopathological progression of AD begins from early degeneration of cholinergic neurones in the nucleus basalis of Meynert and septum. In parallel, the accumulation of intraneuronal β-amyloid and formation of neurofibrillary tangles develop, which ultimately leads to an emergence of senile plaques [182]. Deterioration of neuronal networks begins with synaptic damage and malfunction which affects CNS plasticity; these changes occur prior to the formation of senile plaques and neurofibrillary tangles and prior to neuronal death [251, 255]. In addition, this process suppresses neurogenesis, which further impairs neuronal plasticity [231, 237]. Apart from the cholinergic system, AD pathology impairs other major neurotransmitter systems including noradrenergic, serotonergic and dopaminergic [153, 234, 348].

11.2 Experimental Animal Models of Alzheimer's Disease

Alzheimer's disease, similar to other neurodegenerative diseases, is a specific disease of humans; animals as a rule do not develop AD-like pathology [289]. Experimental study of AD therefore requires the development of animal disease models which are capable of faithful reproduction of single or multiple subsets of neuropathological, histological, cellular, behavioural or biochemical alterations resembling those seen in classical AD [48, 91].

11.2.1 Old Animals

The very first models of AD were represented by aged animals [28, 29, 289]; several species from rodents to primates have demonstrated atrophy and death of basal forebrain neurones expressing choline acetyltransferase or nerve growth factor [60, 78, 248]. In monkeys, alterations of cholinergic neurones were even associated with β-amyloid depositions [289]. In addition, old animals showed not only a cholinergic dysfunction but also a concomitant alteration of other neurotransmitter systems such as the monoaminergic, peptidergic or serotonergic [15, 176, 198, 234].

11.2.2 Lesions

The global lesion models of AD (Table 11.1) relied on destruction of certain brain areas. At the beginning, the electrolytic lesions were used; these caused diffuse damage of several brain areas and lacked specificity [156, 289]. In the majority of global lesion models, the non-selective excitotoxic toxins such as NMDA, ibotenic acid, quisqualic acid, quinolinic acid, colchicine and other alkaloid substances were employed [289]. Injections of these substances triggered cell death with consequent neurological dysfunctions including impaired cognition. The global lesion models also employed injections of alcohol, which is toxic to cholinergic neurons [13, 289]; or injections of β-amyloid peptides, which produces multiple alterations of corticobasal neurones affecting acetylcholine release and cholinergic receptors [88, 211].

Table 11.1 Summary of lesion models of Alzheimer's disease. Modified from [232]

Lesion	Cholinergic	Non-cholinergic	Neuropathology	References
Electrolytic	Yes	Yes	Neuronal death	[156, 293]
Excitotoxins (NMDA, Ibotenic acid, Quisalic acid)	Yes	Yes	Neuronal death	[69, 330]
Quinolinic acid	Yes	Yes	Neuronal death	[37]
Colchicine	Yes	Yes	Neuronal death	[258]
Alkaloids	Yes	Yes	Neuronal death	[65]
AF64A	Yes	No	Neuronal death	[53, 97, 318]
192Ig-G saporin	Yes	No	Neuronal death	[325, 327]
Alcohol	Yes	Yes	Neuronal death	[13]
β-Amyloid	No	No	Cholinergic Dysfunction	[88, 211]

As mentioned above, a loss of cholinergic neurones is a prominent feature of AD [18]. With this in mind, animal models, which employed specific lesioning of cholinergic neurones, were developed. Among these, the most relevant were the rodent models with lesions in the nucleus basalis magnocellularis, which is the equivalent of the nucleus of Meynert in humans [214, 322], in the diagonal band of Broca and the septum [289] (Table 11.1).

Specific cholinergic lesion models used toxins, which affected only cholinergic neurones in the brain regions relevant to AD, including septum, nucleus basalis magnocellularis and the diagonal band of Broca, but did not impair non-cholinergic neurones [289, 325]. For example, the AF64A cholinotoxin, which binds to the high-affinity choline uptake system, was injected. Alternatively, the immunotoxin 192 IgG-saporin that binds selectively and irreversibly to low-affinity nerve growth factor receptor interrupting cholinergic neuronal protein synthesis was employed. Both techniques lead to selective impairment and death of cholinergic neurones [289].

Similarly, the noradrenergic system can be lesioned in rats by the injection of the construct, consisting of antibody against dopamine-β-hydroxylase, the enzyme converting dopamine to noradrenaline, and saporin [215], a ribosome-inactivating plant lectin extracted from *Saponaria officinalis* (Caryophyllaceae) [17, 151]. This technology allows a selective and gradual lesioning of noradrenergic neurones in the brain stem nucleus locus coeruleus, the primary site of noradrenaline production in the CNS [75]. Upon injection into the LC, this immunotoxin binds dopamine-β-hydroxylase, which is not only localised mainly in the cytosol, but also at the plasma membrane surface of noradrenergic neurones [271, 321]. Due to its structure, saporin cannot enter the cell [56], but when coupled to a carrier molecule (for example, an antibody), is able to specifically bind a surface antigen protein (such as dopamine-β-hydroxylase, in this case), the toxin gains access to the cytosol and binds to the ribosomal 60S subunit, interfering with protein synthesis, and soon leading to cell death [326]. In initial anatomical investigations, the immunotoxin, infused into the lateral ventricles of either adult [331] or developing rats [57], produced specific and dose-dependent depletions of locus coeruleus neurones, with no effects on other cholinergic, dopaminergic or serotonergic neuronal populations [153]. The possibility to induce a partial or total noradrenergic loss (by varying the injected dose) makes this immunotoxic approach an ideal model to study events within the noradrenergic projection system, as they occur during age-related demise of locus coeruleus in humans [329].

11.2.3 Transgenic Animals

The AD models described above, although triggering neuronal death with consequent cognitive impairments, did not mimic histopathology and temporal progression of the disease. In the last two decades, an alternative and much more effective approach of using transgenic technologies have produced numerous models of familial forms of AD, which have been widely employed in experimental research. These transgenic animal models replicate several neuropathological features of AD (Table 11.2 and [91, 142, 232]) and they are based on mutated genes isolated from patients with various forms of familial AD. The very first transgenic animal carrying mutant APP and showing an AD-like pathology was developed in 1995 [83]. In this model, known as PDAPP, several pathological hallmarks of AD have been identified, including extracellular β-amyloid deposits, dystrophy of neurites, astrogliosis and memory impairments. Memory impairments, however, did not show any correlation with β-amyloid load [83]. The next transgenic AD mouse model, designated as Tg2576 mice, harboured APP_{swe} (Swedish K670 N/M671L) mutation; this model developed numerous senile plaques in parallel with learning and memory impairments, which begun to develop from 9 months of age onwards [110]. The next generation of transgenic models carried double mutation of APP gene; such a model known as APP23 demonstrated some (~14%) neuronal death in the hippocampus [72, 272]. The next step was to combine mutated APP and PS genes; co-expression of $PS1_{dE9}$ with APP_{Swe} resulted in an AD mouse model characterised by accelerated β-amyloid deposition and memory deficits but without tangle formation [250]. These developments culminated in creation of 5xTG AD mice, designated as Tg6799; these animals carry a single human APP Swedish K670N/M671L double mutation as well as the Florida I716V mutation, and the London V717I mutation, along with PS with double M146L and L286V mutations. These mice develop amyloid depositions as early as 2 months of age [200].

In parallel to the animal models with increased β-amyloid production, the pathological tau models also have been created; the first being produced in 1995. In this model, the hyperphosphorylated tau was accumulated in neuronal somata and dendrites, although neurofibrillary tangles were never developed [90]. The next tauopa-

Table 11.2 Transgenic mouse and rat models of Alzheimer's disease

Transgenic mouse and rat models	Neuropathology	References
APP_{751SL}	Plaques	[32]
APP/Ld/2	Plaques	[183]
APP_{Swe}	Plaques	[72]
APP Swedish, 695 K670N M671L	Plaques	[272]
$PS1_{M146L}$	Diffused plaques	[32]
$APP_{751SL}/PS1_{M146L}$	Plaques	[32]
$APP_{SWE}/PS1_{dE9}$	Plaques	[250]
APPSwedish and $PS1_{M146L}$	Plaques	[115]
APP_{695SWE}	Plaques	[110]
APP_{V717F}	Plaques	[67]
K670N/M671L and V717F	Plaques	[115]
APP Swedish, 695 K670N-M671L and Indiana V717F	Plaques	[72]
APPSwedish and V717F	Plaques	[51]
V337 M	Tangles	[280]
4R/2 N	Tangles	[281]
Tau_{P301L} (4R,2-,3-)	Tangles	[159]
P301L	Tangles	[89]
Tau_{P301L}	Tangles	[12]
P301S/G272V	Tangles	[252]
P301S	Tangles	[6]
G272V, P301L, R406W	Tangles	[72]
Endogenous tau KO	Tangles	[9]
P301L TET-off	Tangles	[224]
7TauTg	Tangles	[113]
Tg2576 × JNPL3 (APP_{SWE})	Plaques and Tangles	[158]
Tg2576 and VLW	Plaques and Tangles	[228]
3xTg-AD	Plaques and Tangles	[202]
Tg478	None	[79]
Tg1116	None	[79]
Tg478/Tg1116	Plaques	[79]
Tg 478/1116/11587	Plaques	[79]
K670M/N671L	Plaques	[132]

thy model, over-expressing Tau_{P301L}, did develop neurofibrillary tangles without β-amyloid pathology and neuronal loss [159].

In 2003, the triple transgenic AD mice (3xTg-AD) was created combining the mutants of the three major implicated genes; these animals harbour the mutant genes for APP_{Swe}, for presenilin $PS1_{M146V}$ and for Tau_{P301L} [201, 202]. These animals demonstrated temporal- and region-specific Aβ and tau pathology, which resembles that seen in the human AD brain. Additionally, the 3xTg-AD animals also displayed plaques and tangles, and also showed reduced long-term potentiation in the hippocampus along with functional and cognitive impairments seen as deficient spatial and long-term memory [201, 202]. These pathological changes progress in an age-related manner; most importantly functional deficits precede the appearance of histological hallmarks. Cognitive deficits in the 3xTg-AD model correlate with the accumulation of intraneuronal Aβ [44, 177]. Moreover, at the cellular level, changes in astroglial subcellular vesicle traffic contribute to the pathophysiology of AD [268, 269].

11.3 Neurodegenerative Diseases and Neuroglia

Neurodegenerative diseases, which affect almost exclusively humans, are chronic disorders that result in a progressive loss of function, structure and number of neural cells, ultimately resulting in atrophy of the brain and profound cognitive deficit. The aetiology of neurodegeneration is complex and multifaceted. Neurodegeneration can have a genetic background or it can be instigated by acute trauma, by chemical poisoning, by metabolic insufficiencies or by infectious attacks, as well as by vascular abnormalities, or by sporadic accumulation of genetic/biochemical errors of unknown nature. At the early stages, neurodegeneration as a rule affects synaptic contacts in the brain tissue, thus causing early cognitive deficits. The early stages of neurodegeneration are of course of specific significance, because during this early phase the pathological process can be arrested or even reversed, thus offering the hope for preventing the cognitive decline.

Cellular and molecular mechanisms of underlying initiation and progression of neurodegenerative diseases are highly complex, which makes it almost impossible to identify a single leading cause. At the level of cellular biochemistry, neurodegeneration is frequently linked to aberrant handling of proteins, which promotes intra- or extracellular accumulation of abnormal proteins such as, for example, β-amyloid, tau or α-synuclein [116]. At a more systemic level, however, neurodegeneration reflects a generalised failure of brain homeostasis, which results in a functional and structural decline in the connectivity of neural networks, thus ultimately destroying information processing. Neurodegeneration starts from functional impairment of synaptic connectivity and synaptic plasticity which leads to a neurotransmission misbalance; these processes stipulate early cognitive deficiency. With further progression of the neurodegenerative process, the structural abnormalities develop, trigger disappearance of synapses and death of neural cells, ultimately resulting in a generalised atrophy of the brain accompanied with profound cognitive deficiency [133, 207, 254, 282].

The neuroglia provides for the birth, maintenance and demise of synapses, as well as for overall homeostasis of the nerve tissue, these functions being summarised in a concept of the astroglial cradle [303, 304]. Thus, these non-neuronal cells likely represent the main cellular element shaping the progression of neurodegenerative processes. The generally acknowledged and prevailing point of view considers neurones as main substrates of neurodegeneration, and it is generally assumed that failures in neuronal protein synthesis and/or direct neuronal damage caused by various factors constitute the leading mechanism of neurodegenerative pathologies. These neurone-centric doctrine has been challenged in the past decade, with considerable attention re-routed to neuroglia, which being primary cells responsible for the brain homeostasis and defences, fundamentally contribute to an overall homeostatic failure promoting neurodegeneration [40, 45, 106, 212, 235, 238, 242, 300, 307, 310, 311, 315].

11.4 Astroglial Atrophy and Astrogliosis in Neurodegenerative Diseases

Astrogliopathology in neurodegenerative diseases includes reactivity and astroglial atrophy, asthenia and loss of function. These processes develop in a stage-specific manner and contribute to pathological progression; frequently astroglial asthenia develops at early stages of the disease, whereas at the advanced stages an emergence of disease-specific lesions (for example, senile plaques) and death of neurones instigates astroglial reactivity [14, 306, 308, 317]. Pathological changes in astroglia evolve in parallel with microglial responses. Microglial reactions, at least in the context of human disease, are represented by either activation (that may contribute to neuroinflammatory progression) or microglial paralysis with loss of neuroprotective capabilities, which all contribute to brain atrophy. Cells of oligodendroglial lineage are also affected, which leads to a failure in myelination and atrophy of brain connectome.

11.4.1 Neurodegeneration Following Toxic Brain Injury

Astrocytes play the leading role in chronic neurodegeneration following the brain poisoning by toxic agents. The core mechanism underlying this astroglial-dependent neurotoxicity, which leads to a substantial neuronal death, is linked to a failure of astroglial glutamate uptake. Glutamate clearance from the extracellular space is mainly accomplished by astroglial Na^+-dependent plasmalemmal glutamate transporters; astrocytes specifically express two types of these glutamate transporters, the excitatory amino acid transporters 1 and 2 (EAAT1 and 2). This glutamate uptake is fundamental for astroglia-mediated neuroprotection against glutamate excitotoxicity; suppressing of astroglial glutamate uptake greatly increases neuronal damage following exposure to glutamate [58]. Astroglial glutamate uptake is usually impaired in neurodegeneration and can be considered as one of the common mechanisms of this process [126].

Exposure to heavy metals triggers neuronal death underlying condition known as heavy metal toxic encephalopathy, which manifests itself by impaired cognition and psychotic symptoms. This neurotoxicity results from astroglial homeostatic failure; heavy metals are accumulated by astroglial cells, thus damaging pathways responsible for glutamate homeostasis and catabolism. In methylmercury-induced encephalopathy known as Minamata disease (the name derives from the Japanese city of Minamata where massive poisoning with methylmercury occurred in 1950s, see [174]). When in astrocytes, methylmercury inhibits glutamate, glutamine and cystine transporters which compromises glutamate homoeostasis [194, 339]. Resulting increase in extracellular glutamate concentration triggers neuronal death underlying clinical symptoms that include cognitive decline, impaired vision and hearing, as well as motor symptoms. Similarly, astrocytes, endowed with capacity manganese transport system, emerge as a main target for manganese toxicity. Again, increased manganese in astroglial cells suppresses astroglial glutamate uptake with subsequent excitotoxic neuronal damage [260]. Similarly, astrocytes appear as a primary target for other heavy metals, such as arsenic, lead and cadmium, which all reduce expression of glial fibrillary acidic protein (GFAP) and trigger astroglial apoptosis, thus reducing astroglial homeostatic presence [223]. In aluminium toxic encephalopathy (symptoms of which include cognitive deficits and speech alterations), astrocytes are again the main targets. Aluminium accumulated by astrocytes impairs plasmalemmal glutamate transporters as well as gap junctions and causes astrocytic death [275]. Likewise, astroglial loss through apoptotic death plays a leading role in the enchephalotoxic damage caused by cypermethrin, a synthetic II pyrethroid insecticide [173].

11.4.2 Wernicke Encephalopathy

Wernicke encephalopathy is a pathoanatomical substrate for Korsakoff syndrome, symptoms of which include ante- and retrograde amnesia, apathy and confabulation [135, 323]. This type of encephalopathy is essentially rapidly progressing malignant thalamo-cortical neurodegeneration. The pathological mechanism of Korsakoff–Wernicke syndrome is primarily associated with acute failure in astroglial glutamate uptake resulting from ~60 to 70% decrease in expression of EAAT1 and EAAT2 glutamate transporters. This remarkable decrease in plasmalemmal glutamate transporters expression has been identified in post-mortem human samples, as well as in the rat thiamine deficiency model of the disease [104, 105]. In addition to decrease in EAAT1/2 expression, astrocytes demonstrated signs of atrophy including decrease in GFAP morphological profiles, as well as decrease in expression of glutamine synthetase (GS) and GAT-3 GABA transporter.

11.4.3 The Human Immunodeficiency Virus-1 (HIV-1)-Associated Dementia (HAD)

In the nervous system, the HIV-1 virus primarily infects and propagates in microglial cells. Microglia contributes to neuronal death by releasing various neurotoxic factors [123, 171], including Nef protein [267]. In HAD, astroglial cells develop both astrodegeneration and reactive astrogliosis. In the basal ganglia, astrocytes undergo a serious loss with the degree of astroglial death correlated with the degree of cognitive impairments [283]. In the entorhinal cortex and in the hippocampus, astrocytes show prominent reactivity [295].

11.4.4 Non-AD Dementia

The non-AD dementia is represented by many disorders including fronto-temporal lobar degeneration, Pick's disease, Cockayne syndrome, juvenile neuronal ceroid lipofuscinosis (JNCL) or Niemann-Pick type C disease. Astroglial contribution to these disorders is complex with signs for astroglial atrophy and astroglial apoptotic death [42, 320] as well as for astroglial reactivity, which is particularly prominent in the frontal and temporal cortices of patients with fronto-temporal dementia [124]. In thalamic dementia, a profound astrogliosis likely represents a key pathophysiological factor [221]. There is evidence indicating uncoupling of astroglial syncytium and aberrant activity of astroglial connexin hemichannels in JNCL [43], whereas early astroglial reactivation was reported in animal models of Niemann-Pick type C disease [222].

11.4.5 Amyotrophic Lateral Sclerosis (ALS)

Astrocytes play fundamental role in the pathogenesis of hereditary familial ALS associated with the mutation of the human superoxide dismutase 1 (hSOD1) gene. In hSOD1/G93A, associated animal models of ALS astrocytes undergo atrophy, pathological remodelling loss of function and cell death. These astroglial changes precede neuronal abnormalities and the emergence of clinical symptoms [241, 242]. The key pathogenetic factor linked to the neurotoxicity is represented by deficient astroglial glutamate uptake. Selective silencing of *hSOD1* gene in astrocytes in animal model delays ALS progression [337].

11.4.6 Parkinson's Disease

The role of neuroglia in emergence and progression of Parkinson's disease remains to be fully elucidated. There are some indications for microglial activation in relevant brain regions; this activation being possibly linked to neurotoxicity [63]. This microglial response, however, can be secondary, being triggered by neuronal death [108]. In 6-hydroxydopamine (6-OHDA) animal model of Parkinson's disease, inhibition of microglial activation was found to be neuroprotective [152]. Astrocytes have been considered to provide neuroprotection to dopaminergic neurones, based on in vitro experiments [180, 181]. In primary neuronal–glial co-cultures, astrocytes were shown to convert L-DOPA, the immediate precursor of dopamine, from neurotoxic to neurotrophic substance, and hence astroglia can be crucial for L-DOPA substitute therapy [179].

11.5 Astrocytes in Alzheimer's Disease

Alzheimer's disease is characterised by progressive neurodegeneration and an occurrence of specific histopathological markers represented by (i) focal extracellular deposits of fibrillar β-amyloid (also called neuritic or senile plaques) in the brain parenchyma and in the walls of blood vessels, and by (ii) intraneuronal accumulation of neurofibrillary tangles composed from abnormal hyperphosphorylated tau filaments. The initial neurodegenerative events in AD appear in the transentorhinal cortex, which subsequently spread to the entorhinal cortex and hippocampus. At later stages of the disease, the neurodegenerative process disseminates through the temporal, frontal, and parietal lobes [284, 285]. At these late stages, the grey matter undergoes severe damage with a profound loss of neurones and synaptic contacts and generalised atrophy of the brain parenchyma; this atrophy includes both white and grey matters. Contribution of neuroglia to the histopathology of Alzheimer's disease has been initially suggested by Alois Alzheimer himself; Fig. 11.1 shows original drawings of Alzheimer depicting pathologically modified glial cells of a senile plaque [8]. The role of astrocytes in the pathogenesis and progression of AD remains to be fully characterised, primarily because of the lack of longitudinal studies assessing the status of astroglia at different stages of the disease. From analyses of human post-mortem tissues, there has been generally agreed that at the late stages of the disease there are prominent reactive astrogliosis and inclusion of astrocytes into senile plaques [106, 193, 235].

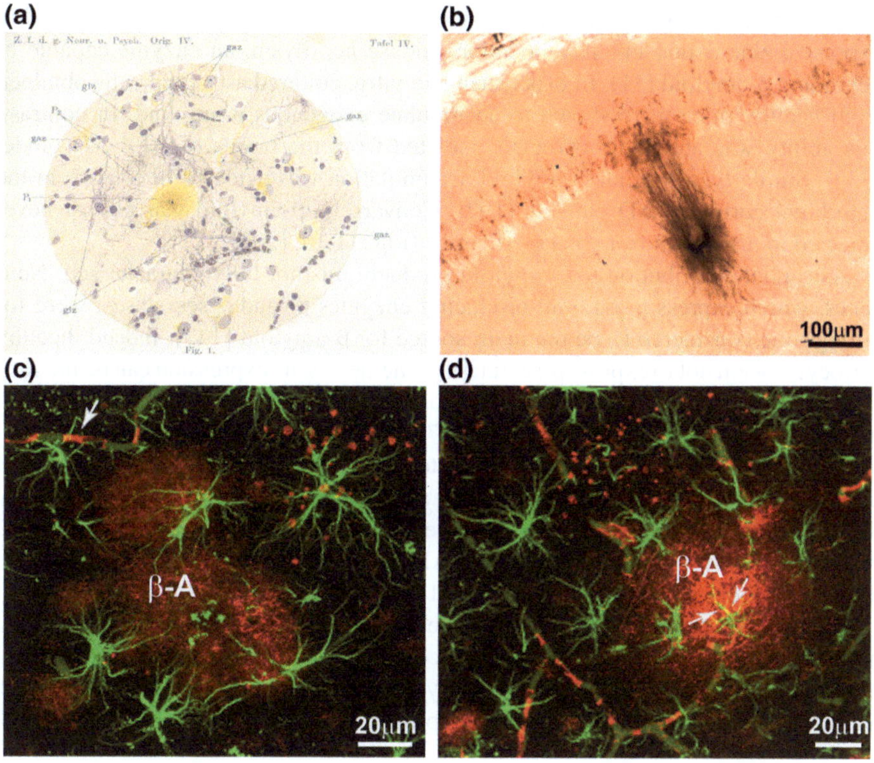

Fig. 11.1 Glial cells in AD **a** Alois Alzheimer´s drawing illustrating the glial reaction (astro- and/or micro-gliosis and hypertrophy) in a pathological brain containing senile plaques. Abbreviations: *gaz*, ganglionic cell—i.e. neurone; *glz*, glial cell, P, central part of the plaque; P_2, peripheral part of the plaque. From [8]. **b** Photomicrograph showing the presence of β-amyloid within the pyramidal neurones of the hippocampal CA1 area as well as the presence of a plaque in 12 months 3xTg-AD mice. **c, d** Confocal images showing GFAP-positive (green) reactive astrocytes surrounding β-amyloid plaques (β-A red; **c**). **d** Reactive astrocytes (green) and an astrocyte showing cytoplasmic β-amyloid accumulation (indicated by arrows; co-localisation is in yellow) near a neuritic plaque (red). Modified and adapted with permission from [235]

11.5.1 Astrocytes and β-Amyloid

The prevailing views on AD pathogenesis associate disease evolution with progressive accumulation of β-amyloid protein in the brain parenchyma and formation of senile plaques [87, 100, 122, 134]. Recently, however, the β-amyloid hypothesis became the subject of extensive criticism [49, 98, 99, 226]. Production of β-amyloid is mostly associated with neurones although there are several reports indicating the role of astrocytes in this process through either direct β-amyloid production or through deficient clearance.

Astroglial contribution to the clearance and degradation of β-amyloid has been suggested some 15 years ago [96, 195]. Reactive astrocytes associated with senile plaques in the transgenic AD mouse model expressing mutant APP were found to express a zinc-dependent metalloendopeptidase neprilysin, an enzyme capable of degrading β-amyloid [11]. In experiments in vitro, cultured astroglial cells obtained from healthy mice were shown to accumulate exogenous β-amyloid. In contrast, this ability was absent in astrocytes isolated from the brains of APP AD model [333]. Similarly, astroglial β-amyloid accumulation was detected in cells from the entorhinal cortex of AD patients [192]. Conversely, β-amyloid was almost never detected in astrocytes from 3xTg-AD mice (Fig. 11.1d, [203]).

Astroglial contribution to β-amyloid production is not fully characterised. Neurones, which express β-amyloid producing enzymes β- and γ-secretases, were for a long time considered to be the main source for β-amyloid [143]. Indeed, healthy astrocytes seem not to express β-secretase; nonetheless, its expression can be induced by conditions of chronic stress or neuroinflammatory environment, thus adding astroglia to amyloidogenesis [33, 82, 117, 157, 205, 243, 342]. Expression of β-secretase was detected in reactive astrocytes emerging following immune lesion of cholinergic septohippocampal afferents or an occlusion of the middle cerebral artery [243]. Similarly, expression of β-secretase was found in reactive astrocytes in AD mice models expressing mutant human APP; these models, for example, included Tg2576, K670N-M671L APP or APP V717I mutations [101, 107, 243]. Of note, an increase in production of APP was characterised in a rat model of chronic neocortical astrogliosis, induced by grafting a foetal cortical tissue in the midbrain of neonatal animals; these chronically activated astrocytes were immunopositive for APP, as well as for another AD-related marker apolipoprotein E4 [166].

11.5.2 Astrogliosis in AD

Astroglial reactivity, generally characterised by an increase in expressions of GFAP, vimentin or s100β protein, has been detected in post-mortem tissues from AD patients [19, 92, 178, 189]. No obvious correlation between GFAP levels, degree of astrogliosis and β-amyloid load was detected [261]. Similarly, no differences in GFAP expression were found between the brains obtained from cognitively sound and demented patients [324]. Reactive, hypertrophic astrocytes, associated with senile plaques and perivascular β-amyloid deposits, are also observed in the brains of AD mice models (Fig. 11.1, [204, 235, 306]). It is important to highlight that astrogliosis in AD is never associated with the scar formation and it does not hamper the physiological non-overlap of astroglial territorial domains. It can be classified therefore as isomorphic or mild astrogliotic response. In the context of AD, the astrogliotic response can be triggered by various molecules, such as β-amyloid, molecules released from damaged cells or certain cytokines and chemokines. Soluble β-amyloid was found to initiate reactive astrogliosis in astrocytes in vitro [64]. This may be associated with certain intracellular signalling events, including, for example, Ca^{2+} signals. Such signals are indeed generated by exposure of cultured astrocytes to β-amyloid [2, 3,

94]. Treatment of cultured astrocytes with β-amyloid also resulted in inhibition of glutamate uptake, which can contribute to pathological progression [168].

11.5.3 Astroglial Atrophy in AD

Pathological changes of astrocytes in the AD pathology are not limited to astrogliotic response; it seems that astrogliosis occurs at later stages of the disease, with reactive astrocytes being mainly associated with senile plaques. Recent studies of transgenic AD mice models revealed a profound astrodegeneration that occurs at the early stages of AD progression [20, 203, 306].

Total number of astrocytes (labelled with antibodies against GFAP, s100β or GS) did not show any age-dependent variations in 3xTg-AD mice of 3–24 months of age [203, 204]. There are complex region- and disease stage-specific morphological changes in astrocytes in 3xTg-AD mice (Figs. 11.2, 11.3 and 11.4). At the early (i.e. pre-plaque) stages of the AD, astrocytes in the entorhinal cortex, prefrontal cortex and hippocampus demonstrate signs of morphological atrophy [139, 203, 204,

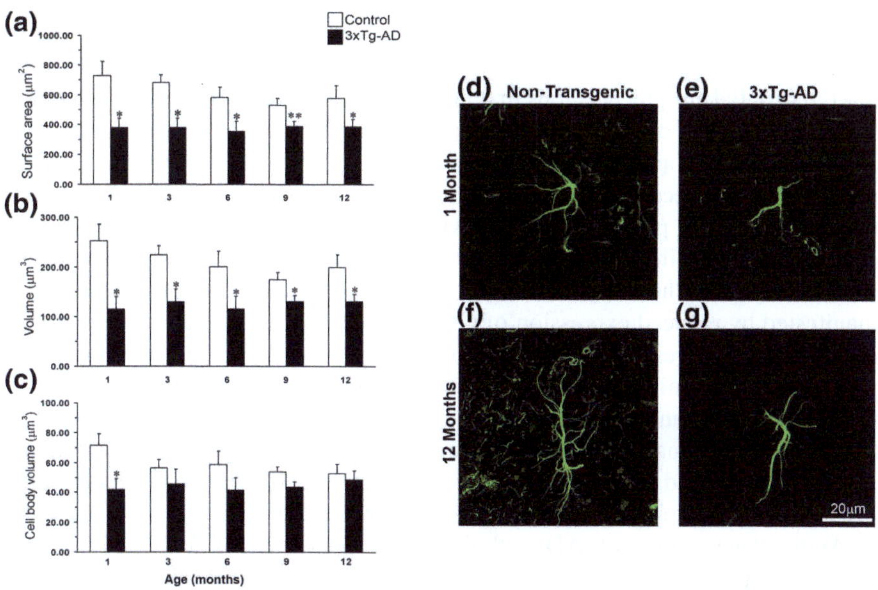

Fig. 11.2 Astroglial atrophy in the entorhinal cortex (EC) of 3xTg-AD mice. Comparison of astrocytic GFAP surface area and volume in the EC of non-Tg and 3xTg-AD animals of different ages. The histograms show a comparison of **a** surface area, **b** total cell volume and **c** somata volume in the EC at the ages of 1, 3, 6, 9 and 12 months between 3xTg-AD and non-Tg animals. Results are means ± S.E.M. (*$p < 0.05$ compared with the age-matched non-Tg control). Confocal micrographs show astrocytic atrophy in 3xTg-AD at 1 month (**e**) and 12 months (**g**) compared with the control animals (**d**, **f**). Reproduced with permission from [338]

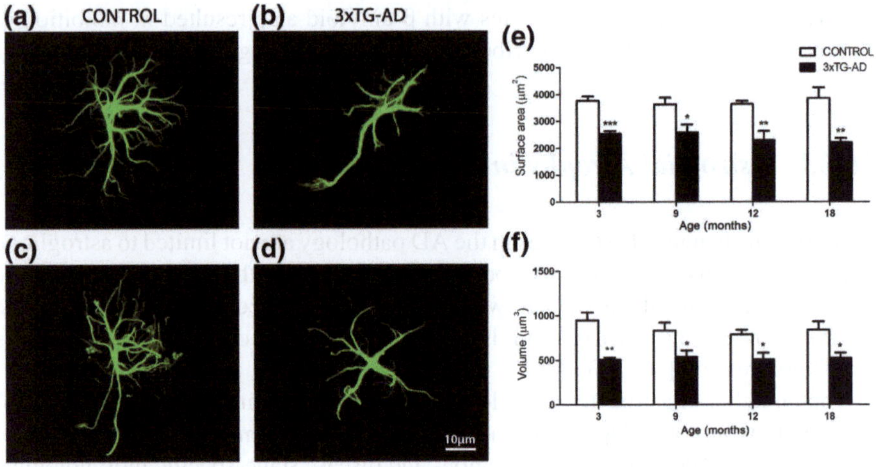

Fig. 11.3 Astroglial atrophy in the prefrontal cortex of 3xTg-AD mice. Confocal images showing morphology of GFAP-positive astrocytes in control non-Tg animals and astrocytic atrophy in the 3xTg-AD animals at 3 months (**a** and **b**, respectively) and 18 months (**c** and **d**, respectively) in the prefrontal cortex. Bar graphs showing the decreases in the surface area and volume (**e**, **f**) in 3xTg-AD mice when compared with control animals. Bars represent mean ± SEM. Reproduced with permission from [139]

338]. Astroglial atrophy develops first in the entorhinal cortex (from 1 month of age, Fig. 11.2); next, it occurs in the prefrontal cortex (3–4 months of age, Fig. 11.3) and finally in the hippocampus (6–9 months of age, Fig. 11.4). Atrophy of GFAP-positive profiles preceded β-amyloid deposition and formation of senile plaques. The reduction in GFAP profiles coincided with the reduced morphological presence of astroglial cells labelled with GS antibodies in the hippocampus and in the prefrontal cortex, but not in the entorhinal cortex. Morphological atrophy of astrocytes was manifested by reduced expression of GFAP-rich cytoskeleton (surface and volume coverage) and decreased somata volume, as well as number and branching of cell processes. Very similar atrophic changes were observed in hippocampal astrocytes from another AD animal model, the mutant APP (PDAPP-J20) mice carrying the Swedish and Indiana APP human mutations [20, 21]. Astroglial atrophy was subsequently confirmed in human material, in astrocytes derived from pluripotent stem cells isolated from patients with family and sporadic AD (Fig. 11.5, [119, 184]).

At the later stages of AD pathology in hippocampi of 3xTG-AD animals (12–18 months; at this time neurofibrillary tangles also start to develop in neurones), formation of plaques and accumulation of extracellular β-amyloid initiates reactive astrogliosis. Numerous hypertrophic astrocytes accumulate exclusively around senile plaques and β-amyloid inundated blood vessels (Figs. 11.6 and 11.7; [203, 230, 316]). This astroglial hypertrophy is characterised by an increased volume and surface of both astrocyte somata and processes, which can increase their size up to 70% (Fig. 11.3). At the same time, astrocytes positioned away from the senile plaques retain their atrophic morphology (Fig. 11.6). In contrast to the hippocampus, accu-

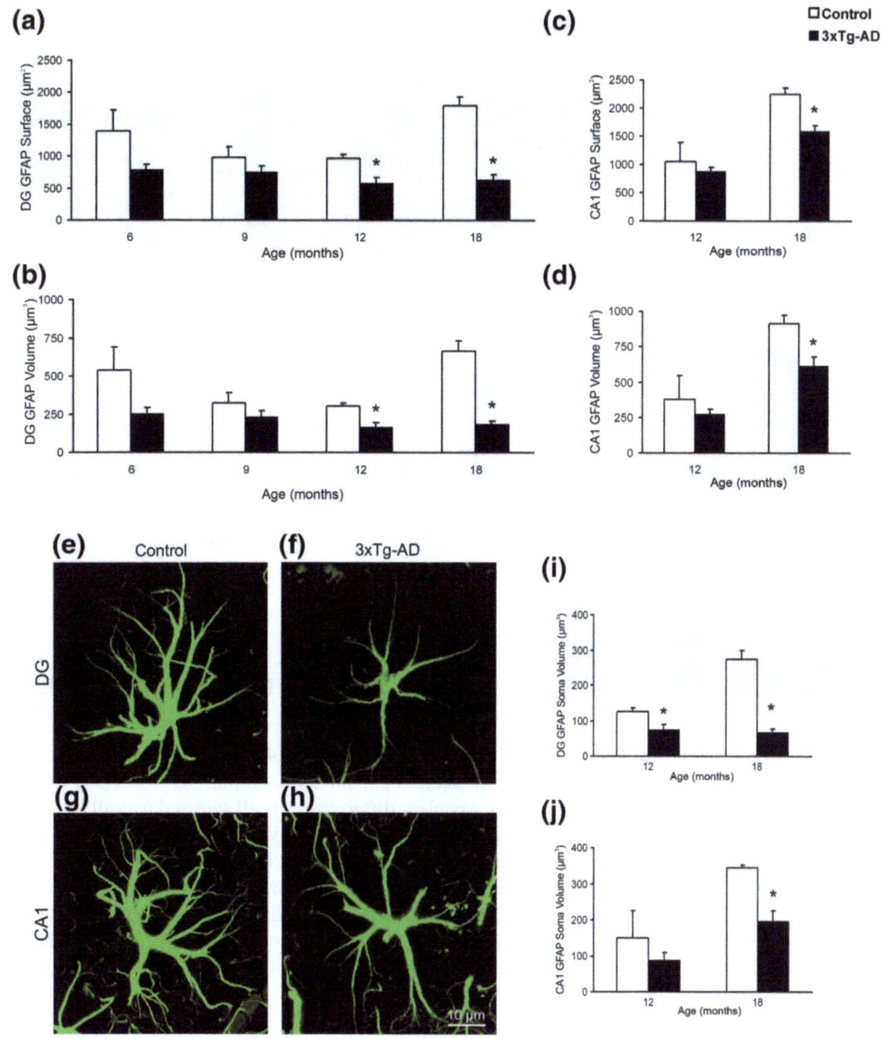

Fig. 11.4 Astroglial atrophy in the hippocampal areas of 3xTg-AD mice. Bar graphs showing the significant decrease in surface area, volume, and soma volume of GFAP-positive astrocytes in the dentate gyrus (DG) (**a**, **b**, **i**) and the CA1 region (**c**, **d**, **j**) of the hippocampus of the 3xTg-AD mice when compared with control animals. Bars represent mean ± SEM ($p < 0.05$). (**g**–**j**). Confocal micrographs illustrating the astrocytic atrophy in 3xTg-AD mice in the DG (**f**) and CA1 (**h**) compared to control animals (**e** and **g**). Reproduced with permission from [203]

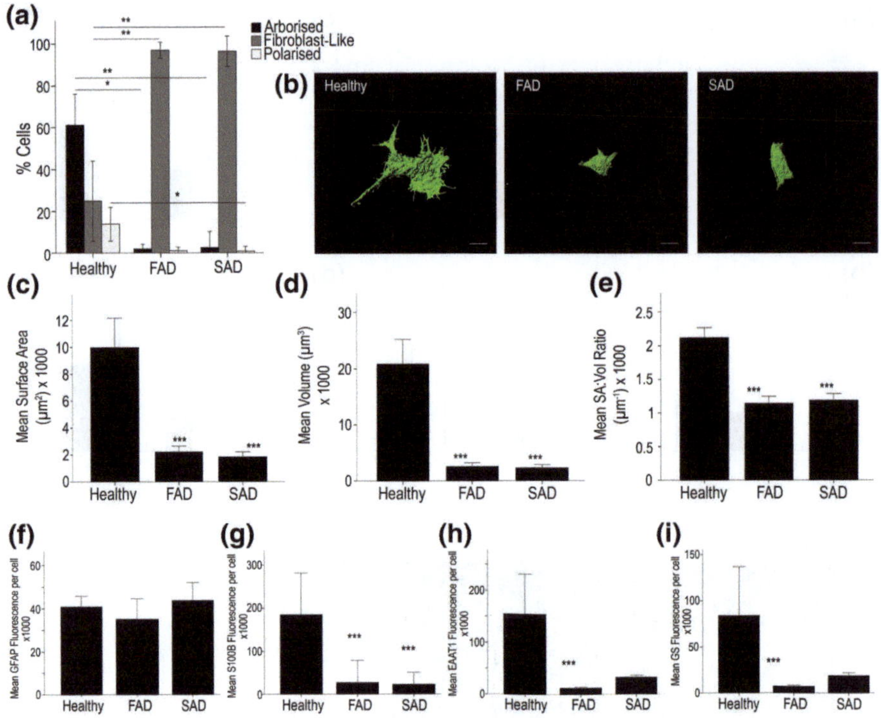

Fig. 11.5 Astrocytes derived from *PSEN1* M146L FAD and *ApoE4*$^{+/+}$ SAD patients exhibit significant atrophy when compared to those from healthy patients. **a** Exemplar 3D isosurface renders constructed from serial confocal z-stacks display clear differences in cell size and overall morphology (**b**). Scale bar = 10 μm. Quantification of cells using these renders by way of surface area (**c**), cell volume (**d**) and SA:Vol ratio (**e**) reveal significant differences in all aspects of cellular morphology between healthy and diseased astrocytes. Quantification of mean fluorescence intensity per immunoreactive cell reveals no significant difference in GFAP staining intensities between AD and control astrocytes (**f**) but S100B, EAAT1 and GS intensities are reduced in both FAD and SAD cells (**g**, **h** and **i**, respectively). Asterisks on graph; ***$p < 0.001$, **$p < 0.005$, *$p < 0.05$. Reproduced from [119]

mulation of β-amyloid and formation of senile plaques do not induce reactivity of astrocytes neither in the entorhinal nor in the prefrontal cortex (Fig. 11.7, [139, 338]). Deficient reactivity of astrocytes may determine the specific vulnerability of different brain regions to AD-type pathology. Atrophy of astrocytes at the early stages of AD may have important functional consequences. The decrease in astroglial complexity may affect synaptic coverage and homoeostatic support as well as functional performance of the neuronal–glial–vascular unit. This in turn can affect connectivity in neural network, reduce synaptic strength and disturb synaptic plasticity thus contributing to cognitive deficits.

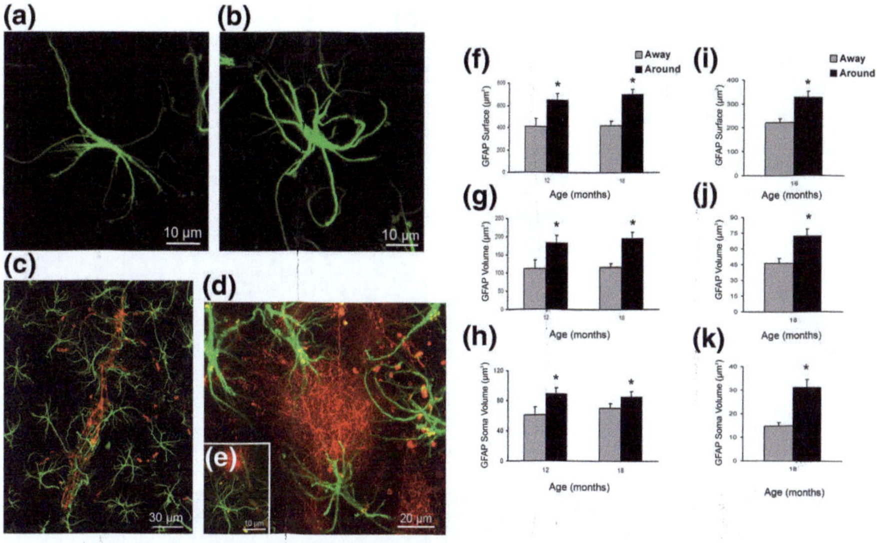

Fig. 11.6 Concomitant astroglial atrophy and astrogliosis at the advanced stages of AD-like pathology in 3xTg-Ad mice. **a, b** Confocal images of hippocampal preparations dually labelled by GFAP and by anti-β amyloid monoclonal antibody illustrating differential changes in GFAP profiles in astrocytes distant to the plaques (**a**) and associated with the β-amyloid plaques (**b**). **c–e** Confocal dual labelling images (GFAP in green and β-amyloid in red) in 3xTg-AD mice showing the accumulation of astrocytes around the β-amyloid plaques and vascular β-amyloid deposits. Astrocytes surrounding β-amyloid plaques (**d, e**) and β-amyloid deposits around a blood vessel (**c**), undergo astrogliosis. **f–k** Bar graphs showing GFAP-positive astrocytic surface area (**f**), volume (**g**) and somata volume (**h**) differences between astrocytes located around the β-amyloid plaques (Aβ) and those distant to the plaques in the CA1 of 3xTg-AD animals. **i–k** Similar astrocytic surface area (**i**), volume (**j**) and somata volume (**k**) differences are observed in the DG at 18 months of age. Bars represent mean 6 SEM ($p < 0.05$). Reproduced with permission from [203]

11.5.4 Loss of Astroglial Homeostatic Support Contributes to Early Cognitive Impairments

Atrophic changes in astrocytes, characterised in several AD animal models as well as in stem cells derived astrocytes appear as general diminution of astroglial territories, of astroglial coverage of neuronal membranes and overall decrease in astroglial homeostatic support. Arguably, this atrophy and loss of function of astrocytes, which occur early in the disease progression, may contribute to the disease pathophysiology. Atrophic astrocytes provide less synaptic coverage with deleterious consequences for synaptic transmission associated with compromised ion and neurotransmitter homeostasis or reduced local metabolic support; astroglial asthenia also results in decreased neuroprotection [232, 238, 316, 317]. All these changes are likely to weaken synaptic transmission and affect synaptic plasticity, thereby being responsible for initial cognitive deficiency observed at the early stages of AD.

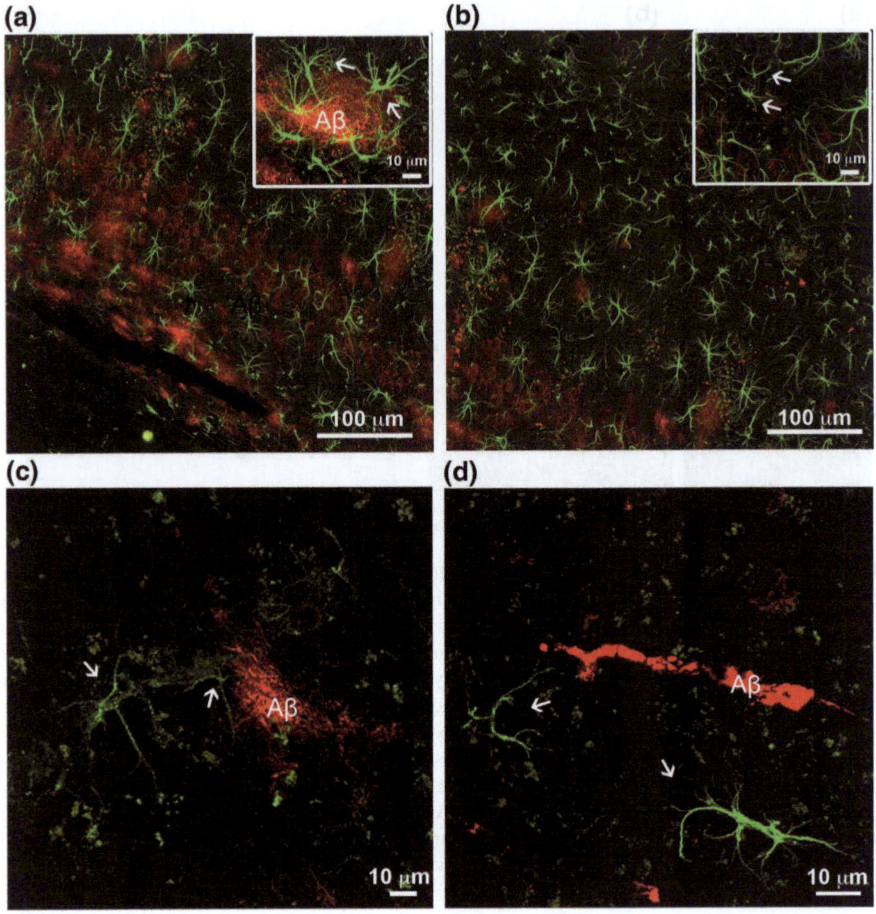

Fig. 11.7 β-Amyloid depositions trigger gliotic response in associated astrocytes in the hippocampus but not in the entorhinal cortex. **a, b** Confocal images of hippocampal preparations labelled by GFAP (green) and β-amyloid (red) illustrating differential changes in GFAP profiles in astrocytes in close association with Aβ plaques (**a**) and atrophic profiles of astrocytes (arrows) distant from β-amyloid deposits (**b**) in 3xTg-AD mice. **c, d** Confocal dual labelling images (GFAP in green and β-amyloid in red) showing the absence of reactive response of astrocytes in the entorhinal cortex of 3xTg-AD mice around perivascular vascular β-amyloid deposits (**c**) and β-amyloid plaques (**d**). Modified and reproduced with permission from [300, 338]

Early cognitive deficits are the very first symptoms of AD, which start to emerge decades before the occurrence of specific histopathology [55, 282]. Loss or impairment of cognitive capacities reflects reduced synaptic connectivity due to decreased synaptic function and synaptic loss [347]. Decrease in number of synapses indeed was found to be the earliest morphological change in AD [282]; and moreover the degree of synaptic loss correlates with the severity of dementia [61, 247]. Atrophy of astroglial perisynaptic processes may indeed underlie synaptic loss at the early stages of AD. Furthermore, astrocytes are fundamental for synaptogenesis and synaptic maintenance; furthermore, astroglial plasmalemmal transporters control local concentrations of ions and neurotransmitters, most notably glutamate, that may contribute to local excitotoxicity [73, 292, 303]. Astroglial asthenia also impairs metabolic support accomplished by lactate shuttle [213]. Astrocytes are also critical for maintaining normal neurotransmission by supplying neurones with glutamine that is an indispensable precursor for both glutamate and GABA. Impairment of all these fundamental functions associated with astroglial atrophy and loss of function may be considered as a primary cause for distorted synaptic connectivity and early cognitive deficits in AD [300, 306, 316].

11.5.5 Neurovascular Unit in AD

Clinical evolution of AD is almost invariably associated with vascular deficiency and pathologies of the blood–brain barrier [276, 277]. It is well documented that the blood flow is significantly reduced in the brains of patients with AD, with these vascular deficits being prominent already at the early stages of the disease [24, 345]. These functional deficits stem from substantial remodelling of vascularisation in the brains altered by AD pathology [74]. Brain vessels are controlled by both neuronal and astroglial inputs [112, 346]. Astrocytes are central integrating elements of neurovascular units that bridge brain parenchyma with local circulation. By secreting various agents astrocytes target pericytes, vascular smooth muscle cells and endothelial cells, thus contributing to functional hyperaemia and regulating the blood–brain barrier [191, 279, 346]. Astroglial atrophy as well as reactivity may differentially remodel the neurovascular unit, even that can occur at both early and late stages of the disease and can contribute to cognitive abnormalities and neuronal damage.

11.5.6 AD and Astroglial Metabolic Support

Metabolic deficiency of the brain is a common feature of AD. Functional brain imaging demonstrated a progressive loss of utilisation of glucose in patients with different stages of AD; deficits in brain metabolism are present already at the very early stages of the disease, having thus diagnostic significance [187, 229]. Exposure of cultured astrocytes to β-amyloid impairs cellular metabolism, although both decrease [209, 264] and increase [5] of glucose utilisation were detected. Likewise, both decrease [34, 160] and increase [31, 264] of the activity of glucose metabolism enzymes were described in post-mortem AD brains.

11.5.7 Deficient Astroglial Reactivity Defines Susceptibility of Brain Tissue to AD Pathology

Astroglial atrophy and asthenia in AD also lead to a loss of their defensive function [300]. As has been alluded previously, in experiments on 3xTg-AD mice, reactive astrocytes were accumulated around senile plaques and form perivascular β-amyloid deposits [203, 204]. These hypertrophic astrocytes are specifically associated with extracellular β-amyloid deposits, whereas astrocytes distant to the plaques remain atrophic (so in this sense astroglial atrophy emerges at the early stages of AD and is complimented by astrogliosis at later stages, when specific lesions develop). In contrast, in entorhinal and prefrontal cortices, extracellular β-amyloid accumulation does not trigger astrogliotic response (Fig. 11.7, [139, 338]) indicating failure of astroglial neuroprotection.

There are several lines of evidence demonstrating that reactive astrocytes are neuroprotective in the context of AD. For example, the Tg2576 mice (that harbour the APP_{Swe} mutation—see [110]) demonstrate early and prominent astroglial reactivity which correlates with relatively slow development of AD. Furthermore, senile plaques in these animals are resembling human β-amyloid deposit being represented by fleecy, granular, cored and diffused amyloid plaques [336], Incidentally, the Tg2576 mice display certain similarities with the prodromal stage of AD known in humans as mild cognitive impairment [16]. Astroglial capabilities to mount astrogliotic response change with age. The density of reactive astrocytes changes with age. In old Tg2576 mice, GFAP staining demonstrated prevalence of atrophic astrocytes with fewer reactive astroglial cells, which may be related with increased AD pathology in ageing [300]. Inhibition of reactive astrogliosis in the AD mouse model significantly increased β-amyloid load and exacerbated pathological progression [137].

The in vivo brain imaging of astrocytes uses PET detection of ^{11}C-deuterium-L-deprenyl (^{11}C-DED) that binds to MAO-B in the astrocytes [81]. When using a multi-tracer PET detecting ^{11}C-PIB (marker of fibrillar β-amyloid), ^{18}F-FDG (marker cerebral glucose metabolism) and ^{11}C-DED (marker of astrogliosis), the highest binding of ^{11}C-DED (which reflects maximal reactivity of astrocytes) was observed in patients with mild cognitive impairment (MCI) and high levels of fibrillar amyloid plaques in the brain (PIB+) reflecting prodromal AD [46]. Decrease in astroglial reactivity parallels the switch from MCI to full-blown AD with senile dementia again demonstrating the neuroprotective role of astrogliotic remodelling (Fig. 11.8 and [300]).

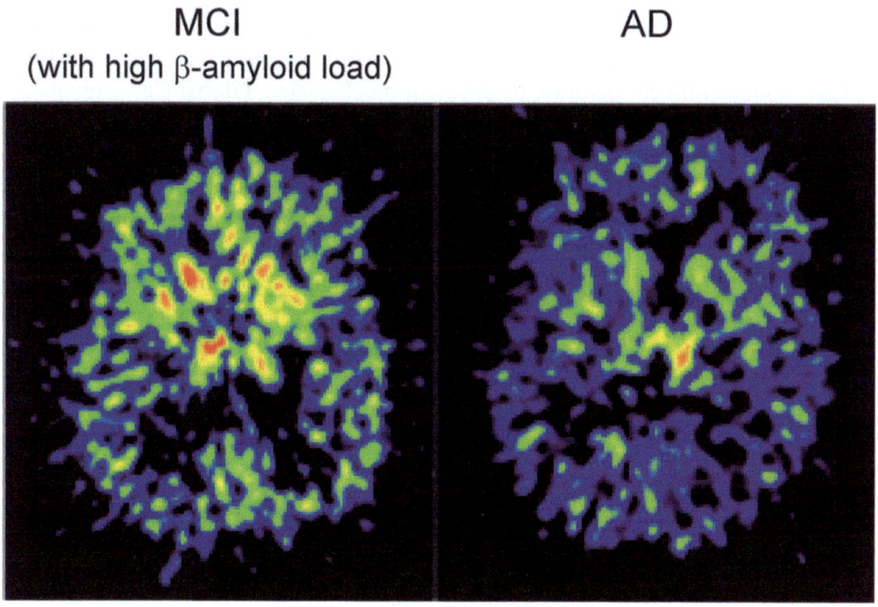

Fig. 11.8 Failure in astroglial reactivity defines the switch between mild cognitive impairment and senility in AD. Prominent astrogliosis in the brain of patient with mild cognitive impairment associated with high β-amyloid load (*Left panel*) in comparison with patient with Alzheimer's disease (*Right panel*). Representative images of ^{11}C-d-deprenyl binding (that reflects MAO-B expression in astrocytes) were obtained by position emission tomography. The MCI patient also showed high presence of fibrillar amyloid plaque as measured with ^{11}C-PIB (the status that could be identified as a prodromal AD). The PET scans show sagittal sections of the brain at the level of basal ganglia. Colour scale indicates red = very high, yellow = moderately high, green = high, blue = low ^{11}C-d-deprenyl binding. Photo courtesy of A. Nordberg, Karolinska institutet. Reproduced with permission from [300]

11.6 Astroglial Calcium Signalling in AD

11.6.1 Ionic Signalling as a Substrate of Astroglial Excitability

Astroglial excitability is based on spatially and temporally controlled fluctuations of intracellular concentration of ions, most notably of Ca^{2+} and Na^{+}, although recently the signalling role for K^{+} and Cl^{-} begun to be considered [130, 240, 246, 256, 313, 314, 328]. Astroglial Ca^{2+} signalling is the most studied; astroglial Ca^{2+} responses have been discovered in the late 1980s [71, 299], and are implicated in various signalling functions.

Physiological stimulation has been demonstrated to trigger Ca^{2+} signals in astrocytes in vitro, in situ and in vivo [22, 66, 129, 131, 256]. Astroglial calcium signalling has a spatio-temporal hierarchical organisation: at the cellular level, astrocytic Ca^{2+}

signals are classified into local Ca^{2+} microdomains, intracellular propagating waves, global Ca^{2+} signals and Ca^{2+} oscillations [95, 196, 259, 265]. These distinct forms of Ca^{2+} signals reflect operation of different mechanisms. Global Ca^{2+} signals and propagating Ca^{2+} waves originate from Ca^{2+} release from the endoplasmic reticulum Ca^{2+} store; this release is primarily mediated by inositol 1,4,5 trisphosphate ($InsP_3$) receptor type 2, ($InsP_3R2$). Local Ca^{2+} microdomains in contrast are often generated by Ca^{2+} entry through ionotropic receptors, transient receptor potential channels, store-operated Ca^{2+} entry (SOCE) or reversed Na^+/Ca^{2+} exchanger [305]. Astroglial Ca^{2+} signals regulate several cellular processes, including secretion, metabolism and astroglial reactivity. Astroglial Na^+ signalling is much less characterised, although basic parameters of Na^+ transients evoked by physiological stimulation have been described in experiments in cultured cells and in astrocytes in brain slices [86, 127, 128, 148–150, 227, 344]. Astroglial Na^+ signals regulate many homeostatic plasmalemmal transporters, thus coordinating neuronal activity with astroglial support [130, 240].

11.6.2 Aberrant Calcium Signalling in AD

The fundamental role of Ca^{2+} in regulation of cellular survival and cell death inspired the "calcium hypothesis of ageing and neurodegeneration" formulated by Zaven Khachatirian [125] who based this hypothesis on experimental studies of Philipp W. Landfield [146, 147]. This Ca^{2+} hypothesis posits that ageing neurones experience increased Ca^{2+} influx during depolarisation, which elevates cytosolic Ca^{2+} concentration ($[Ca^{2+}]_i$), thus triggering excitotoxicity. Subsequent studies revealed that physiological neuronal ageing is associated with much subtle alterations of neuronal Ca^{2+} extrusion, which, although capable of handling normal Ca^{2+} loads, fail to clear excessive Ca^{2+} influx. This deficit in Ca^{2+} handling stipulates higher vulnerability of old neurones to the periods of high activity [286, 287, 312, 334]. In neurodegenerative diseases (including AD), Ca^{2+} homeostatic machinery is, however, seriously compromised, and hence these disorders have been regarded as "chronic calciumopathies" [273, 274]. Almost nothing is known about changes in Ca^{2+} homeostatic machinery, resting Ca^{2+} handling, and Ca^{2+} signalling in aged astrocytes. There are several isolated reports demonstrating a decrease in evoked astrocytic Ca^{2+} signals in mice aged 16–21 months, when compared to adult animals [144, 145].

11.6.3 Exposure to β-Amyloid Disturbs Astroglial Ca^{2+} Signalling

Whether β-amyloid is indeed a causal factor in AD or a mere epiphenomenon, exposure to it affects astroglial Ca^{2+} dynamics. Experimental studies in vitro in primary astroglial cultures demonstrated acute effects of β-amyloid on Ca^{2+} signalling. Resting $[Ca^{2+}]_i$ significantly (2–3 times) increased in astrocytes exposed to β-amyloid (in concentrations ranging between 100 nM and 5 μM) for several hours [103, 161]. These findings, however, have not been universally confirmed; several investigations found that incubations of cultured astrocytes with 100–200 nM of β-amyloid (or its toxic fragment β-amyloid$_{25\text{-}35}$) for 48–72 h did not significantly change resting $[Ca^{2+}]_i$ [47, 288].

Acute exposure to β-amyloid triggered oscillations of $[Ca^{2+}]_i$ transients in cultured astrocytes and in astrocytes in organotypic slices [2–4, 52, 114, 163]. These acute effects, however, were not always observed and several studies have not noticed such acute effects [47, 288]. Treatment of cultured astrocytes with 1 μM of β-amyloid$_{1\text{-}40}$ induced $[Ca^{2+}]_i$ elevations only in 17% of all the cells, whereas application of β-amyloid$_{25\text{-}35}$ triggered Ca^{2+} signals in 36% of all astrocytes [114]. In primary cultured rat newborn astrocytes, application of 1 μM of β-amyloid$_{25\text{-}35}$ induced $[Ca^{2+}]_i$ transients in 27% of primary cultured rat newborn astrocytes; at 2–5 μM ~60% of astrocytes responded with $[Ca^{2+}]_i$ transients [270]. Of note, low concentrations of β-amyloid apparently stimulate astroglial Ca^{2+}-permeable α7 nicotinic cholinoreceptors, which resulted in Ca^{2+} influx and generation of Ca^{2+} responses [154, 216].

11.6.4 Pathological Ca^{2+} Signalling in AD Astrocytes In Vitro

Analysis of $[Ca^{2+}]_i$ dynamics in astrocytes isolated from several mouse models of AD also demonstrated aberrant Ca^{2+} signalling [162, 163]. Abnormally large Ca^{2+} signals have been detected in astrocytes isolated from newborn 3xTg-AD mice, indicating intrinsic alterations of Ca^{2+} homeostatic cascades [239]. Astrocytes isolated from 3xTg-AD mice in particular showed increased store-operated calcium entry (SOCE) [239]. Cultures of astrocytes isolated from 3xTg-AD animals also demonstrated aberrant kinetics of ATP-induced Ca^{2+} signals and Ca^{2+} oscillations [269]. Further analysis revealed that these aberrations are most likely associated with expression of mutant PS1 presenilins residing in the endoplasmic reticulum [269]. In the Tg5469 AD mouse which over-expressed APP, the SOCE was not changed, whereas deletion of APP caused an inhibition of store-operated Ca^{2+} entry [164]. This inhibition may be associated with down-regulation of expression of either TRPC1 or Orai 1 channels.

11.6.5 Pathological Ca^{2+} Signalling in Astrocytes In Vivo

Imaging astroglial $[Ca^{2+}]_i$ dynamics in vivo in the brains of AD animal models reliably demonstrated aberrant, hyperactive $[Ca^{2+}]_i$ dynamics, which is fundamentally similar to neuronal hyperexcitability routinely observed in AD-like experimental pathology [350]. Aberrant hyperactive $[Ca^{2+}]_i$ oscillations have been observed in reactive astrocytes associated with senile plaques. High levels of resting $[Ca^{2+}]_i$, pathological Ca^{2+} oscillations and long-projecting propagating Ca^{2+} waves have been identified in plaque-associated astrocytes in the brains of APP/PS1 mice [138]. Emergence of astroglial Ca^{2+} hyperactivity was also suggested to be linked with abnormal purinergic signalling in reactive astrocytes. There are claims that reactive astrocytes release excessive amounts of ATP through connexin hemichannels. This ATP, acting in autocrine manner, activates astroglial P2Y purinoceptors, which mediate pathological Ca^{2+} signalling [62]. An increased frequency of astroglial Ca^{2+} oscillations was also observed in AD animals in the pre-plaque stage, and these abnormal $[Ca^{2+}]_i$ dynamics coincided with the instability of vascular tone probably indicating that astrocytes in their ability to regulate local blood flow [278].

11.6.6 Astroglial Ca^{2+} Signalling Toolkit Is Remodelled in AD

The AD as a chronic pathology leads to a substantial remodelling of astroglial Ca^{2+} signalling toolkit. Chronic exposure of cultured astrocytes to β-amyloid as well in vivo AD pathology (in model animals) changes expression of various components of Ca^{2+} homeostatic/signalling system; these molecules include, for example, metabotropic and ionotropic receptors, intracellular Ca^{2+} channels, store-operated Ca^{2+} channels and Ca^{2+} sensors [162, 163, 309].

Exposure of astroglial cultures to 10–30 μM β-amyloid$_{1\text{-}40}$ for 48–72 h resulted in an increase of the amplitude of $[Ca^{2+}]_i$ transients in response to stimulation of metabotropic glutamate receptor mGluR5. This augmentation of metabotropic Ca^{2+} signalling was a consequence of an up-regulated expression of mGluR5 detected at both mRNA and protein levels [47]. This was corroborated in another series of experiments which demonstrated that 24–72 h exposure of cultured astrocytes to 100 nM–20 μM of oligomeric β-amyloid increased expression of mGluR5 [93, 94, 161]. This up-regulation of mGluR5 expression was suppressed by the inhibitors of calcineurin and Nf-κB (nuclear factor κ-light-chain-enhancer of activated B cells) [161]. Similar increase in expression of mGluR5 was detected in astrocytes in the animal AD models and in post-mortem human tissues. Increased levels of mGluR5 protein were found in the post-mortem hippocampal preparations obtained from AD patients at advanced (Braak V–VI) stages of the disease [47, 161]. Incubation of astrocytes with nanomolar (0.1–100 nM) concentrations of β-amyloid$_{1\text{-}42}$ for 24–72 h increased the expression of several subunits of nicotinic cholinoreceptors including α7nAChR, α 4nAChR and β2nAChR [335]. Similarly, increased levels of α7nAChR were identified in the post-mortem brain tissue of patients with sporadic AD and familial AD associated with the Swedish APP mutation [341].

Another important class of molecules affected by AD progression is represented by intracellular Ca^{2+} release channels. Exposure of cultured astrocytes to 125 nM of Tat-ProADAM10$_{709\text{-}729}$ peptide (this peptide inhibits production of β-amyloid$_{1\text{-}40}$ and β-amyloid$_{1\text{-}42}$) for 72 h leads to an increased expression of $InsP_3R1$ [93]. Similarly, up-regulation of expression of $InsP_3R1$ and $InsP_3R2$ mRNA was detected in astrocytes in vitro which were exposed to 100 nM oligomeric β-amyloid$_{1\text{-}42}$ [161]. Pathological remodelling of Ca^{2+} homeostatic and signalling cascades differ between different brain regions. Expression of InsP3R1 is increased in healthy hippocampal astrocytes exposed to β-amyloid, but remains unchanged in astrocytes from the entorhinal cortex [94]. However, β-amyloid did not affect expression of $InsP_3R1$ in astrocytes isolated from 3xTg-AD animals, indicating that exogenous β-amyloid and over-expression of mutated AD-related genes share common molecular pathways that cause deregulation of Ca^{2+} homeostasis. In post-mortem studies, however, generalised decrease in the expression of $InsP_3Rs$ was detected in all brain regions including frontal, parietal and entorhinal cortices and the hippocampus [102, 141, 340]. These studies did not, however, discriminate between cell types. Many other components of Ca^{2+} signalling system are affected by AD; these include components calpain-10 [85], NFAT (Nuclear factor of activated T-cells) [1], NF-κB [93], calcineurin [93, 199], L-type calcium channels [59] and store-operated Ca^{2+} channels [239]. All in all 32 genes associated with Ca^{2+} signalling were found to be affected in the transcriptome of astrocytes microdissected from patients with different Braak stages of AD. It appeared that expression of several isoforms of calmodulin kinase CaMKII, two isoforms of calmodulin, plasma membrane Ca^{2+}-ATPases, ryanodine receptors and InsP3Rs, was decreased at advanced (Braak V–VI) stage when compared with early (Braak I–II) stage of the disease [262].

11.6.7 Ca^{2+} Release and Astroglial Reactivity

As has been alluded before, astrogliosis is a prominent component in certain brain regions in the context of AD; reactive astrocytes associate themselves with senile plaques in human tissue and in the brains of AD animal models, arguably forming a defensive barrier protecting neural networks [106, 306]. Suppression of astrogliotic response (for instance, by genetic deletion of GFAP and vimentin) exacerbates β-amyloid load and facilitates plaques dissemination [137]. Astroglial reactivity, however, is different in different regions of the brain. Prominent astroglial reactivity is observed in the hippocampus, whereas the emergence of senile plaques and β-amyloid depositions does not trigger astrogliosis in entorhinal and prefrontal cortices of AD mice models. Underlying molecular mechanisms might be linked to a deficient Ca^{2+} signalling in astrocytes from different brain regions.

In the AD context, one of the most relevant signals instigating astroglial reactivity is β-amyloid, and indeed exposure of astroglia to β-amyloid in vitro or in situ triggers astrogliosis [4, 306]. As has been described above, β-amyloid also evokes $[Ca^{2+}]_i$ elevation. It appears that β-amyloid-induced Ca^{2+} signals originate from Ca^{2+} release from the endoplasmic reticulum Ca^{2+} store and these Ca^{2+} signals are directly linked to the initiation of astrogliotic response. Suppression of Ca^{2+} release from the ER with pharmacological tools effectively inhibits astrogliosis induced by β-amyloid in both cultured astrocytes and astroglia in organotypic slices [4]. The causal role of

Ca^{2+} release in astroglial reactivity was directly demonstrated: deletion of $InsP_3R2$ effectively suppressed astrogliotic activation [120]. Sensitivity of astrocytes from different brain regions to β-amyloid is different: β-amyloid up-regulates expression of molecules providing for ER Ca^{2+} release in hippocampal but not in entorhinal astrocytes [94]. This may explain the absence of astrogliotic defensive response in astrocytes from cortical regions, which renders these parts of the brain vulnerable to the AD pathology [162, 300].

11.7 AD Pathology Affects Astroglial Vesicular Trafficking and Secretion

Astrocytes are secretory cells, being a part of CNS-wide "gliocrine" system [301]. Astrocytes are known to secrete ~200 molecules, many of which are released through exocytosis of secretory vesicles [349]. Intracellular astroglial vesicles are also fundamental for delivery of various molecules [298] such as ion channels, membrane receptors and transporters, as well as major histocompatibility complex II (MHC-II, [296]) and EAAT 2 [266], to the plasma membrane. Vesicular traffic is controlled by sophisticated molecular cascades, which in turn are regulated by increases in $[Ca^{2+}]_i$ [220, 266]. Changes in $[Ca^{2+}]_i$ differentially regulate motility of distinct vesicles types. Increases in $[Ca^{2+}]_i$ reduce the motility of vesicles carrying peptides, such as atrial natriuretic peptide, while accelerating motility of vesicles containing vesicular L-glutamate transporter VGLUT1 [218–220]. Proteolytic enzymes stored in endolysosomes may contribute to the development of AD. One of these proteases is represented by the insulin-degrading enzyme (IDE), which, when secreted into the extracellular space, may degrade β-amyloid. Astrocytes are the main cell type which produces and releases IDE [68, 263]. It has been hypothesised that in AD the capacity of secreting IDE is reduced, leading to an increase in β-amyloid, which involves a reduction in autophagy-based lysosomal secretion of IDE [263].

Astrocytes from 3xTg-AD mice demonstrated an aberrant vesicular traffic. Spontaneous mobility of peptidergic and endolysosomal vesicles as well as the ATP-evoked, Ca^{2+}-dependent, vesicle mobility was all diminished in AD astrocytes (Fig. 11.9). Transfection of healthy rat astrocytes to express familial AD-associated mutated presenilin 1 (PS1M146V) caused very similar impairment of peptidergic vesicle trafficking. The stimulation-dependent peptide secretion from single vesicles was less efficient in 3xTg-AD and PS1M146V-expressing astrocytes than in healthy controls. The impaired vesicle dynamics and reduced evoked secretion of the signalling peptides both may contribute to the development of AD [269].

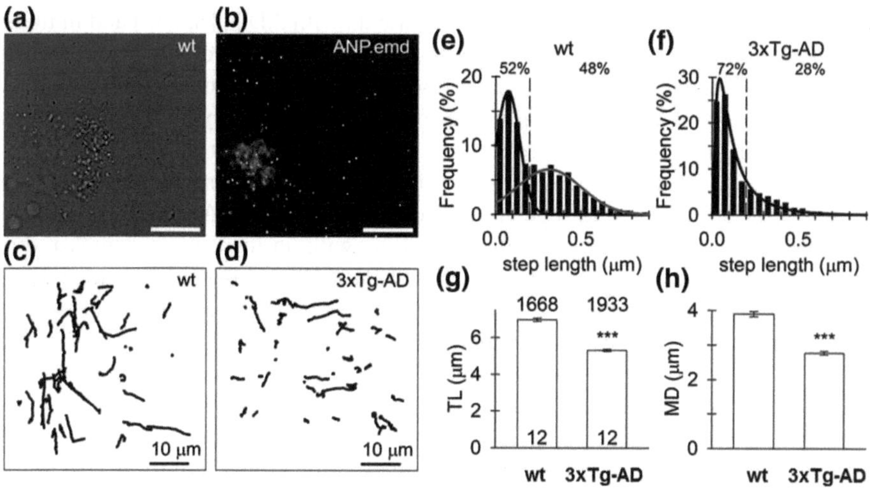

Fig. 11.9 Decreased spontaneous mobility of peptidergic vesicles in 3xTg-AD astrocytes. **a** Live cultured wild-type (wt) astrocyte under DIC optics and **b** the confocal image of the same cell expressing fluorescent peptide atrial natriuretic peptide-emerald green (ANP.emd), stored in individual vesicles, observed as bright fluorescent puncta; scale bars, 10 μm. **c** Vesicle tracks (N = 50) obtained in a 15-s epoch of imaging representative control (wt) and **d** 3xTg-AD astrocytes expressing ANP.emd, respectively. Note less elongated vesicle tracks in the 3xTg-AD astrocyte. **e**, **f** Frequency histogram of the step length in spontaneously moving vesicles in wt ($N = 5025$, **e**) and 3xTg-AD ($N = 5072$, **f**) astrocytes. The data were fitted with the function $f = a \times \exp(-0.5 \times (x/x_0)/b)^2/x$, where $a = 17.88 \pm 0.00$, $b = 0.07 \pm 0.00\ \mu m^{-0.5}$, $x_0 = 0.07 \pm 0.00\ \mu m$ (black curve) and $a = 6.53 \pm 0.13$, $b = 0.19 \pm 0.01\ \mu m^{-0.5}$, $x_0 = 0.31 \pm 0.01\ \mu m$ (grey curve) in wt astrocyte, and with the function $f = a \times \exp(-0.5 \times (\ln x/x_0)/b)^2/x$, where $a = 1.96 \pm 0.04$, $b = 0.92 \pm 0.02\ \mu m^{-0.5}$, $x_0 = 0.10 \pm 0.00\ \mu m$ (black curve) in 3xTg-AD astrocyte. The vertical dashed line indicates the step length of 0.2 μm obtained close to the intersection of distributions (black and grey curve) in wt astrocytes to discriminate small (<0.2 μm) from large (≥0.2 μm) steps. Note the higher proportion (%) of smaller steps lengths in the 3xTg-AD astrocyte indicated by the absence of the second mode distribution seen in wt astrocytes. **g** Track length (TL), **h** maximal displacement (MD),note substantially diminished TL, MD in 3xTg-AD astrocytes. The numbers above the top of the bars (mean ± SEM) indicate the number of vesicles analysed; the numbers at the bottom of the bars indicate the number of cells analysed; "***"—indicates p values < 0.001. Modified with permission from [269]

11.8 GABAergic Astrocytes in AD

In the healthy young CNS, astrocytes contribute to GABAergic transmission through (i) supplying glutamine, needed for GABA biosynthesis in neuronal terminals and (ii) removing ~20% of all released GABA by dedicated plasmalemmal transporters GAT-1 and GAT-3. After being transported into the astrocytes, most of GABA is catabolised by GABA transaminase (GABA-T) to succinate, which is subsequently utilised for production of ATP [253, 305, 319]. Due to this energy-oriented catabolism, the concentration of GABA in the cytosol of astrocytes is rather low. Ageing and neurodegeneration, however, significantly affect astroglial GABA metabolism; concentration of GABA in astrocytes in elderly [155], in patients with AD [118, 332] and in transgenic AD models [41, 118, 332], is significantly higher. This increase is particularly prominent in reactive astrocytes associated with senile plaques in AD animal models; intracellular GABA concentration in these AD reactive astrocytes is several times higher than in age-matched controls and is very similar to neuronal GABA content [41, 118, 332]. These changes in astrocytic GABA content in reactive astrocytes are accompanied with an up-regulation of expression of GABA producing enzyme glutamic acid decarboxylase GAD67 as well as with an increase in expression of astroglia-specific monoaminoxidase-B (MAO-B) [118]. At the same time, expression of glutamine synthetase is specifically down-regulated in reactive astrocytes surrounding senile plaques in the hippocampus and prefrontal cortex of 3xTg-AD mice (Fig. 11.10 and [204]). Thus reactive astrocytes acquire machinery to synthetase GABA either from glutamate (through GAD67 and increased glutamate availability due to the loss of glutamine synthetase) or from putrescine through MAO-B pathway [84]. Furthermore, there is an increased glutamatergic neuronal activity around senile plaques [206]; this conceivably increases astroglial glutamate uptake and availability of cytosolic glutamate for conversion to GABA cytosolic glutamate concentration glutamate transport into astrocytes Astroglial GABA may potentially be released from astrocytes by diffusion through Bestrophin-1 Cl^- channels or through reversed GAT3 transporters (Fig. 11.11, [84]). The emergence of GABAergic astrocytes may represent yet another defensive response; as GABA release from astroglia may counteract neuronal hyperexcitability by an increase of tonic inhibition [84].

11.9 Astrocytes as Therapeutic Targets in AD

Neuroglia is yet to be considered as a fundamental target for novel therapeutic agents for neurological disorders and neurodegenerative diseases in particular. It is conceivable that by modulating the status of astrocytes, by reversing or halting astrocytes degeneration and asthenia or by modulating astroglial reactivity, the course of AD can be altered and the disease can be delayed or cognitive alterations reversed. Several possible strategies that may affect astroglial pathology have recently emerged.

11.9.1 Lifestyle Changes May Reverse Astrodegeneration

Recent experiments have demonstrated that environmental modifications such as sensory stimulation, dieting or usage of food supplements may affect AD progression and at the same time change astroglia morphology and revert astroglial atrophy. Experiments on APP and 3xTg-AD mice models revealed that chronic exposure of these animals to physical activity and/or to enriched environment reversed morphological atrophy of astrocytes, increased GFAP expression and normalised GFAP-positive astroglial profiles (Fig. 11.12); most importantly, these astroglia-specific

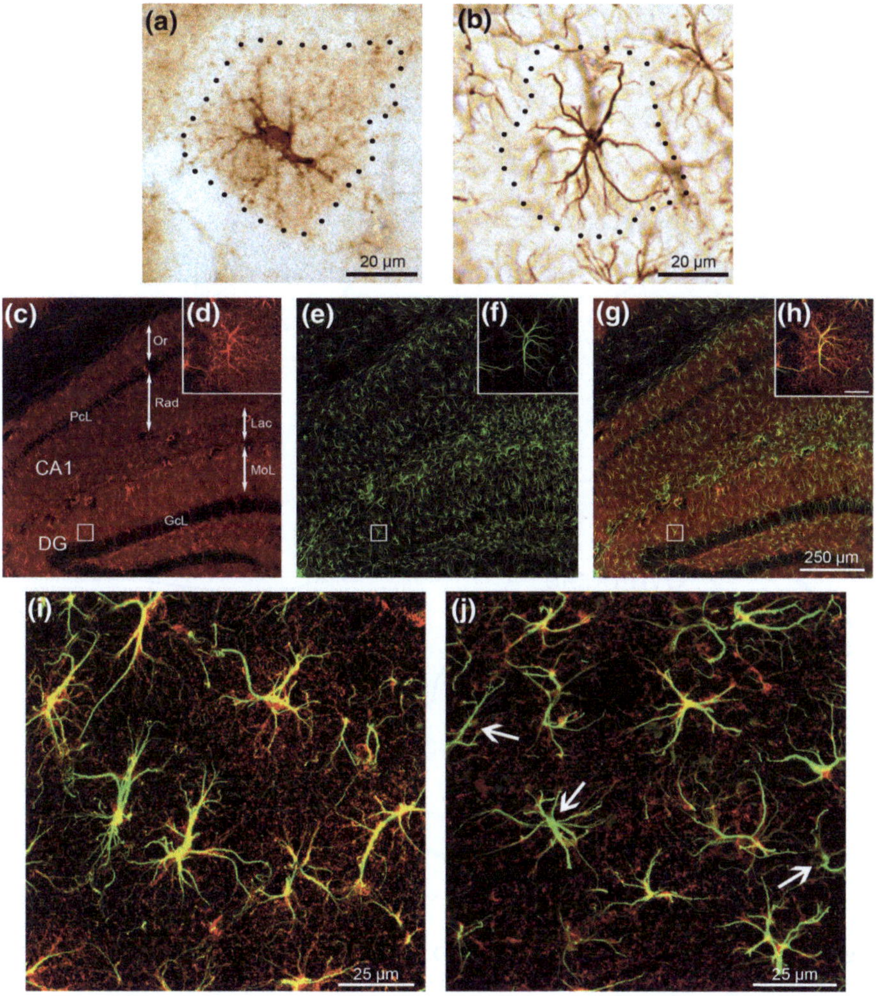

Fig. 11.10 Down-regulation of glutamine synthetase (GS) expression in hippocampal astrocytes in 3xTg-AD mice. **a**, **b** Light microscopy images of GS—(**a**) and GFAP—(**b**) positive astrocytes. **c**, **e**, **g** Confocal images of hippocampal preparation labelled for GS (**c**, red), GFAP (**e**, green) and their co-localisation (**g**, yellow). **d**, **f**, **h** High magnification confocal images illustrating the co-expression of GS and GFAP. **i**, **j** Ubiquitous co-expression of GS and GFAP in wild-type control mice (**i**) and down-regulation of GS expression (astrocytes lacking GS are indicated by arrows) in 3xTg-AD mice (**j**). DG, dentate gyrus; GcL, granule cell layer; MoL, molecular layer; Lac, stratum lacunosum moleculare; Or, stratum oriens; PcL, pyramidal layer; Rad, stratum radiatum. Modified and reproduced with permission from [204]

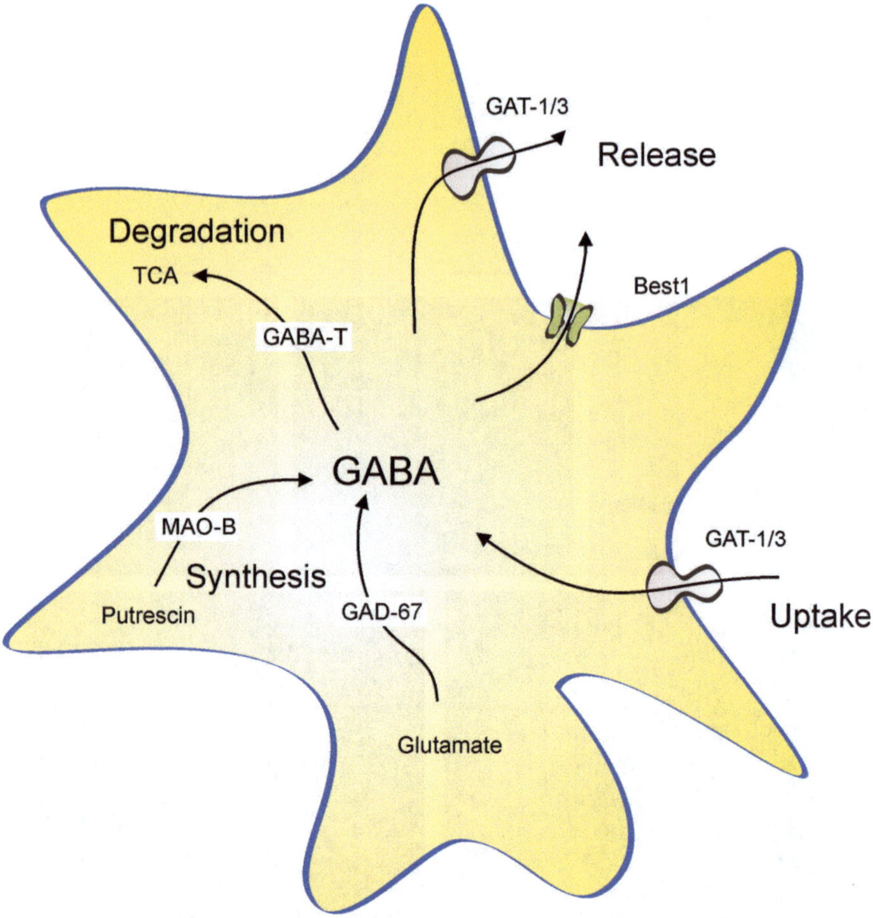

Fig. 11.11 GABAergic reactive astrocytes in AD. See text for explanation. Abbreviations: GAT1/3 GABA transporters 1 (SLC6A1) and 3 (SLC6A11); Best1—bestrophin 1 anion channel 1; GABA-T—GABA transaminase; TCA—tricarboxylic acid (Krebs) cycle; MAO-B—Monoamine oxidase B; GAD67—glutamate decarboxylase. Modified from [84]

changes developed in parallel to a decrease in β-amyloid load [20, 236]. Incidentally, environmental stimulation also improved neurogenesis which is impaired in the AD [231, 233]. Another AD model, the 5xFAD mice chronically treated with polyunsaturated fatty acid 2-hydroxy-docosahexaenoic acid similarly rescued astroglial atrophy, restored adult neurogenesis and improved cognitive performance [76]. Treatment with specific diets may also affect ageing and AD progression. It is well appreciated that caloric restriction exerts prominent positive effect of lifespan of several species and may boost cognitive resilience of the brain [80, 169, 170, 175]. It appeared that caloric restriction induces growth of astroglial perisynaptic processes, thus extending synaptic coverage, preventing glutamate spillover, improving K^+ buffering and glu-

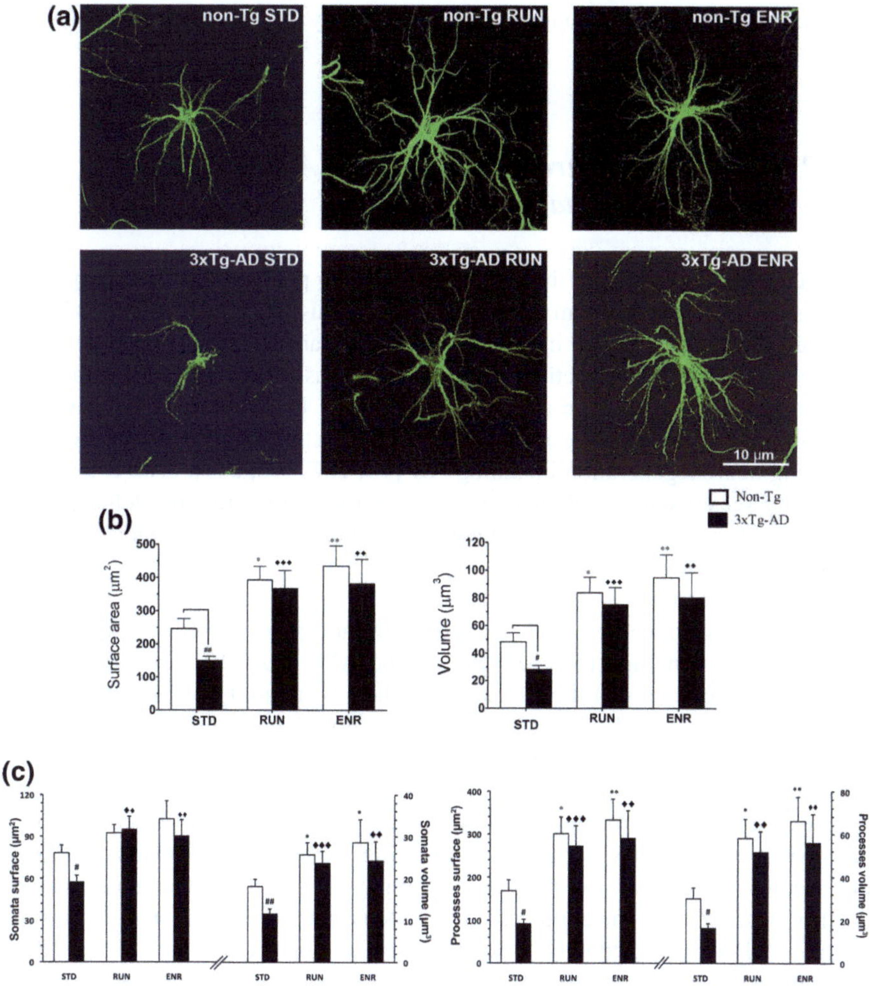

Fig. 11.12 Environmental stimulation (enriched environment, ENR and physical activity, RUN) reverse morphological atrophy of astrocytes seen in the dentate gyrus isolates from 3xTg-AD mice. GFAP-immunoreactivity of astrocytes in the DG of non-Tg and 3xTg-AD animals housed in different conditions. **a** High magnification of representative confocal micrographs showing the astrocytic morphology in mice housed in standard conditions (STD), RUN and ENR. Scale bars, 10 μm. Note the morphological changes of the astrocytes from both genotypes induced by the different living conditions. **b** Histograms showing difference of surface area and volume of GFAP-positive astrocytes in the DG of non-Tg and 3xTg-AD mice housed under different housing conditions. **c** Histograms showing differences in surface area and volume of GFAP-immunoreactivity of astrocytic cell bodies and processes detected between non-Tg and 3xTg-AD mice housed under different housing conditions. Bars represent means ± S.E.M., #$p < 0.05$, ##$p < 0.01$ compared with non-Tg animals in same housing environment; *$p < 0.05$, **$p < 0.01$ compared with non-Tg mice housed under STD; ••$p < 0.01$ and •••$p < 0.001$ compared with 3xTg-AD mice housed under STD. Reproduced with permission from [236]

tamate uptake from the synaptic cleft, thus ultimately enhancing synaptic plasticity [217].

11.9.2 Preventing Neurodegeneration by Adrenergic Astroglial Excitation

Noradrenergic innervation of the CNS is provided by projections of adrenergic neurones localised in the brainstem nucleus locus coeruleus. This small nucleus is located near the fourth ventricle and, in humans, comprises around 50,000 neurones [188]. Diffuse innervation by projections of locus coeruleus neurones reaches practically all parts of the brain and the spinal cord [25]. The locus coeruleus neurones are vulnerable to oxidative stress; apparently, they are lost in ageing and they are first to die during neurodegeneration including AD [75, 167, 190, 249]. Astrocytes, being universally sensitive to noradrenaline, represent the major target for deficient noradrenergic innervation and interfering with astroglial adrenergic mechanisms may be therapeutically relevant [348].

Astroglial sensitivity to noradrenaline, released from locus coeruleus neuronal projections, is mediated by both α- and β-adrenoceptors linked, respectively, to cytosolic Ca^{2+} signalling [66, 131] and cyclic AMP (cAMP) cascades [297]. In the in vivo experiments in the awake mice, the vast majority of astrocytes generated synchronous $[Ca^{2+}]_i$ signals in response to noradrenaline released from locus coeruleus projections [22, 66, 210]; of note neurones did not generate Ca^{2+} responses to the same stimulation [208]. This difference reflects upon much higher density of adrenoceptors in astrocytes when compared to neurones [10]. Degeneration of locus coeruleus neurones associated with ageing most certainly impairs adrenoceptors–medicated astroglial excitability, which may be linked to the cognitive decline [348]. Consequently, preventing death of locus coeruleus neurones or boosting astroglial adrenergic excitability may represent a valid therapeutic strategy [348]. Alternative possibilities may involve drugs, such as deprenyl, that limit noradrenaline catabolism in astrocytes.

Transcranial direct current stimulation (tDCS) was used with positive effects including memory enhancements, accelerated motor function rehabilitation, alleviation of depressive symptoms and decelerated progression of cognitive impairments in AD patients [140, 197]. The mechanism of action of tDCS is astroglial and noradrenergic. It has been revealed that tDCS induces a massive increase in astroglial $[Ca^{2+}]_i$ which has been suppressed by the ablation of noradrenergic neurones or by the inhibition of α_1-adrenoceptors [186]. Alternative possibilities may involve drugs that limit noradrenaline catabolism in astrocytes such as, for example, deprenil.

11.10 Conclusions

Astroglial contributions to the pathophysiology of AD are complex and range from early astroglial atrophy, which limits homeostatic support and may cause synaptic weakness and early cognitive decline, to astroglial reactivity, which seems to protect the CNS against AD-associated pathology and limits the spread of β-amyloid load. Astroglia may also undergo pathological remodelling, in which astrocytes may acquire new functions such as, for example, secreting GABA. Specific manipulations with astroglia may represent a valid therapeutic approach for treating neurodegenerative disorders including AD.

Acknowledgments VP's work is supported by a grant from the National Institute of General Medical Sciences of the National Institutes of Health (R01GM123971). VP is an Honorary Professor at University of Rijeka, Croatia.

References

1. Abdul HM, Sama MA, Furman JL, Mathis DM, Beckett TL, Weidner AM, Patel ES, Baig I, Murphy MP, LeVine H 3rd, Kraner SD, Norris CM (2009) Cognitive decline in Alzheimer's disease is associated with selective changes in calcineurin/NFAT signaling. J Neurosci 29:12957–12969
2. Abramov AY, Canevari L, Duchen MR (2003) Changes in intracellular calcium and glutathione in astrocytes as the primary mechanism of amyloid neurotoxicity. J Neurosci 23:5088–5095
3. Abramov AY, Canevari L, Duchen MR (2004) β-Amyloid peptides induce mitochondrial dysfunction and oxidative stress in astrocytes and death of neurons through activation of NADPH oxidase. J Neurosci 24:565–575
4. Alberdi E, Wyssenbach A, Alberdi M, Sanchez-Gomez MV, Cavaliere F, Rodriguez JJ, Verkhratsky A, Matute C (2013) Ca2+-dependent endoplasmic reticulum stress correlates with astrogliosis in oligomeric amyloid β-treated astrocytes and in a model of Alzheimer's disease. Aging Cell 12:292–302
5. Allaman I, Gavillet M, Belanger M, Laroche T, Viertl D, Lashuel HA, Magistretti PJ (2010) Amyloid-β aggregates cause alterations of astrocytic metabolic phenotype: impact on neuronal viability. J Neurosci 30:3326–3338
6. Allen B, Ingram E, Takao M, Smith MJ, Jakes R, Virdee K, Yoshida H, Holzer M, Craxton M, Emson PC, Atzori C, Migheli A, Crowther RA, Ghetti B, Spillantini MG, Goedert M (2002) Abundant tau filaments and nonapoptotic neurodegeneration in transgenic mice expressing human P301S tau protein. J Neurosci 22:9340–9351
7. Alzheimer A (1907) Über eine eigenartige Erkrankung der Hirnrinde. Allg Z Psychiat Psych-Gericht Med 64:146–148
8. Alzheimer A (1910) Beiträge zur Kenntnis der pathologischen Neuroglia und ihrer Beziehungen zu den Abbauvorgängen im Nervengewebe. In: Nissl F, Alzheimer A (eds) Histologische und histopathologische Arbeiten über die Grosshirnrinde mit besonderer Berücksichtigung der pathologischen Anatomie der Geisteskrankheiten. Jena, Gustav Fischer, pp 401–562
9. Andorfer C, Kress Y, Espinoza M, de Silva R, Tucker KL, Barde YA, Duff K, Davies P (2003) Hyperphosphorylation and aggregation of tau in mice expressing normal human tau isoforms. J Neurochem 86:582–590

10. Aoki C (1992) β-adrenergic receptors: astrocytic localization in the adult visual cortex and their relation to catecholamine axon terminals as revealed by electron microscopic immunocytochemistry. J Neurosci 12:781–792
11. Apelt J, Ach K, Schliebs R (2003) Aging-related down-regulation of neprilysin, a putative β-amyloid-degrading enzyme, in transgenic Tg2576 Alzheimer-like mouse brain is accompanied by an astroglial upregulation in the vicinity of β-amyloid plaques. Neurosci Lett 339:183–186
12. Arendash GW, Lewis J, Leighty RE, McGowan E, Cracchiolo JR, Hutton M, Garcia MF (2004) Multi-metric behavioral comparison of APPsw and P301L models for Alzheimer's disease: linkage of poorer cognitive performance to tau pathology in forebrain. Brain Res 1012:29–41
13. Arendt T (1994) Impairment in memory function and neurodegenerative changes in the cholinergic basal forebrain system induced by chronic intake of ethanol. J Neural Transm Suppl 44:173–187
14. Arranz AM, De Strooper B (2019) The role of astroglia in Alzheimer's disease: pathophysiology and clinical implications. Lancet Neurol 18:406–414
15. Arranz B, Blennow K, Ekman R, Eriksson A, Mansson JE, Marcusson J (1996) Brain monoaminergic and neuropeptidergic variations in human aging. J Neural Transm (Vienna) 103:101–115
16. Ashe KH, Zahs KR (2010) Probing the biology of Alzheimer's disease in mice. Neuron 66:631–645
17. Barthelemy I, Martineau D, Ong M, Matsunami R, Ling N, Benatti L, Cavallaro U, Soria M, Lappi DA (1993) The expression of saporin, a ribosome-inactivating protein from the plant Saponaria officinalis, in Escherichia coli. J Biol Chem 268:6541–6548
18. Bartus RT, Dean RL 3rd, Beer B, Lippa AS (1982) The cholinergic hypothesis of geriatric memory dysfunction. Science 217:408–414
19. Beach TG, McGeer EG (1988) Lamina-specific arrangement of astrocytic gliosis and senile plaques in Alzheimer's disease visual cortex. Brain Res 463:357–361
20. Beauquis J, Pavia P, Pomilio C, Vinuesa A, Podlutskaya N, Galvan V, Saravia F (2013) Environmental enrichment prevents astroglial pathological changes in the hippocampus of APP transgenic mice, model of Alzheimer's disease. Exp Neurol 239:28–37
21. Beauquis J, Vinuesa A, Pomilio C, Pavia P, Galvan V, Saravia F (2014) Neuronal and glial alterations, increased anxiety, and cognitive impairment before hippocampal amyloid deposition in PDAPP mice, model of Alzheimer's disease. Hippocampus 24:257–269
22. Bekar LK, He W, Nedergaard M (2008) Locus coeruleus α-adrenergic-mediated activation of cortical astrocytes in vivo. Cereb Cortex 18:2789–2795
23. Bekris LM, Yu CE, Bird TD, Tsuang DW (2010) Genetics of Alzheimer disease. J Geriatr Psychiatry Neurol 23:213–227
24. Bell RD, Zlokovic BV (2009) Neurovascular mechanisms and blood-brain barrier disorder in Alzheimer's disease. Acta Neuropathol 118:103–113
25. Benarroch EE (2009) The locus ceruleus norepinephrine system: functional organization and potential clinical significance. Neurology 73:1699–1704
26. Berchtold NC, Cotman CW (1998) Evolution in the conceptualization of dementia and Alzheimer's disease: Greco-Roman period to the 1960s. Neurobiol Aging 19:173–189
27. Berrios GE (1990) Alzheimer's disease: A conceptual history. Int J Geriatric Psychiat 5:355–365
28. Bertoni-Freddari C, Giuli C, Pieri C, Paci D (1986) Quantitative investigation of the morphological plasticity of synaptic junctions in rat dentate gyrus during aging. Brain Res 366:187–192
29. Biegon A, Greenberger V, Segal M (1986) Quantitative histochemistry of brain acetylcholinesterase and learning rate in the aged rat. Neurobiol Aging 7:215–217
30. Bielschowsky M (1903) Die Ziele bei Impregnation der Neurofibrillen. Neurol Centralbl 22:997–1006

31. Bigl M, Bruckner MK, Arendt T, Bigl V, Eschrich K (1999) Activities of key glycolytic enzymes in the brains of patients with Alzheimer's disease. J Neural Transm 106:499–511
32. Blanchard V, Moussaoui S, Czech C, Touchet N, Bonici B, Planche M, Canton T, Jedidi I, Gohin M, Wirths O, Bayer TA, Langui D, Duyckaerts C, Tremp G, Pradier L (2003) Time sequence of maturation of dystrophic neurites associated with Abeta deposits in APP/PS1 transgenic mice. Exp Neurol 184:247–263
33. Blasko I, Veerhuis R, Stampfer-Kountchev M, Saurwein-Teissl M, Eikelenboom P, Grubeck-Loebenstein B (2000) Costimulatory effects of interferon-gamma and interleukin-1beta or tumor necrosis factor alpha on the synthesis of Abeta1-40 and Abeta1-42 by human astrocytes. Neurobiol Dis 7:682–689
34. Blass JP, Sheu RK, Gibson GE (2000) Inherent abnormalities in energy metabolism in Alzheimer disease. Interaction with cerebrovascular compromise. Ann NY Acad Sci 903:204–221
35. Blennow K, de Leon MJ, Zetterberg H (2006) Alzheimer's disease. Lancet 368:387–403
36. Blocq P, Marinesco G (1892) Sur les lesions et la pathogenie de l'epilepsie dite essentielle. Semaine Medical 12:445–446
37. Boegman RJ, el-Defrawy SR, Jhamandas K, Beninger RJ, Ludwin SK (1985) Quinolinic acid neurotoxicity in the nucleus basalis antagonized by kynurenic acid. Neurobiol Aging 6:331–336
38. Boller F, Forbes MM (1998) History of dementia and dementia in history: an overview. J Neurol Sci 158:125–133
39. Braak H, Braak E (1991) Neuropathological stageing of Alzheimer-related changes. Acta Neuropathol 82:239–259
40. Brambilla L, Martorana F, Rossi D (2013) Astrocyte signaling and neurodegeneration: new insights into CNS disorders. Prion 7:28–36
41. Brawek B, Chesters R, Klement D, Muller J, Lerdkrai C, Hermes M, Garaschuk O (2018) A bell-shaped dependence between amyloidosis and GABA accumulation in astrocytes in a mouse model of Alzheimer's disease. Neurobiol Aging 61:187–197
42. Broe M, Kril J, Halliday GM (2004) Astrocytic degeneration relates to the severity of disease in frontotemporal dementia. Brain 127:2214–2220
43. Burkovetskaya M, Karpuk N, Xiong J, Bosch M, Boska MD, Takeuchi H, Suzumura A, Kielian T (2014) Evidence for aberrant astrocyte hemichannel activity in Juvenile Neuronal Ceroid Lipofuscinosis (JNCL). PLoS ONE 9:e95023
44. Carroll JC, Rosario ER, Chang L, Stanczyk FZ, Oddo S, LaFerla FM, Pike CJ (2007) Progesterone and estrogen regulate Alzheimer-like neuropathology in female 3xTg-AD mice. J Neurosci 27:13357–13365
45. Carter SF, Herholz K, Rosa-Neto P, Pellerin L, Nordberg A, Zimmer ER (2019) Astrocyte biomarkers in Alzheimer's disease. Trends Mol Med 25:77–95
46. Carter SF, Scholl M, Almkvist O, Wall A, Engler H, Langstrom B, Nordberg A (2012) Evidence for astrocytosis in prodromal Alzheimer disease provided by ^{11}C-deuterium-L-deprenyl: a multitracer PET paradigm combining ^{11}C-Pittsburgh compound B and ^{18}F-FDG. J Nucl Med 53:37–46
47. Casley CS, Lakics V, Lee HG, Broad LM, Day TA, Cluett T, Smith MA, O'Neill MJ, Kingston AE (2009) Up-regulation of astrocyte metabotropic glutamate receptor 5 by amyloid-β peptide. Brain Res
48. Cassel JC, Mathis C, Majchrzak M, Moreau PH, Dalrymple-Alford JC (2008) Coexisting cholinergic and parahippocampal degeneration: a key to memory loss in dementia and a challenge for transgenic models? Neurodegener Dis 5:304–317
49. Castellani RJ, Lee HG, Siedlak SL, Nunomura A, Hayashi T, Nakamura M, Zhu X, Perry G, Smith MA (2009) Reexamining Alzheimer's disease: evidence for a protective role for amyloid-beta protein precursor and amyloid-beta. J Alzheimers Dis 18:447–452
50. Celsus AC (1935–1938) De Medicina. With an english translation by W. G. Spencer. William Heinemann Ltd., London

51. Chishti MA, Yang DS, Janus C, Phinney AL, Horne P, Pearson J, Strome R, Zuker N, Loukides J, French J, Turner S, Lozza G, Grilli M, Kunicki S, Morissette C, Paquette J, Gervais F, Bergeron C, Fraser PE, Carlson GA, George-Hyslop PS, Westaway D (2001) Early-onset amyloid deposition and cognitive deficits in transgenic mice expressing a double mutant form of amyloid precursor protein 695. J Biol Chem 276:21562–21570
52. Chow SK, Yu D, Macdonald CL, Buibas M, Silva GA (2010) Amyloid β-peptide directly induces spontaneous calcium transients, delayed intercellular calcium waves and gliosis in rat cortical astrocytes. ASN Neuro 2:e00026
53. Chrobak JJ, Hanin I, Schmechel DE, Walsh TJ (1988) AF64A-induced working memory impairment: behavioral, neurochemical and histological correlates. Brain Res 463:107–117
54. Cicero (2003) On old age. In: On the good life, pp 160–194. Folio Society, London
55. Coleman P, Federoff H, Kurlan R (2004) A focus on the synapse for neuroprotection in Alzheimer disease and other dementias. Neurology 63:1155–1162
56. Contestabile A, Stirpe F (1993) Ribosome-inactivating proteins from plants as agents for suicide transport and immunolesioning in the nervous system. Eur J Neurosci 5:1292–1301
57. Coradazzi M, Gulino R, Garozzo S, Leanza G (2010) Selective lesion of the developing central noradrenergic system: short- and long-term effects and reinnervation by noradrenergic-rich tissue grafts. J Neurochem 114:761–771
58. Danbolt NC (2001) Glutamate uptake. Progr Neurobiol 65:1–105
59. Daschil N, Geisler S, Obermair GJ, Humpel C (2014) Short- and long-term treatment of mouse cortical primary astrocytes with beta-amyloid differentially regulates the mRNA expression of L-type calcium channels. Pharmacology 93:24–31
60. Decker MW (1987) The effects of aging on hippocampal and cortical projections of the forebrain cholinergic system. Brain Res 434:423–438
61. DeKosky ST, Scheff SW (1990) Synapse loss in frontal cortex biopsies in Alzheimer's disease: correlation with cognitive severity. Ann Neurol 27:457–464
62. Delekate A, Fuchtemeier M, Schumacher T, Ulbrich C, Foddis M, Petzold GC (2014) Metabotropic P2Y1 receptor signalling mediates astrocytic hyperactivity in vivo in an Alzheimer's disease mouse model. Nat Commun 5:5422
63. Depboylu C, Stricker S, Ghobril JP, Oertel WH, Priller J, Hoglinger GU (2012) Brain-resident microglia predominate over infiltrating myeloid cells in activation, phagocytosis and interaction with T-lymphocytes in the MPTP mouse model of Parkinson disease. Exp Neurol 238:183–191
64. DeWitt DA, Perry G, Cohen M, Doller C, Silver J (1998) Astrocytes regulate microglial phagocytosis of senile plaque cores of Alzheimer's disease. Exp Neurol 149:329–340
65. Di Patre PL, Abbamondi A, Bartolini L, Pepeu G (1989) GM1 ganglioside counteracts cholinergic and behavioral deficits induced in the rat by intracerebral injection of vincristine. Eur J Pharmacol 162:43–50
66. Ding F, O'Donnell J, Thrane AS, Zeppenfeld D, Kang H, Xie L, Wang F, Nedergaard M (2013) α1-Adrenergic receptors mediate coordinated Ca2+ signaling of cortical astrocytes in awake, behaving mice. Cell Calcium 54:387–394
67. Dodart JC, Mathis C, Saura J, Bales KR, Paul SM, Ungerer A (2000) Neuroanatomical abnormalities in behaviorally characterized APP(V717F) transgenic mice. Neurobiol Dis 7:71–85
68. Dorfman VB, Pasquini L, Riudavets M, Lopez-Costa JJ, Villegas A, Troncoso JC, Lopera F, Castano EM, Morelli L (2010) Differential cerebral deposition of IDE and NEP in sporadic and familial Alzheimer's disease. Neurobiol Aging 31:1743–1757
69. Dunnett SB, Everitt BJ, Robbins TW (1991) The basal forebrain-cortical cholinergic system: interpreting the functional consequences of excitotoxic lesions. Trends Neurosci 14:494–501
70. Duyckaerts C, Delatour B, Potier MC (2009) Classification and basic pathology of Alzheimer disease. Acta Neuropathol 118:5–36
71. Enkvist MO, Holopainen I, Akerman KE (1989) Glutamate receptor-linked changes in membrane potential and intracellular Ca^{2+} in primary rat astrocytes. Glia 2:397–402

72. Eriksen JL, Janus CG (2007) Plaques, tangles, and memory loss in mouse models of neurodegeneration. Behav Genet 37:79–100
73. Eroglu C, Barres BA (2010) Regulation of synaptic connectivity by glia. Nature 468:223–231
74. Farkas E, Luiten PG (2001) Cerebral microvascular pathology in aging and Alzheimer's disease. Prog Neurobiol 64:575–611
75. Feinstein DL, Kalinin S, Braun D (2016) Causes, consequences, and cures for neuroinflammation mediated via the locus coeruleus: noradrenergic signaling system. J Neurochem 139(Suppl 2):154–178
76. Fiol-deRoque MA, Gutierrez-Lanza R, Torres M, Terés S, Barceló P, Rial RV, Verkhratsky A, Escribá PV, Busquets X, Rodríguez JJ (2013) Cognitive recovery and restoration of cell proliferation in the dentate gyrus in the 5XFAD transgenic mice model of Alzheimer's disease following 2-hydroxy-DHA treatment. Biogerontology (in press)
77. Fischer O (1907) Miliäre Nekrosen mit drusigen wucherungen der neurofibrillen, eine regelmässige veränderung der hirnrinde bei seniler demenz. Monatsschr Psychiatr Neurol 22:361–372
78. Fischer W, Chen KS, Gage FH, Bjorklund A (1992) Progressive decline in spatial learning and integrity of forebrain cholinergic neurons in rats during aging. Neurobiol Aging 13:9–23
79. Flood DG, Lin YG, Lang DM, Trusko SP, Hirsch JD, Savage MJ, Scott RW, Howland DS (2009) A transgenic rat model of Alzheimer's disease with extracellular Aβ deposition. Neurobiol Aging 30:1078–1090
80. Fontana L, Partridge L, Longo VD (2010) Extending healthy life span–from yeast to humans. Science 328:321–326
81. Fowler JS, Volkow ND, Wang GJ, Logan J, Pappas N, Shea C, MacGregor R (1997) Age-related increases in brain monoamine oxidase B in living healthy human subjects. Neurobiol Aging 18:431–435
82. Frost GR, Li YM (2017) The role of astrocytes in amyloid production and Alzheimer's disease. Open Biol 7
83. Games D, Adams D, Alessandrini R, Barbour R, Berthelette P, Blackwell C, Carr T, Clemens J, Donaldson T, Gillespie F et al (1995) Alzheimer-type neuropathology in transgenic mice overexpressing V717F beta-amyloid precursor protein. Nature 373:523–527
84. Garaschuk O, Verkhratsky A (2019) GABAergic astrocytes in Alzheimer's disease. Aging (Albany NY) 11:1602–1604
85. Garwood C, Faizullabhoy A, Wharton SB, Ince PG, Heath PR, Shaw PJ, Baxter L, Gelsthorpe C, Forster G, Matthews FE, Brayne C, Simpson JE (2013) Calcium dysregulation in relation to Alzheimer-type pathology in the ageing brain. Neuropathol Appl Neurobiol 39:788–799
86. Gerkau NJ, Kafitz KW, Rose CR (2019) Imaging of local and global sodium signals in astrocytes. Methods Mol Biol 1938:187–202
87. Gerlai R (2001) Alzheimer's disease: beta-amyloid hypothesis strengthened! Trends Neurosci 24:199
88. Giovannini MG, Scali C, Prosperi C, Bellucci A, Vannucchi MG, Rosi S, Pepeu G, Casamenti F (2002) Beta-amyloid-induced inflammation and cholinergic hypofunction in the rat brain in vivo: involvement of the p38MAPK pathway. Neurobiol Dis 11:257–274
89. Gotz J, Chen F, van Dorpe J, Nitsch RM (2001) Formation of neurofibrillary tangles in P301 l tau transgenic mice induced by Abeta 42 fibrils. Science 293:1491–1495
90. Gotz J, Probst A, Spillantini MG, Schafer T, Jakes R, Burki K, Goedert M (1995) Somatodendritic localization and hyperphosphorylation of tau protein in transgenic mice expressing the longest human brain tau isoform. EMBO J 14:1304–1313
91. Gotz J, Streffer JR, David D, Schild A, Hoerndli F, Pennanen L, Kurosinski P, Chen F (2004) Transgenic animal models of Alzheimer's disease and related disorders: histopathology, behavior and therapy. Mol Psychiatry 9:664–683
92. Griffin WS, Stanley LC, Ling C, White L, MacLeod V, Perrot LJ, White CL 3rd, Araoz C (1989) Brain interleukin 1 and S-100 immunoreactivity are elevated in Down syndrome and Alzheimer disease. Proc Natl Acad Sci USA 86:7611–7615

93. Grolla AA, Fakhfouri G, Balzaretti G, Marcello E, Gardoni F, Canonico PL, DiLuca M, Genazzani AA, Lim D (2013) Aβ leads to Ca2+ signaling alterations and transcriptional changes in glial cells. Neurobiol Aging 34:511–522
94. Grolla AA, Sim JA, Lim D, Rodriguez JJ, Genazzani AA, Verkhratsky A (2013) Amyloid-β and Alzheimer's disease type pathology differentially affects the calcium signalling toolkit in astrocytes from different brain regions. Cell Death Dis 4:e623
95. Grosche J, Matyash V, Moller T, Verkhratsky A, Reichenbach A, Kettenmann H (1999) Microdomains for neuron-glia interaction: parallel fiber signaling to Bergmann glial cells. Nat Neurosci 2:139–143
96. Guenette SY (2003) Astrocytes: a cellular player in Aβ clearance and degradation. Trends Mol Med 9:279–280
97. Hanin I (1996) The AF64A model of cholinergic hypofunction: an update. Life Sci 58:1955–1964
98. Hardy J (2006) Has the amyloid cascade hypothesis for Alzheimer's disease been proved? Curr Alzheimer Res 3:71–73
99. Hardy J (2009) The amyloid hypothesis for Alzheimer's disease: a critical reappraisal. J Neurochem 110:1129–1134
100. Hardy J, Selkoe DJ (2002) The amyloid hypothesis of Alzheimer's disease: progress and problems on the road to therapeutics. Science 297:353–356
101. Hartlage-Rubsamen M, Zeitschel U, Apelt J, Gartner U, Franke H, Stahl T, Gunther A, Schliebs R, Penkowa M, Bigl V, Rossner S (2003) Astrocytic expression of the Alzheimer's disease β-secretase (BACE1) is stimulus-dependent. Glia 41:169–179
102. Haug H, Eggers R (1991) Morphometry of the human cortex cerebri and corpus striatum during aging. Neurobiol Aging 12:336–338; discussion 352–335
103. Haughey NJ, Mattson MP (2003) Alzheimer's amyloid β-peptide enhances ATP/gap junction-mediated calcium-wave propagation in astrocytes. Neuromolecular Med 3:173–180
104. Hazell AS (2009) Astrocytes are a major target in thiamine deficiency and Wernicke's encephalopathy. Neurochem Int 55:129–135
105. Hazell AS, Sheedy D, Oanea R, Aghourian M, Sun S, Jung JY, Wang D, Wang C (2009) Loss of astrocytic glutamate transporters in Wernicke encephalopathy. Glia 58:148–156
106. Heneka MT, Rodriguez JJ, Verkhratsky A (2010) Neuroglia in neurodegeneration. Brain Res Rev (in press)
107. Heneka MT, Sastre M, Dumitrescu-Ozimek L, Dewachter I, Walter J, Klockgether T, Van Leuven F (2005) Focal glial activation coincides with increased BACE1 activation and precedes amyloid plaque deposition in APPV717I transgenic mice. J Neuroinflam 2:22
108. Henry V, Paille V, Lelan F, Brachet P, Damier P (2009) Kinetics of microglial activation and degeneration of dopamine-containing neurons in a rat model of Parkinson disease induced by 6-hydroxydopamine. J Neuropathol Exp Neurol 68:1092–1102
109. Hodges JR (2006) Alzheimer's centennial legacy: origins, landmarks and the current status of knowledge concerning cognitive aspects. Brain 129:2811–2822
110. Hsiao K, Chapman P, Nilsen S, Eckman C, Harigaya Y, Younkin S, Yang F, Cole G (1996) Correlative memory deficits, Abeta elevation, and amyloid plaques in transgenic mice. Science 274:99–102
111. Humphry GM (1889) Old age. Macmillan & Bowes, Cambridge
112. Iadecola C, Nedergaard M (2007) Glial regulation of the cerebral microvasculature. Nat Neurosci 10:1369–1376
113. Ishihara T, Higuchi M, Zhang B, Yoshiyama Y, Hong M, Trojanowski JQ, Lee VM (2001) Attenuated neurodegenerative disease phenotype in tau transgenic mouse lacking neurofilaments. J Neurosci 21:6026–6035
114. Jalonen TO, Charniga CJ, Wielt DB (1997) β-Amyloid peptide-induced morphological changes coincide with increased K+ and Cl- channel activity in rat cortical astrocytes. Brain Res 746:85–97

115. Janus C, Pearson J, McLaurin J, Mathews PM, Jiang Y, Schmidt SD, Chishti MA, Horne P, Heslin D, French J, Mount HT, Nixon RA, Mercken M, Bergeron C, Fraser PE, St George-Hyslop P, Westaway D (2000) A beta peptide immunization reduces behavioural impairment and plaques in a model of Alzheimer's disease. Nature 408:979–982
116. Jellinger KA (2008) Neuropathological aspects of Alzheimer disease, Parkinson disease and frontotemporal dementia. Neurodegener Dis 5:118–121
117. Jin SM, Cho HJ, Kim YW, Hwang JY, Mook-Jung I (2012) Abeta-induced Ca(2+) influx regulates astrocytic BACE1 expression via calcineurin/NFAT4 signals. Biochem Biophys Res Commun 425:649–655
118. Jo S, Yarishkin O, Hwang YJ, Chun YE, Park M, Woo DH, Bae JY, Kim T, Lee J, Chun H, Park HJ, Lee DY, Hong J, Kim HY, Oh SJ, Park SJ, Lee H, Yoon BE, Kim Y, Jeong Y, Shim I, Bae YC, Cho J, Kowall NW, Ryu H, Hwang E, Kim D, Lee CJ (2014) GABA from reactive astrocytes impairs memory in mouse models of Alzheimer's disease. Nat Med 20:886–896
119. Jones VC, Atkinson-Dell R, Verkhratsky A, Mohamet L (2017) Aberrant iPSC-derived human astrocytes in Alzheimer's disease. Cell Death Dis 8:e2696
120. Kanemaru K, Kubota J, Sekiya H, Hirose K, Okubo Y, Iino M (2013) Calcium-dependent N-cadherin up-regulation mediates reactive astrogliosis and neuroprotection after brain injury. Proc Natl Acad Sci USA 110:11612–11617
121. Karenberg A, Forstl H (2006) Dementia in the Greco-Roman world. J Neurol Sci 244:5–9
122. Karran E, Mercken M, De Strooper B (2011) The amyloid cascade hypothesis for Alzheimer's disease: an appraisal for the development of therapeutics. Nat Rev Drug Discov 10:698–712
123. Kaul M, Lipton SA (2006) Mechanisms of neuronal injury and death in HIV-1 associated dementia. Curr HIV Res 4:307–318
124. Kersaitis C, Halliday GM, Kril JJ (2004) Regional and cellular pathology in frontotemporal dementia: relationship to stage of disease in cases with and without Pick bodies. Acta Neuropathol 108:515–523
125. Khachaturian ZS (1987) Hypothesis on the regulation of cytosol calcium concentration and the aging brain. Neurobiol Aging 8:345–346
126. Kim K, Lee SG, Kegelman TP, Su ZZ, Das SK, Dash R, Dasgupta S, Barral PM, Hedvat M, Diaz P, Reed JC, Stebbins JL, Pellecchia M, Sarkar D, Fisher PB (2011) Role of excitatory amino acid transporter-2 (EAAT2) and glutamate in neurodegeneration: opportunities for developing novel therapeutics. J Cell Physiol 226:2484–2493
127. Kirischuk S, Kettenmann H, Verkhratsky A (1997) Na+/Ca2+ exchanger modulates kainate-triggered Ca2+ signaling in Bergmann glial cells in situ. FASEB J 11:566–572
128. Kirischuk S, Kettenmann H, Verkhratsky A (2007) Membrane currents and cytoplasmic sodium transients generated by glutamate transport in Bergmann glial cells. Pflugers Arch 454:245–252
129. Kirischuk S, Moller T, Voitenko N, Kettenmann H, Verkhratsky A (1995) ATP-induced cytoplasmic calcium mobilization in Bergmann glial cells. J Neurosci 15:7861–7871
130. Kirischuk S, Parpura V, Verkhratsky A (2012) Sodium dynamics: another key to astroglial excitability? Trends Neurosci 35:497–506
131. Kirischuk S, Tuschick S, Verkhratsky A, Kettenmann H (1996) Calcium signalling in mouse Bergmann glial cells mediated by α1-adrenoreceptors and H1 histamine receptors. Eur J Neurosci 8:1198–1208
132. Kloskowska E, Pham TM, Nilsson T, Zhu S, Oberg J, Codita A, Pedersen LA, Pedersen JT, Malkiewicz K, Winblad B, Folkesson R, Benedikz E (2010) Cognitive impairment in the Tg6590 transgenic rat model of Alzheimer's disease. J Cell Mol Med 14:1816–1823
133. Knight RA, Verkhratsky A (2010) Neurodegenerative diseases: failures in brain connectivity? Cell Death Differ 17:1069–1070
134. Korczyn AD (2008) The amyloid cascade hypothesis. Alzheimers Dement 4:176–178
135. Korsakoff SS (1889) Корсаков, С.С. Психическое расстройство в сочетании с множественным невритом (psychosis polineuritica, s. cerebropathia psychica toxaemica). English translation: Korsakoff SS. Psychic disorder in conjunction with multiple neuritis, Translated from Russian by M. Victor and P. Yakovlev, Neurology (1955), 5:394–406. *Мед обозр* 32:3–18

136. Kraepelin E (1910) Psychiatrie: Ein Lehrbuch fuer Studierende und Arzte. Johann Ambrosius Barth, Leipzig
137. Kraft AW, Hu X, Yoon H, Yan P, Xiao Q, Wang Y, Gil SC, Brown J, Wilhelmsson U, Restivo JL, Cirrito JR, Holtzman DM, Kim J, Pekny M, Lee JM (2013) Attenuating astrocyte activation accelerates plaque pathogenesis in APP/PS1 mice. FASEB J 27:187–198
138. Kuchibhotla KV, Lattarulo CR, Hyman BT, Bacskai BJ (2009) Synchronous hyperactivity and intercellular calcium waves in astrocytes in Alzheimer mice. Science 323:1211–1215
139. Kulijewicz-Nawrot M, Verkhratsky A, Chvatal A, Sykova E, Rodriguez JJ (2012) Astrocytic cytoskeletal atrophy in the medial prefrontal cortex of a triple transgenic mouse model of Alzheimer's disease. J Anat 221:252–262
140. Kuo MF, Paulus W, Nitsche MA (2014) Therapeutic effects of non-invasive brain stimulation with direct currents (tDCS) in neuropsychiatric diseases. Neuroimage 85(Pt 3):948–960
141. Kurumatani T, Fastbom J, Bonkale WL, Bogdanovic N, Winblad B, Ohm TG, Cowburn RF (1998) Loss of inositol 1,4,5-trisphosphate receptor sites and decreased PKC levels correlate with staging of Alzheimer's disease neurofibrillary pathology. Brain Res 796:209–221
142. LaFerla FM, Green KN (2012) Animal models of Alzheimer disease. Cold Spring Harb Perspect Med 2
143. Laird FM, Cai H, Savonenko AV, Farah MH, He K, Melnikova T, Wen H, Chiang HC, Xu G, Koliatsos VE, Borchelt DR, Price DL, Lee HK, Wong PC (2005) BACE1, a major determinant of selective vulnerability of the brain to amyloid-beta amyloidogenesis, is essential for cognitive, emotional, and synaptic functions. J Neurosci 25:11693–11709
144. Lalo U, Palygin O, North RA, Verkhratsky A, Pankratov Y (2011) Age-dependent remodelling of ionotropic signalling in cortical astroglia. Aging Cell 10:392–402
145. Lalo U, Rasooli-Nejad S, Pankratov Y (2014) Exocytosis of gliotransmitters from cortical astrocytes: implications for synaptic plasticity and aging. Biochem Soc Trans 42:1275–1281
146. Landfield PW (1987) 'Increased calcium-current' hypothesis of brain aging. Neurobiol Aging 8:346–347
147. Landfield PW, Pitler TA (1984) Prolonged Ca2+ -dependent afterhyperpolarizations in hippocampal neurons of aged rats. Science 226:1089–1092
148. Langer J, Gerkau NJ, Derouiche A, Kleinhans C, Moshrefi-Ravasdjani B, Fredrich M, Kafitz KW, Seifert G, Steinhauser C, Rose CR (2017) Rapid sodium signaling couples glutamate uptake to breakdown of ATP in perivascular astrocyte endfeet. Glia 65:293–308
149. Langer J, Rose CR (2009) Synaptically induced sodium signals in hippocampal astrocytes in situ. J Physiol 587:5859–5877
150. Langer J, Stephan J, Theis M, Rose CR (2012) Gap junctions mediate intercellular spread of sodium between hippocampal astrocytes in situ. Glia 60:239–252
151. Lappi DA, Esch FS, Barbieri L, Stirpe F, Soria M (1985) Characterization of a Saponaria officinalis seed ribosome-inactivating protein: immunoreactivity and sequence homologies. Biochem Biophys Res Commun 129:934–942
152. Lazzarini M, Martin S, Mitkovski M, Vozari RR, Stuhmer W, Bel ED (2013) Doxycycline restrains glia and confers neuroprotection in a 6-OHDA Parkinson model. Glia 61:1084–1100
153. Leanza G, Gulino R, Zorec R (2018) Noradrenergic hypothesis linking neurodegeneration-based cognitive decline and astroglia. Front Mol Neurosci 11:254
154. Lee L, Kosuri P, Arancio O (2014) Picomolar amyloid-β peptides enhance spontaneous astrocyte calcium transients. J Alzheimers Dis 38:49–62
155. Lee M, Schwab C, McGeer PL (2011) Astrocytes are GABAergic cells that modulate microglial activity. Glia 59:152–165
156. Lescaudron L, Stein DG (1999) Differences in memory impairment and response to GM1 ganglioside treatment following electrolytic or ibotenic acid lesions of the nucleus basalis magnocellularis. Restor Neurol Neurosci 15:25–37
157. Leuba G, Wernli G, Vernay A, Kraftsik R, Mohajeri MH, Saini KD (2005) Neuronal and non-neuronal quantitative BACE immunocytochemical expression in the entorhinohippocampal and frontal regions in Alzheimer's disease. Dement Geriatr Cogn Disord 19:171–183

158. Lewis J, Dickson DW, Lin WL, Chisholm L, Corral A, Jones G, Yen SH, Sahara N, Skipper L, Yager D, Eckman C, Hardy J, Hutton M, McGowan E (2001) Enhanced neurofibrillary degeneration in transgenic mice expressing mutant tau and APP. Science 293:1487–1491
159. Lewis J, McGowan E, Rockwood J, Melrose H, Nacharaju P, Van Slegtenhorst M, Gwinn-Hardy K, Paul Murphy M, Baker M, Yu X, Duff K, Hardy J, Corral A, Lin WL, Yen SH, Dickson DW, Davies P, Hutton M (2000) Neurofibrillary tangles, amyotrophy and progressive motor disturbance in mice expressing mutant (P301L) tau protein. Nat Genet 25:402–405
160. Liang WS, Reiman EM, Valla J, Dunckley T, Beach TG, Grover A, Niedzielko TL, Schneider LE, Mastroeni D, Caselli R, Kukull W, Morris JC, Hulette CM, Schmechel D, Rogers J, Stephan DA (2008) Alzheimer's disease is associated with reduced expression of energy metabolism genes in posterior cingulate neurons. Proc Natl Acad Sci U S A 105:4441–4446
161. Lim D, Iyer A, Ronco V, Grolla AA, Canonico PL, Aronica E, Genazzani AA (2013) Amyloid β deregulates astroglial mGluR5-mediated calcium signaling via calcineurin and Nf-κB. Glia 61:1134–1145
162. Lim D, Rodriguez-Arellano JJ, Parpura V, Zorec R, Zeidan-Chulia F, Genazzani AA, Verkhratsky A (2016) Calcium signalling toolkits in astrocytes and spatio-temporal progression of Alzheimer's disease. Curr Alzheimer Res 13:359–369
163. Lim D, Ronco V, Grolla AA, Verkhratsky A, Genazzani AA (2014) Glial calcium signalling in Alzheimer's disease. Rev Physiol Biochem Pharmacol 167:45–65
164. Linde CI, Baryshnikov SG, Mazzocco-Spezzia A, Golovina VA (2011) Dysregulation of Ca^{2+} signaling in astrocytes from mice lacking amyloid precursor protein. Am J Physiol Cell Physiol 300:C1502–1512
165. Lyketsos CG, Olin J (2002) Depression in Alzheimer's disease: overview and treatment. Biol Psychiatry 52:243–252
166. Martins RN, Taddei K, Kendall C, Evin G, Bates KA, Harvey AR (2001) Altered expression of apolipoprotein E, amyloid precursor protein and presenilin-1 is associated with chronic reactive gliosis in rat cortical tissue. Neuroscience 106:557–569
167. Mather M, Harley CW (2016) The locus coeruleus: essential for maintaining cognitive function and the aging brain. Trends Cogn Sci 20:214–226
168. Matos M, Augusto E, Oliveira CR, Agostinho P (2008) Amyloid-β peptide decreases glutamate uptake in cultured astrocytes: involvement of oxidative stress and mitogen-activated protein kinase cascades. Neuroscience 156:898–910
169. Mattison JA, Colman RJ, Beasley TM, Allison DB, Kemnitz JW, Roth GS, Ingram DK, Weindruch R, de Cabo R, Anderson RM (2017) Caloric restriction improves health and survival of rhesus monkeys. Nat Commun 8:14063
170. Mattson MP (2012) Energy intake and exercise as determinants of brain health and vulnerability to injury and disease. Cell Metab 16:706–722
171. Mattson MP, Haughey NJ, Nath A (2005) Cell death in HIV dementia. Cell Death Differ 12(Suppl 1):893–904
172. Maurer K, Volk S, Gerbaldo H (1997) Auguste D and Alzheimer's disease. Lancet 349:1546–1549
173. Maurya SK, Rai A, Rai NK, Deshpande S, Jain R, Mudiam MK, Prabhakar YS, Bandyopadhyay S (2012) Cypermethrin induces astrocyte apoptosis by the disruption of the autocrine/paracrine mode of epidermal growth factor receptor signaling. Toxicol Sci 125:473–487
174. McAlpine D, Araki S (1958) Minamata disease: an unusual neurological disorder caused by contaminated fish. Lancet 2:629–631
175. McCay CM, Crowell MF, Maynard LA (1935) The effect of retarded growth upon the length of life span and upon the ultimate body size. J Nutr 10:63–79
176. McEntee WJ, Crook TH (1991) Serotonin, memory, and the aging brain. Psychopharmacology 103:143–149
177. McKee AC, Carreras I, Hossain L, Ryu H, Klein WL, Oddo S, LaFerla FM, Jenkins BG, Kowall NW, Dedeoglu A (2008) Ibuprofen reduces Abeta, hyperphosphorylated tau and memory deficits in Alzheimer mice. Brain Res 1207:225–236

178. Meda L, Baron P, Scarlato G (2001) Glial activation in Alzheimer's disease: the role of Abeta and its associated proteins. Neurobiol Aging 22:885–893
179. Mena MA, Casarejos MJ, Carazo A, Paino CL, Garcia de Yebenes J (1996) Glia conditioned medium protects fetal rat midbrain neurones in culture from L-DOPA toxicity. NeuroReport 7:441–445
180. Mena MA, de Bernardo S, Casarejos MJ, Canals S, Rodriguez-Martin E (2002) The role of astroglia on the survival of dopamine neurons. Mol Neurobiol 25:245–263
181. Mena MA, Garcia de Yebenes J (2008) Glial cells as players in parkinsonism: the "good," the "bad," and the "mysterious" glia. Neuroscientist 14:544–560
182. Mesulam M, Shaw P, Mash D, Weintraub S (2004) Cholinergic nucleus basalis tauopathy emerges early in the aging-MCI-AD continuum. Ann Neurol 55:815–828
183. Moechars D, Lorent K, De Strooper B, Dewachter I, Van Leuven F (1996) Expression in brain of amyloid precursor protein mutated in the alpha-secretase site causes disturbed behavior, neuronal degeneration and premature death in transgenic mice. EMBO J 15:1265–1274
184. Mohamet L, Jones VC, Dayanithi G, Verkhratsky A (2018) Pathological human astroglia in Alzheimer's disease: opening new horizons with stem cell technology. Future Neurol 13:87–99
185. Mohs RC (2005) The clinical syndrome of Alzheimer's disease: aspects particularly relevant to clinical trials. Genes Brain Behav 4:129–133
186. Monai H, Ohkura M, Tanaka M, Oe Y, Konno A, Hirai H, Mikoshiba K, Itohara S, Nakai J, Iwai Y, Hirase H (2016) Calcium imaging reveals glial involvement in transcranial direct current stimulation-induced plasticity in mouse brain. Nat Commun 7:11100
187. Mosconi L, Pupi A, De Leon MJ (2008) Brain glucose hypometabolism and oxidative stress in preclinical Alzheimer's disease. Ann NY Acad Sci 1147:180–195
188. Mouton PR, Pakkenberg B, Gundersen HJ, Price DL (1994) Absolute number and size of pigmented locus coeruleus neurons in young and aged individuals. J Chem Neuroanat 7:185–190
189. Mrak RE, Griffin WS (2005) Glia and their cytokines in progression of neurodegeneration. Neurobiol Aging 26:349–354
190. Mravec B, Lejavova K, Cubinkova V (2014) Locus (coeruleus) minoris resistentiae in pathogenesis of Alzheimer's disease. Curr Alzheimer Res 11:992–1001
191. Mulligan SJ, MacVicar BA (2004) Calcium transients in astrocyte endfeet cause cerebrovascular constrictions. Nature 431:195–199
192. Nagele RG, D'Andrea MR, Lee H, Venkataraman V, Wang HY (2003) Astrocytes accumulate A β 42 and give rise to astrocytic amyloid plaques in Alzheimer disease brains. Brain Res 971:197–209
193. Nagele RG, Wegiel J, Venkataraman V, Imaki H, Wang KC, Wegiel J (2004) Contribution of glial cells to the development of amyloid plaques in Alzheimer's disease. Neurobiol Aging 25:663–674
194. Ni M, Li X, Rocha JB, Farina M, Aschner M (2012) Glia and methylmercury neurotoxicity. J Toxicol Environ Health A 75:1091–1101
195. Nicoll JA, Weller RO (2003) A new role for astrocytes: β-amyloid homeostasis and degradation. Trends Mol Med 9:281–282
196. Nimmerjahn A, Mukamel EA, Schnitzer MJ (2009) Motor behavior activates Bergmann glial networks. Neuron 62:400–412
197. Nitsche MA, Cohen LG, Wassermann EM, Priori A, Lang N, Antal A, Paulus W, Hummel F, Boggio PS, Fregni F, Pascual-Leone A (2008) Transcranial direct current stimulation: state of the art 2008. Brain Stimul 1:206–223
198. Noristani HN, Olabarria M, Verkhratsky A, Rodriguez JJ (2010) Serotonin fibre sprouting and increase in serotonin transporter immunoreactivity in the CA1 area of hippocampus in a triple transgenic mouse model of Alzheimer's disease. Eur J Neurosci 32:71–79
199. Norris CM, Kadish I, Blalock EM, Chen KC, Thibault V, Porter NM, Landfield PW, Kraner SD (2005) Calcineurin triggers reactive/inflammatory processes in astrocytes and is upregulated in aging and Alzheimer's models. J Neurosci 25:4649–4658
200. Oakley H, Cole SL, Logan S, Maus E, Shao P, Craft J, Guillozet-Bongaarts A, Ohno M, Disterhoft J, Van Eldik L, Berry R, Vassar R (2006) Intraneuronal beta-amyloid aggregates,

neurodegeneration, and neuron loss in transgenic mice with five familial Alzheimer's disease mutations: potential factors in amyloid plaque formation. J Neurosci 26:10129–10140
201. Oddo S, Caccamo A, Kitazawa M, Tseng BP, LaFerla FM (2003) Amyloid deposition precedes tangle formation in a triple transgenic model of Alzheimer's disease. Neurobiol Aging 24:1063–1070
202. Oddo S, Caccamo A, Shepherd JD, Murphy MP, Golde TE, Kayed R, Metherate R, Mattson MP, Akbari Y, LaFerla FM (2003) Triple-transgenic model of Alzheimer's disease with plaques and tangles: intracellular Aβ and synaptic dysfunction. Neuron 39:409–421
203. Olabarria M, Noristani HN, Verkhratsky A, Rodriguez JJ (2010) Concomitant astroglial atrophy and astrogliosis in a triple transgenic animal model of Alzheimer's disease. Glia 58:831–838
204. Olabarria M, Noristani HN, Verkhratsky A, Rodriguez JJ (2011) Age-dependent decrease in glutamine synthetase expression in the hippocampal astroglia of the triple transgenic Alzheimer's disease mouse model: mechanism for deficient glutamatergic transmission? Mol Neurodegener 6:55
205. Orre M, Kamphuis W, Osborn LM, Jansen AHP, Kooijman L, Bossers K, Hol EM (2014) Isolation of glia from Alzheimer's mice reveals inflammation and dysfunction. Neurobiol Aging 35:2746–2760
206. Ovsepian SV, O'Leary VB, Zaborszky L, Ntziachristos V, Dolly JO (2018) Amyloid plaques of Alzheimer's disease as hotspots of glutamatergic activity. Neuroscientist. https://doi.org/10.1177/1073858418791128
207. Palop JJ, Mucke L (2010) Amyloid-beta-induced neuronal dysfunction in Alzheimer's disease: from synapses toward neural networks. Nat Neurosci 13:812–818
208. Pankratov Y, Lalo U (2015) Role for astroglial α1-adrenoreceptors in gliotransmission and control of synaptic plasticity in the neocortex. Front Cell Neurosci 9:230
209. Parpura-Gill A, Beitz D, Uemura E (1997) The inhibitory effects of beta-amyloid on glutamate and glucose uptakes by cultured astrocytes. Brain Res 754:65–71
210. Paukert M, Agarwal A, Cha J, Doze VA, Kang JU, Bergles DE (2014) Norepinephrine controls astroglial responsiveness to local circuit activity. Neuron 82:1263–1270
211. Pavia J, Alberch J, Alvarez I, Toledano A, de Ceballos ML (2000) Repeated intracerebroventricular administration of beta-amyloid(25-35) to rats decreases muscarinic receptors in cerebral cortex. Neurosci Lett 278:69–72
212. Pekny M, Pekna M, Messing A, Steinhauser C, Lee JM, Parpura V, Hol EM, Sofroniew MV, Verkhratsky A (2016) Astrocytes: a central element in neurological diseases. Acta Neuropathol 131:323–345
213. Pellerin L, Magistretti PJ (2012) Sweet sixteen for ANLS. J Cereb Blood Flow Metab 32:1152–1166
214. Pepeu G, Marconcini Pepeu I (1994) Dysfunction of the brain cholinergic system during aging and after lesions of the nucleus basalis of Meynert. J Neural Transm Suppl 44:189–194
215. Picklo MJ, Wiley RG, Lappi DA, Robertson D (1994) Noradrenergic lesioning with an anti-dopamine beta-hydroxylase immunotoxin. Brain Res 666:195–200
216. Pirttimaki TM, Codadu NK, Awni A, Pratik P, Nagel DA, Hill EJ, Dineley KT, Parri HR (2013) α7 Nicotinic receptor-mediated astrocytic gliotransmitter release: Aβ effects in a preclinical Alzheimer's mouse model. PLoS ONE 8:e81828
217. Plata A, Popov A, Denisov P, Bychkov M, Brazhe A, Lyukmanova E, Natalia Lazareva N, Verkhratsky A, Semyanov A (2019) An astrocytic basis of caloric restriction action on the brain plasticity. BioRxiv
218. Potokar M, Kreft M, Pangrsic T, Zorec R (2005) Vesicle mobility studied in cultured astrocytes. Biochem Biophys Res Commun 329:678–683
219. Potokar M, Stenovec M, Gabrijel M, Li L, Kreft M, Grilc S, Pekny M, Zorec R (2010) Intermediate filaments attenuate stimulation-dependent mobility of endosomes/lysosomes in astrocytes. Glia 58:1208–1219
220. Potokar M, Vardjan N, Stenovec M, Gabrijel M, Trkov S, Jorgacevski J, Kreft M, Zorec R (2013) Astrocytic vesicle mobility in health and disease. Int J Mol Sci 14:11238–11258

221. Potts R, Leech RW (2005) Thalamic dementia: an example of primary astroglial dystrophy of Seitelberger. Clin Neuropathol 24:271–275
222. Pressey SN, Smith DA, Wong AM, Platt FM, Cooper JD (2012) Early glial activation, synaptic changes and axonal pathology in the thalamocortical system of Niemann-Pick type C1 mice. Neurobiol Dis 45:1086–1100
223. Rai A, Maurya SK, Sharma R, Ali S (2013) Down-regulated GFAPα: a major player in heavy metal induced astrocyte damage. Toxicol Mech Methods 23:99–107
224. Ramsden M, Kotilinek L, Forster C, Paulson J, McGowan E, SantaCruz K, Guimaraes A, Yue M, Lewis J, Carlson G, Hutton M, Ashe KH (2005) Age-dependent neurofibrillary tangle formation, neuron loss, and memory impairment in a mouse model of human tauopathy (P301L). J Neurosci 25:10637–10647
225. Redlich E (1898) Ueber miliare Sklerose der Hirnrinde bei seniler Atrophie. J Psychiat Neurol 17:208–216
226. Reitz C (2012) Alzheimer's disease and the amyloid cascade hypothesis: a critical review. Int J Alzheimers Dis 2012:369808
227. Reyes RC, Verkhratsky A, Parpura V (2013) TRPC1-mediated Ca2+ and Na+ signalling in astroglia: differential filtering of extracellular cations. Cell Calcium 54:120–125
228. Ribe EM, Perez M, Puig B, Gich I, Lim F, Cuadrado M, Sesma T, Catena S, Sanchez B, Nieto M, Gomez-Ramos P, Moran MA, Cabodevilla F, Samaranch L, Ortiz L, Perez A, Ferrer I, Avila J, Gomez-Isla T (2005) Accelerated amyloid deposition, neurofibrillary degeneration and neuronal loss in double mutant APP/tau transgenic mice. Neurobiol Dis 20:814–822
229. Rodriguez-Vieitez E, Saint-Aubert L, Carter SF, Almkvist O, Farid K, Scholl M, Chiotis K, Thordardottir S, Graff C, Wall A, Langstrom B, Nordberg A (2016) Diverging longitudinal changes in astrocytosis and amyloid PET in autosomal dominant Alzheimer's disease. Brain 139:922–936
230. Rodriguez JJ, Butt AM, Gardenal E, Parpura V, Verkhratsky A (2016) Complex and differential glial responses in Alzheimer's disease and ageing. Curr Alzheimer Res 13:343–358
231. Rodriguez JJ, Jones VC, Tabuchi M, Allan SM, Knight EM, LaFerla FM, Oddo S, Verkhratsky A (2008) Impaired adult neurogenesis in the dentate gyrus of a triple transgenic mouse model of Alzheimer's disease. PLoS ONE 3:e2935
232. Rodríguez JJ, Matute C, Verkhratsky AI (ed)(2011) Neuroglia in Alzheimer's disease. In: Schemes E, Spray DC (eds) Astrocytes: wiring the brain, pp 311–337. Taylor & Francis Inc
233. Rodriguez JJ, Noristani HN, Olabarria M, Fletcher J, Somerville TD, Yeh CY, Verkhratsky A (2011) Voluntary running and environmental enrichment restores impaired hippocampal neurogenesis in a triple transgenic mouse model of Alzheimer's disease. Curr Alzheimer Res 8:707–717
234. Rodriguez JJ, Noristani HN, Verkhratsky A (2012) The serotonergic system in ageing and Alzheimer's disease. Prog Neurobiol 99:15–41
235. Rodriguez JJ, Olabarria M, Chvatal A, Verkhratsky A (2009) Astroglia in dementia and Alzheimer's disease. Cell Death Differ 16:378–385
236. Rodriguez JJ, Terzieva S, Olabarria M, Lanza RG, Verkhratsky A (2013) Enriched environment and physical activity reverse astrogliodegeneration in the hippocampus of AD transgenic mice. Cell Death Dis 4:e678
237. Rodriguez JJ, Verkhratsky A (2011) Neurogenesis in Alzheimer's disease. J Anat 219:78–89
238. Rodriguez JJ, Verkhratsky A (2011) Neuroglial roots of neurodegenerative diseases? Mol Neurobiol 43:87–96
239. Ronco V, Grolla AA, Glasnov TN, Canonico PL, Verkhratsky A, Genazzani AA, Lim D (2014) Differential deregulation of astrocytic calcium signalling by amyloid-β, TNFα, IL-1β and LPS. Cell Calcium 55:219–229
240. Rose CR, Verkhratsky A (2016) Principles of sodium homeostasis and sodium signalling in astroglia. Glia
241. Rossi D, Brambilla L, Valori CF, Roncoroni C, Crugnola A, Yokota T, Bredesen DE, Volterra A (2008) Focal degeneration of astrocytes in amyotrophic lateral sclerosis. Cell Death Differ 15:1691–1700

242. Rossi D, Volterra A (2009) Astrocytic dysfunction: insights on the role in neurodegeneration. Brain Res Bull 80:224–232
243. Rossner S, Lange-Dohna C, Zeitschel U, Perez-Polo JR (2005) Alzheimer's disease beta-secretase BACE1 is not a neuron-specific enzyme. J Neurochem 92:226–234
244. Roth M, Tomlinson BE, Blessed G (1966) Correlation between scores for dementia and counts of 'senile plaques' in cerebral grey matter of elderly subjects. Nature 209:109–110
245. Roth M, Tomlinson BE, Blessed G (1967) The relationship between quantitative measures of dementia and of degenerative changes in the cerebral grey matter of elderly subjects. Proc R Soc Med 60:254–260
246. Rusakov DA (2015) Disentangling calcium-driven astrocyte physiology. Nat Rev Neurosci 16:226–233
247. Samuel W, Masliah E, Hill LR, Butters N, Terry R (1994) Hippocampal connectivity and Alzheimer's dementia: effects of synapse loss and tangle frequency in a two-component model. Neurology 44:2081–2088
248. Sani S, Traul D, Klink A, Niaraki N, Gonzalo-Ruiz A, Wu CK, Geula C (2003) Distribution, progression and chemical composition of cortical amyloid-beta deposits in aged rhesus monkeys: similarities to the human. Acta Neuropathol 105:145–156
249. Satoh A, Iijima KM (2019) Roles of tau pathology in the locus coeruleus (LC) in age-associated pathophysiology and Alzheimer's disease pathogenesis: potential strategies to protect the LC against aging. Brain Res 1702:17–28
250. Savonenko A, Xu GM, Melnikova T, Morton JL, Gonzales V, Wong MP, Price DL, Tang F, Markowska AL, Borchelt DR (2005) Episodic-like memory deficits in the APPswe/PS1dE9 mouse model of Alzheimer's disease: relationships to beta-amyloid deposition and neurotransmitter abnormalities. Neurobiol Dis 18:602–617
251. Scheff SW, Price DA, Schmitt FA, Mufson EJ (2006) Hippocampal synaptic loss in early Alzheimer's disease and mild cognitive impairment. Neurobiol Aging 27:1372–1384
252. Schindowski K, Bretteville A, Leroy K, Begard S, Brion JP, Hamdane M, Buee L (2006) Alzheimer's disease-like tau neuropathology leads to memory deficits and loss of functional synapses in a novel mutated tau transgenic mouse without any motor deficits. Am J Pathol 169:599–616
253. Schousboe A, Wellendorph P, Frolund B, Clausen RP, Krogsgaard-Larsen P (2017) Astrocytic GABA transporters: pharmacological properties and targets for antiepileptic drugs. Adv Neurobiol 16:283–296
254. Selkoe DJ (2001) Alzheimer's disease: genes, proteins, and therapy. Physiol Rev 81:741–766
255. Selkoe DJ (2002) Alzheimer's disease is a synaptic failure. Science 298:789–791
256. Semyanov A (2019) Spatiotemporal pattern of calcium activity in astrocytic network. Cell Calcium 78:15–25
257. Shastry BS, Giblin FJ (1999) Genes and susceptible loci of Alzheimer's disease. Brain Res Bull 48:121–127
258. Shaughnessy LW, Barone S Jr, Mundy WR, Herr DW, Tilson HA (1994) Comparison of intracranial infusions of colchicine and ibotenic acid as models of neurodegeneration in the basal forebrain. Brain Res 637:15–26
259. Shigetomi E, Kracun S, Sofroniew MV, Khakh BS (2010) A genetically targeted optical sensor to monitor calcium signals in astrocyte processes. Nat Neurosci 13:759–766
260. Sidoryk-Wegrzynowicz M, Aschner M (2013) Role of astrocytes in manganese mediated neurotoxicity. BMC Pharmacol Toxicol 14:23
261. Simpson JE, Ince PG, Lace G, Forster G, Shaw PJ, Matthews F, Savva G, Brayne C, Wharton SB (2010) Astrocyte phenotype in relation to Alzheimer-type pathology in the ageing brain. Neurobiol Aging 31:578–590
262. Simpson JE, Ince PG, Shaw PJ, Heath PR, Raman R, Garwood CJ, Gelsthorpe C, Baxter L, Forster G, Matthews FE, Brayne C, Wharton SB (2011) Microarray analysis of the astrocyte transcriptome in the aging brain: relationship to Alzheimer's pathology and APOE genotype. Neurobiol Aging 32:1795–1807

263. Son SM, Cha MY, Choi H, Kang S, Choi H, Lee MS, Park SA, Mook-Jung I (2016) Insulin-degrading enzyme secretion from astrocytes is mediated by an autophagy-based unconventional secretory pathway in Alzheimer disease. Autophagy 12:784–800
264. Soucek T, Cumming R, Dargusch R, Maher P, Schubert D (2003) The regulation of glucose metabolism by HIF-1 mediates a neuroprotective response to amyloid beta peptide. Neuron 39:43–56
265. Srinivasan R, Huang BS, Venugopal S, Johnston AD, Chai H, Zeng H, Golshani P, Khakh BS (2015) Ca^{2+} signaling in astrocytes from $Ip3r2^{-/-}$ mice in brain slices and during startle responses *in vivo*. Nat Neurosci 18:708–717
266. Stenovec M, Kreft M, Grilc S, Pangrsic T, Zorec R (2008) EAAT2 density at the astrocyte plasma membrane and Ca^{2+}-regulated exocytosis. Mol Membr Biol 25:203–215
267. Stenovec M, Lasic E, Dominkus PP, Bobnar ST, Zorec R, Lenassi M, Kreft M (2019) Slow release of HIV-1 protein nef from vesicle-like structures is inhibited by cytosolic calcium elevation in single human microglia. Mol Neurobiol 56:102–118
268. Stenovec M, Trkov Bobnar S, Smolic T, Kreft M, Parpura V, Zorec R (2018) Presenilin PS1E9 disrupts mobility of secretory organelles in rat astrocytes. Acta Physiol (Oxf) 223:e13046
269. Stenovec M, Trkov S, Lasic E, Terzieva S, Kreft M, Rodriguez Arellano JJ, Parpura V, Verkhratsky A, Zorec R (2016) Expression of familial Alzheimer disease presenilin 1 gene attenuates vesicle traffic and reduces peptide secretion in cultured astrocytes devoid of pathologic tissue environment. Glia 64:317–329
270. Stix B, Reiser G (1998) β-amyloid peptide 25–35 regulates basal and hormone-stimulated Ca2+ levels in cultured rat astrocytes. Neurosci Lett 243:121–124
271. Studelska DR, Brimijoin S (1989) Partial isolation of two classes of dopamine beta-hydroxylase-containing particles undergoing rapid axonal transport in rat sciatic nerve. J Neurochem 53:622–631
272. Sturchler-Pierrat C, Abramowski D, Duke M, Wiederhold KH, Mistl C, Rothacher S, Ledermann B, Burki K, Frey P, Paganetti PA, Waridel C, Calhoun ME, Jucker M, Probst A, Staufenbiel M, Sommer B (1997) Two amyloid precursor protein transgenic mouse models with Alzheimer disease-like pathology. Proc Natl Acad Sci USA 94:13287–13292
273. Stutzmann GE (2007) The pathogenesis of Alzheimers disease is it a lifelong "calciumopathy"? Neuroscientist 13:546–559
274. Stutzmann GE, Mattson MP (2011) Endoplasmic reticulum Ca^{2+} handling in excitable cells in health and disease. Pharmacol Rev 63:700–727
275. Suarez-Fernandez MB, Soldado AB, Sanz-Medel A, Vega JA, Novelli A, Fernandez-Sanchez MT (1999) Aluminum-induced degeneration of astrocytes occurs via apoptosis and results in neuronal death. Brain Res 835:125–136
276. Sweeney MD, Montagne A, Sagare AP, Nation DA, Schneider LS, Chui HC, Harrington MG, Pa J, Law M, Wang DJJ, Jacobs RE, Doubal FN, Ramirez J, Black SE, Nedergaard M, Benveniste H, Dichgans M, Iadecola C, Love S, Bath PM, Markus HS, Salman RA, Allan SM, Quinn TJ, Kalaria RN, Werring DJ, Carare RO, Touyz RM, Williams SCR, Moskowitz MA, Katusic ZS, Lutz SE, Lazarov O, Minshall RD, Rehman J, Davis TP, Wellington CL, Gonzalez HM, Yuan C, Lockhart SN, Hughes TM, Chen CLH, Sachdev P, O'Brien JT, Skoog I, Pantoni L, Gustafson DR, Biessels GJ, Wallin A, Smith EE, Mok V, Wong A, Passmore P, Barkof F, Muller M, Breteler MMB, Roman GC, Hamel E, Seshadri S, Gottesman RF, van Buchem MA, Arvanitakis Z, Schneider JA, Drewes LR, Hachinski V, Finch CE, Toga AW, Wardlaw JM, Zlokovic BV (2019) Vascular dysfunction-The disregarded partner of Alzheimer's disease. Alzheimers Dement 15:158–167
277. Sweeney MD, Zhao Z, Montagne A, Nelson AR, Zlokovic BV (2019) Blood-brain barrier: from physiology to disease and back. Physiol Rev 99:21–78
278. Takano T, Han X, Deane R, Zlokovic B, Nedergaard M (2007) Two-photon imaging of astrocytic Ca^{2+} signaling and the microvasculature in experimental mice models of Alzheimer's disease. Ann N Y Acad Sci 1097:40–50
279. Takano T, Tian GF, Peng W, Lou N, Libionka W, Han X, Nedergaard M (2006) Astrocyte-mediated control of cerebral blood flow. Nat Neurosci 9:260–267

280. Tanemura K, Murayama M, Akagi T, Hashikawa T, Tominaga T, Ichikawa M, Yamaguchi H, Takashima A (2002) Neurodegeneration with tau accumulation in a transgenic mouse expressing V337M human tau. J Neurosci 22:133–141
281. Tatebayashi Y, Miyasaka T, Chui DH, Akagi T, Mishima K, Iwasaki K, Fujiwara M, Tanemura K, Murayama M, Ishiguro K, Planel E, Sato S, Hashikawa T, Takashima A (2002) Tau filament formation and associative memory deficit in aged mice expressing mutant (R406W) human tau. Proc Natl Acad Sci USA 99:13896–13901
282. Terry RD (2000) Cell death or synaptic loss in Alzheimer disease. J Neuropathol Exp Neurol 59:1118–1119
283. Thompson KA, McArthur JC, Wesselingh SL (2001) Correlation between neurological progression and astrocyte apoptosis in HIV-associated dementia. Ann Neurol 49:745–752
284. Thompson PM, Hayashi KM, de Zubicaray G, Janke AL, Rose SE, Semple J, Herman D, Hong MS, Dittmer SS, Doddrell DM, Toga AW (2003) Dynamics of gray matter loss in Alzheimer's disease. J Neurosci 23:994–1005
285. Thompson PM, Hayashi KM, Dutton RA, Chiang MC, Leow AD, Sowell ER, De Zubicaray G, Becker JT, Lopez OL, Aizenstein HJ, Toga AW (2007) Tracking Alzheimer's disease. Ann NY Acad Sci 1097:183–214
286. Toescu EC, Verkhratsky A (2007) The importance of being subtle: small changes in calcium homeostasis control cognitive decline in normal aging. Aging Cell 6:267–273
287. Toescu EC, Verkhratsky A, Landfield PW (2004) Ca2+ regulation and gene expression in normal brain aging. Trends Neurosci 27:614–620
288. Toivari E, Manninen T, Nahata AK, Jalonen TO, Linne ML (2011) Effects of transmitters and amyloid-beta peptide on calcium signals in rat cortical astrocytes: Fura-2AM measurements and stochastic model simulations. PLoS ONE 6:e17914
289. Toledano A, Alvarez MI (2004) Lesions and dysfunctions of the nucleus basalis as Alzheimer's disease models: general and critical overview and analysis of the long-term changes in several excitotoxic models. Curr Alzheimer Res 189–214
290. Tomlinson BE, Blessed G, Roth M (1970) Observations on the brains of demented old people. J Neurol Sci 11:205–242
291. Torack R (1996) The early history of senile dementia. In: Reisberg B (ed) Alzheimer's disease: the standard reference. The Free Press, New York, pp 23–28
292. Ullian EM, Christopherson KS, Barres BA (2004) Role for glia in synaptogenesis. Glia 47:209–216
293. Vale-Martinez A, Guillazo-Blanch G, Marti-Nicolovius M, Nadal R, Arevalo-Garcia R, Morgado-Bernal I (2002) Electrolytic and ibotenic acid lesions of the nucleus basalis magnocellularis interrupt long-term retention, but not acquisition of two-way active avoidance, in rats. Exp Brain Res 142:52–66
294. Van Cauwenberghe C, Van Broeckhoven C, Sleegers K (2016) The genetic landscape of Alzheimer disease: clinical implications and perspectives. Genet Med 18:421–430
295. Vanzani MC, Iacono RF, Caccuri RL, Troncoso AR, Berria MI (2006) Regional differences in astrocyte activation in HIV-associated dementia. Medicina (B Aires) 66:108–112
296. Vardjan N, Gabrijel M, Potokar M, Svajger U, Kreft M, Jeras M, de Pablo Y, Faiz M, Pekny M, Zorec R (2012) IFN-γ-induced increase in the mobility of MHC class II compartments in astrocytes depends on intermediate filaments. J Neuroinflammation 9:144
297. Vardjan N, Kreft M, Zorec R (2014) Dynamics of β-adrenergic/cAMP signaling and morphological changes in cultured astrocytes. Glia 62:566–579
298. Vardjan N, Verkhratsky A, Zorec R (2015) Pathologic potential of astrocytic vesicle traffic: new targets to treat neurologic diseases? Cell Transplant 24:599–612
299. Verkhratsky A, Kettenmann H (1996) Calcium signalling in glial cells. Trends Neurosci 19:346–352
300. Verkhratsky A, Marutle A, Rodriguez-Arellano JJ, Nordberg A (2015) Glial asthenia and functional paralysis: a new perspective on neurodegeneration and Alzheimer's disease. Neuroscientist 21:552–568

301. Verkhratsky A, Matteoli M, Parpura V, Mothet JP, Zorec R (2016) Astrocytes as secretory cells of the central nervous system: idiosyncrasies of vesicular secretion. EMBO J 35:239–257
302. Verkhratsky A, Mattson MP, Toescu EC (2004) Aging in the mind. Trends Neurosci 27:577–578
303. Verkhratsky A, Nedergaard M (2014) Astroglial cradle in the life of the synapse. Philos Trans R Soc Lond B Biol Sci 369:20130595
304. Verkhratsky A, Nedergaard M (2016) The homeostatic astroglia emerges from evolutionary specialization of neural cells. Philos Trans R Soc Lond B Biol Sci 371
305. Verkhratsky A, Nedergaard M (2018) Physiology of astroglia. Physiol Rev 98:239–389
306. Verkhratsky A, Olabarria M, Noristani HN, Yeh CY, Rodriguez JJ (2010) Astrocytes in Alzheimer's disease. Neurotherapeutics 7:399–412
307. Verkhratsky A, Parpura V (2016) Astrogliopathology in neurological, neurodevelopmental and psychiatric disorders. Neurobiol Dis 85:254–261
308. Verkhratsky A, Parpura V, Rodriguez JJ (2014) Neurodegeneration and neuroglia: emphasis on astroglia in Alzheimer's disease. In: Parpura V, Verkhratsky A (eds) Pathological potential of neuroglia: possible new targets for medical intervention. Springer, Heidelberg, pp 264–291
309. Verkhratsky A, Rodriguez-Arellano JJ, Parpura V, Zorec R (2017) Astroglial calcium signalling in Alzheimer's disease. Biochem Biophys Res Commun 483:1005–1012
310. Verkhratsky A, Rodriguez JJ, Parpura V (2013) Astroglia in neurological diseases. Future Neurol 8:149–158
311. Verkhratsky A, Sofroniew MV, Messing A, deLanerolle NC, Rempe D, Rodriguez JJ, Nedergaard M (2012) Neurological diseases as primary gliopathies: a reassessment of neurocentrism. ASN Neuro 4
312. Verkhratsky A, Toescu EC (1998) Calcium and neuronal ageing. Trends Neurosci 21:2–7
313. Verkhratsky A, Trebak M, Perocchi F, Khananshvili D, Sekler I (2018) Crosslink between calcium and sodium signalling. Exp Physiol 103:157–169
314. Verkhratsky A, Untiet V, Rose CR (2019) Ionic signalling in astroglia beyond calcium. J Physiol
315. Verkhratsky A, Zorec R, Parpura V (2017) Stratification of astrocytes in healthy and diseased brain. Brain Pathol 27:629–644
316. Verkhratsky A, Zorec R, Rodriguez JJ, Parpura V (2016) Astroglia dynamics in ageing and Alzheimer's disease. Curr Opin Pharmacol 26:74–79
317. Verkhratsky A, Zorec R, Rodriguez JJ, Parpura V (2017) Neuroglia: functional paralysis and reactivity in Alzheimer's disease and other neurodegenerative pathologies. Adv Neurobiol 15:427–449
318. Waite JJ, Chen AD, Wardlow ML, Wiley RG, Lappi DA, Thal LJ (1995) 192 immunoglobulin G-saporin produces graded behavioral and biochemical changes accompanying the loss of cholinergic neurons of the basal forebrain and cerebellar Purkinje cells. Neuroscience 65:463–476
319. Walls AB, Waagepetersen HS, Bak LK, Schousboe A, Sonnewald U (2015) The glutamine-glutamate/GABA cycle: function, regional differences in glutamate and GABA production and effects of interference with GABA metabolism. Neurochem Res 40:402–409
320. Weidenheim KM, Dickson DW, Rapin I (2009) Neuropathology of Cockayne syndrome: evidence for impaired development, premature aging, and neurodegeneration. Mech Ageing Dev 130:619–636
321. Weinshilboum RM (1978) Serum dopamine beta-hydroxylase. Pharmacol Rev 30:133–166
322. Wellman CL, Pelleymounter MA (1999) Differential effects of nucleus basalis lesions in young adult and aging rats. Neurobiol Aging 20:381–393
323. Wernicke C (1881–1883) Lehrbuch der Gehirnkrankheiten für Aerzte und Studirende. Theodor Fischer, Kassel und Berlin
324. Wharton SB, O'Callaghan JP, Savva GM, Nicoll JA, Matthews F, Simpson JE, Forster G, Shaw PJ, Brayne C, Ince PG (2009) Population variation in glial fibrillary acidic protein levels in brain ageing: relationship to Alzheimer-type pathology and dementia. Dement Geriatr Cogn Disord 27:465–473

325. Wiley RG (1992) Neural lesioning with ribosome-inactivating proteins: suicide transport and immunolesioning. Trends Neurosci 15:285–290
326. Wiley RG, Kline IR (2000) Neuronal lesioning with axonally transported toxins. J Neurosci Methods 103:73–82
327. Wiley RG, Oeltmann TN, Lappi DA (1991) Immunolesioning: selective destruction of neurons using immunotoxin to rat NGF receptor. Brain Res 562:149–153
328. Wilson CS, Mongin AA (2019) The signaling role for chloride in the bidirectional communication between neurons and astrocytes. Neurosci Lett 689:33–44
329. Wilson RS, Nag S, Boyle PA, Hizel LP, Yu L, Buchman AS, Schneider JA, Bennett DA (2013) Neural reserve, neuronal density in the locus ceruleus, and cognitive decline. Neurology 80:1202–1208
330. Winkler J, Thal LJ (1995) Effects of nerve growth factor treatment on rats with lesions of the nucleus basalis magnocellularis produced by ibotenic acid, quisqualic acid, and AMPA. Exp Neurol 136:234–250
331. Wrenn CC, Picklo MJ, Lappi DA, Robertson D, Wiley RG (1996) Central noradrenergic lesioning using anti-DBH-saporin: anatomical findings. Brain Res 740:175–184
332. Wu Z, Guo Z, Gearing M, Chen G (2014) Tonic inhibition in dentate gyrus impairs long-term potentiation and memory in an Alzheimer's [corrected] disease model. Nat Commun 5:4159
333. Wyss-Coray T, Loike JD, Brionne TC, Lu E, Anankov R, Yan F, Silverstein SC, Husemann J (2003) Adult mouse astrocytes degrade amyloid-β in vitro and in situ. Nat Med 9:453–457
334. Xiong J, Verkhratsky A, Toescu EC (2002) Changes in mitochondrial status associated with altered Ca2+ homeostasis in aged cerebellar granule neurons in brain slices. J Neurosci 22:10761–10771
335. Xiu J, Nordberg A, Zhang JT, Guan ZZ (2005) Expression of nicotinic receptors on primary cultures of rat astrocytes and up-regulation of the α7, α4 and β2 subunits in response to nanomolar concentrations of the β-amyloid peptide1-42. Neurochem Int 47:281–290
336. Yamaguchi H, Sugihara S, Ogawa A, Saido TC, Ihara Y (1998) Diffuse plaques associated with astroglial amyloid beta protein, possibly showing a disappearing stage of senile plaques. Acta Neuropathol 95:217–222
337. Yamanaka K, Chun SJ, Boillee S, Fujimori-Tonou N, Yamashita H, Gutmann DH, Takahashi R, Misawa H, Cleveland DW (2008) Astrocytes as determinants of disease progression in inherited amyotrophic lateral sclerosis. Nat Neurosci 11:251–253
338. Yeh CY, Vadhwana B, Verkhratsky A, Rodriguez JJ (2011) Early astrocytic atrophy in the entorhinal cortex of a triple transgenic animal model of Alzheimer's disease. ASN Neuro 3:271–279
339. Yin Z, Milatovic D, Aschner JL, Syversen T, Rocha JB, Souza DO, Sidoryk M, Albrecht J, Aschner M (2007) Methylmercury induces oxidative injury, alterations in permeability and glutamine transport in cultured astrocytes. Brain Res 1131:1–10
340. Young LT, Kish SJ, Li PP, Warsh JJ (1988) Decreased brain [^{3}H]inositol 1,4,5-trisphosphate binding in Alzheimer's disease. Neurosci Lett 94:198–202
341. Yu WF, Guan ZZ, Bogdanovic N, Nordberg A (2005) High selective expression of α7 nicotinic receptors on astrocytes in the brains of patients with sporadic Alzheimer's disease and patients carrying Swedish APP 670/671 mutation: a possible association with neuritic plaques. Exp Neurol 192:215–225
342. Zhao J, O'Connor T, Vassar R (2011) The contribution of activated astrocytes to Abeta production: implications for Alzheimer's disease pathogenesis. J Neuroinflammation 8:150
343. Zhu CW, Sano M (2006) Economic considerations in the management of Alzheimer's disease. Clin Interv Aging 1:143–154
344. Ziemens D, Oschmann F, Gerkau NJ, Rose CR (2019) Heterogeneity of activity-induced sodium transients between astrocytes of the mouse hippocampus and neocortex: mechanisms and consequences. J Neurosci 39:2620–2634
345. Zlokovic BV (2008) The blood-brain barrier in health and chronic neurodegenerative disorders. Neuron 57:178–201

346. Zonta M, Angulo MC, Gobbo S, Rosengarten B, Hossmann KA, Pozzan T, Carmignoto G (2003) Neuron-to-astrocyte signaling is central to the dynamic control of brain microcirculation. Nat Neurosci 6:43–50
347. Zorec R, Parpura V, Vardjan N, Verkhratsky A (2017) Astrocytic face of Alzheimer's disease. Behav Brain Res 322:250–257
348. Zorec R, Parpura V, Verkhratsky A (2018) Preventing neurodegeneration by adrenergic astroglial excitation. FEBS J 285:3645–3656
349. Zorec R, Verkhratsky A, Rodriguez JJ, Parpura V (2016) Astrocytic vesicles and gliotransmitters: Slowness of vesicular release and synaptobrevin2-laden vesicle nanoarchitecture. Neuroscience 323:67–75
350. Zott B, Busche MA, Sperling RA, Konnerth A (2018) What happens with the circuit in Alzheimer's disease in mice and humans? Annu Rev Neurosci 41:277–297

Chapter 12
Oligodendroglial Cells in Alzheimer's Disease

Arthur M. Butt, Irene Chacon De La Rocha and Andrea Rivera

Abstract Oligodendrocytes form the myelin that ensheaths CNS axons, which is essential for rapid neuronal signalling and underpins the massive computing power of the human brain. Oligodendrocytes and myelin also provide metabolic and trophic support for axons and their disruption results in axonal demise and neurodegeneration, which are key features of Alzheimer's disease (AD). Notably, the brain has a remarkable capacity for regenerating oligodendrocytes, which is the function of adult oligodendrocyte progenitor cells (OPCs) or NG2-glia. White matter loss is often among the earliest brain changes in AD, preceding the tangles and plaques that characterize neuronal deficits. The underlying causes of myelin loss include oxidative stress, neuroinflammation and excitotoxicity, associated with accumulation of Aβ and tau hyperphosphorylation, pathological hallmarks of AD. Moreover, there is evidence that NG2-glia are disrupted in AD, which may be associated with disruption of synaptic signalling. This has led to the hypothesis that a vicious cycle of myelin loss and failure of regeneration from NG2-glia plays a key role in AD. Therapies that target NG2-glia are likely to have positive effects on myelination and neuroprotection in AD.

Keywords Oligodendrocyte · Oligodendrocyte precursor cell · OPC · NG2-glia · Myelin · Axon

12.1 Introduction

Alzheimer's disease (AD) is characterized by a loss of neurones and synapses, with an associated progressive decline in cognitive function and dementia [56]. The pathological hallmarks of AD are deposition of amyloid-β (Aβ) plaques and neurofibrillary tangles of hyperphosphorylated tau, although their specific roles in neuronal demise remain unclear [52]. AD is classified as a neurodegenerative disorder, but glial cells

A. M. Butt (✉) · I. C. De La Rocha · A. Rivera
School of Pharmacy and Biomedical Science, University of Portsmouth, St. Michael's Building, White Sawn Road, Portsmouth PO1 2DT, UK
e-mail: Arthur.butt@port.ac.uk

A. Verkhratsky et al. (eds.), *Neuroglia in Neurodegenerative Diseases*,
Advances in Experimental Medicine and Biology 1175,
https://doi.org/10.1007/978-981-13-9913-8_12

are intricately involved in disease progression [46]. Indeed, disruption of white matter (WM) is a major element in AD and, in many cases, myelin disruption may precede overt neuropathology [6]. Indeed, loss of myelin can result in axonal and neuronal degeneration [51]. The causes of myelin disruption in AD are likely to be complex and to include oxidative stress, together with glutamate, iron and metabolic dyshomeostasis [42, 55, 60]. Furthermore, Aβ may be directly toxic to oligodendrocytes and their progenitors [19, 30, 43]. In addition, astrocytes and microglia are markedly altered in AD (Chaps. 12 and 14), which will impact on oligodendrocyte degenerative changes and regeneration from oligodendrocyte progenitor cells (OPCs), or NG2-glia, which are responsible for the life-long generation of oligodendrocytes (Chap. 5). Studies in human post-mortem tissue and mouse models provide evidence that NG2-glia are altered in AD, which may be indicative of reduced regenerative capacity [21, 43, 62]. The self-renewal and differentiation of NG2-glia is modulated by synaptic activity [13], and is considered to underpin adult oligodendrogenesis and experience-dependent or 'adaptive' myelination, which plays a critical role in neuronal network remodelling and learning [25, 28, 38, 58]. On top of this, myelin is required for the integrity and survival of axons [51], due at least in part to axonal metabolic support provided by oligodendrocytes [1]. Thus, neurones and their axons, together with oligodendrocytes and their myelin, are interdependent functional units, whereby loss or disruption of one affects the others. On top of this, a decline in NG2-glia self-renewal results in a vicious cycle of neuronal disruption, myelin loss, and failure of regeneration (Fig. 12.1) [45]. Hence, novel therapies for rejuvenating oligodendrogenesis in the ageing brain have the potential for neuroprotection in AD.

12.2 Evidence of WM Disruption in Human AD

Magnetic resonance imaging (MRI) and post-mortem studies have demonstrated reduced WM volume and alterations of WM microstructure in AD [6, 41]. WM abnormalities correlate with phosphorylated tau 181/β-amyloid 42 [17], and it appears late myelinated regions are more vulnerable to myelin breakdown and AD pathology compared to areas that meylinate earlier, referred to as 'neuropathologic retrogenesis' [5, 9]. Moreover, analysis of human samples enriched for high AD risk (APOE ε4 and parental history of AD) suggests that WM degeneration is an early pathological feature of AD [27]. Indeed, WM changes are detectable in preclinical AD and may precede overt neurodegenerative changes [23], suggesting oligodendrocyte disruption and myelin loss may be primary events in AD pathology [41]. Nonetheless, some degree of WM abnormalities in AD is associated with cortical neurodegeneration [36], and a number of studies have emphasized the importance of vascular disease in the development of WM abnormalities in AD [10].

Post-mortem analyses support widespread abnormalities in myelin and oligodendrocytes in AD [41], and genomic analyses have identified that oligodendrocyte genes are dysregulated in AD and are associated with AD risk variants, such as BIN1 and

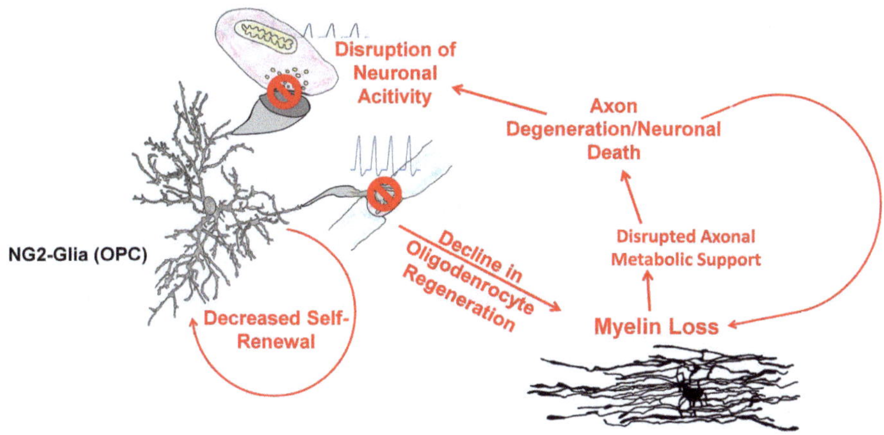

Fig. 12.1 Disruption of NG2-glia has a negative impact on myelin replacement in AD. NG2-glial cells regenerate myelinating oligodendrocytes throughout life, which is essential for replacing myelin lost through pathology and providing new myelin in response to new life experiences. NG2-glia contact neurons at synapses, the sites of neurotransmission, and nodes of Ranvier, the sites of action potential propagation along axons. Neuronal activity helps drive self-renewal and differentiation of NG2-glia, termed 'adaptive myelination', which is important for neural circuit remodelling and learning. Disruption of synaptic signalling is a key factor in AD and would have adverse effects on NG2-glia, which are disrupted in AD. In this scenario, loss of myelin and oligodendrocytes results in disruption of their metabolic and trophic support for axons, resulting in axonal demise and neurodegeneration, all of which are aggravated by reduced regenerativae capacity of NG2-glia . This vicious cycle may be a key factor in the pathogenesis of AD and therapies that promote oligodendrocyte regeneration and myelination are likely to have important neuroprotective effects, and vice versa

GOT2 [37]. In the human brain, mature oligodendrocytes are the main cells expressing BIN1, which has diverse functions in membrane remodelling, implicating it in myelin disruption in AD [16]. Myelin degeneration has been demonstrated throughout frontal and temporal lobes in AD [29], and biochemical analyses of post-mortem WM revealed decreased myelin basic protein (MBP), myelin proteolipid protein (PLP), cyclic nucleotide phosphohydrolase (CNPase) and cholesterol in AD [47]. Post-mortem analyses have demonstrated a loss of Olig2+ oligodendrocyte lineage cells and NG2-glia in AD [7, 43]. Myelin injury in AD cortex is associated with axon degeneration and amyloid plaques [61], and focal loss of oligodendrocytes in AD is associated with Aβ plaque cores [40]. A recent study has indicated that Olig2+ and NG2-glia associated with Aβ plaques exhibited a 'senescence-like' phenotype and suggested a role for Aβ-induced OPC cell senescence in cognitive deficits in AD [62]. Thus, myelin deficiencies are evident in AD and are associated with senescence of NG2-glia and loss of oligodendrocytes.

12.3 Oligodendrocyte and Myelin Changes in Animal Models of AD

Studies in animal models of AD support human evidence that oligodendrocyte disruption and myelin loss is an early event in AD pathology [46]. A number of studies provide evidence of Aβ toxicity in oligodendrocytes [18, 19], and myelin disruption correlates with the earliest appearance of Aβ accumulation in the 3×Tg-AD mouse model [20]. Similarly, studies in the APP/PS1 mouse model of AD demonstrate oligodendrocyte differentiation is disrupted [57], together with downregulation of MBP, shrinkage of the corpus callosum, increased NG2-glia and behavioural deficits [21]. Recent studies in the 5xFAD mouse model of AD demonstrate myelin deficits occur at an early stage and progress with ageing [24], with evidence that subcellular accumulation of Aβ drives axonopathy and myelinopathy [14]. However, there are contradictions between mouse models and human AD, for example, Olig2+ cells and NG2-glia are increased in APP/PS1 mice, but are decreased in human AD pathology [7, 43]. In addition, in vitro studies indicate NG2-glia engulf Aβ peptides and degrade them by autophagy [32], and provide evidence that Aβ promotes oligodendrocyte differentiation, maturation and survival [44]. Thus, Aβ may not directly cause oligodendrocyte loss and myelin disruption in AD, but instead may be related to OPC-mediated repair mechanisms.

12.4 Mechanisms of Oligodendrocyte and Myelin Disruption in AD

The molecular mechanisms leading to myelin loss in AD have not been elucidated, but are likely to include oxidative stress, neuroinflammation and excitotoxicity [41]. Multiple studies implicate Aβ in oligodendrocyte dysfunction and myelin breakdown, which could be due to direct Aβ toxicity in oligodendroglial cells or oxidative stress [19, 31, 59], but equally could be related to OPC-mediated repair mechanisms [44]. It is important to note that clinical trials targeting the removal of Aβ plaques did not prevent progressive neurodegeneration and cognitive decline in AD patients [26]. Moreover, early changes in oligodendrocytes and myelin precede Aβ deposition in the 3xTg mouse [20]. Hence, the primary importance of Aβ as causative in oligodendrocyte and myelin pathology is no less clear than its role in neuronal pathology [50]. It also seems that WM disruption and myelin loss in AD is associated with tau hyperphosphorylation [35], and maybe triggered by the formation of NFTs, with oxidative stress as a common factor [11].

A feature that is common to both AD and oligodendrocyte pathology is dysregulation of glutamate signalling and Ca^{2+} dyshomeostasis [22, 46]. Oligodendrocytes and myelin express ligand-gated channels that are permeable to Ca^{2+}, including glutamate receptors (Chap. 5). Prolonged activation of NMDAR triggers oligodendrocyte death and myelin destruction [39, 49], and this can be partly offset by the NMDAR receptor blocker memantine [4]. Interestingly, NMDAR contributes to synaptic dysfunction in AD and this can be alleviated by treatment with memantine [33]. These studies raise the possibility that the effects of memantine on oligodendrocytes and myelin may play an important role in its neuroprotective effects in AD. Moreover, pathological changes in oligodendrocytes and/or myelin may compromise their trophic support of axons [1]. Hence, therapies that protect oligodendrocytes are also likely to be neuroprotective.

In addition to its role in pathology, glutamatergic signalling is also implicated in the regulation of NG2-glia self-renewal and differentiation (Chap. 5). Glutamate released from electrically active axons acting on AMPAR promotes proliferation and differentiation of NG2-glia [12, 53], which is important for neural circuit remodelling and is gradually lost with ageing [25]. Altered glutamatergic synaptic signalling is a major component of AD [15], which would impact on oligodendrogenesis [45]. In addition, ablation of NG2-glia causes deficits in glutamatergic neurotransmission and depressive-like behaviour in mice [8, 48]. Thus, the changes in NG2-glia observed in AD [7, 21, 43] are likely to impact upon information processing in multiple ways.

The pivotal role of Glycogen Synthase Kinase-3 (GSK-3β) in the formation of Aβ plaques and NFTs has identified GSK-3β as a key factor in AD progression and a relevant therapeutic target [34]. In addition, activation of Wnt signalling through inhibition of GSK-3β, a key negative regulator of Wnt signalling, is able to protect against Aβ toxicity and ameliorate cognitive performance in AD [54]. It is noteworthy, therefore, that we have identified a persistent role for Wnt signalling in driving oligodendrogenesis in the adult brain [3]. Moreover, Wnt signalling declines in ageing and targeting Wnt signalling with GSK3β inhibitors can rejuvenate the regenerative capacity of the ageing brain [2]. These studies indicate that manipulating the GSK3β-Wnt signalling pathway may be a potential treatment for promoting myelination and neuroprotection in AD.

12.5 Concluding Remarks

In summary, oligodendrocyte and myelin disturbances are pathological features in AD and may even precede and predict overt neuropathology [5, 6]. Although the molecular mechanisms leading to myelin loss in AD have not been elucidated [41], there is an age-related decline in myelination that is accelerated in AD and the loss of axonal trophic support provided by oligodendrocytes is likely to contribute to neurodegenerative changes. It seems unlikely that it will be possible to determine unequivocally whether myelin loss is secondary to or is a primary contributor to neuronal demise. However, this may be a moot point, since oligodendrocytes and the axons they myelinate are interdependent units, and loss or disruption of one will have adverse effects on the other. In addition, myelin loss will be aggravated by the apparent age-related decline in NG2-glial cell regenerative capacity, most likely associated with the disruption of neuronal signalling [45]. Thus, novel therapies for rejuvenating oligodendrogenesis in the ageing brain, such as targeting GSK3β-Wnt signalling, have the potential for neuroprotection in AD.

Acknowledgements The authors are supported by grants from the BBSRC, MRC and Multiple Sclerosis Society of the UK.

References

1. Alexandra IM, Constanze D, Klaus-Armin N (2018) An emerging role of dysfunctional axon-oligodendrocyte coupling in neurodegenerative diseases. Dialogues Clin Neurosci 20:283–292
2. Azim K, Angonin D, Marcy G, Pieropan F, Rivera A, Donega V, Cantu C, Williams G, Berninger B, Butt AM, Raineteau O (2017) Pharmacogenomic identification of small molecules for lineage specific manipulation of subventricular zone germinal activity. PLoS Biol 15:e2000698
3. Azim K, Fischer B, Hurtado-Chong A, Draganova K, Cantu C, Zemke M, Sommer L, Butt A, Raineteau O (2014) Persistent Wnt/beta-catenin signaling determines dorsalization of the postnatal subventricular zone and neural stem cell specification into oligodendrocytes and glutamatergic neurons. Stem Cells 32:1301–1312
4. Bakiri Y, Hamilton NB, Káradóttir R, Attwell D (2008) Testing NMDA receptor block as a therapeutic strategy for reducing ischaemic damage to CNS white matter. Glia 56:233–240
5. Bartzokis G (2004) Age-related myelin breakdown: a developmental model of cognitive decline and Alzheimer's disease. Neurobiol Aging 25, 5–18; author reply 49–62
6. Bartzokis G (2011) Alzheimer's disease as homeostatic responses to age-related myelin breakdown. Neurobiol Aging 32:1341–1371
7. Behrendt G, Baer K, Buffo A, Curtis MA, Faull RL, Rees MI, Gotz M, Dimou L (2013) Dynamic changes in myelin aberrations and oligodendrocyte generation in chronic amyloidosis in mice and men. Glia 61:273–286
8. Birey F, Kloc M, Chavali M, Hussein I, Wilson M, Christoffel DJ, Chen T, Frohman MA, Robinson JK, Russo SJ, Maffei A, Aguirre A (2015) Genetic and stress-induced loss of NG2 Glia triggers emergence of depressive-like behaviors through reduced secretion of FGF2. Neuron 88:941–956
9. Braak H, Braak E (1996) Development of Alzheimer-related neurofibrillary changes in the neocortex inversely recapitulates cortical myelogenesis. Acta Neuropathol 92:197–201
10. Brown WR, Thore CR (2011) Review: cerebral microvascular pathology in ageing and neurodegeneration. Neuropathol Appl Neurobiol 37:56–74
11. Cai Z, Xiao M (2016) Oligodendrocytes and Alzheimer's disease. Int J Neurosci 126:97–104
12. Chen TJ, Kula B, Nagy B, Barzan R, Gall A, Ehrlich I, Kukley M (2018) In vivo regulation of oligodendrocyte precursor cell proliferation and differentiation by the AMPA-receptor subunit GluA2. Cell Rep 25:852–861.e7
13. Chorghay Z, Káradóttir RT, Ruthazer ES (2018) White matter plasticity keeps the brain in tune: axons conduct while glia wrap. Front Cell Neurosc 12:428–428
14. Chu T-H, Cummins K, Sparling JS, Tsutsui S, Brideau C, Nilsson KPR, Joseph JT, Stys PK (2017) Axonal and myelinic pathology in 5xFAD Alzheimer's mouse spinal cord. PLoS ONE 12:e0188218–e0188218
15. Crimins JL, Pooler A, Polydoro M, Luebke JI, Spires-Jones TL (2013) The intersection of amyloid beta and tau in glutamatergic synaptic dysfunction and collapse in Alzheimer's disease. Ageing Res Rev 12:757–763

16. de Rossi P, Buggia-Prevot V, Clayton BL, Vasquez JB, van Sanford C, Andrew RJ, Lesnick R, Botte A, Deyts C, Salem S, Rao E, Rice RC, Parent A, Kar S, Popko B, Pytel P, Estus S, Thinakaran G (2016) Predominant expression of Alzheimer's disease-associated BIN1 in mature oligodendrocytes and localization to white matter tracts. Mol Neurodegener 11:59
17. Dean DC, III, Hurley SA, Kecskemeti SR, O'Grady JP, Canda C, Davenport-Sis NJ, Carlsson CM, Zetterberg H, Blennow K, Asthana S, Sager MA, Johnson SC, Alexander AL, Bendlin BB (2017) Association of amyloid pathology with myelin alteration in preclinical alzheimer disease. JAMA Neurol 74:41–49
18. Desai MK, Guercio BJ, Narrow WC, Bowers WJ (2011) An Alzheimer's disease-relevant presenilin-1 mutation augments amyloid-beta-induced oligodendrocyte dysfunction. Glia 59:627–640
19. Desai MK, Mastrangelo MA, Ryan DA, Sudol KL, Narrow WC, Bowers WJ (2010) Early oligodendrocyte/myelin pathology in Alzheimer's disease mice constitutes a novel therapeutic target. Am J Pathol 177:1422–1435
20. Desai MK, Sudol KL, Janelsins MC, Mastrangelo MA, Frazer ME, Bowers WJ (2009) Triple-transgenic Alzheimer's disease mice exhibit region-specific abnormalities in brain myelination patterns prior to appearance of amyloid and tau pathology. Glia 57:54–65
21. Dong YX, Zhang HY, Li HY, Liu PH, Sui Y, Sun XH (2018) Association between Alzheimer's disease pathogenesis and early demyelination and oligodendrocyte dysfunction. Neural Regen Res 13:908–914
22. Fern RF, Matute C, Stys PK (2014) White matter injury: ischemic and nonischemic. Glia 62:1780–1789
23. Fischer FU, Wolf D, Scheurich A, Fellgiebel A (2015) Altered whole-brain white matter networks in preclinical Alzheimer's disease. Neuroimage Clin 8:660–666
24. Gu L, Wu D, Tang X, Qi X, Li X, Bai F, Chen X, Ren Q, Zhang Z (2018) Myelin changes at the early stage of 5XFAD mice. Brain Res Bull 137:285–293
25. Hill RA, Li AM, Grutzendler J (2018) Lifelong cortical myelin plasticity and age-related degeneration in the live mammalian brain. Nat Neurosci 21:683–695
26. Holmes C, Boche D, Wilkinson D, Yadegarfar G, Hopkins V, Bayer A, Jones RW, Bullock R, Love S, Neal JW, Zotova E, Nicoll JA (2008) Long-term effects of Abeta42 immunisation in Alzheimer's disease: follow-up of a randomised, placebo-controlled phase I trial. Lancet 372:216–223
27. Hoy AR, Ly M, Carlsson CM, Okonkwo OC, Zetterberg H, Blennow K, Sager MA, Asthana S, Johnson SC, Alexander AL, Bendlin BB (2017) Microstructural white matter alterations in preclinical Alzheimer's disease detected using free water elimination diffusion tensor imaging. PLoS ONE 12:e0173982
28. Hughes EG, Orthmann-Murphy JL, Langseth AJ, Bergles DE (2018) Myelin remodeling through experience-dependent oligodendrogenesis in the adult somatosensory cortex. Nat Neurosci 21:696–706
29. Ihara M, Polvikoski TM, Hall R, Slade JY, Perry RH, Oakley AE, Englund E, O'Brien JT, Ince PG, Kalaria RN (2010) Quantification of myelin loss in frontal lobe white matter in vascular dementia, Alzheimer's disease, and dementia with Lewy bodies. Acta Neuropathol 119:579–589
30. Jantaratnotai N, Ryu JK, Kim SU, McLarnon JG (2003) Amyloid beta peptide-induced corpus callosum damage and glial activation in vivo. NeuroReport 14:1429–1433
31. Lee JT, Xu J, Lee JM, Ku G, Han X, Yang DI, Chen S, Hsu CY (2004) Amyloid-beta peptide induces oligodendrocyte death by activating the neutral sphingomyelinase-ceramide pathway. J Cell Biol 164:123–131
32. Li W, Tang Y, Fan Z, Meng Y, Yang G, Luo J, Ke ZJ (2013) Autophagy is involved in oligodendroglial precursor-mediated clearance of amyloid peptide. Mol Neurodegener 8:27
33. Liu J, Chang L, Song Y, Li H, Wu Y (2019) The Role of NMDA Receptors in Alzheimer's Disease. Front Neurosci 13:43–43
34. Llorens-Martin M, Jurado J, Hernandez F, Avila J (2014) GSK-3beta, a pivotal kinase in Alzheimer disease. Front Mol Neurosci 7:46

35. McAleese KE, Firbank M, Dey M, Colloby SJ, Walker L, Johnson M, Beverley JR, Taylor JP, Thomas AJ, O'Brien JT, Attems J (2015) Cortical tau load is associated with white matter hyperintensities. Acta Neuropathol Commun 3:60
36. McAleese KE, Walker L, Graham S, Moya ELJ, Johnson M, Erskine D, Colloby SJ, Dey M, Martin-Ruiz C, Taylor JP, Thomas AJ, McKeith IG, de Carli C, Attems J (2017) Parietal white matter lesions in Alzheimer's disease are associated with cortical neurodegenerative pathology, but not with small vessel disease. Acta Neuropathol 134:459–473
37. McKenzie AT, Moyon S, Wang M, Katsyv I, Song WM, Zhou X, Dammer EB, Duong DM, Aaker J, Zhao Y, Beckmann N, Wang P, Zhu J, Lah JJ, Seyfried NT, Levey AI, Katsel P, Haroutunian V, Schadt EE, Popko B, Casaccia P, Zhang B (2017) Multiscale network modeling of oligodendrocytes reveals molecular components of myelin dysregulation in Alzheimer's disease. Mol Neurodegener 12:82
38. McKenzie IA, Ohayon D, Li H, de Faria JP, Emery B, Tohyama K, Richardson WD (2014) Motor skill learning requires active central myelination. Science 346:318–322
39. Micu I, Jiang Q, Coderre E, Ridsdale A, Zhang L, Woulfe J, Yin X, Trapp BD, McRory JE, Rehak R, Zamponi GW, Wang W, Stys PK (2006) NMDA receptors mediate calcium accumulation in myelin during chemical ischaemia. Nature 439:988–992
40. Mitew S, Kirkcaldie MT, Halliday GM, Shepherd CE, Vickers JC, Dickson TC (2010) Focal demyelination in Alzheimer's disease and transgenic mouse models. Acta Neuropathol 119:567–577
41. Nasrabady SE, Rizvi B, Goldman JE, Brickman AM (2018) White matter changes in Alzheimer's disease: a focus on myelin and oligodendrocytes. Acta neuropathologica communications 6:22–22
42. Ndayisaba A, Kaindlstorfer C, Wenning GK (2019) Iron in neurodegeneration—cause or consequence? Front Neurosci 13:180
43. Nielsen HM, Ek D, Avdic U, Orbjorn C, Hansson O, Veerhuis R, Rozemuller AJ, Brun A, Minthon L, Wennstrom M (2013) NG2 cells, a new trail for Alzheimer's disease mechanisms? Acta Neuropathol Commun 1:7
44. Quintela-Lopez T, Ortiz-Sanz C, Serrano-Regal MP, Gaminde-Blasco A, Valero J, Baleriola J, Sanchez-Gomez MV, Matute C, Alberdi E (2019) Abeta oligomers promote oligodendrocyte differentiation and maturation via integrin beta1 and Fyn kinase signaling. Cell Death Dis 10:445
45. Rivera A, Vanzuli I, Arellano JJ, Butt A (2016) Decreased regenerative capacity of oligodendrocyte progenitor cells (NG2-Glia) in the ageing brain: a vicious cycle of synaptic dysfunction, myelin loss and neuronal disruption? Curr Alzheimer Res 13:413–418
46. Rodriguez JJ, Butt AM, Gardenal E, Parpura V, Verkhratsky A (2016) Complex and differential glial responses in Alzheimer's disease and ageing. Curr Alzheimer Res 13:343–358
47. Roher AE, Weiss N, Kokjohn TA, Kuo YM, Kalback W, Anthony J, Watson D, Luehrs DC, Sue L, Walker D, Emmerling M, Goux W, Beach T (2002) Increased A beta peptides and reduced cholesterol and myelin proteins characterize white matter degeneration in Alzheimer's disease. Biochemistry 41:11080–11090
48. Sakry D, Neitz A, Singh J, Frischknecht R, Marongiu D, Biname F, Perera SS, Endres K, Lutz B, Radyushkin K, Trotter J, Mittmann T (2014) Oligodendrocyte precursor cells modulate the neuronal network by activity-dependent ectodomain cleavage of glial NG2. PLoS Biol 12:e1001993
49. Salter MG, Fern R (2005) NMDA receptors are expressed in developing oligodendrocyte processes and mediate injury. Nature 438:1167–1171
50. Selkoe DJ, Hardy J (2016) The amyloid hypothesis of Alzheimer's disease at 25 years. EMBO Mol Med 8:595–608
51. Stassart RM, Mobius W, Nave KA, Edgar JM (2018) The Axon-Myelin unit in development and degenerative disease. Front Neurosci 12:467
52. Thal DR, Rub U, Orantes M, Braak H (2002) Phases of A beta-deposition in the human brain and its relevance for the development of AD. Neurology 58:1791–1800

53. Wake H, Lee PR, Fields RD (2011) Control of local protein synthesis and initial events in myelination by action potentials. Science 333:1647–1651
54. Wan W, Xia S, Kalionis B, Liu L, Li Y (2014) The role of Wnt signaling in the development of Alzheimer's disease: a potential therapeutic target? Biomed Res Int 2014:301575
55. Wang R, Reddy PH (2017) Role of glutamate and NMDA receptors in Alzheimer's disease. J Alzheimers Dis 57:1041–1048
56. Weller J, Budson A (2018) Current understanding of Alzheimer's disease diagnosis and treatment [version 1; peer review: 2 approved]. F1000Research 7
57. Wu Y, Ma Y, Liu Z, Geng Q, Chen Z, Zhang Y (2017) Alterations of myelin morphology and oligodendrocyte development in early stage of Alzheimer's disease mouse model. Neurosci Lett 642:102–106
58. Xiao L, Ohayon D, McKenzie IA, Sinclair-Wilson A, Wright JL, Fudge AD, Emery B, Li H, Richardson WD (2016) Rapid production of new oligodendrocytes is required in the earliest stages of motor-skill learning. Nat Neurosci 19:1210–1217
59. Xu J, Chen S, Ahmed SH, Chen H, Ku G, Goldberg MP, Hsu CY (2001) Amyloid-beta peptides are cytotoxic to oligodendrocytes. J Neurosci 21:Rc118
60. Yin F, Sancheti H, Patil I, Cadenas E (2016) Energy metabolism and inflammation in brain aging and Alzheimer's disease. Free Radic Biol Med 100:108–122
61. Zhan X, Jickling GC, Ander BP, Stamova B, Liu D, Kao PF, Zelin MA, Jin LW, Decarli C, Sharp FR (2015) Myelin basic protein associates with AbetaPP, Abeta1-42, and amyloid plaques in cortex of Alzheimer's disease brain. J Alzheimers Dis 44:1213–1229
62. Zhang P, Kishimoto Y, Grammatikakis I, Gottimukkala K, Cutler RG, Zhang S, Abdelmohsen K, Bohr VA, Misra Sen J, Gorospe M, Mattson MP (2019) Senolytic therapy alleviates Abeta-associated oligodendrocyte progenitor cell senescence and cognitive deficits in an Alzheimer's disease model. Nat Neurosci 22:719–728

Chapter 13
Microglia in Parkinson's Disease

Margaret S. Ho

Abstract Microglia are the most abundant immune cells in the central nervous system (CNS), where they interact with neurons and exhibit a wide array of functions in physiological and pathological conditions. Physiologically, microglia mediate synaptic pruning and remodeling crucial for neural circuits and brain connectivity. In pathological conditions such as neurodegeneration in the Parkinson's disease (PD), microglia are activated, migrated to the injury site, and prone to engulf debris, sense pathology, and secrete possible pro- and anti-inflammatory factors. Microglia mediate responses such as inflammation and phagocytosis associated with neurodegeneration and are pivotal players in exacerbating or relieving disease progression. This chapter provides an overview on microglial function in the neurodegenerative disease—Parkinson's disease (PD). An overview on the pathology of PD will first be given, followed by discussion on receptors and signaling pathways involved in microglia-mediated inflammation and phagocytosis. Mechanism of how microglia contribute to PD by inflammation, phagocytosis of α-Synuclein (α-Syn), and interaction with PD genes will also be discussed.

Keywords Microglia · Neuroinflammation · Parkinson's disease · Phagocytosis · Alpha-synuclein

13.1 Introduction

As the most abundant immune cells residing in the central nervous system (CNS), microglia are small cells intertwining with neurons both physically and functionally, exhibiting a wide array of functions in physiological and pathological conditions. Microglia display differential density in various brain regions, with different combinations of markers underlying their regional identity and distinct functional roles [37]. This distributional difference, dynamic behavior, and unique cellular features

M. S. Ho (✉)
School of Life Science and Technology, ShanghaiTech University, #B416, L Building, #230 Haike Road, Pudong New District, Shanghai 201210, China
e-mail: margareth@shanghaitech.edu.cn

A. Verkhratsky et al. (eds.), *Neuroglia in Neurodegenerative Diseases*, Advances in Experimental Medicine and Biology 1175,
https://doi.org/10.1007/978-981-13-9913-8_13

have made them significant brain cells that receive substantial attention and merit in-depth exploration.

Microglia in physiological conditions mediate a variety of brain functions such as synaptic pruning and remodeling. Neuron-microglia bidirectional signaling is particularly crucial for neural circuits and brain connectivity [98, 114, 142]. Upon pathological trigger, microglia migrate to the injured site and act as a double-edged sword to relieve or exacerbate the injury. Decades of study have indicated that microglia doing these jobs are in two major states, resting and activated, distinguishable by the forms of their morphology. While microglia constantly surf around the environments and sense pathology in their resting state [93], they transit to an activated state once the nervous system is under detrimental attack and becomes pathological. Transition from the resting to activated state requires complex regulation, thus allowing microglial activation to be under tight control [55]. Despite this common bipartite categorization, it is generally believed that different targets and receptors tune microglial responses in a continuous manner and multiple forms of activation state exist [43, 116, 137].

Intriguingly, activation of microglia is often associated with neurodegeneration, a degenerative process underlying the ultimate pathology of neurodegenerative diseases. Distinct from resting microglia, activated microglia are often of amoeboid morphology, short processes, enlarged soma, and de novo expression of cell surface receptors. They are prone to engulf debris, sense pathology, and secrete possible pro- and anti-inflammatory factors that exacerbate or relieve disease progression. Thus, the activation profile of microglia is often an important indicator for and reflects neuronal dysfunction in neurodegenerative diseases. In this chapter, we will discuss microglial function in the neurodegenerative disease Parkinson's disease (PD). An overview of the pathology of PD will first be given, followed by a discussion on receptors and signaling pathways involved in microglia-mediated inflammation and phagocytosis. How microglia contribute to the occurrence of PD pathological hallmarks such as dopaminergic (DA) neuron death and formation of α-Synuclein (α-Syn)-containing Lewy bodies (LB) aggregates and mechanisms pertaining to PD gene function will also be discussed.

13.2 Parkinson's Disease

As the second most common neurodegenerative disorder, PD is clinically characterized by symptoms such as resting tremor, bradykinesia, postural instability, accompanying non-motor symptoms like cognitive impairment and autonomic dysfunction. Inside the brain, a series of neuropathological changes appears throughout the course of PD development, ultimately leading to the diagnostic hallmark: the aggregation of intracellular inclusions named Lewy bodies (LBs) and Lewy Neurites (LNs) and the loss of dopaminergic (DA) neurons in the *substantia nigra pars compacta* (*SNpc*).

This progressive brain pathology can be staged by the LB appearance in different regions of the brain [12, 13], with initial detection of LB in the periphery such as dorsal motor nucleus of the glossopharyngeal and vagal nerves or the olfactory bulb [74], followed by the appearance in the *SNpc* DA neurons in mid-stage, then the rostral propagation to other parts of the brain.

While most of the PD cases are sporadic, studies on rare familial cases offer the strength of identifying possible genetic causes for PD. The central component for LB and LN, α-Synuclein (α-Syn), is the gene product from *SNCA* (*PARK1*)—the first *PARK* gene identified in the studies of rare familial PD cases. Not only that genome-wide association (GWAS) studies have identified *SNCA* SNPs as risk variants for sporadic PD [112, 118], the missense *SNCA* mutation A53T [104], along with many others and duplications in the *SNCA* locus, have all been shown to associate with PD [2, 18, 66, 73, 101, 119, 141]. Interestingly, PD-associated mutations of α-Syn confer differential self-aggregation properties [19, 21, 28, 34, 35, 38], implicating that mutant α-Syn with altered propensities are potentially toxic and more prone for aggregation in disease conditions.

13.3 Microglia in PD

How microglia contribute to PD pathology remains to be an important area of study for researchers centering on the perspective of non-cell-autonomous regulation of neurodegeneration. Although no affirmative connection between DA neuron death and microglial activation has been established yet, microglial activation is thought to be a significant part of the disease process integrated either as a cause or consequence [11, 56]. Some of the earlier studies have provided evidence that microglia are involved in PD. First, the human leukocyte antigen gene (HLA-DRA) expressed specifically in microglia has been identified by a genome-wide association (GWAS) study as a genetic risk factor for late-onset PD [40]. p.R47H variant of microglial triggering receptor expressed on myeloid cells-2 (TREM-2) is also associated with PD [106]. These results suggest that microglia-specific regulation of PD progression exists. In addition, positron emission tomography (PET) studies show that microgliosis is an early and sustained response of PD [6, 33, 122]. Reactive microglia have also been detected in toxin-induced and transgenic mouse models of PD [22, 46, 84, 111]. Brains with an injection of the Gram-negative bacterial endotoxin lipopolysaccharide (LPS), a toxin that specifically induces microgliosis, show signs of DA neuron loss in the substantia nigra (SN). In toto, these findings suggest that microglial activation correlates with PD progression and induces DA neuron toxicity and death. Finally, studies on regional density of microglia revealed that they are prominently distributed in SN and striatum, where microglia exhibit a region- and stage-specific release of cytokines and mediators, thereby affecting DA neuron death [37, 76]. Taken together, these observations indicate a pivotal role for microglia in PD disease progression and raise interests in studying their functional roles during PD pathology (Fig. 13.1).

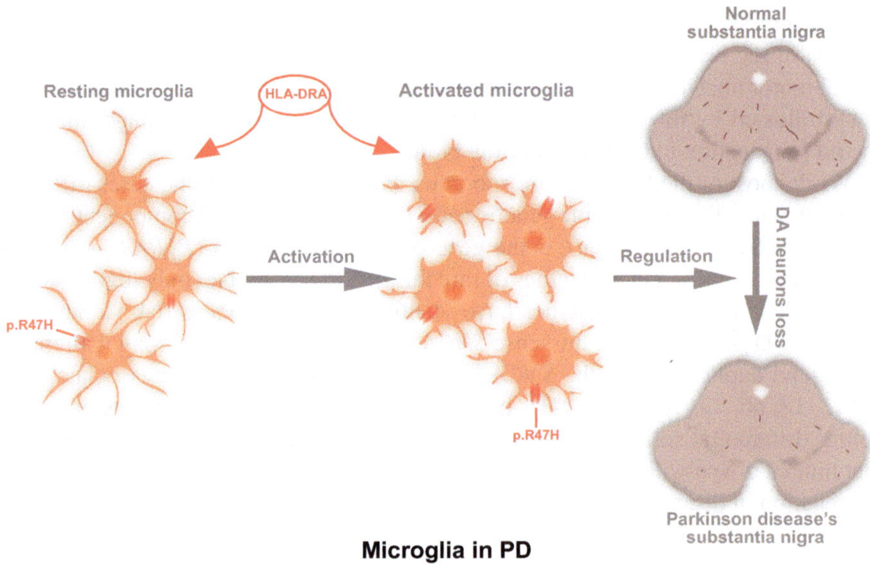

Fig. 13.1 Microglia in PD. Microglia transit from resting state to activating state when the nervous system undergoes pathological attack such as DA neuron loss in PD. A number of observations have dictated a pivotal role for microglia during this process, such as the identification of a microglial specific gene HLA-DRA by GWAS associated with PD, p.R47H variant of TREM-2 associated with PD, and microgliosis as an early and sustained response of PD shown by PET studies

13.4 Microglial Receptors in PD

Similar to other immune cells, CNS microglia express pattern recognition receptors (PRRs) that respond to pathogen-associated molecular patterns (PAMPs) and recognize invading pathogens for host defense immune mechanisms. One type of microglial PRRs, toll-like receptors (TLRs) [132, 136], are single-pass transmembrane proteins with an N-terminal extracellular ligand recognition domain carrying leucine-rich repeats [83] and the C-terminal intracellular Toll-interleukin 1 receptor (TIR) domains transforming extracellular recognition into an intracellular response [15, 48, 140]. TIR domains interact with adaptor molecules such as MyD88, TRIF, and TRAM in response to various stimuli. For instance, α-Syn or Pam3CSK4, a synthetic triacylated lipopeptide, activates TLR2-mediated downstream signaling via the adaptor MyD88 and a co-receptor, either TLR1 or TLR6. Upon α-Syn activation of the TLR1/2 receptor, interaction between MyD88 and TIR domain first activates the kinase activity of the interleukin-1 receptor-associated kinase (IRAK) complex [75], which in turn interacts with and activates the TNF receptor-associated factor 6 (TRAF6) via its K63-linked auto-ubiquitination. These sequential events lead to the activation of the transforming growth factor β-activated kinase-1 (TAK1) complex and the release of IKKs, which mediate IκBα degradation and the ultimate production of pro-inflammatory cytokines through MAPK activation and the nuclear translocation of NF-κB, JNK, and p38 (Figs. 13.1 and 13.2) [57, 58, 125].

Microglial TLR1/2 has been shown to be central to the α-Syn pathogenesis: an important therapeutical target for analysis [4]. First, the expression level of microglial TLR2 is elevated in patients of incidental Lewy Body disease (iLBD) which equals to a prodromal Braak stage 1–3, suggesting that elevated TLR2 level correlates early microglial activation response in PD [25]. Next, α-Syn activates microglial TLR1/2 in different experimental systems including BV-2 microglia, primary mouse microglia, or human microglia [7, 23, 60]. Similarly, medium from α-Syn overexpressing SH-SY5Y cells containing oligomeric α-Syn activates microglia in a TLR2-dependent manner [60, 61]. Upon TLR1/2 activation, microglia release pro-inflammatory cytokines tumor necrosis factor-alpha (TNFα) and interleukin (IL)-1β in a MyD88-dependent manner [23, 124, 143]. On the other hand, activated microglia also release anti-inflammatory cytokines, pointing to a diverse functional output upon α-Syn activation of microglial TLR1/2. Taken together, these results suggest that microglial TLR1/2 is a direct α-Syn target and the subsequent TLR1/2-activated signaling pathways participate in PD progression by releasing cytokines that tune the degree of neuroinflammation.

Furthermore, mice lacking the fractalkine receptor CX3CR1 show extensive loss of tyrosine-hydroxylase (TH)-positive neurons in the MPTP-induced PD mouse model [17]. CXCL-CX3CR1 signaling is also involved in a 6-hydroxydopamine (6-OHDA) rat model of PD, where CX3CL1 was found to suppress microglial activation and reduce neuronal loss [96]. It has also been shown that mice lacking CX3CR1 exhibit reduced α-Syn-mediated inflammatory response and microglial phagocytosis, further strengthening the importance of CXCL-CX3CR1 signaling in PD [128].

13.5 Microglia-Mediated Neuroinflammation in PD

The very first evidence that inflammation is involved in PD came from the observation that pro-inflammatory mediators such as TNFα, IL-1β, and IL-6 were detected in elevated levels in the cerebral spinal fluid (CSF) and brains of PD patients, particularly in the striatum [87, 88, 91]. The elevation of cytokine levels is part of the microglial activation and recruitment (microgliosis) process that starts early, accompanies neurodegeneration, and persists throughout the course of PD [50, 51, 65]. When activated microglia induces neuroinflammation, they either exhibit the M1 neurotoxic phenotype or the M2 neuroprotective phenotype [42, 63, 89, 105]. In the scenario of activated M1-like microglia, these cells often adopt an amoeboid morphology, are highly capable to phagocytose and remove apoptotic cell debris, and release massive pro-inflammatory factors such as IL-1β, IL-12, TNFα, and inducible nitric oxide synthase (iNOS). Microglial release of these factors often couples with DA neuron loss in PD. On the contrary, M2-like activated microglia are of thin cellular bodies and ramified processes, and secrete anti-inflammatory cytokines including IL-4, IL-13, IL-10, TGFβ, and neurotrophic insulin-like growth factor 1(IGF-1) to

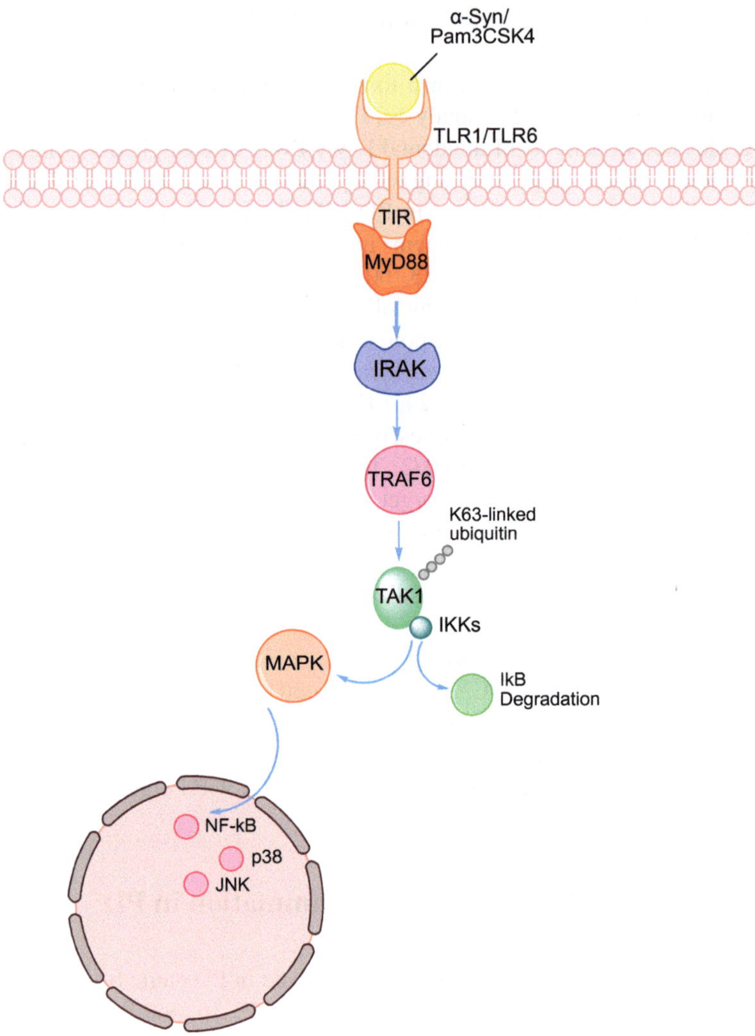

Fig. 13.2 Microglial receptors in PD. An example of microglial receptor TLR1/6 and its downstream pathway was illustrated. TIR domains of TLRs interact with adaptor molecules such as MyD88, TRIF, and TRAM in response to various stimuli. Ligands such as α-Syn or Pam3CSK4 activate TLR2-mediated downstream signaling via the adaptor MyD88 and a co-receptor, either TLR1 or TLR6. Upon α-Syn activation of the TLR1/2 receptor, interaction between MyD88 and TIR domain first activates the kinase activity of the interleukin-1 receptor-associated kinase (IRAK) complex, which in turn interacts with and activates the TNF receptor-associated factor 6 (TRAF6) via its K63-linked auto-ubiquitination. These sequential events lead to the activation of the transforming growth factor β-activated kinase-1 (TAK1) complex and the release of IKKs, which mediate IκBα degradation and the ultimate production of pro-inflammatory cytokines through MAPK activation and the nuclear translocation of NF-κB, JNK, and p38

ease inflammation and accelerate repair. Thus, microglia-mediated inflammation has double-sided effects in terms of relieving and exacerbating disease progression [139]. At one end, the inflammatory response might be beneficial by promoting neuron survival, whereas, on the other hand, the production of neurotoxic factors might also enhance the neurodegeneration. It is noteworthy to mention that microglia-released pro- and anti-inflammatory molecules coexist in the early stage of PD and their expression profiles change over time, suggesting that dynamic regulation of microgliosis correlates with PD progression [102, 113].

Interestingly, prominent microgliosis is detected in various toxin-based models of PD such as 6-OHDA, MPTP, and rotenone [80, 81, 94, 120, 133, 138] as well as transgenic models of PD based on α-Syn. Microgliosis in α-Syn transgenic models occurs early in the stage and precedes DA neuron death, suggesting that cell death is not necessarily a prerequisite for microglial activation [71, 85, 123]. Based on these findings, it has been suggested that signals inducing microgliosis and inflammation might be released from the toxic α-Syn protein aggregates or the degenerated neurons, making an increasing number of microglial cells reactive and migrate to the injury site to defend the progressively degenerating environment.

13.6 Microglial Activation by α-Syn

Microglia-mediated inflammation is regulated by PD risk factors such as DJ-1, LRRK2, and α-Syn. For instance, lacking LRRK2 attenuates inflammation via inhibiting p38 MAPK and NF-κB pathways [59, 86]. α-Syn, as mentioned above, positively regulates microglial inflammatory responses [124, 143]. These findings provide the molecular link between microglia-mediated neuroinflammation and PD pathology. Given that some of these factors might be neuronal specific, bidirectional signaling between neurons and microglia is therefore established as an extremely important theme in PD disease progression.

During PD pathology, α-Syn is secreted to the extracellular space from neurons and detected in the extracellular biological fluids in PD patients [79, 129]. Clear evidence shows that extracellular α-Syn directly activates microglia [124, 143]. This activation has significant consequences. First, it is part of a key event for fully shifting activated microglia to exhibit a pro-inflammatory phenotype [3, 127]. Next, α-Syn-induced microglial activation promotes α-Syn phagocytosis via microglial FcγR receptor and subsequently activates a series of pro-inflammatory events such as nuclear translocation of NFγB p65 and elevated release of cytokines, potentiating the loss of DA neurons and chronic neurodegeneration in PD [16, 64, 69, 72, 124].

Results from studies on the form of α-Syn that activates microglia were contradictory. Different forms of α-Syn exhibit different effects on microglial phagocytosis and inflammatory activation. Pathogenic form of α-Syn, such as α-Syn^{A53T}, triggers pro-inflammatory microglial response and impairs phagocytosis [46, 108]. In a different study, however, α-Syn^{A53T} is implicated in promoting phagocytosis [109]. Physiological α-Syn, on the other hand, inhibits inflammation yet promotes phagocytosis [3]. It is generally believed that monomeric α-Syn promotes phagocytosis whereas oligomeric α-Syn acts in an opposite way [100], yet other studies also indicate enhanced microglial phagocytosis by fibrillar and C-terminal truncated α-Syn [29]. Aggregated α-Syn has been shown to inhibit microglial phagocytosis by activating SHP-1 via interaction with FcγRIIB, and is more potent in mediating microglial release of TNFα and IL-1β [20, 47]. Taken together, α-Syn conformation and its pathogenic form play pivotal roles in regulating microglial phagocytosis and subsequent activated inflammatory response, accompanying neurodegeneration in PD.

In addition to the aforementioned TLRs, α-Syn interacts with a number of different microglial receptors for potentiating phagocytosis and inflammatory responses. For instance, in response to α-Syn, the Prostaglandin E receptor subtype 2 (EP2) regulates α-Syn phagocytosis and CD11b-mediated microglial activation [54]. α-Syn also interacts with CD11b to activate NOX2 through Erk1/2 kinase activation and RhoA-dependent pathway to direct microglial migration [49, 134]. Furthermore, α-Syn, in its monomeric or pathogenic mutant form, interacts with the scavenger receptor CD36 to regulate microglial activation and TNFα release [123, 124]. Another receptor associated with α-Syn is the protease-activated receptor (PAR-1), working in a paracrine manner initiated by the secretion of matrix metalloproteinases [69]. The microglial purinergic P2X7 receptor is also implicated in α-Syn-mediated microglial activation via PHOX activation [53]. Taken together, these receptors receive signals from α-Syn and trigger different cascades of signaling pathways within microglia to activate inflammatory responses, creating the diversity in microglial outputs upon pathological trigger as PD progresses.

13.7 Microglial Phagocytosis in PD

Microglia mediate phagocytosis of apoptotic cells, unfolded proteins, or neuronal debris, a process carried out by the resting microglia in the developing brain or the reactive microglia in pathological conditions such as PD [116, 117]. Interestingly, phagocytosis has been considered beneficial associated with the anti-inflammatory function of microglia, raising the interest in studying cellular machinery mediating this process. A list of receptors has been shown to mediate microglial phagocytosis, including TLRs, the scavenger receptors CD14, TAM (Tyro3, Axl, and Mer) receptor, and TREM-2 [31, 41, 121, 131]. First, microglial phagocytosis of α-Syn is impaired in the absence of TLR4, suggesting TLR4 is involved in microglial α-Syn uptake. Alternation of TLR4 signaling modulates pro-inflammatory responses and ROS production, and promotes neurodegeneration [29, 121], whereas treatment of TLR-4 agonist also protects the survival of transgenic α-Syn overexpressing mice [131]. These findings suggest that TLR4 promotes microglial clearance of α-Syn, thus playing a beneficial role in controlling α-Syn spread and PD progression.

TREM-2 is another microglia-specific receptor that mediates phagocytosis of apoptotic neurons [126]. Alternation in TREM-2 expression affects phagocytosis and subsequent microglia-mediated inflammatory responses by regulating pro-inflammatory gene transcription. TREM-2 is considered neuroprotective as increased levels of TREM-2 enhances microglial phagocytosis and decreases pro-inflammatory responses by regulating TLR4-mediated activation of NF-κB signaling [107].

Expression of the scavenger receptor Mannose Receptor C-Type 1 (MRC1) is decreased in an MPTP mice PD model, suggesting MRC1-mediated microglial phagocytosis is crucial for PD progression. Like TREM2, increased MRC1 expression (thus MRC1-mediated phagocytosis) is also beneficial as the increased MRC expression is part of the peroxisome proliferator-activated receptor gamma (PPARγ)-mediated mechanism of neuroprotection [68]. It is noteworthy mentioning that despite the supporting evidence from TREM-2 and MRC1 that microglial phagocytosis is beneficial for PD, other evidence has suggested phagocytosis contributes to neurodegeneration [5]. For instance, loss of TAM phagocytic receptor slightly extended survival of α-Syn^{A53T} overexpressing mice, suggesting TAM-mediated microglial phagocytosis promotes neurodegeneration and accelerates animal death [31] (Fig. 13.3).

Microglial phagocytosis is also regulated by PD risk factors such as DJ-1, α-Syn, and LRRK2 [20, 78, 82, 92, 100]. For instance, DJ-1 regulates microglial phagocy-

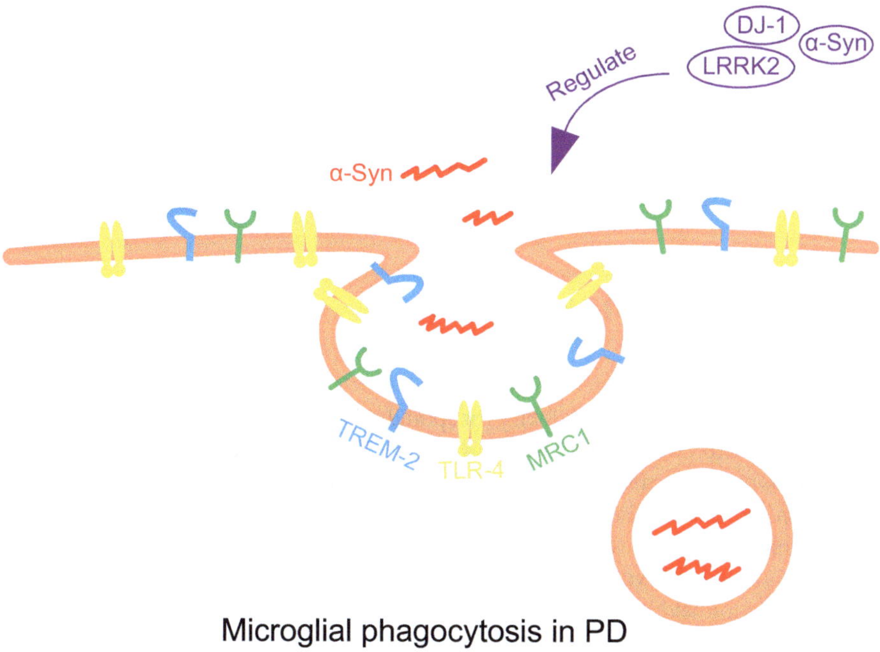

Fig. 13.3 Microglial phagocytosis in PD. Using α-Syn as an example, the receptor-mediated microglial phagocytosis were illustrated. TLR4, TREM-2, and MRC1 are receptors that mediate microglial phagocytosis during PD. This process is under tight regulation by other PD factors such as LRRK2, DJ-1, and α-Syn

tosis of α-Syn via autophagy and in an LC3 (microtubule-associated protein 1A/1B-light chain 3)-dependent (LAP) manner [52, 92], whereas α-Syn is both a regulator and a substrate for microglial phagocytosis. How α-Syn contributes to microglial activation and phagocytosis is discussed above, and microglial phagocytosis of α-Syn is summarized in the next section.

13.8 Microglial Phagocytosis of α-Syn

Previous studies have indicated that microglial phagocytose α-Syn [10, 70]. Some of the receptors functioning in microglial phagocytosis and activation, like TLR2 and TLR4, have been implicated in α-Syn uptake and α-Syn-mediated activation [29, 60, 121, 130]. While TLR4 is required for both microglial activation and phagocytosis of α-Syn [29, 121], TLR2 mainly receives signals from oligomeric α-Syn, but not monomeric or fibrillar α-Syn [60]. TLR2 activation is also crucial in neurons to decrease the uptake and autophagy of α-Syn, promoting neuronal α-Syn accumulation [26, 62]. These findings suggest that signaling cascade initiated by TL2 might be different in neurons and glia, and contribute differently to the disease. Moreover, microglia have been observed in vitro to uptake α-Syn-containing exosomes released by oligodendrocytes via macropinocytosis [30]. In addition to phagocytosis and macropinocytosis, other clathrin-independent routes such as monosialoganglioside (GM1)-dependent lipid rafts have also been shown to mediate microglial uptake of α-Syn [99]. Reduced expression of DJ-1, another PD risk factor, reduces cell surface lipid raft expression in microglia and impairs their ability to uptake soluble α-Syn [92].

13.9 Microglia, LRRK2 and PD

Mutations in the leucine-rich repeat kinase 2 (LRRK2) gene are the most common cause of familial PD and a risk factor for sporadic PD [44, 67, 97]. In the immune system, LRRK2 expression in monocytes is increased upon inflammation and the release of pro-inflammatory mediators like IL-1β, TNFα, and IFNγ [32]. Upon microglial activation by LPS in the brain, LRRK2 expression is also increased in a TLR4-dependent manner [86]. It is generally believed that LRRK function correlates with its phosphorylation level on Serine residue 935 (Ser935) [27, 115], a site in which phosphorylation level is crucial for microglial activation and induces inflammatory response in PD. It has also been shown that PD-associated mutations of LRRK2 at position R1441 reduce PKA-mediated LRRK2 phosphorylation and prevent its binding with the adaptor protein 14-3-3 binding [90], suggesting that mutations in LRRK2 associated with PD could affect LRRK2 phosphorylation, hence its kinase activity and cellular function.

One of the major LRRK2 funtions is to regulate the autophagy/lysosome degradation pathway. LRRK2 is localized on the autophagosome vesicles as shown by immune-electron microscopy and biochemical approaches [1, 36, 115]. Membrane localization of LRRK2 on autophagic and lysosome-related vesicles indicates that LRRK2 plays a pivotal role in regulating their function [8, 9]. LRRK2 also interacts with membrane proteins on these vesicles such as Rab7 (late endosomes) and Lamp2A [24, 45, 77, 95]. These results suggest that LRRK2 is involved in different steps of autophagy/lysosome pathway and its activity alters the degradative activity, development, or final maturation of these different vesicles.

Interestingly, the PD-associated LRRK2 mutation, G2019S, in its kinase domain results in an upregulation of LRRK2 kinase activity [39, 135] and has been implicated in autophagic dysfunction. Cells expressing $LRRK2^{G2019S}$ consistently exhibit increased autophagic vesicles or marker expression [14, 103, 110], possibly due to an increase in autophagic flux or an arrest in autophagosome/lysosome fusion. It is possible that defects in the autophagy/lysosome pathway caused by increased LRRK2 activity result in insufficient degradation of accumulated protein such as α-Syn, disrupting α-Syn proteostasis and underlying the mechanism of PD.

Acknowledgements My sincere apologies to colleagues in the field whose work I was not able to mention because of space limitations. Work in the MH laboratory is supported by grants from the National Basic Research Program of China (973 Program 2013CB945602) and the National Natural Science Foundation of China (31871039). We would like to thank Honglei Wang for the illustrations and all members of the MH laboratory for comments and suggestions on the manuscript.

References

1. Alegre-Abarrategui J, Christian H, Lufino MM, Mutihac R, Venda LL, Ansorge O, Wade-Martins R (2009) LRRK2 regulates autophagic activity and localizes to specific membrane microdomains in a novel human genomic reporter cellular model. Hum Mol Genet 18:4022–4034
2. Appel-Cresswell S, Vilarino-Guell C, Encarnacion M, Sherman H, Yu I, Shah B, Weir D, Thompson C, Szu-Tu C, Trinh J et al (2013) Alpha-synuclein p. H50Q, a novel pathogenic mutation for Parkinson's disease. Mov Disord 28:811–813
3. Austin SA, Floden AM, Murphy EJ, Combs CK (2006) Alpha-synuclein expression modulates microglial activation phenotype. J Neurosci 26:10558–10563
4. Babcock AA, Wirenfeldt M, Holm T, Nielsen HH, Dissing-Olesen L, Toft-Hansen H, Millward JM, Landmann R, Rivest S, Finsen B et al (2006) Toll-like receptor 2 signaling in response to brain injury: an innate bridge to neuroinflammation. J Neurosci 26:12826–12837
5. Barcia C, Ros CM, Ros-Bernal F, Gomez A, Annese V, Carrillo-de Sauvage MA, Yuste JE, Campuzano CM, de Pablos V, Fernandez-Villalba E et al (2013) Persistent phagocytic characteristics of microglia in the substantia nigra of long-term Parkinsonian macaques. J Neuroimmunol 261:60–66
6. Bartels AL, Willemsen AT, Doorduin J, de Vries EF, Dierckx RA, Leenders KL (2010) [11C]-PK11195 PET: quantification of neuroinflammation and a monitor of anti-inflammatory treatment in Parkinson's disease? Parkinsonism Relat Disord 16:57–59
7. Beraud D, Hathaway HA, Trecki J, Chasovskikh S, Johnson DA, Johnson JA, Federoff HJ, Shimoji M, Mhyre TR, Maguire-Zeiss KA (2013) Microglial activation and antioxi-

dant responses induced by the Parkinson's disease protein alpha-synuclein. J Neuroimmune Pharmacol 8:94–117

8. Berger Z, Smith KA, Lavoie MJ (2010) Membrane localization of LRRK2 is associated with increased formation of the highly active LRRK2 dimer and changes in its phosphorylation. Biochemistry 49:5511–5523
9. Biskup S, Moore DJ, Celsi F, Higashi S, West AB, Andrabi SA, Kurkinen K, Yu SW, Savitt JM, Waldvogel HJ et al (2006) Localization of LRRK2 to membranous and vesicular structures in mammalian brain. Ann Neurol 60:557–569
10. Bliederhaeuser C, Grozdanov V, Speidel A, Zondler L, Ruf WP, Bayer H, Kiechle M, Feiler MS, Freischmidt A, Brenner D et al (2016) Age-dependent defects of alpha-synuclein oligomer uptake in microglia and monocytes. Acta Neuropathol 131:379–391
11. Block ML, Zecca L, Hong JS (2007) Microglia-mediated neurotoxicity: uncovering the molecular mechanisms. Nat Rev Neurosci 8:57–69
12. Braak H, Del Tredici K, Rub U, de Vos RA, Jansen Steur EN, Braak E (2003) Staging of brain pathology related to sporadic Parkinson's disease. Neurobiol Aging 24:197–211
13. Braak H, Muller CM, Rub U, Ackermann H, Bratzke H, de Vos RA, Del Tredici K (2006) Pathology associated with sporadic Parkinson's disease–where does it end? J Neural Transm Suppl 89–97
14. Bravo-San Pedro JM, Niso-Santano M, Gomez-Sanchez R, Pizarro-Estrella E, Aiastui-Pujana A, Gorostidi A, Climent V, Lopez de Maturana R, Sanchez-Pernaute R, Lopez de Munain A et al (2013) The LRRK2 G2019S mutant exacerbates basal autophagy through activation of the MEK/ERK pathway. Cell Mol Life Sci 70:121–136
15. Brown V, Brown RA, Ozinsky A, Hesselberth JR, Fields S (2006) Binding specificity of Toll-like receptor cytoplasmic domains. Eur J Immunol 36:742–753
16. Cao S, Standaert DG, Harms AS (2012) The gamma chain subunit of Fc receptors is required for alpha-synuclein-induced pro-inflammatory signaling in microglia. J Neuroinflammation 9:259
17. Cardona AE, Pioro EP, Sasse ME, Kostenko V, Cardona SM, Dijkstra IM, Huang D, Kidd G, Dombrowski S, Dutta R et al (2006) Control of microglial neurotoxicity by the fractalkine receptor. Nat Neurosci 9:917–924
18. Chartier-Harlin MC, Kachergus J, Roumier C, Mouroux V, Douay X, Lincoln S, Levecque C, Larvor L, Andrieux J, Hulihan M et al (2004) Alpha-synuclein locus duplication as a cause of familial Parkinson's disease. Lancet 364:1167–1169
19. Choi W, Zibaee S, Jakes R, Serpell LC, Davletov B, Crowther RA, Goedert M (2004) Mutation E46 K increases phospholipid binding and assembly into filaments of human alpha-synuclein. FEBS Lett 576:363–368
20. Choi YR, Kang SJ, Kim JM, Lee SJ, Jou I, Joe EH, Park SM (2015) FcgammaRIIB mediates the inhibitory effect of aggregated alpha-synuclein on microglial phagocytosis. Neurobiol Dis 83:90–99
21. Conway KA, Harper JD, Lansbury PT (1998) Accelerated in vitro fibril formation by a mutant alpha-synuclein linked to early-onset Parkinson disease. Nat Med 4:1318–1320
22. Czlonkowska A, Kohutnicka M, Kurkowska-Jastrzebska I, Czlonkowski A (1996) Microglial reaction in MPTP (1-methyl-4-phenyl-1,2,3,6-tetrahydropyridine) induced Parkinson's disease mice model. Neurodegeneration 5:137–143
23. Daniele SG, Beraud D, Davenport C, Cheng K, Yin H, Maguire-Zeiss KA (2015) Activation of MyD88-dependent TLR1/2 signaling by misfolded alpha-synuclein, a protein linked to neurodegenerative disorders. Sci Signal 8:ra45
24. Dodson MW, Zhang T, Jiang C, Chen S, Guo M (2012) Roles of the Drosophila LRRK2 homolog in Rab7-dependent lysosomal positioning. Hum Mol Genet 21:1350–1363
25. Doorn KJ, Moors T, Drukarch B, van de Berg W, Lucassen PJ, van Dam AM (2014) Microglial phenotypes and toll-like receptor 2 in the substantia nigra and hippocampus of incidental Lewy body disease cases and Parkinson's disease patients. Acta Neuropathol Commun 2:90
26. Dzamko N, Gysbers A, Perera G, Bahar A, Shankar A, Gao J, Fu Y, Halliday GM (2017) Toll-like receptor 2 is increased in neurons in Parkinson's disease brain and may contribute to alpha-synuclein pathology. Acta Neuropathol 133:303–319

27. Dzamko N, Inesta-Vaquera F, Zhang J, Xie C, Cai H, Arthur S, Tan L, Choi H, Gray N, Cohen P et al (2012) The IkappaB kinase family phosphorylates the Parkinson's disease kinase LRRK2 at Ser935 and Ser910 during Toll-like receptor signaling. PLoS ONE 7:e39132
28. Fares MB, Ait-Bouziad N, Dikiy I, Mbefo MK, Jovicic A, Kiely A, Holton JL, Lee SJ, Gitler AD, Eliezer D et al (2014) The novel Parkinson's disease linked mutation G51D attenuates in vitro aggregation and membrane binding of alpha-synuclein, and enhances its secretion and nuclear localization in cells. Hum Mol Genet 23:4491–4509
29. Fellner L, Irschick R, Schanda K, Reindl M, Klimaschewski L, Poewe W, Wenning GK, Stefanova N (2013) Toll-like receptor 4 is required for alpha-synuclein dependent activation of microglia and astroglia. Glia 61:349–360
30. Fitzner D, Schnaars M, van Rossum D, Krishnamoorthy G, Dibaj P, Bakhti M, Regen T, Hanisch UK, Simons M (2011) Selective transfer of exosomes from oligodendrocytes to microglia by macropinocytosis. J Cell Sci 124:447–458
31. Fourgeaud L, Traves PG, Tufail Y, Leal-Bailey H, Lew ED, Burrola PG, Callaway P, Zagorska A, Rothlin CV, Nimmerjahn A et al (2016) TAM receptors regulate multiple features of microglial physiology. Nature 532:240–244
32. Gardet A, Benita Y, Li C, Sands BE, Ballester I, Stevens C, Korzenik JR, Rioux JD, Daly MJ, Xavier RJ et al (2010) LRRK2 is involved in the IFN-gamma response and host response to pathogens. J Immunol 185:5577–5585
33. Gerhard A, Pavese N, Hotton G, Turkheimer F, Es M, Hammers A, Eggert K, Oertel W, Banati RB, Brooks DJ (2006) In vivo imaging of microglial activation with [11C](R)-PK11195 PET in idiopathic Parkinson's disease. Neurobiol Dis 21:404–412
34. Ghosh D, Mondal M, Mohite GM, Singh PK, Ranjan P, Anoop A, Ghosh S, Jha NN, Kumar A, Maji SK (2013) The Parkinson's disease-associated H50Q mutation accelerates alpha-synuclein aggregation in vitro. Biochemistry 52:6925–6927
35. Ghosh D, Sahay S, Ranjan P, Salot S, Mohite GM, Singh PK, Dwivedi S, Carvalho E, Banerjee R, Kumar A et al (2014) The newly discovered Parkinson's disease associated Finnish mutation (A53E) attenuates alpha-synuclein aggregation and membrane binding. Biochemistry 53:6419–6421
36. Gomez-Suaga P, Luzon-Toro B, Churamani D, Zhang L, Bloor-Young D, Patel S, Woodman PG, Churchill GC, Hilfiker S (2012) Leucine-rich repeat kinase 2 regulates autophagy through a calcium-dependent pathway involving NAADP. Hum Mol Genet 21:511–525
37. Grabert K, Michoel T, Karavolos MH, Clohisey S, Baillie JK, Stevens MP, Freeman TC, Summers KM, McColl BW (2016) Microglial brain region-dependent diversity and selective regional sensitivities to aging. Nat Neurosci 19:504–516
38. Greenbaum EA, Graves CL, Mishizen-Eberz AJ, Lupoli MA, Lynch DR, Englander SW, Axelsen PH, Giasson BI (2005) The E46 K mutation in alpha-synuclein increases amyloid fibril formation. J Biol Chem 280:7800–7807
39. Greggio E, Jain S, Kingsbury A, Bandopadhyay R, Lewis P, Kaganovich A, van der Brug MP, Beilina A, Blackinton J, Thomas KJ et al (2006) Kinase activity is required for the toxic effects of mutant LRRK2/dardarin. Neurobiol Dis 23:329–341
40. Hamza TH, Zabetian CP, Tenesa A, Laederach A, Montimurro J, Yearout D, Kay DM, Doheny KF, Paschall J, Pugh E et al (2010) Common genetic variation in the HLA region is associated with late-onset sporadic Parkinson's disease. Nat Genet 42:781–785
41. Han J, Wang M, Ren M, Lou H (2017) Contributions of triggering-receptor-expressed-on-myeloid-cells-2 to neurological diseases. Int J Neurosci 127:368–375
42. Hanisch UK (2013) Functional diversity of microglia—how heterogeneous are they to begin with? Front Cell Neurosci 7:65
43. Hanisch UK, Kettenmann H (2007) Microglia: active sensor and versatile effector cells in the normal and pathologic brain. Nat Neurosci 10:1387–1394
44. Healy DG, Falchi M, O'Sullivan SS, Bonifati V, Durr A, Bressman S, Brice A, Aasly J, Zabetian CP, Goldwurm S et al (2008) Phenotype, genotype, and worldwide genetic penetrance of LRRK2-associated Parkinson's disease: a case-control study. Lancet Neurol 7:583–590

45. Higashi S, Moore DJ, Yamamoto R, Minegishi M, Sato K, Togo T, Katsuse O, Uchikado H, Furukawa Y, Hino H et al (2009) Abnormal localization of leucine-rich repeat kinase 2 to the endosomal-lysosomal compartment in lewy body disease. J Neuropathol Exp Neurol 68:994–1005
46. Hoenen C, Gustin A, Birck C, Kirchmeyer M, Beaume N, Felten P, Grandbarbe L, Heuschling P, Heurtaux T (2016) Alpha-synuclein proteins promote pro-inflammatory cascades in microglia: stronger effects of the A53T mutant. PLoS ONE 11:e0162717
47. Hoffmann A, Ettle B, Bruno A, Kulinich A, Hoffmann AC, von Wittgenstein J, Winkler J, Xiang W, Schlachetzki JC (2016) Alpha-synuclein activates BV2 microglia dependent on its aggregation state. Biochem Biophys Res Commun 479:881–886
48. Horng T, Barton GM, Flavell RA, Medzhitov R (2002) The adaptor molecule TIRAP provides signalling specificity for Toll-like receptors. Nature 420:329–333
49. Hou L, Bao X, Zang C, Yang H, Sun F, Che Y, Wu X, Li S, Zhang D, Wang Q (2018) Integrin CD11b mediates alpha-synuclein-induced activation of NADPH oxidase through a Rho-dependent pathway. Redox Biol 14:600–608
50. Hunot S, Boissiere F, Faucheux B, Brugg B, Mouatt-Prigent A, Agid Y, Hirsch EC (1996) Nitric oxide synthase and neuronal vulnerability in Parkinson's disease. Neuroscience 72:355–363
51. Imamura K, Hishikawa N, Sawada M, Nagatsu T, Yoshida M, Hashizume Y (2003) Distribution of major histocompatibility complex class II-positive microglia and cytokine profile of Parkinson's disease brains. Acta Neuropathol 106:518–526
52. Janda E, Boi L, Carta AR (2018) Microglial phagocytosis and its regulation: a therapeutic target in Parkinson's disease? Front Mol Neurosci 11:144
53. Jiang T, Hoekstra J, Heng X, Kang W, Ding J, Liu J, Chen S, Zhang J (2015) P2X7 receptor is critical in alpha-synuclein–mediated microglial NADPH oxidase activation. Neurobiol Aging 36:2304–2318
54. Jin J, Shie FS, Liu J, Wang Y, Davis J, Schantz AM, Montine KS, Montine TJ, Zhang J (2007) Prostaglandin E2 receptor subtype 2 (EP2) regulates microglial activation and associated neurotoxicity induced by aggregated alpha-synuclein. J Neuroinflammation 4:2
55. Joers V, Tansey MG, Mulas G, Carta AR (2017) Microglial phenotypes in Parkinson's disease and animal models of the disease. Prog Neurobiol 155:57–75
56. Kannarkat GT, Boss JM, Tansey MG (2013) The role of innate and adaptive immunity in Parkinson's disease. J Parkinsons Dis 3:493–514
57. Kawai T, Akira S (2007) Signaling to NF-kappaB by Toll-like receptors. Trends Mol Med 13:460–469
58. Kawai T, Akira S (2010) The role of pattern-recognition receptors in innate immunity: update on Toll-like receptors. Nat Immunol 11:373–384
59. Kim B, Yang MS, Choi D, Kim JH, Kim HS, Seol W, Choi S, Jou I, Kim EY, Joe EH (2012) Impaired inflammatory responses in murine Lrrk2-knockdown brain microglia. PLoS ONE 7:e34693
60. Kim C, Ho DH, Suk JE, You S, Michael S, Kang J, Joong Lee S, Masliah E, Hwang D, Lee HJ et al (2013) Neuron-released oligomeric alpha-synuclein is an endogenous agonist of TLR2 for paracrine activation of microglia. Nat Commun 4:1562
61. Kim C, Lee HJ, Masliah E, Lee SJ (2016) Non-cell-autonomous neurotoxicity of alpha-synuclein through microglial toll-like receptor 2. Exp Neurobiol 25:113–119
62. Kim C, Rockenstein E, Spencer B, Kim HK, Adame A, Trejo M, Stafa K, Lee HJ, Lee SJ, Masliah E (2015) Antagonizing neuronal toll-like receptor 2 prevents synucleinopathy by activating autophagy. Cell Rep 13:771–782
63. Kim CC, Nakamura MC, Hsieh CL (2016) Brain trauma elicits non-canonical macrophage activation states. J Neuroinflammation 13:117
64. Klegeris A, Pelech S, Giasson BI, Maguire J, Zhang H, McGeer EG, McGeer PL (2008) Alpha-synuclein activates stress signaling protein kinases in THP-1 cells and microglia. Neurobiol Aging 29:739–752

65. Knott C, Stern G, Wilkin GP (2000) Inflammatory regulators in Parkinson's disease: iNOS, lipocortin-1, and cyclooxygenases-1 and -2. Mol Cell Neurosci 16:724–739
66. Kruger R, Kuhn W, Muller T, Woitalla D, Graeber M, Kosel S, Przuntek H, Epplen JT, Schols L, Riess O (1998) Ala30Pro mutation in the gene encoding alpha-synuclein in Parkinson's disease. Nat Genet 18:106–108
67. Kumari U, Tan EK (2009) LRRK2 in Parkinson's disease: genetic and clinical studies from patients. FEBS J 276:6455–6463
68. Lecca D, Janda E, Mulas G, Diana A, Martino C, Angius F, Spolitu S, Casu MA, Simbula G, Boi L et al (2018) Boosting phagocytosis and anti-inflammatory phenotype in microglia mediates neuroprotection by PPARgamma agonist MDG548 in Parkinson's disease models. Br J Pharmacol
69. Lee EJ, Woo MS, Moon PG, Baek MC, Choi IY, Kim WK, Junn E, Kim HS (2010) Alpha-synuclein activates microglia by inducing the expressions of matrix metalloproteinases and the subsequent activation of protease-activated receptor-1. J Immunol 185:615–623
70. Lee HJ, Suk JE, Bae EJ, Lee SJ (2008) Clearance and deposition of extracellular alpha-synuclein aggregates in microglia. Biochem Biophys Res Commun 372:423–428
71. Lee MK, Stirling W, Xu Y, Xu X, Qui D, Mandir AS, Dawson TM, Copeland NG, Jenkins NA, Price DL (2002) Human alpha-synuclein-harboring familial Parkinson's disease-linked Ala-53 –> Thr mutation causes neurodegenerative disease with alpha-synuclein aggregation in transgenic mice. Proc Natl Acad Sci USA 99:8968–8973
72. Lee SB, Park SM, Ahn KJ, Chung KC, Paik SR, Kim J (2009) Identification of the amino acid sequence motif of alpha-synuclein responsible for macrophage activation. Biochem Biophys Res Commun 381:39–43
73. Lesage S, Anheim M, Letournel F, Bousset L, Honore A, Rozas N, Pieri L, Madiona K, Durr A, Melki R et al (2013) G51D alpha-synuclein mutation causes a novel parkinsonian-pyramidal syndrome. Ann Neurol 73:459–471
74. Lewy FH (1912) Paralysis agitans. I. Pathologische anatomie. In: Lewandowsky M (ed). Springer, Berlin
75. Lin SC, Lo YC, Wu H (2010) Helical assembly in the MyD88-IRAK4-IRAK2 complex in TLR/IL-1R signalling. Nature 465:885–890
76. Lopez Gonzalez I, Garcia-Esparcia P, Llorens F, Ferrer I (2016) Genetic and transcriptomic profiles of inflammation in neurodegenerative diseases: Alzheimer, Parkinson, Creutzfeldt-Jakob and Tauopathies. Int J Mol Sci 17:206
77. MacLeod DA, Rhinn H, Kuwahara T, Zolin A, Di Paolo G, McCabe BD, Marder KS, Honig LS, Clark LN, Small SA et al (2013) RAB7L1 interacts with LRRK2 to modify intraneuronal protein sorting and Parkinson's disease risk. Neuron 77:425–439
78. Maekawa T, Sasaoka T, Azuma S, Ichikawa T, Melrose HL, Farrer MJ, Obata F (2016) Leucine-rich repeat kinase 2 (LRRK2) regulates alpha-synuclein clearance in microglia. BMC Neurosci 17:77
79. Majbour NK, Vaikath NN, Eusebi P, Chiasserini D, Ardah M, Varghese S, Haque ME, Tokuda T, Auinger P, Calabresi P et al (2016) Longitudinal changes in CSF alpha-synuclein species reflect Parkinson's disease progression. Mov Disord 31:1535–1542
80. Manocha GD, Floden AM, Puig KL, Nagamoto-Combs K, Scherzer CR, Combs CK (2017) Defining the contribution of neuroinflammation to Parkinson's disease in humanized immune system mice. Mol Neurodegener 12:17
81. Marinova-Mutafchieva L, Sadeghian M, Broom L, Davis JB, Medhurst AD, Dexter DT (2009) Relationship between microglial activation and dopaminergic neuronal loss in the substantia nigra: a time course study in a 6-hydroxydopamine model of Parkinson's disease. J Neurochem 110:966–975
82. Marker DF, Puccini JM, Mockus TE, Barbieri J, Lu SM, Gelbard HA (2012) LRRK2 kinase inhibition prevents pathological microglial phagocytosis in response to HIV-1 Tat protein. J Neuroinflammation 9:261
83. Matsushima N, Tanaka T, Enkhbayar P, Mikami T, Taga M, Yamada K, Kuroki Y (2007) Comparative sequence analysis of leucine-rich repeats (LRRs) within vertebrate toll-like receptors. BMC Genom 8:124

84. McGeer PL, McGeer EG, Kawamata T, Yamada T, Akiyama H (1991) Reactions of the immune system in chronic degenerative neurological diseases. Can J Neurol Sci 18:376–379
85. Miller RM, Kiser GL, Kaysser-Kranich T, Casaceli C, Colla E, Lee MK, Palaniappan C, Federoff HJ (2007) Wild-type and mutant alpha-synuclein induce a multi-component gene expression profile consistent with shared pathophysiology in different transgenic mouse models of PD. Exp Neurol 204:421–432
86. Moehle MS, Webber PJ, Tse T, Sukar N, Standaert DG, DeSilva TM, Cowell RM, West AB (2012) LRRK2 inhibition attenuates microglial inflammatory responses. J Neurosci 32:1602–1611
87. Mogi M, Harada M, Kondo T, Riederer P, Inagaki H, Minami M, Nagatsu T (1994) Interleukin-1 beta, interleukin-6, epidermal growth factor and transforming growth factor-alpha are elevated in the brain from parkinsonian patients. Neurosci Lett 180:147–150
88. Mogi M, Harada M, Riederer P, Narabayashi H, Fujita K, Nagatsu T (1994) Tumor necrosis factor-alpha (TNF-alpha) increases both in the brain and in the cerebrospinal fluid from parkinsonian patients. Neurosci Lett 165:208–210
89. Morganti JM, Riparip LK, Rosi S (2016) Call off the Dog(ma): M1/M2 polarization is concurrent following traumatic brain injury. PLoS ONE 11:e0148001
90. Muda K, Bertinetti D, Gesellchen F, Hermann JS, von Zweydorf F, Geerlof A, Jacob A, Ueffing M, Gloeckner CJ, Herberg FW (2014) Parkinson-related LRRK2 mutation R1441C/G/H impairs PKA phosphorylation of LRRK2 and disrupts its interaction with 14-3-3. Proc Natl Acad Sci USA 111:E34–43
91. Nagatsu T, Mogi M, Ichinose H, Togari A (2000) Cytokines in Parkinson's disease. J Neural Transm Suppl 143–151
92. Nash Y, Schmukler E, Trudler D, Pinkas-Kramarski R, Frenkel D (2017) DJ-1 deficiency impairs autophagy and reduces alpha-synuclein phagocytosis by microglia. J Neurochem 143:584–594
93. Nimmerjahn A, Kirchhoff F, Helmchen F (2005) Resting microglial cells are highly dynamic surveillants of brain parenchyma in vivo. Science 308:1314–1318
94. Ojha S, Javed H, Azimullah S, Haque ME (2016) Beta-caryophyllene, a phytocannabinoid attenuates oxidative stress, neuroinflammation, glial activation, and salvages dopaminergic neurons in a rat model of Parkinson disease. Mol Cell Biochem 418:59–70
95. Orenstein SJ, Kuo SH, Tasset I, Arias E, Koga H, Fernandez-Carasa I, Cortes E, Honig LS, Dauer W, Consiglio A et al (2013) Interplay of LRRK2 with chaperone-mediated autophagy. Nat Neurosci 16:394–406
96. Pabon MM, Bachstetter AD, Hudson CE, Gemma C, Bickford PC (2011) CX3CL1 reduces neurotoxicity and microglial activation in a rat model of Parkinson's disease. J Neuroinflammation 8:9
97. Paisan-Ruiz C, Nath P, Washecka N, Gibbs JR, Singleton AB (2008) Comprehensive analysis of LRRK2 in publicly available Parkinson's disease cases and neurologically normal controls. Hum Mutat 29:485–490
98. Paolicelli RC, Bolasco G, Pagani F, Maggi L, Scianni M, Panzanelli P, Giustetto M, Ferreira TA, Guiducci E, Dumas L et al (2011) Synaptic pruning by microglia is necessary for normal brain development. Science 333:1456–1458
99. Park JY, Kim KS, Lee SB, Ryu JS, Chung KC, Choo YK, Jou I, Kim J, Park SM (2009) On the mechanism of internalization of alpha-synuclein into microglia: roles of ganglioside GM1 and lipid raft. J Neurochem 110:400–411
100. Park JY, Paik SR, Jou I, Park SM (2008) Microglial phagocytosis is enhanced by monomeric alpha-synuclein, not aggregated alpha-synuclein: implications for Parkinson's disease. Glia 56:1215–1223
101. Pasanen P, Myllykangas L, Siitonen M, Raunio A, Kaakkola S, Lyytinen J, Tienari PJ, Poyhonen M, Paetau A (2014) Novel alpha-synuclein mutation A53E associated with atypical multiple system atrophy and Parkinson's disease-type pathology. Neurobiol Aging 35(2180):e2181–2185

102. Pisanu A, Lecca D, Mulas G, Wardas J, Simbula G, Spiga S, Carta AR (2014) Dynamic changes in pro- and anti-inflammatory cytokines in microglia after PPAR-gamma agonist neuroprotective treatment in the MPTPp mouse model of progressive Parkinson's disease. Neurobiol Dis 71:280–291
103. Plowey ED, Cherra SJ 3rd, Liu YJ, Chu CT (2008) Role of autophagy in G2019S-LRRK2-associated neurite shortening in differentiated SH-SY5Y cells. J Neurochem 105:1048–1056
104. Polymeropoulos MH, Lavedan C, Leroy E, Ide SE, Dehejia A, Dutra A, Pike B, Root H, Rubenstein J, Boyer R et al (1997) Mutation in the alpha-synuclein gene identified in families with Parkinson's disease. Science 276:2045–2047
105. Ransohoff RM (2016) A polarizing question: do M1 and M2 microglia exist? Nat Neurosci 19:987–991
106. Rayaprolu S, Mullen B, Baker M, Lynch T, Finger E, Seeley WW, Hatanpaa KJ, Lomen-Hoerth C, Kertesz A, Bigio EH et al (2013) TREM2 in neurodegeneration: evidence for association of the p.R47H variant with frontotemporal dementia and Parkinson's disease. Mol Neurodegener 8:19
107. Ren M, Guo Y, Wei X, Yan S, Qin Y, Zhang X, Jiang F, Lou H (2018) TREM2 overexpression attenuates neuroinflammation and protects dopaminergic neurons in experimental models of Parkinson's disease. Exp Neurol 302:205–213
108. Rojanathammanee L, Murphy EJ, Combs CK (2011) Expression of mutant alpha-synuclein modulates microglial phenotype in vitro. J Neuroinflammation 8:44
109. Roodveldt C, Labrador-Garrido A, Gonzalez-Rey E, Fernandez-Montesinos R, Caro M, Lachaud CC, Waudby CA, Delgado M, Dobson CM, Pozo D (2010) Glial innate immunity generated by non-aggregated alpha-synuclein in mouse: differences between wild-type and Parkinson's disease-linked mutants. PLoS ONE 5:e13481
110. Sanchez-Danes A, Richaud-Patin Y, Carballo-Carbajal I, Jimenez-Delgado S, Caig C, Mora S, Di Guglielmo C, Ezquerra M, Patel B, Giralt A et al (2012) Disease-specific phenotypes in dopamine neurons from human iPS-based models of genetic and sporadic Parkinson's disease. EMBO Mol Med 4:380–395
111. Sanchez-Guajardo V, Febbraro F, Kirik D, Romero-Ramos M (2010) Microglia acquire distinct activation profiles depending on the degree of alpha-synuclein neuropathology in a rAAV based model of Parkinson's disease. PLoS ONE 5:e8784
112. Satake W, Nakabayashi Y, Mizuta I, Hirota Y, Ito C, Kubo M, Kawaguchi T, Tsunoda T, Watanabe M, Takeda A et al (2009) Genome-wide association study identifies common variants at four loci as genetic risk factors for Parkinson's disease. Nat Genet 41:1303–1307
113. Sawada M, Imamura K, Nagatsu T (2006) Role of cytokines in inflammatory process in Parkinson's disease. J Neural Transm Suppl 373–381
114. Schafer DP, Lehrman EK, Kautzman AG, Koyama R, Mardinly AR, Yamasaki R, Ransohoff RM, Greenberg ME, Barres BA, Stevens B (2012) Microglia sculpt postnatal neural circuits in an activity and complement-dependent manner. Neuron 74:691–705
115. Schapansky J, Nardozzi JD, Felizia F, LaVoie MJ (2014) Membrane recruitment of endogenous LRRK2 precedes its potent regulation of autophagy. Hum Mol Genet 23:4201–4214
116. Sierra A, Abiega O, Shahraz A, Neumann H (2013) Janus-faced microglia: beneficial and detrimental consequences of microglial phagocytosis. Front Cell Neurosci 7:6
117. Sierra A, Encinas JM, Deudero JJ, Chancey JH, Enikolopov G, Overstreet-Wadiche LS, Tsirka SE, Maletic-Savatic M (2010) Microglia shape adult hippocampal neurogenesis through apoptosis-coupled phagocytosis. Cell Stem Cell 7:483–495
118. Simon-Sanchez J, Schulte C, Bras JM, Sharma M, Gibbs JR, Berg D, Paisan-Ruiz C, Lichtner P, Scholz SW, Hernandez DG et al (2009) Genome-wide association study reveals genetic risk underlying Parkinson's disease. Nat Genet 41:1308–1312
119. Singleton AB, Farrer M, Johnson J, Singleton A, Hague S, Kachergus J, Hulihan M, Peuralinna T, Dutra A, Nussbaum R et al (2003) Alpha-synuclein locus triplication causes Parkinson's disease. Science 302:841
120. Smeyne RJ, Breckenridge CB, Beck M, Jiao Y, Butt MT, Wolf JC, Zadory D, Minnema DJ, Sturgess NC, Travis KZ et al (2016) Assessment of the effects of MPTP and paraquat on

dopaminergic neurons and microglia in the substantia nigra pars compacta of C57BL/6 mice. PLoS ONE 11:e0164094
121. Stefanova N, Fellner L, Reindl M, Masliah E, Poewe W, Wenning GK (2011) Toll-like receptor 4 promotes alpha-synuclein clearance and survival of nigral dopaminergic neurons. Am J Pathol 179:954–963
122. Stokholm MG, Iranzo A, Ostergaard K, Serradell M, Otto M, Svendsen KB, Garrido A, Vilas D, Borghammer P, Santamaria J et al (2017) Assessment of neuroinflammation in patients with idiopathic rapid-eye-movement sleep behaviour disorder: a case-control study. Lancet Neurol 16:789–796
123. Su X, Federoff HJ, Maguire-Zeiss KA (2009) Mutant alpha-synuclein overexpression mediates early proinflammatory activity. Neurotox Res 16:238–254
124. Su X, Maguire-Zeiss KA, Giuliano R, Prifti L, Venkatesh K, Federoff HJ (2008) Synuclein activates microglia in a model of Parkinson's disease. Neurobiol Aging 29:1690–1701
125. Symons A, Beinke S, Ley SC (2006) MAP kinase kinase kinases and innate immunity. Trends Immunol 27:40–48
126. Takahashi K, Rochford CD, Neumann H (2005) Clearance of apoptotic neurons without inflammation by microglial triggering receptor expressed on myeloid cells-2. J Exp Med 201:647–657
127. Theodore S, Cao S, McLean PJ, Standaert DG (2008) Targeted overexpression of human alpha-synuclein triggers microglial activation and an adaptive immune response in a mouse model of Parkinson disease. J Neuropathol Exp Neurol 67:1149–1158
128. Thome AD, Standaert DG, Harms AS (2015) Fractalkine signaling regulates the inflammatory response in an alpha-synuclein model of Parkinson disease. PLoS ONE 10:e0140566
129. Tokuda T, Qureshi MM, Ardah MT, Varghese S, Shehab SA, Kasai T, Ishigami N, Tamaoka A, Nakagawa M, El-Agnaf OM (2010) Detection of elevated levels of alpha-synuclein oligomers in CSF from patients with Parkinson disease. Neurology 75:1766–1772
130. Valdinocci D, Radford RA, Siow SM, Chung RS, Pountney DL (2017) Potential modes of intercellular alpha-synuclein transmission. Int J Mol Sci 18
131. Venezia S, Refolo V, Polissidis A, Stefanis L, Wenning GK, Stefanova N (2017) Toll-like receptor 4 stimulation with monophosphoryl lipid A ameliorates motor deficits and nigral neurodegeneration triggered by extraneuronal alpha-synucleinopathy. Mol Neurodegener 12:52
132. Vezzani A, Maroso M, Balosso S, Sanchez MA, Bartfai T (2011) IL-1 receptor/Toll-like receptor signaling in infection, inflammation, stress and neurodegeneration couples hyperexcitability and seizures. Brain Behav Immun 25:1281–1289
133. Walsh S, Finn DP, Dowd E (2011) Time-course of nigrostriatal neurodegeneration and neuroinflammation in the 6-hydroxydopamine-induced axonal and terminal lesion models of Parkinson's disease in the rat. Neuroscience 175:251–261
134. Wang S, Chu CH, Stewart T, Ginghina C, Wang Y, Nie H, Guo M, Wilson B, Hong JS, Zhang J (2015) alpha-Synuclein, a chemoattractant, directs microglial migration via H_2O_2-dependent Lyn phosphorylation. Proc Natl Acad Sci USA 112:E1926–1935
135. West AB, Moore DJ, Biskup S, Bugayenko A, Smith WW, Ross CA, Dawson VL, Dawson TM (2005) Parkinson's disease-associated mutations in leucine-rich repeat kinase 2 augment kinase activity. Proc Natl Acad Sci USA 102:16842–16847
136. West AP, Koblansky AA, Ghosh S (2006) Recognition and signaling by toll-like receptors. Annu Rev Cell Dev Biol 22:409–437
137. Wolf SA, Boddeke HW, Kettenmann H (2017) Microglia in physiology and disease. Annu Rev Physiol 79:619–643
138. Wu DC, Jackson-Lewis V, Vila M, Tieu K, Teismann P, Vadseth C, Choi DK, Ischiropoulos H, Przedborski S (2002) Blockade of microglial activation is neuroprotective in the 1-methyl-4-phenyl-1,2,3,6-tetrahydropyridine mouse model of Parkinson disease. J Neurosci 22:1763–1771
139. Wyss-Coray T, Mucke L (2002) Inflammation in neurodegenerative disease–a double-edged sword. Neuron 35:419–432

140. Xu Y, Tao X, Shen B, Horng T, Medzhitov R, Manley JL, Tong L (2000) Structural basis for signal transduction by the Toll/interleukin-1 receptor domains. Nature 408:111–115
141. Zarranz JJ, Alegre J, Gomez-Esteban JC, Lezcano E, Ros R, Ampuero I, Vidal L, Hoenicka J, Rodriguez O, Atares B et al (2004) The new mutation, E46 K, of alpha-synuclein causes Parkinson and Lewy body dementia. Ann Neurol 55:164–173
142. Zhan Y, Paolicelli RC, Sforazzini F, Weinhard L, Bolasco G, Pagani F, Vyssotski AL, Bifone A, Gozzi A, Ragozzino D et al (2014) Deficient neuron-microglia signaling results in impaired functional brain connectivity and social behavior. Nat Neurosci 17:400–406
143. Zhang W, Wang T, Pei Z, Miller DS, Wu X, Block ML, Wilson B, Zhang W, Zhou Y, Hong JS et al (2005) Aggregated alpha-synuclein activates microglia: a process leading to disease progression in Parkinson's disease. FASEB J 19:533–542

Chapter 14
Astrocytes in Huntington's Disease

Michelle Gray

Abstract Huntington's disease (HD) is a dominantly inherited neurodegenerative disease that results in motor, cognitive and psychiatric dysfunction. It is caused by a polyglutamine repeat expansion mutation in the widely expressed HTT protein. The clinical manifestations of HD have been largely attributed to the neurodegeneration of specific neuronal cell types in the brain. However, it has become clear that other cell types, including astrocytes, play important roles in the pathogenesis of HD. The mutant HTT (mHTT) protein is present in neuronal and non-neuronal cell types throughout the nervous system. Studies designed to understand the contribution of mHTT expression in non-neuronal cell types to HD pathogenesis has lagged considerably behind those focused on neurons. However, the role of astrocytes in HD has received more attention over the last 5–10 years. In this chapter we present an overview of HD and our current understanding of astrocytic involvement in this disease. We describe the neuropathological features of HD and provide evidence of morphological and molecular changes in mHTT expressing astrocytes. We review data from animal models and HD patients that implicate mHTT expressing astrocytes to the progression of HD.

Keywords Astrocytes · Huntingtin · Excitotoxicity · BDNF · Cholesterol

14.1 Introduction

Huntington's Disease (HD) is a progressive and fatal neurodegenerative disorder characterized by motor dysfunction, psychiatric disturbances, and cognitive impairment for which we have no neuroprotective therapies. The age of onset of HD is in the mid-forties, with death usually occurring 15–20 years after diagnosis [47]. George Huntington published the first clinical description of HD in 1872 in his paper

M. Gray (✉)
Department of Neurology and Center for Neurodegeneration and Experimental Therapeutics, University of Alabama at Birmingham, 1720 2nd Ave S,
CIRC 425B, Birmingham AL 35294, USA
e-mail: mccgray@uabmc.edu

A. Verkhratsky et al. (eds.), *Neuroglia in Neurodegenerative Diseases*,
Advances in Experimental Medicine and Biology 1175,
https://doi.org/10.1007/978-981-13-9913-8_14

"On Chorea" in which he described sufferers of this disease quite accurately [64]. He described a "*hereditary* chorea" where chorea is defined by "*dancing* propensities", and "with no loss of volition attending these contractions….the will is there, but its power to perform is deficient." He noted that the disease appeared hereditary in nature, with a tendency to insanity and suicide and that it manifested in adult life [64]. Remarkably, George Huntington published this classic work just 6 years after Gregor Mendel published his famous study of pea plants (in 1866) without any knowledge of it, but he was clearly able to identify its pattern of inheritance.

These descriptions of the disease, which now bares Huntington's name, have stood the test of time with more information having been added over time. To date, the clinical diagnosis of HD is based on the development of the classic movement deficit, chorea, which can present in combination with other movement deficits including dystonia and bradykinesia. While the motor symptoms are the most visible symptoms of HD, cognitive and behavioral problems are also very prominent in this disease. The cognitive changes can manifest early in the disease process and can include deficits in emotional recognition, time production, and speed of initiating thought processes [59]. HD patients also display changes in learning and working memory [122]. They have difficulties in learning new information and in memory recall [94].

14.2 Huntington's Disease Genetics

HD is one of the most common single-gene dominantly inherited neurodegenerative disorders. While it is rare, meta-analysis of studies of prevalence worldwide shows it affects 2–3 persons per 100,000. In areas with populations largely of European descent, the overall prevalence is 5–6 persons per 100,000 [103]. HD is caused by a repeat expansion in the highly conserved huntingtin (*HTT*) gene [52]. The triplet repeat, cytosine–adenosine–guanine (CAG), is found in exon1 of the gene and encodes glutamine. The disease allele results in the production of an expanded polyglutamine (polyQ) stretch in the N-terminal portion of the large 3,144 amino acid protein [52, 149]. When someone carries a CAG stretch with less than 35 repeats, there is no risk for getting HD. Alleles with CAG repeat lengths in the 36–40 range are incompletely penetrant. Those persons with CAG repeat lengths in this range will likely not develop HD symptoms; but if they do, it tends to be at a very advanced age [104, 112, 121]. When the CAG repeat expansion is greater than 40, the carriers will develop HD [4, 52, 149].

The average age at onset in HD is in the mid-forties and is inversely correlated to the length of the CAG repeat expansion in HD [36, 149]. The CAG repeat length can explain nearly half of the variability in the age at onset for HD [142] with additional modifiers present in the genome accounting for another significant portion. Recently, age at onset genetic modifiers have been identified that are involved in DNA repair and DNA damage mechanisms including those that could contribute to increasing the length of the CAG repeat in somatic cells and germ cells (mismatch and base excision repair) [26, 48, 73].

14.3 Huntington's Disease Neuropathology

While the mutant protein is found in cells throughout the nervous system, the classic HD neuropathology is characterized by degeneration of the γ-aminobutyric acid (GABA) ergic medium spiny neurons (MSNs) in the striatum, with the vast majority of these cells degenerating at advanced stages of the disease (Fig. 14.1), [135]. HD disease grades are determined postmortem-based on the degree of striatal degeneration, with modest degeneration assigned as Grade 0 and the most severe atrophy and degeneration assigned to Grade 4 with ~95% of neurons lost in the striatum [137]. Grade 0 brains are determined microscopically since no sizeable changes are observed at a macroscopic (whole brain) level. At this stage, neuronal loss (30–40%) is observed in the tail of the caudate nucleus. In Grade 1 tissue, the body and tail of the caudate nucleus degenerate and then the tissue progresses in a caudal to rostral and dorsal to the ventral gradient. There are also differences between the interspersed MSNs based on neurochemical and anatomical features, with the MSNs expressing enkaphalin and dopamine receptor D2/Drd2 (and projecting to the globus pallidus externa) being lost earlier than MSNs expressing dopamine receptor D1/Drd1 and substance P (projecting to the globus pallidus interna and the substantia nigra pars reticulata) loss after Drd2 MSNs [109, 137]. While the MSNs are the most degen-

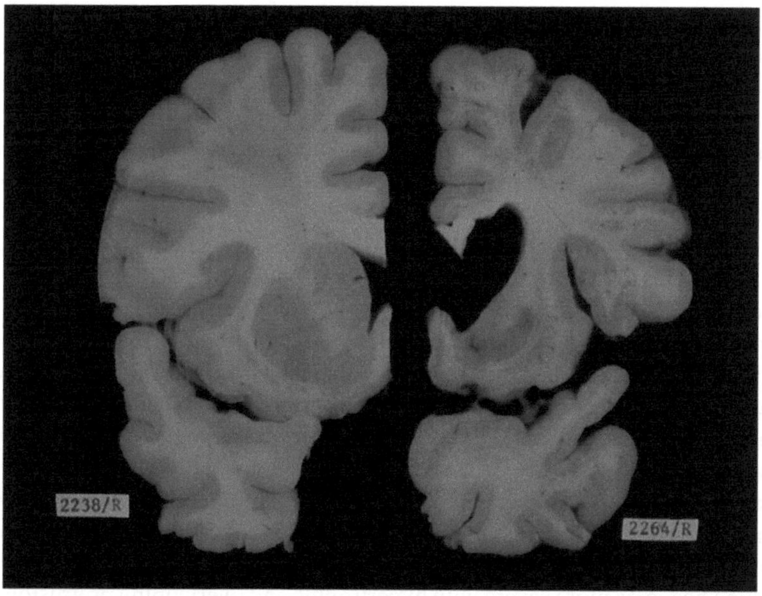

Fig. 14.1 Postmortem human brain at the level of the caudate-putamen. Coronal brain sections taken through the caudate-putamen of a normal (left) and a Huntington's disease patient (right). The Huntington's disease brain on the right shows degeneration of the caudate nucleus adjacent to the lateral ventricle, which has enlarged in response to the striatal atrophy. Courtesy of J.-P. Vonsattel. Reproduced from Alexi et al. (2000), with permission from Elsevier

erated cell type in the striatum, there is increasing evidence from postmortem HD patient tissue that parvalbumin-positive interneurons in the caudate and putamen are decreased in number as disease progresses [108]. Interestingly, prior to a decrease in number, these cells appear shrunken and have decreased expression of PARV [108]. A recent study of aged transgenic HD monkeys (5 years) revealed a significant decrease in the number parvalbumin-positive interneurons in the caudate nucleus and putamen as compared to control [71].

Magnetic resonance imaging, positron emission tomography, and computed tomography have been used to image brains of presymptomatic *HTT* mutation carriers. Atrophy of the striatum was shown prior to overt motor symptoms and a clinical diagnosis of HD [8]. These studies also helped to confirm degeneration of extra-striatal regions, with the cortex as the next most affected brain region in HD. There is significant cortical atrophy that can be detected early in disease [110, 111] and degeneration of cortical pyramidal neurons especially those in cortical layers III, V, and VI, including those that project directly to the striatum [56, 57]. The atrophy of these structures seems to appear long before the onset of overt motor dysfunction as the imaging studies demonstrate atrophy prior to clinical diagnosis. Although HD affects most prominently the MSNs in the striatum, there is also significant atrophy of other brain regions as disease progresses, including the nucleus accumbens, globus pallidus, thalamus, and parts of the hypothalamus [70, 125, 133].

One prominent feature found upon neuropathological examination of HD patient tissue is the presence of astrogliosis [39, 137] (Fig. 14.2). In neurodegeneration, it is generally believed that astrogliosis is a response to dysfunction or death of neurons. In HD patient brains, there is a significant increase in the number of reactive astrocytes as disease grade (neuropathological severity) increases (0–4). In Grade 0 brains, there is no significant astrogliosis in the caudate nucleus [136, 137]. The reactive astrogliosis can be seen most predominantly in the tail of the caudate nucleus at Grade 1 coincident with neuronal loss. In another study, astrogliosis was assessed in striatal tissue from all disease grades using GFAP immunohistochemistry. These studies reveal that increased GFAP immunoreactivity is present throughout the striatum in all disease grades. Furthermore, the GFAP level seems to increase as a larger number of astrocytes are expressing GFAP as disease grade increases [39]. The pattern of astrogliosis in the striatal tissue from these patients seems to follow the pattern of neurodegeneration, where it is first seen in the dorsal striatum and then in the ventral striatum (nucleus accumbens). Since the dorsal striatal astrogliosis is so early in the disease process (Grade 0–1), at least as far as obvious neuropathological changes are seen, this suggests that there is likely a cell-autonomous change in the astrocyte. The increase in GFAP staining intensity exists with characteristic reactive astrocytic phenotypes including hypertrophic cell bodies. As the disease progresses, the reactive phenotype becomes more severe with hypertrophic and overlapping protrusions from the astrocytes (Fig. 14.2) [39]. Neuronal loss and reactive astrogliosis are present predominantly in the tail of the caudate nucleus.

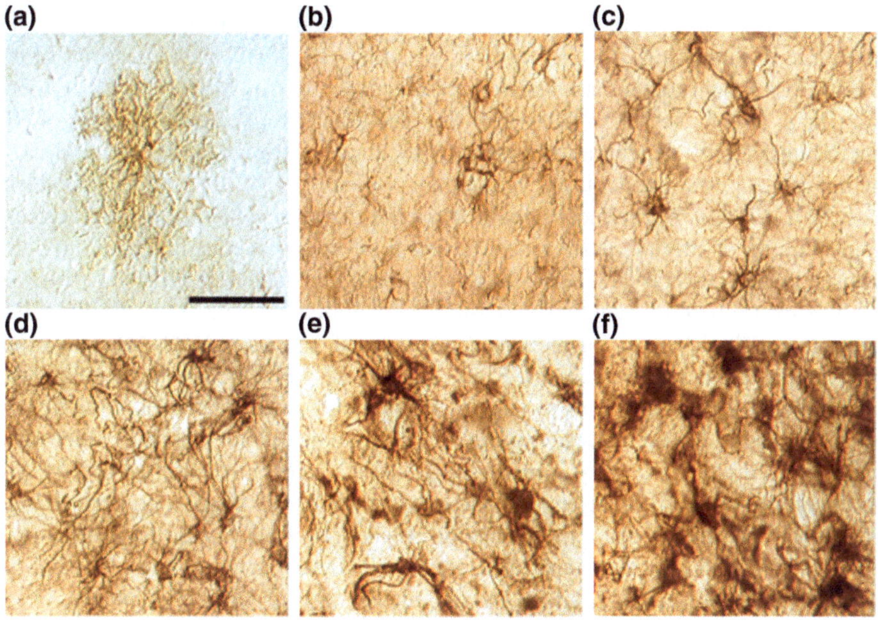

Fig. 14.2 Astrogliosis in HD patient striatal tissue. GFAP immunohistochemistry in 50 μm tissue sections from the caudate nucleus in non-neurological control (**a**), and in increasingly severe HD specimens, Grades 0–4, Grade 0 (**b**), Grade 1 (**c**), Grade 2 (**d**), Grade 3 (**e**), and Grade 4 (**f**) HD subjects. Normal astrocytes present as faintly GFAP-stained cells with a short lace-like branching pattern distributed symmetrically around the cell soma. With increasing disease progression, there was greater GFAP immunoreactivity, twisting and thickened arbors, with larger somal size. The degree of astrogliosis became so great as to mask their individual appearance. The magnification bar in (**a**) represents 100 μm and is the same in all photomicrographs. Reproduced with permission from [39]; Oxford University Press

14.4 Excitotoxicity in Huntington's Disease

14.4.1 Excitotoxicity and Neurons

There are multiple mechanisms that may contribute to toxicity in HD; one of these is excitotoxicity. Excitotoxicity leading to neuronal dysfunction and death is caused by excessive activation of glutamate-gated N-methyl D-aspartate receptors (NMDARs) due to increased exposure to glutamate. This leads to Ca^{2+} overload and mitochondria energy failure [27]. This mechanism had been hypothesized for HD many decades ago. In HD this mechanism has primarily focused on the corticostriatal synapse, with the pre- and postsynaptic neuron receiving the most attention. The MSNs in the striatum receive glutamatergic excitatory input from both the cortex and thalamus [43, 44]. The excitotoxicity hypothesis of HD pathogenesis is supported by the existence of these extensive inputs and the presence of high densities of glutamatergic receptors in striatal neurons [1, 10, 72]. Many of the initial studies in HD used excitotoxins to mimic HD pathology. One of the first rodent models of HD used injections of the excitotoxin kainic acid into the striatum to selectively destroy MSNs [28, 87]. Quinolinic acid, a selective NMDAR agonist was also used to replicate the features of HD including selective degeneration and morphological changes in MSNs including loss of dendritic spines in rodents and nonhuman primates [11, 42, 113, 147]. This selective NMDAR agonist produced specific toxicities for striatal MSNs, without causing degeneration of striatal interneurons, further reinforcing the idea that excitotoxicity caused by activation of NMDARs as an important mechanism in HD.

The excitotoxicity hypothesis is further supported by data from multiple HD mouse models. There is an increased response of MSNs to NMDAR activation. When quinolinic acid was injected into the YAC72 and YAC128 mice, there was a significant difference in lesion size as compared to wild-type mice, although as the disease progressed in these models, this phenotype did not persist in older mice [49, 147]. Furthermore, there is increased glutamate level in the striatum of YAC128 mice upon cortical stimulation [67], although other studies suggest no such change in young YAC128 mice not yet displaying behavioral or neuropathological features of HD [30, 91]. These data support an altered NMDAR function early in the course of disease in these mouse models, and that altered striatal NMDAR signaling likely contributes to the deficits seen in HD.

14.4.2 Excitotoxicity and Astrocytes

The glutamate transporter is critical for regulating glutamate levels at the synapse. The uptake of glutamate and its conversion into glutamine reduces the level of glutamate in the synaptic space. In support of the excitotoxicity hypothesis, HD brains show a decrease in the level of the excitatory amino acid transporter 2 (EAAT2; human)/Glutamate transporter 1 (GLT-1; rodent) [7, 29]. In situ hybridization studies on HD brain tissue reveal a decrease in EAAT2 mRNA labeling that correlated with disease severity [7]. In the tissue, a decrease in the number of cells expressing EAAT2 mRNA can be seen in the remaining tissue of both the caudate and putamen, although the putamen seemed to have a greater decrease. In addition, immunohistochemistry performed on Grade 0 to Grade 4 tissues with an EAAT2/GLT-1 antibody revealed a grade-dependent decrease in protein levels [39]. This data reveals that there is a loss of EAAT2 early in the disease process, which can implicate the transporter as a primary component in the initiation of disease. To properly maintain synaptic function and glutamate neurotransmission, there must be coordinated activity of the pre- and postsynaptic cells, but also the astrocyte. The role of this transporter in astrocytes is to remove glutamate from the synaptic cleft after it is released [31, 86]. With excitotoxicity as one of the proposed toxic mechanisms in HD, due to increased levels of glutamate in striatal tissue that is hypothesized to come from the corticostriatal inputs, the ability to efficiently uptake glutamate from the synaptic space is of vital importance.

14.5 Huntingtin Expression and Mutant Huntingtin Aggregation

The *HTT* RNA is found throughout the nervous system. In situ hybridization used to localize the RNA found the transcript throughout the brain. Neurons express a small amount more *HTT* RNA than other cell types. It is interesting that the neurons most susceptible to neurodegeneration in HD, the MSNs, do not display higher levels of *HTT* RNA expression than neurons in other brain regions. It can be found in nonneuronal cell types as well. The HTT protein is found in all of these cell types with the levels similar in all of the cell types as a whole [72, 115, 117, 119, 148].

Like many other neurodegenerative diseases, another hallmark of HD is the progressive aggregation or inclusion body formation of mutant HTT (mHTT). These aggregates/inclusions were initially identified in R6 mouse models harboring an expanding CAG repeat stretch within exon1 of a human *HTT* transgene [32]. These inclusions were identified as neuronal intranuclear inclusions containing HTT and ubiquitin prior to neurological phenotypic development in these animals. Subsequently, these inclusions were identified in the neurons of HD patients [35]. The initial descriptions of these inclusions in HD tissue, largely identified them as neuronal in nature and cytoplasmic/neuropil with a few intranuclear locations, and found primarily in gray matter (with EM48 antibody) [55]. The largest number of inclusions identified in patient tissue in those studies were observed in the deeper cortical layers, with many fewer and much smaller inclusions found in the striatum. More recent analysis of inclusions and aggregates in postmortem tissue from HD patients with an antibody that recognizes aggregated mHTT (S829), showed that mHTT nuclear inclusions are found in all nervous system cell types—neurons, astrocytes, oligodendrocytes, and microglia [65].

14.5.1 Expression and mHTT Aggregation in Human Astrocytes

RNA in situ hybridization for *HTT* identified positive signals in astrocytes from normal brain tissue [72]. Furthermore, brain tissue from HD patients that was stained with antibodies to HTT and glial fibrillary acidic protein (GFAP) revealed the presence of HTT in astrocytes [119], although to a lesser degree than what is seen in neurons. Astrocytes in various brain regions including the striatum (caudate nucleus and putamen) as well as white matter from HD patients also contained mHTT positive aggregates [18, 39, 65, 118, 119]. Recent work has confirmed with the S829 antibody that the frequency of nuclear inclusions in astrocytes was much lower than that observed in neurons. They also identified nuclear inclusions in the astrocytes that were less frequent than neuronal nuclear inclusions [65].

14.5.2 Expression and mHTT Aggregation in Mouse Astrocytes

The HTT protein is also found in astrocytes from mice [18, 74]. The normal function of this protein in astrocytes remains to be completely elucidated. However, in mouse *Hdh* (encoding endogenous mouse huntingtin) knock-out neural stem cells treated using a neuronal differentiation protocol, there was a significant increase in the number of GFAP positive cells and a decrease in the number of microtubule-associated protein 2 positive neurons when compared to control cultures. This finding suggests that wild-type HTT is involved in controlling the differentiation of neuronal and glial cells and that production of neurons from neural stem cells requires a normal level of wild-type HTT [25]. This data is in line with previous studies showing that wild-type HTT plays a role in central nervous system development and neuronal survival [82, 107]. The wild-type HTT expressed in these stem cells could be acting in an instructive or repressive role, to promote neuronal fate and/or repress the glial fate. However, the exact role of wild-type HTT in astrocytes in the nervous system will need to be further assessed in conditional knock-in mouse models, where one can specifically reduce the expression of endogenous HTT only in astrocytes. Mutant HTT positive aggregates are found in astrocytes in the brains of HD mouse models. These mice contain aggregates in the white matter as well as the gray matter [105, 118, 146]. Like the aggregation found in neurons in these mice, aggregation in the astrocytes also appears to be progressive, with the number of mHTT aggregates increasing as the animal ages. The aggregates are found not only in the striatum and cortex of these mice but also in the corpus callosum [19, 118]. In addition, in a mouse model expressing a fragment of m*HTT* only in astrocytes, driven by the human GFAP promoter, there are aggregates in the astrocytes in the cortex, striatum, brainstem, and spinal cord [18]. Further analysis of various mouse models expressing mHTT using the S829 antibody reveals the presence of nuclear inclusions in astrocytes in those models [65].

14.6 Cell Autonomous and Non-cell Autonomous Toxicity in Huntington's Disease

The most prominent area of neurodegeneration in HD is the striatum. The dysfunction of the MSNs in this region and their ultimate degeneration is at the core of the motor abnormalities that exist in this disease. The striatum is the central input area of the basal ganglia receiving excitatory glutamatergic input from both the cortex and thalamus, and dopaminergic input from the substantia nigra [17, 51, 143]. Although the mHTT protein is expressed throughout the nervous system, much of the focus of HD research has been in the striatum. However, a series of studies in mice demonstrate the importance of other cell types in HD and their effect on neuropathological and behavioral manifestations of disease phenotypes in mice. There is clear evidence for non-cell autonomous mechanisms of toxicity in HD. In a mouse model with inducible expression of an m*HTT*-exon1 fragment (the Rosa/HD mouse), induction of expression throughout the brain (using Nestin-Cre) in neurons and glia results in neuropathological abnormalities including gliosis and neurodegenerative changes in cortex and striatum. However, when the expression of mHTT was restricted to the cortex (using Emx1-Cre), or the striatum alone (using Dlx5/6-Cre), there were no significant behavioral or neuropathological changes at the ages examined [53, 54]. In another model conditionally expressing a different m*HTT* fragment in medium spiny neurons under the control of the Darpp32 promoter (DE5 mice), there are late-onset motor abnormalities but no evidence of neurodegenerative changes [20], whereas mice expressing this fragment throughout the brain showed extensive neuropathological changes [146].

14.7 Astrocyte Changes in Huntington's Disease Mouse Models

14.7.1 Astrogliosis in Mutant Huntingtin Expressing Models

The examination of astrogliosis, mHTT aggregation, and excitotoxicity has been performed in many different mouse models expressing different forms of the mutated *HTT* gene. These mouse models have not only revealed phenotypes that are observed in HD patients but they have also uncovered mechanisms that were not previously implicated for astrocytic contribution to HD pathogenesis. Progressive astrogliosis has been observed in postmortem tissue of increasing grade from HD patient brains. The reactive astrocyte phenotype has also been observed in many of the mouse models expressing mHTT. Neuropathologically, these models display varying degrees of pathological changes, with regional atrophy, cellular atrophy, dark neuron degenerative changes, and mHTT aggregation. There are multiple mouse models that express some form of the mutant protein, either a full-length or truncated protein by either a knock-in or transgenic approach. Many of these models also display some degree of astrogliosis. In the RosaHD/Nestin-Cre model, expressing mHTT throughout the nervous system in both neurons and glia, there is significant astrogliosis in the cortex and striatum [54]. Studies of astrogliosis in models where the mHTT protein is expressed in the cortex with RosaHD/Emx1-Cre and striatum with RosaHD/Dlx5/6-Cre revealed no significant astrogliosis when mHTT was restricted to neurons [53, 54]. The data from these models suggested that cell–cell interactions were important for the development of the reactive astrocyte phenotype.

In the mouse model HTT171-82Q, where mHTT expression was targeted to astrocytes in the striatum through lentiviral expression, there was an increase in GFAP staining and astrocytes exhibiting a reactive phenotype. This phenotype increased in severity as the animal aged, with an increased astrocyte soma size [39]. This indicates a cell-autonomous effect of the mHTT protein within astrocytes as this reactive phenotype is not due to the expression of mHTT in neurons, thus reinforcing the idea that reactive gliosis is not merely a consequence or response to sick or degenerating neurons in neurodegenerative diseases. In the knock-in mouse model, *Hdh* CAG 150, there is extensive astrogliosis in the striatum [80].

One of the most widely used mHTT expressing mouse models, R6/2, which express an exon1 fragment with an expanded CAG repeat, does not exhibit significant astrogliosis as indicated by morphological changes in the cortex nor striatum as the animal ages [85, 126]. There is also not a significant upregulation of the GFAP level in the mice. Additional analysis of the somatosensory cortex specifically did not reveal significant astrogliosis [138]. There is no significant astrogliosis identified in the knock-in Q175 model nor in the human mHTT expressing YAC128 or BACHD models [50, 60, 126].

Multiple monkey models have been generated for use in understanding HD pathogenesis [23, 145]. A study describing the characterization of two monkeys created via lentiviral gene transfer was recently published. One monkey contained exons 1–10 of the m*HTT* gene with 70 CAG repeats under the control of a minimal human *HTT* promoter and the other monkey contained human *HTT* exon1 with 29 CAG repeats and is regulated by the human polyubiquitin C promoter. A study of these two types of monkeys at 5 years of age revealed significant astrogliosis in the model [71]. The caudate nucleus and putamen contained an increase in astrocytes that were heavily GFAP-stained. The increase was more pronounced in the monkey that contained only exon1 with 29 CAG repeats. The control monkey did not have much GFAP positive staining in these regions (Fig. 14.3).

Although it is not completely understood why astrogliosis is not a striking feature in all of the animal models expressing mHTT, there is astrogliosis in some of the models. The lack of significant astrogliosis does not negate the possibility that the astrocytes are not completely normal in those models. Nonetheless, some of the mouse models demonstrate that the reactive astrocyte phenotype can be elicited from the expression of mHTT specifically in the astrocytes or observed when mHTT expression is also found in neurons; therefore, the mHTT protein is able to elicit both cell-autonomous and non-cell autonomous phenotypes.

14.7.2 GLT-1 in mHTT Expressing Mouse Models

Based on the decrease in the level of the astrocyte-specific glutamate transporter EAAT2/GLT-1, this has been a major focus of studies aimed at understanding the astrocytic contribution to HD. The loss of GLT-1 in HD could be of critical consequence given the important role of GLT-1 in maintaining glutamate homeostasis at the excitatory synapse. Much of this work has been performed in multiple mouse models expressing various forms of the mHTT protein. There is a reduction in the level of *GLT-1* mRNA in the R6/2 mice, which express an exon1 fragment with an expanded CAG repeat when compared to littermates [12, 118, 126]. The decrease in *GLT-1* mRNA levels seem progressive as a decrease can be seen in R6/2 animals

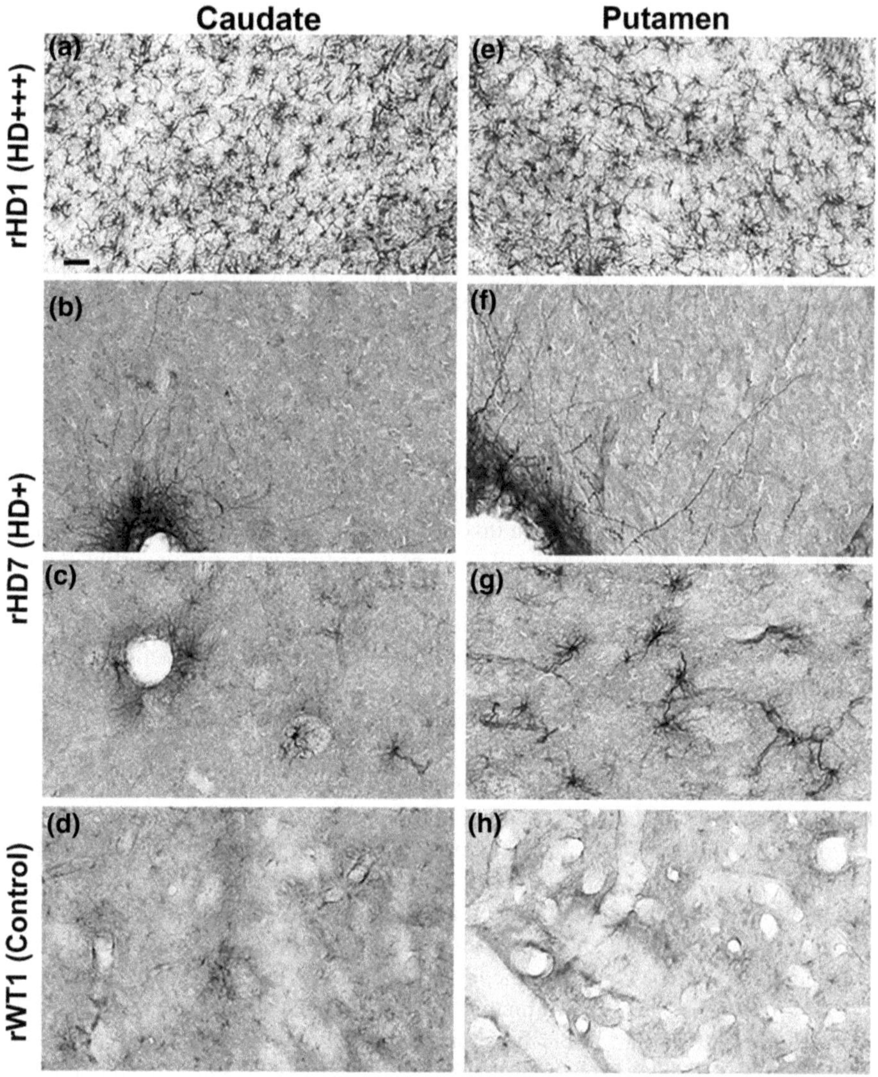

Fig. 14.3 Astrogliosis in the brains of 5 year old mHTT expressing monkeys. (**a**–**h**) GFAP labeling in the caudate nucleus (CD; **a**–**d**) and putamen (PU; **e**–**h**) of the two transgenic HD monkeys (rHD1, rHD7) and the control animal (rWT1). There is a large density of GFAP labeled astrocytes in the CD and PU of the most severely affected HD monkey (rHD1-**a**, **e**) compared with the least affected HD monkey (rHD7; **b**, **c**; **f**, **g**) and the control (**d**, **h**). Scale bar in A: 30 μm in A and E; 20 μm in (**b**–**d** and **f**–**h**). Edits to the figure legend were made to decrease space. Reproduced from [71], under a Creative Commons Attribution 4.0 International License. https://creativecommons.org/licenses/by/4.0/legalcode

in 6-week-old animals, in both cortex and striatum and further declines at 12 weeks [12]. In addition to mRNA levels in the R6/2 mice, there is a significant reduction in the protein level of GLT-1 during 11–12 weeks of age [12, 100], although there is a likely decrease in protein levels early as well; it does not reach statistical significance [118]. Glutamate uptake is decreased in the striatum of the R6/2 transgenic model [12, 38, 79, 118]. Another mouse model, the *Hdh* CAG 150 knock-in mouse, did not show a significant difference in the *GLT-1* mRNA at 9 months of age although there was a slight reduction in *GLT-1* mRNA levels [118]. This result is likely given the slower progression in this full-length mHTT mouse model that exhibits no obvious neuropathological changes at this age [80].

The YAC128 model expresses full-length human mHTT and recapitulates aspects of HD neuropathology and behavior that becomes progressively worse as the animal ages [60, 120]. There does not appear to be a decrease in the levels of GLT-1 protein in the brain of these mice even as the disease progresses. Interestingly, glutamate uptake was decreased in the striatum of these mice as early as 3 months of age; it is also seen at 12 months in the cortex [63]. EAAT2 was identified in a proteomics study of palmitoylation [68], which involves the thioesterification of palmitic acid to cysteine residues and functions in tethering proteins to membranes or sorting them to lipid microdomains. Palmitoylation was reduced on GLT-1 in the brains of YAC128 mice [63]. The decrease in palmitoylation of GLT-1 was found to impair glutamate uptake, without affecting is localization to the membrane. Thus, exactly how palmitoylation affects the function of this receptor is still up for debate; nonetheless, these data provide two mechanisms for decreased GLT-1 levels and activity in HD mouse models.

BACHD mice express full-length human mHTT from a Bacterial Artificial chromosome that has been modified to contain a mixed CAA–CAG repeat. This is a slowly progressing mouse model with behavioral and neuropathological changes that are similar to those observed in HD patients [50]. The neuropathological changes manifest at about 1 year of age with a decrease in the striatal volume and robust mHTT aggregation. Western blot analysis of protein extracted from the cortex and striatum revealed no significant change in the levels of GLT-1 at 6 or 12 months [144].

The knock-in line zQ175 (human HTT exon1) is a slowly progressing mHTT expressing mouse model. Robust phenotypic changes are observed at 1 year of age. Examination of GLT-1 expression in this line revealed a significant reduction of the GLT-1 protein at 10–12 months of age [126]. Interestingly, *GLT-1* mRNA was only reduced in homozygous Q175 mice at 10 months of age [126]. Another study revealed a significant reduction in the level of the *GLT-1* transcript at about 10 months of age in these mice [88]. In addition, glutamate transport activity is altered in the zQ175 mice that are thought to critically depend on or be caused by a decrease in Kir4.1 conductance in the HD models [37]. It is interesting, though not surprising, that the degree of GLT-1 loss or dysfunction is variable across the models expressing different forms of mHTT. Some of these models express full-length mHTT (knock-in or transgenic) driven by endogenous promotor regions that give rise to a much slower progressing HD-like phenotype, while others express truncated forms of mHTT driven by minimal *HTT* gene promotors or other neuronal promotors that seem to produce a more rapid progression of HD-like phenotypes. The differences in the genetic construct used in these various models could contribute to some of the variability observed in the animals.

Mice generated using the human GFAP promoter to specifically express an mHTT exon1 fragment carrying 160Q repeat in astrocytes (GFAP-HD) [18]. There is a significant decrease in the level of GLT-1 protein in the brains and cultured astrocytes from these mice. As a consequence of the decrease in the levels of *GLT-1*, there is also decreased glutamate uptake in the GFAP-HD mice [18]. In mice that expressed an mHTT fragment containing 82 CAG repeats (viral HTT-82Q) specifically in astrocytes using lentiviral vectors, there was a decrease in the level of *GLT-1* mRNA in the striatum of these mice 12 weeks after injection [39]. The level of the GLT-1 protein that was assessed by immunohistochemistry in the mHTT positive astrocytes showed a significant and progressive decrease in the striatum. Interestingly, there was also a decrease in GLAST, but this was not significant until late in the disease process in this animal. Glutamate transport was also decreased in the astrocyte HTT171-82Q expressing mice, and no decrease in glutamate transport or GLT-1 levels was observed with neuronal expression of the HTT171-82Q [39]. Perhaps the most interesting observation from these mice was the decrease in the levels of two neuronal proteins DARPP-32 and GluN2B subunits of NMDARs, thus indicating that the presence of mHTT in astrocytes is likely sufficient to alter neuronal activity and function. The exact mechanism whereby the mHTT in the astrocytes caused the decrease in the levels of these proteins is unclear but implicates the inability of GLT-1 to properly buffer extracellular glutamate as a possible mechanism for decreased expression of DARPP-32 and GluN2B in neurons.

The mechanism for decrease in GLT-1 levels in the mouse models includes a change in Sp1-dependent transcription of *GLT-1* and palmitoylation of GLT-1. In mHTT expressing astrocytes from GFAP-HD transgenic mice, there is a reduction of the transcription factor Sp1 occupancy at the *GLT-1* promoter as compared to littermate controls [18]. This reduction in this model is likely due to the increased association of mHTT with Sp1 as shown by more mHTT precipitating with Sp1 than with HTT with a polyQ repeat in the normal range [18].

With the alterations in GLT-1, studies have been performed to determine if increasing the expression of GLT-1 would alleviate phenotypes caused by the presence of mHTT. When *GLT-1* was overexpressed by using a lentiviral vector in astrocytes also expressing HTT-171-82Q, the level of GLT-1 increased and the reactive astrocyte previously seen in those mice significantly decreased [39]. Furthermore, the use of ceftriaxone, a β-lactam antibiotic in R6/2 mice raised GLT-1 levels and reversed the glutamate uptake deficit in these mice. Ceftriaxone improved some of the motor deficits found in the R6/2 mice [90]. However, use of ceftriaxone must be approached with caution as it has been found to affect long-term depression in the hippocampus [96] and impairs prepulse inhibition [13]. Nonetheless, while the appropriate approach to take is yet to be determined, increasing the levels of GLT-1 in HD may have beneficial effects on glutamate uptake deficit and motor impairment.

Another study using primary cultured astrocytes expressing an N-terminal fragment of mHTT (HTT-552), observed a significant decrease in the level *GLT-1* mRNA as well as the GLT-1 protein. The authors tested whether enhancing autophagy with rapamycin, a known autophagy activator, would have beneficial effects on clearing mHTT protein and thus increasing GLT-1 protein levels. When the cultures were treated with rapamycin, there was a significant increase in the levels of GLT-1 and also increased glutamate uptake in the astrocytes [24].

14.8 Glutamate Release from Mutant HTT Expressing Astrocytes

Much of the study in HD on astrocytes centers around the decrease in EAAT2/GLT-1 level, the ability of the astrocyte to take up excess glutamate from the synaptic cleft, and on how that may contribute to excitotoxicity. However, there has been a lack of appreciation for the ability of the astrocyte to release glutamate and whether that ability is changed in HD. The astrocyte is the only cell in the brain that can synthesize glutamate de novo [58]. They are capable of releasing glutamate through various mechanisms, including but not limited to Ca^{2+}-dependent vesicular exocytosis [99]. Glutamate released from astrocytes has been shown to act on extrasynaptic NMDARs to modulate neuronal excitability and synaptic transmission [5, 41]. Based on this ability of astrocytes to modulate neuronal excitability, it is possible that a change in this function due to the presence of mHTT in astrocytes could contribute to changes in the activity at the most critical excitatory synapses in the brains of HD patients.

To date, only one study has been performed to determine whether mHTT expression in astrocytes had any effect on the levels of Ca^{2+}-dependent glutamate release [74]. This study used cultured solitary astrocytes from the full-length human mHTT expressing BACHD mouse. This study employed a previously validated system where astrocytes were first grown in culture flasks and purified solitary astrocytes were then mechanically stimulated resulting in glutamate release [62]. Mechanical stimulation of astrocytes allows one to assess the exocytotic release of glutamate [93]. Mechanical stimulation of full-length mHTT containing cortical astrocytes from BACHD mice resulted in an increase in the level of glutamate that was released into the extracellular space near the solitary astrocytes as compared to wild-type astrocytes [74]. Astrocytes exhibiting Ca^{2+}-dependent exocytotic release of glutamate do so by increased cytosolic Ca^{2+} responses which are usually in proportion to the level of glutamate released from these cells [97]. While these BACHD cells responded to mechanical stimulation with a rise in cytosolic Ca^{2+} levels, there was not a significant difference between wild-type and full-length mHTT expressing cortical astrocytes in cytosolic Ca^{2+} levels upon mechanical stimulation.

Multiple mechanisms could account for the change in levels of glutamate released from astrocytes. These include the trafficking of glutamate containing vesicles and glutamate synthesizing machinery within these cells. However, analysis of glutamate containing vesicles revealed no change in mHTT expressing astrocytes when compared to wild-type astrocyte, suggesting that other mechanisms are involved in causing the increase in levels of glutamate released from these astrocytes.

Glutamate can be synthesized in astrocytes de novo as a by-product of the tricarboxylic acid cycle. In addition, a decrease in the enzyme glutamine synthetase could result in an increase in the levels of glutamate present intracellularly in these astrocytes, which ultimately could contribute to an increase in the amount released. Further examination of these mHTT expressing astrocytes revealed an increase in the level of the mitochondria resident enzyme pyruvate carboxylase and no change in the glutamine synthetase. The increase in pyruvate carboxylase provides a mechanism for the increase in glutamate released from these cortical astrocytes. The exact mechanism leading to the increase in pyruvate carboxylase remains to be elucidated and validated in patient samples.

14.9 BDNF and mHTT Expressing Astrocytes

The motor abnormalities originate due to degeneration of the MSNs in the striatum. These cells receive trophic support from the cortex in the form of the brain-derived neurotrophic factor (BDNF) [3]. A lack of BDNF trophic support can increase a neuron's susceptibility to cell death and it is crucial for the survival of MSNs [9, 21]. The MSN does not produce BDNF, but receives it from other sources. There is a decrease of BDNF levels in HD patient striatum that is also recapitulated in some of the animal models expressing mHTT [149–152]. This decrease is attributed not only to transcriptional dysregulation caused by mHTT in the cortex but also to a decrease in BDNF transport to the striatum [46, 77]. Astrocytes play important roles in providing trophic support in the nervous system. They are able to synthesize and release neurotrophic factors including glial-derived neurotrophic factor (GDNF), ciliary neurotrophic factor (CNTF), and brain-derived neurotrophic factor (BDNF) [16, 78, 92]. Astrocytes can release BDNF through exocytosis [99].

Astrocyte expression of mHTT impairs multiple normal aspects of astrocyte function. The expression of mHTT (100Q) in astrocytes by adenoviral vector in cultured rat primary astrocytes caused a decrease in the level of BDNF found in the astrocyte-conditioned media as compared to conditioned media from cells expressing HTT with a short (18Q) polyQ with a retention of BDNF within the astrocytes [139]. Analysis of the BDNF transcript level revealed a decrease in the transcript, thus indicating that the presence of mHTT repressed BDNF transcription in the astrocytes [140]. Using two mouse models: one is a full-length knock-in mouse model where mHTT is expressed in all cells of the nervous system, including astrocytes; and another is where expression of a truncated mHTT fragment is driven by the GFAP promoter. Hong et al. showed a reduction in secreted BDNF from mHTT expressing astrocytes and it was shown to likely be due to impaired docking of BDNF containing vesicles in astrocytes [61]. Thus, mHTT expression in astrocytes is capable of interfering with multiple mechanisms that would contribute to a decrease of BDNF levels in astrocytes.

There clearly is a deficit in BDNF generated and released by mHTT containing astrocytes; thus, these cells are contributing to the BDNF deficit observed in HD patient brains suggesting that these cells are good targets for therapeutic intervention. Adenoviral astrocyte specific expression of BDNF using the GFAP promotor injected into the striatum of R6/2 mHTT expressing model there was a delay in the onset of motor abnormalities in the model [6]. In addition, glatiramer acetate, which is known to increase BDNF expression in T cells [69, 81, 124] was used to treat R6/2 mice and YAC128 mice and increased the expression of BDNF in the brains of the mice and also decreased the neurodegenerative changes observed in these mice [106]. Together these studies implicate another mechanism whereby mHTT in astrocytes could contribute to the pathogenesis of HD.

14.10 Cholesterol Changes in Huntington's Disease

Brain cholesterol is required for nervous system development and to maintain proper nervous system function [34]. The most cholesterol-rich organ in the body is the brain and all cholesterol found in the brain is synthesized in the central nervous system [128]. In the adult brain, astrocytes are essential for de novo cholesterol synthesis [15, 101]. Cholesterol homeostasis is maintained in the brain by de novo synthesis and export of excess cholesterol from the brain to the plasma in the form of 24-hydroxycholesterol (24-OHC) [15, 83]. In humans, mutations in genes that affect cholesterol synthesis or transport can result in neurodegeneration and abnormal brain formation [102, 134]. The mutations in these genes also impair cognitive function [102].

Studies in HD patient postmortem tissue have revealed changes in cholesterol metabolism with increased striatal neuronal cholesterol membrane accumulation [132]. There is a decrease in the enzymes involved in cholesterol synthesis in post-mortem HD brain tissue. The levels of mRNA expression for genes involved in cholesterol synthesis, including HMG-CoAR and 7-DHC reductase and lanosterol 14 a-demethylase, are reduced in postmortem striatal and cortical tissue [132]. One group reported an increase in cholesterol levels in postmortem HD brains [33]. This contradictory data is likely due to the differences in methodology used to assess cholesterol content in tissue. Plasma levels of cholesterol are decreased in late-stage HD patient plasma [75]. Plasma levels of 24-OHC change as the disease worsens in HD patients. In plasma from premanifest HD patients (prior to overt motor deficits), the 24-OHC levels are normal as compared to controls. However, in HD patients that display motor deficits, the level of 24-OHC levels are reduced [75, 76]. In general, there is a decrease in the level of cholesterol synthesis enzymes in postmortem HD brain tissue and also a reduction in the level of 24-OHC that is transported out of the nervous system.

There is significant data from mHTT expressing mouse models that demonstrate changes in cholesterol levels in the brain and plasma. While there are numerous studies now focused on cholesterol and its contribution to the pathogenesis observed in HD, some of this work is contradictory. However, it must be recognized that this work has been performed in different animal models with various polyQ lengths, with mHTT driven from different promoters, and with different timeframes for HD-like disease manifestation. Thus, while some of the data seems contradictory, it highlights that the cholesterol story in HD is highly complicated and in need of further investigation to ascertain what is truly happening during disease progression. The R6/2, YAC72, YAC128, HdhQ111/111, and zQ175 [116, 129, 130], all have reduced brain cholesterol levels. Another study in YAC72 mice suggests they have increased cholesterol striatal neuron accumulation with increased cholesterol content [127]. There is data from mouse models that clearly show reduced expression of genes involved in cholesterol synthesis, like HMG-CoAR, CYP51, and 7-DHC-reductase, just like what is observed in HD patient brain [132]. The effect on the expression of these genes may be due to the interaction of HTT with specific transcription factors like Sp1, which is known to coordinate sterol regulatory element-binding protein (SREBP) that activates gene transcription when cholesterol levels are low [114]. Plasma levels of 24-OHC are decreased in zQ175, YAC128, and R6/2 mice as the disease progresses in those mHTT expressing models [116]. It has been shown that mHTT expressing astrocytes in a culture that secreted less cholesterol into the medium and conditioned media from these astrocytes (and media depleted of lipoproteins) did not support neurite outgrowth nor synaptic activity when compared to cholesterol supplementation or conditioned media from normal astrocytes [131]. Due to the significant amount of data in support of cholesterol changes in HD, a study using cholesterol-loaded nanoparticles that were able to cross the blood–brain barrier were injected intraperitonealy, and localized to glial and neuronal cells [131]. The nanoparticles improved electrophysiological deficits, cognitive dysfunction, and increased decreased synaptic proteins [131].

14.11 Genetic Manipulation of Astrocytic mHTT Expression in Mice

Numerous studies have been performed in mice expressing mHTT specifically in astrocytes as well as studies focused on their contribution to HD pathogenesis in the context of mHTT expressing neurons. An important study to demonstrate that astrocytes expressing mHTT were important for neurodegeneration in HD was performed in cell culture. Co-culturing of adenoviral-mediated expression of a fragment of mHTT in astrocytes in culture with normal neurons, resulted in increased neuronal cell death in culture [118]. Studies in a mouse model where mHTT with 160Q was expressed under the control of the human glial fibrillary acidic protein (GFAP) promoter, revealed abnormal neurological phenotypes as the animals aged and mHTT aggregation within astrocytes. Furthermore, those mice had motor dysfunction, a decrease in body weight, and premature death [18]. These mice also had a decrease in the level of GLT-1 attributed to a decreased interaction of the transcription factor Sp1 with the promoter region of the GLT-1. Interestingly, breeding a transgenic mouse with a GFAP promoter-driven fragment of mHTT with 98Qs to the transgenic mouse model N171-82Q that contains the mHTT fragment only in neurons revealed an exacerbation of the HD-like phenotypes observed in the N171-82Q model, thus revealing that the mHTT expressing astrocytes are important targets in HD [19].

Another mouse model expressing a fragment of mHTT was developed using lentiviral vectors with a pseudotype (Mokola) that drove expression specifically in striatal astrocytes. This model showed increased astrogliosis with a concomitant decrease in the expression level of GLT-1 and GLAST. There was also a reduction in dopamine and adenosine 3′,5′-monophosphate-regulated phosphoprotein (32 kDa) DARPP-32, a protein highly enriched in MSNs, and a marker of neuronal loss and dysfunction [39]. This study reveals non-cell autonomous toxicity of mHTT expressing astrocytes on the MSNs in this model. A study using AAV2/5 viral vectors expressing mHTT N171-82Q with the Gfa2(B)3 promoter or the Chicken-β Actin (CBA) in the striatum of mice, revealed that when mHTT was expressed alone in astrocytes, less severe motor phenotypes were observed in the mouse than when mHTT was expressed in both neurons and astrocytes [89]. Using this system, the authors demonstrated minimal changes in the mice when the vectors were targeted to astrocytes in the striatum. Interestingly, when both neurons and astrocytes expressed mHTT, the number of mHTT aggregates was increased and the level of DARPP-32 decrease was exacerbated [89]. This work suggests that the mHTT expressing astrocytes within the striatum are contributors to the disease although mostly in the context of mHTT expressing astrocytes.

The BACHD mouse model is a conditional transgenic full-length human mHTT expressing mouse model with exon1 that contains the expanded repeat flanked by loxP sites [50]. The genetic design of the model allows assessment of cell-type-specific contributions to HD pathogenesis. It has been used to explore the necessity of fl-mHTT expression within the vulnerable MSN and cortical pyramidal neuron population in HD [141]. None of the other models are designed to be able to assess whether mHTT expression in astrocytes is necessary for HD phenotypic progression when modulated only in astrocytes and not changed in neurons. A study using this BACHD model bred to two different GFAP-CreER^{T2} lines [22, 45], reveals that the Cre expression is sufficient to decrease mHTT levels in the brains of the BACHD mice [144]. Using a reporter mouse model, it showed the recombination was highly selective for astrocytes in the cortex and striatum [84]. The fl-mHTT was decreased in the animals after weaning and behavioral assessments were taken as the phenotype progressed in the animals from 2–12 months of age. This study revealed the reducing mHTT in the astrocytes at 2 mths did not affect the onset of the behavioral phenotypes in the BACHD model. However, the decrease of mHTT in the astrocytes decreased disease progression. The BACHD/GFAP-CreER^{T2} expressing mice showed behavioral phenotypes that were better than the BACHD mice at 6 and 12 mths of age, indicating a slowing of the behavioral phenotypes. In addition, the BACHD/GFAPCreER^{T2} mouse brain weight increased and the striatal volume significantly improved over what was observed in the BACHD mice alone. In agreement with the change in the neuropathology observed in the mice, the electrophysiological deficits displayed by the BACHD MSNs at 12 mths were restored to normal in the MSNs in the BACHD/GFAPCreER^{T2} mice. Together, the behavioral, neuropathological, and electrophysiological rescue observed in the BACHD/GFAP-CreER^{T2} mice, reveal the significant contribution to the disease progression in this model. This has important implications for therapeutic development in HD. If targeting disease onset is not feasible, targeting mechanisms at play in astrocytes may slow the progression of disease in HD patients.

14.12 Concluding Remarks

The importance of astrocytes to HD pathogenesis is starting to gain more focus and acceptance. While the presence of reactive astrocytes has been found in postmortem tissue and seems to progress with age, whether this phenotype is truly an intrinsic phenotype or a response to the neuronal changes in HD brains remain to be completely determined. Interestingly, many of the mouse models expressing mHTT do not fully recapitulate this neuropathological feature. The astrocyte functions in and impacts a variety of processes in the brain. The uptake of excess glutamate from the synaptic space by astrocytes is critical to proper neural circuit function and its decreased uptake in HD could contribute to disease pathogenesis. Both the levels of the receptor responsible for this uptake into astrocytes and posttranslational modification of the receptor are likely important mechanisms that could contribute to excitotoxicity. Mouse models expressing mHTT do not all demonstrate a broad decrease in the level of GLT-1. However, it is important to note that changes on the microcircuit level may not be as obvious when using techniques such as Western blot to assess protein levels of this highly expressed protein. However, if more sophisticated imaging modalities are employed, one may be able to visualize cell-specific changes. The identification of an increase in the levels of glutamate released from mHTT expressing cortical astrocytes also provides another mechanism whereby mHTT in astrocytes could lead to excitotoxicity. However, another study suggests that there is no change in astrocyte glutamate release in the striatum of a different mouse model using a glutamate sensor to measure glutamate levels [66]. Although neurotransmitter uptake/levels have been widely studied, some of the additional processes that astrocytes are involved in have not been fully explored in HD although increased efforts along these lines are underway. These processes including the release of glutamate [74], astrocyte roles in the inflammatory response [14, 123], and sleep (also abnormal in HD) [40, 95], need further exploration in HD. These studies will require the use of advanced imaging modalities, manipulation of normal HTT and mHTT in both cell and animal models, and novel methodologies for introducing agents specifically to astrocytes. Many of these studies can likely be performed using the many genetic models of expressing mHTT in all the cell types or only in astrocytes that have been generated to date. However, proper performance and interpretation of these studies will require the cooperation of expert glial biologists and those focused on the elucidation of pathogenic mechanisms in HD.

Note: This chapter contains content from "The role of astrocytes in Huntington's disease" published in "Pathological Potential of Neuroglia [98]".

Acknowledgements The authors' work is supported by grants from the Dixon Family Foundation, the Hereditary Disease Foundation and the National Institutes of Health.

References

1. Albin RL, Young AB, Penney JB, Handelin B, Balfour R, Anderson KD, Markel DS, Tourtellotte WW, Reiner A (1990) Abnormalities of striatal projection neurons and N-methyl-D-aspartate receptors in presymptomatic Huntington's disease. N Engl J Med 322:1293–1298
2. Alexi T, Borlongan CV, Faull RL, Williams CE, Clark RG, Gluckman PD, Hughes PE (2000) Neuroprotective strategies for basal ganglia degeneration: Parkinson's and Huntington's diseases. Prog Neurobiol 60:409–470
3. Altar CA, Cai N, Bliven T, Juhasz M, Conner JM, Acheson AL, Lindsay RM, Wiegand SJ (1997) Anterograde transport of brain-derived neurotrophic factor and its role in the brain. Nature 389:856–860
4. Andrew SE, Goldberg YP, Kremer B, Telenius H, Theilmann J, Adam S, Starr E, Squitieri F, Lin B, Kalchman MA et al (1993) The relationship between trinucleotide (CAG) repeat length and clinical features of Huntington's disease. Nat Genet 4:398–403
5. Araque A, Sanzgiri RP, Parpura V, Haydon PG (1999) Astrocyte-induced modulation of synaptic transmission. Can J Physiol Pharmacol 77:699–706
6. Arregui L, Benitez JA, Razgado LF, Vergara P, Segovia J (2011) Adenoviral astrocyte-specific expression of BDNF in the striata of mice transgenic for Huntington's disease delays the onset of the motor phenotype. Cell Mol Neurobiol 31:1229–1243
7. Arzberger T, Krampfl K, Leimgruber S, Weindl A (1997) Changes of NMDA receptor subunit (NR1, NR2B) and glutamate transporter (GLT1) mRNA expression in Huntington's disease–an in situ hybridization study. J Neuropathol Exp Neurol 56:440–454
8. Aylward EH, Codori AM, Rosenblatt A, Sherr M, Brandt J, Stine OC, Barta PE, Pearlson GD, Ross CA (2000) Rate of caudate atrophy in presymptomatic and symptomatic stages of Huntington's disease. Mov Disord 15:552–560
9. Baquet ZC, Gorski JA, Jones KR (2004) Early striatal dendrite deficits followed by neuron loss with advanced age in the absence of anterograde cortical brain-derived neurotrophic factor. J Neurosci 24:4250–4258
10. Beal MF (1994) Huntington's disease, energy, and excitotoxicity. Neurobiol Aging 15:275–276
11. Beal MF, Ferrante RJ, Swartz KJ, Kowall NW (1991) Chronic quinolinic acid lesions in rats closely resemble Huntington's disease. J Neurosci 11:1649–1659
12. Behrens PF, Franz P, Woodman B, Lindenberg KS, Landwehrmeyer GB (2002) Impaired glutamate transport and glutamate-glutamine cycling: downstream effects of the Huntington mutation. Brain 125:1908–1922
13. Bellesi M, Melone M, Gubbini A, Battistacci S, Conti F (2009) GLT-1 upregulation impairs prepulse inhibition of the startle reflex in adult rats. Glia 57:703–713
14. Benveniste EN (1992) Inflammatory cytokines within the central nervous system: sources, function, and mechanism of action. Am J Physiol 263:C1–16
15. Bjorkhem I, Lutjohann D, Breuer O, Sakinis A, Wennmalm A (1997) Importance of a novel oxidative mechanism for elimination of brain cholesterol. Turnover of cholesterol and 24(S)-hydroxycholesterol in rat brain as measured with 18O2 techniques in vivo and in vitro. J Biol Chem 272:30178–30184

16. Blondel O, Collin C, McCarran WJ, Zhu S, Zamostiano R, Gozes I, Brenneman DE, McKay RD (2000) A glia-derived signal regulating neuronal differentiation. J Neurosci 20:8012–8020
17. Bolam JP, Hanley JJ, Booth PA, Bevan MD (2000) Synaptic organisation of the basal ganglia. J Anat 196(Pt 4):527–542
18. Bradford J, Shin JY, Roberts M, Wang CE, Li XJ, Li S (2009) Expression of mutant huntingtin in mouse brain astrocytes causes age-dependent neurological symptoms. Proc Natl Acad Sci USA 106:22480–22485
19. Bradford J, Shin JY, Roberts M, Wang CE, Sheng G, Li S, Li XJ (2010) Mutant huntingtin in glial cells exacerbates neurological symptoms of Huntington disease mice. J Biol Chem 285:10653–10661
20. Brown TB, Bogush AI, Ehrlich ME (2008) Neocortical expression of mutant huntingtin is not required for alterations in striatal gene expression or motor dysfunction in a transgenic mouse. Hum Mol Genet 17:3095–3104
21. Canals JM, Pineda JR, Torres-Peraza JF, Bosch M, Martin-Ibanez R, Munoz MT, Mengod G, Ernfors P, Alberch J (2004) Brain-derived neurotrophic factor regulates the onset and severity of motor dysfunction associated with enkephalinergic neuronal degeneration in Huntington's disease. J Neurosci 24:7727–7739
22. Casper KB, Jones K, McCarthy KD (2007) Characterization of astrocyte-specific conditional knockouts. Genesis 45:292–299
23. Chan AW, Xu Y, Jiang J, Rahim T, Zhao D, Kocerha J, Chi T, Moran S, Engelhardt H, Larkin K et al (2014) A two years longitudinal study of a transgenic Huntington disease monkey. BMC Neurosci 15:36
24. Chen LL, Wu JC, Wang LH, Wang J, Qin ZH, Difiglia M, Lin F (2012) Rapamycin prevents the mutant huntingtin-suppressed GLT-1 expression in cultured astrocytes. Acta Pharmacol Sin 33:385–392
25. Conforti P, Camnasio S, Mutti C, Valenza M, Thompson M, Fossale E, Zeitlin S, MacDonald ME, Zuccato C, Cattaneo E (2013) Lack of huntingtin promotes neural stem cells differentiation into glial cells while neurons expressing huntingtin with expanded polyglutamine tracts undergo cell death. Neurobiol Dis 50:160–170
26. Consortium, G.M.O.H.D.G.-H (2015) Identification of genetic factors that modify clinical onset of Huntington's disease. Cell 162:516–526
27. Coyle JT, Puttfarcken P (1993) Oxidative stress, glutamate, and neurodegenerative disorders. Science 262:689–695
28. Coyle JT, Schwarcz R (1976) Lesion of striatal neurones with kainic acid provides a model for Huntington's chorea. Nature 263:244–246
29. Cross AJ, Slater P, Reynolds GP (1986) Reduced high-affinity glutamate uptake sites in the brains of patients with Huntington's disease. Neurosci Lett 67:198–202
30. Cummings DM, Cepeda C, Levine MS (2010) Alterations in striatal synaptic transmission are consistent across genetic mouse models of Huntington's disease. ASN Neuro 2:e00036
31. Danbolt NC (2001) Glutamate uptake. Prog Neurobiol 65:1–105
32. Davies SW, Turmaine M, Cozens BA, DiFiglia M, Sharp AH, Ross CA, Scherzinger E, Wanker EE, Mangiarini L, Bates GP (1997) Formation of neuronal intranuclear inclusions underlies the neurological dysfunction in mice transgenic for the HD mutation. Cell 90:537–548
33. del Toro D, Xifro X, Pol A, Humbert S, Saudou F, Canals JM, Alberch J (2010) Altered cholesterol homeostasis contributes to enhanced excitotoxicity in Huntington's disease. J Neurochem 115:153–167
34. Dietschy JM, Turley SD (2004) Thematic review series: brain Lipids. Cholesterol metabolism in the central nervous system during early development and in the mature animal. J Lipid Res 45:1375–1397
35. DiFiglia M, Sapp E, Chase KO, Davies SW, Bates GP, Vonsattel JP, Aronin N (1997) Aggregation of huntingtin in neuronal intranuclear inclusions and dystrophic neurites in brain. Science 277:1990–1993
36. Duyao M, Ambrose C, Myers R, Novelletto A, Persichetti F, Frontali M, Folstein S, Ross C, Franz M, Abbott M et al (1993) Trinucleotide repeat length instability and age of onset in Huntington's disease. Nat Genet 4:387–392

37. Dvorzhak A, Vagner T, Kirmse K, Grantyn R (2016) Functional indicators of glutamate transport in single striatal astrocytes and the influence of Kir4.1 in normal and Huntington mice. J Neurosci 36:4959–4975
38. Estrada-Sanchez AM, Montiel T, Segovia J, Massieu L (2009) Glutamate toxicity in the striatum of the R6/2 Huntington's disease transgenic mice is age-dependent and correlates with decreased levels of glutamate transporters. Neurobiol Dis 34:78–86
39. Faideau M, Kim J, Cormier K, Gilmore R, Welch M, Auregan G, Dufour N, Guillermier M, Brouillet E, Hantraye P et al (2010) In vivo expression of polyglutamine-expanded huntingtin by mouse striatal astrocytes impairs glutamate transport: a correlation with Huntington's disease subjects. Hum Mol Genet 19:3053–3067
40. Fellin T, Ellenbogen JM, De Pitta M, Ben-Jacob E, Halassa MM (2012) Astrocyte regulation of sleep circuits: experimental and modeling perspectives. Front Comput Neurosci 6:65
41. Fellin T, Pascual O, Gobbo S, Pozzan T, Haydon PG, Carmignoto G (2004) Neuronal synchrony mediated by astrocytic glutamate through activation of extrasynaptic NMDA receptors. Neuron 43:729–743
42. Ferrante RJ, Kowall NW, Cipolloni PB, Storey E, Beal MF (1993) Excitotoxin lesions in primates as a model for Huntington's disease: histopathologic and neurochemical characterization. Exp Neurol 119:46–71
43. Fonnum F, Soreide A, Kvale I, Walker J, Walaas I (1981) Glutamate in cortical fibers. Adv Biochem Psychopharmacol 27:29–41
44. Fonnum F, Storm-Mathisen J, Divac I (1981) Biochemical evidence for glutamate as neurotransmitter in corticostriatal and corticothalamic fibres in rat brain. Neuroscience 6:863–873
45. Ganat YM, Silbereis J, Cave C, Ngu H, Anderson GM, Ohkubo Y, Ment LR, Vaccarino FM (2006) Early postnatal astroglial cells produce multilineage precursors and neural stem cells in vivo. J Neurosci 26:8609–8621
46. Gauthier LR, Charrin BC, Borrell-Pages M, Dompierre JP, Rangone H, Cordelieres FP, De Mey J, MacDonald ME, Lessmann V, Humbert S et al (2004) Huntingtin controls neurotrophic support and survival of neurons by enhancing BDNF vesicular transport along microtubules. Cell 118:127–138
47. Gomez-Tortosa E, MacDonald ME, Friend JC, Taylor SA, Weiler LJ, Cupples LA, Srinidhi J, Gusella JF, Bird ED, Vonsattel JP et al (2001) Quantitative neuropathological changes in presymptomatic Huntington's disease. Ann Neurol 49:29–34
48. Goold R, Flower M, Moss DH, Medway C, Wood-Kaczmar A, Andre R, Farshim P, Bates GP, Holmans P, Jones L et al (2019) FAN1 modifies Huntington's disease progression by stabilizing the expanded HTT CAG repeat. Hum Mol Genet 28:650–661
49. Graham RK, Pouladi MA, Joshi P, Lu G, Deng Y, Wu NP, Figueroa BE, Metzler M, Andre VM, Slow EJ et al (2009) Differential susceptibility to excitotoxic stress in YAC128 mouse models of Huntington disease between initiation and progression of disease. J Neurosci 29:2193–2204
50. Gray M, Shirasaki DI, Cepeda C, Andre VM, Wilburn B, Lu XH, Tao J, Yamazaki I, Li SH, Sun YE et al (2008) Full-length human mutant huntingtin with a stable polyglutamine repeat can elicit progressive and selective neuropathogenesis in BACHD mice. J Neurosci 28:6182–6195
51. Graybiel AM (1990) Neurotransmitters and neuromodulators in the basal ganglia. Trends Neurosci 13:244–254
52. Group, T.H.s.D.C.R (1993) A novel gene containing a trinucleotide repeat that is expanded and unstable on Huntington's disease chromosomes. The Huntington's disease collaborative research group. Cell 72:971–983
53. Gu X, Andre VM, Cepeda C, Li SH, Li XJ, Levine MS, Yang XW (2007) Pathological cell-cell interactions are necessary for striatal pathogenesis in a conditional mouse model of Huntington's disease. Mol Neurodegener 2:8
54. Gu X, Li C, Wei W, Lo V, Gong S, Li SH, Iwasato T, Itohara S, Li XJ, Mody I et al (2005) Pathological cell-cell interactions elicited by a neuropathogenic form of mutant Huntingtin contribute to cortical pathogenesis in HD mice. Neuron 46:433–444

55. Gutekunst CA, Li SH, Yi H, Mulroy JS, Kuemmerle S, Jones R, Rye D, Ferrante RJ, Hersch SM, Li XJ (1999) Nuclear and neuropil aggregates in Huntington's disease: relationship to neuropathology. J Neurosci Off J Soc Neurosci 19:2522–2534
56. Halliday GM, McRitchie DA, Macdonald V, Double KL, Trent RJ, McCusker E (1998) Regional specificity of brain atrophy in Huntington's disease. Exp Neurol 154:663–672
57. Hedreen JC, Peyser CE, Folstein SE, Ross CA (1991) Neuronal loss in layers V and VI of cerebral cortex in Huntington's disease. Neurosci Lett 133:257–261
58. Hertz L, Dringen R, Schousboe A, Robinson SR (1999) Astrocytes: glutamate producers for neurons. J Neurosci Res 57:417–428
59. Hinton SC, Paulsen JS, Hoffmann RG, Reynolds NC, Zimbelman JL, Rao SM (2007) Motor timing variability increases in preclinical Huntington's disease patients as estimated onset of motor symptoms approaches. J Int Neuropsychol Soc 13:539–543
60. Hodgson JG, Agopyan N, Gutekunst CA, Leavitt BR, LePiane F, Singaraja R, Smith DJ, Bissada N, McCutcheon K, Nasir J et al (1999) A YAC mouse model for Huntington's disease with full-length mutant huntingtin, cytoplasmic toxicity, and selective striatal neurodegeneration. Neuron 23:181–192
61. Hong Y, Zhao T, Li XJ, Li S (2016) Mutant Huntingtin impairs BDNF release from astrocytes by disrupting conversion of Rab3a-GTP into Rab3a-GDP. J Neurosci 36:8790–8801
62. Hua X, Malarkey EB, Sunjara V, Rosenwald SE, Li WH, Parpura V (2004) C(a2+)-dependent glutamate release involves two classes of endoplasmic reticulum Ca(2+) stores in astrocytes. J Neurosci Res 76:86–97
63. Huang K, Kang MH, Askew C, Kang R, Sanders SS, Wan J, Davis NG, Hayden MR (2010) Palmitoylation and function of glial glutamate transporter-1 is reduced in the YAC128 mouse model of Huntington disease. Neurobiol Dis 40:207–215
64. Huntington G (2003) On chorea. George Huntington, M.D. J Neuropsychiatry Clin Neurosci 15:109–112
65. Jansen AH, van Hal M, Op den Kelder IC, Meier RT, de Ruiter AA, Schut MH, Smith DL, Grit C, Brouwer N, Kamphuis W et al (2017) Frequency of nuclear mutant huntingtin inclusion formation in neurons and glia is cell-type-specific. Glia 65:50–61
66. Jiang R, Diaz-Castro B, Looger LL, Khakh BS (2016) Dysfunctional calcium and glutamate signaling in striatal astrocytes from Huntington's disease model mice. J Neurosci 36:3453–3470
67. Joshi PR, Wu NP, Andre VM, Cummings DM, Cepeda C, Joyce JA, Carroll JB, Leavitt BR, Hayden MR, Levine MS et al (2009) Age-dependent alterations of corticostriatal activity in the YAC128 mouse model of Huntington disease. J Neurosci 29:2414–2427
68. Kang R, Wan J, Arstikaitis P, Takahashi H, Huang K, Bailey AO, Thompson JX, Roth AF, Drisdel RC, Mastro R et al (2008) Neural palmitoyl-proteomics reveals dynamic synaptic palmitoylation. Nature 456:904–909
69. Kerschensteiner M, Gallmeier E, Behrens L, Leal VV, Misgeld T, Klinkert WE, Kolbeck R, Hoppe E, Oropeza-Wekerle RL, Bartke I et al (1999) Activated human T cells, B cells, and monocytes produce brain-derived neurotrophic factor in vitro and in inflammatory brain lesions: a neuroprotective role of inflammation? J Exp Med 189:865–870
70. Kremer HP, Roos RA, Dingjan GM, Bots GT, Bruyn GW, Hofman MA (1991) The hypothalamic lateral tuberal nucleus and the characteristics of neuronal loss in Huntington's disease. Neurosci Lett 132:101–104
71. Lallani SB, Villalba RM, Chen Y, Smith Y, Chan AWS (2019) Striatal interneurons in transgenic nonhuman primate model of Huntington's Disease. Sci Rep 9:3528
72. Landwehrmeyer GB, McNeil SM, Dure LSt, Ge P, Aizawa H, Huang Q, Ambrose CM, Duyao MP, Bird ED, Bonilla E et al (1995) Huntington's disease gene: regional and cellular expression in brain of normal and affected individuals. Ann Neurol 37:218–230
73. Lee JM, Chao MJ, Harold D, Abu Elneel K, Gillis T, Holmans P, Jones L, Orth M, Myers RH, Kwak S et al (2017) A modifier of Huntington's disease onset at the MLH1 locus. Hum Mol Genet 26:3859–3867

74. Lee W, Reyes RC, Gottipati MK, Lewis K, Lesort M, Parpura V, Gray M (2013) Enhanced Ca-dependent glutamate release from astrocytes of the BACHD Huntington's disease mouse model. Neurobiol Dis 58C:192–199
75. Leoni V, Mariotti C, Nanetti L, Salvatore E, Squitieri F, Bentivoglio AR, Bandettini di Poggio M, Piacentini S, Monza D, Valenza M et al (2011) Whole body cholesterol metabolism is impaired in Huntington's disease. Neurosci Lett 494:245–249
76. Leoni V, Mariotti C, Tabrizi SJ, Valenza M, Wild EJ, Henley SM, Hobbs NZ, Mandelli ML, Grisoli M, Bjorkhem I et al (2008) Plasma 24S-hydroxycholesterol and caudate MRI in pre-manifest and early Huntington's disease. Brain 131:2851–2859
77. Li JL, Hayden MR, Almqvist EW, Brinkman RR, Durr A, Dode C, Morrison PJ, Suchowersky O, Ross CA, Margolis RL et al (2003) A genome scan for modifiers of age at onset in Huntington disease: The HD MAPS study. Am J Hum Genet 73:682–687
78. Liberto CM, Albrecht PJ, Herx LM, Yong VW, Levison SW (2004) Pro-regenerative properties of cytokine-activated astrocytes. J Neurochem 89:1092–1100
79. Lievens JC, Woodman B, Mahal A, Spasic-Boscovic O, Samuel D, Kerkerian-Le Goff L, Bates GP (2001) Impaired glutamate uptake in the R6 Huntington's disease transgenic mice. Neurobiol Dis 8:807–821
80. Lin CH, Tallaksen-Greene S, Chien WM, Cearley JA, Jackson WS, Crouse AB, Ren S, Li XJ, Albin RL, Detloff PJ (2001) Neurological abnormalities in a knock-in mouse model of Huntington's disease. Hum Mol Genet 10:137–144
81. Linker RA, Lee DH, Demir S, Wiese S, Kruse N, Siglienti I, Gerhardt E, Neumann H, Sendtner M, Luhder F et al (2010) Functional role of brain-derived neurotrophic factor in neuroprotective autoimmunity: therapeutic implications in a model of multiple sclerosis. Brain 133:2248–2263
82. Lo Sardo V, Zuccato C, Gaudenzi G, Vitali B, Ramos C, Tartari M, Myre MA, Walker JA, Pistocchi A, Conti L et al (2012) An evolutionary recent neuroepithelial cell adhesion function of huntingtin implicates ADAM10-Ncadherin. Nat Neurosci 15:713–721
83. Lund EG, Guileyardo JM, Russell DW (1999) cDNA cloning of cholesterol 24-hydroxylase, a mediator of cholesterol homeostasis in the brain. Proc Natl Acad Sci USA 96:7238–7243
84. Madisen L, Zwingman TA, Sunkin SM, Oh SW, Zariwala HA, Gu H, Ng LL, Palmiter RD, Hawrylycz MJ, Jones AR et al (2010) A robust and high-throughput Cre reporting and characterization system for the whole mouse brain. Nat Neurosci 13:133–140
85. Mangiarini L, Sathasivam K, Seller M, Cozens B, Harper A, Hetherington C, Lawton M, Trottier Y, Lehrach H, Davies SW et al (1996) Exon 1 of the HD gene with an expanded CAG repeat is sufficient to cause a progressive neurological phenotype in transgenic mice. Cell 87:493–506
86. Maragakis NJ, Rothstein JD (2001) Glutamate transporters in neurologic disease. Arch Neurol 58:365–370
87. McGeer EG, McGeer PL (1976) Duplication of biochemical changes of Huntington's chorea by intrastriatal injections of glutamic and kainic acids. Nature 263:517–519
88. Menalled LB, Kudwa AE, Miller S, Fitzpatrick J, Watson-Johnson J, Keating N, Ruiz M, Mushlin R, Alosio W, McConnell K et al (2012) Comprehensive behavioral and molecular characterization of a new knock-in mouse model of Huntington's disease: zQ175. PLoS ONE 7:e49838
89. Meunier C, Merienne N, Jolle C, Deglon N, Pellerin L (2016) Astrocytes are key but indirect contributors to the development of the symptomatology and pathophysiology of Huntington's disease. Glia 64:1841–1856
90. Miller BR, Dorner JL, Shou M, Sari Y, Barton SJ, Sengelaub DR, Kennedy RT, Rebec GV (2008) Up-regulation of GLT1 expression increases glutamate uptake and attenuates the Huntington's disease phenotype in the R6/2 mouse. Neuroscience 153:329–337
91. Milnerwood AJ, Raymond LA (2007) Corticostriatal synaptic function in mouse models of Huntington's disease: early effects of huntingtin repeat length and protein load. J Physiol 585:817–831

92. Miyamoto N, Maki T, Shindo A, Liang AC, Maeda M, Egawa N, Itoh K, Lo EK, Lok J, Ihara M et al (2015) Astrocytes promote oligodendrogenesis after white matter damage via brain-derived neurotrophic factor. J Neurosci 35:14002–14008
93. Montana V, Ni Y, Sunjara V, Hua X, Parpura V (2004) Vesicular glutamate transporter-dependent glutamate release from astrocytes. J Neurosci Off J Soc Neurosci 24:2633–2642
94. Montoya A, Price BH, Menear M, Lepage M (2006) Brain imaging and cognitive dysfunctions in Huntington's disease. J Psychiatry Neurosci 31:21–29
95. Morton AJ (2013) Circadian and sleep disorder in Huntington's disease. Exp Neurol 243:34–44
96. Omrani A, Melone M, Bellesi M, Safiulina V, Aida T, Tanaka K, Cherubini E, Conti F (2009) Up-regulation of GLT-1 severely impairs LTD at mossy fibre–CA3 synapses. J Physiol 587:4575–4588
97. Parpura V, Haydon PG (2000) Physiological astrocytic calcium levels stimulate glutamate release to modulate adjacent neurons. Proc Natl Acad Sci U S A 97:8629–8634
98. Parpura V, Verkhratskiĭ AN (2014) Pathological potential of neuroglia: possible new targets for medical intervention. Springer, New York
99. Parpura V, Zorec R (2010) Gliotransmission: exocytotic release from astrocytes. Brain Res Rev 63:83–92
100. Petr GT, Schultheis LA, Hussey KC, Sun Y, Dubinsky JM, Aoki C, Rosenberg PA (2013) Decreased expression of GLT-1 in the R6/2 model of Huntington's disease does not worsen disease progression. Eur J Neurosci 38:2477–2490
101. Pfrieger FW (2003) Role of cholesterol in synapse formation and function. Biochim Biophys Acta 1610:271–280
102. Porter FD, Herman GE (2011) Malformation syndromes caused by disorders of cholesterol synthesis. J Lipid Res 52:6–34
103. Pringsheim T, Wiltshire K, Day L, Dykeman J, Steeves T, Jette N (2012) The incidence and prevalence of Huntington's disease: a systematic review and meta-analysis. Mov Disord 27:1083–1091
104. Quarrell OW, Rigby AS, Barron L, Crow Y, Dalton A, Dennis N, Fryer AE, Heydon F, Kinning E, Lashwood A et al (2007) Reduced penetrance alleles for Huntington's disease: a multi-centre direct observational study. J Med Genet 44:e68
105. Reddy PH, Williams M, Charles V, Garrett L, Pike-Buchanan L, Whetsell WO Jr, Miller G, Tagle DA (1998) Behavioural abnormalities and selective neuronal loss in HD transgenic mice expressing mutated full-length HD cDNA. Nat Genet 20:198–202
106. Reick C, Ellrichmann G, Tsai T, Lee DH, Wiese S, Gold R, Saft C, Linker RA (2016) Expression of brain-derived neurotrophic factor in astrocytes—beneficial effects of glatiramer acetate in the R6/2 and YAC128 mouse models of Huntington's disease. Exp Neurol 285:12–23
107. Reiner A, Dragatsis I, Zeitlin S, Goldowitz D (2003) Wild-type huntingtin plays a role in brain development and neuronal survival. Mol Neurobiol 28:259–276
108. Reiner A, Shelby E, Wang H, Demarch Z, Deng Y, Guley NH, Hogg V, Roxburgh R, Tippett LJ, Waldvogel HJ et al (2013) Striatal parvalbuminergic neurons are lost in Huntington's disease: implications for dystonia. Mov Disord 28:1691–1699
109. Richfield EK, Maguire-Zeiss KA, Vonkeman HE, Voorn P (1995) Preferential loss of preproenkephalin versus preprotachykinin neurons from the striatum of Huntington's disease patients. Ann Neurol 38:852–861
110. Rosas HD, Hevelone ND, Zaleta AK, Greve DN, Salat DH, Fischl B (2005) Regional cortical thinning in preclinical Huntington disease and its relationship to cognition. Neurology 65:745–747
111. Rosas HD, Koroshetz WJ, Chen YI, Skeuse C, Vangel M, Cudkowicz ME, Caplan K, Marek K, Seidman LJ, Makris N et al (2003) Evidence for more widespread cerebral pathology in early HD: an MRI-based morphometric analysis. Neurology 60:1615–1620
112. Rubinsztein DC, Leggo J, Coles R, Almqvist E, Biancalana V, Cassiman JJ, Chotai K, Connarty M, Crauford D, Curtis A et al (1996) Phenotypic characterization of individuals with 30-40 CAG repeats in the Huntington disease (HD) gene reveals HD cases with 36 repeats and apparently normal elderly individuals with 36-39 repeats. Am J Hum Genet 59:16–22

113. Sanberg PR, Calderon SF, Giordano M, Tew JM, Norman AB (1989) The quinolinic acid model of Huntington's disease: locomotor abnormalities. Exp Neurol 105:45–53
114. Sanchez HB, Yieh L, Osborne TF (1995) Cooperation by sterol regulatory element-binding protein and Sp1 in sterol regulation of low density lipoprotein receptor gene. J Biol Chem 270:1161–1169
115. Schilling G, Sharp AH, Loev SJ, Wagster MV, Li SH, Stine OC, Ross CA (1995) Expression of the Huntington's disease (IT15) protein product in HD patients. Hum Mol Genet 4:1365–1371
116. Shankaran M, Di Paolo E, Leoni V, Caccia C, Ferrari Bardile C, Mohammed H, Di Donato S, Kwak S, Marchionini D, Turner S et al (2017) Early and brain region-specific decrease of de novo cholesterol biosynthesis in Huntington's disease: a cross-validation study in Q175 knock-in mice. Neurobiol Dis 98:66–76
117. Sharp AH, Loev SJ, Schilling G, Li SH, Li XJ, Bao J, Wagster MV, Kotzuk JA, Steiner JP, Lo A et al (1995) Widespread expression of Huntington's disease gene (IT15) protein product. Neuron 14:1065–1074
118. Shin JY, Fang ZH, Yu ZX, Wang CE, Li SH, Li XJ (2005) Expression of mutant huntingtin in glial cells contributes to neuronal excitotoxicity. J Cell Biol 171:1001–1012
119. Singhrao SK, Thomas P, Wood JD, MacMillan JC, Neal JW, Harper PS, Jones AL (1998) Huntingtin protein colocalizes with lesions of neurodegenerative diseases: an investigation in Huntington's, Alzheimer's, and Pick's diseases. Exp Neurol 150:213–222
120. Slow EJ, van Raamsdonk J, Rogers D, Coleman SH, Graham RK, Deng Y, Oh R, Bissada N, Hossain SM, Yang YZ et al (2003) Selective striatal neuronal loss in a YAC128 mouse model of Huntington disease. Hum Mol Genet 12:1555–1567
121. Snell RG, MacMillan JC, Cheadle JP, Fenton I, Lazarou LP, Davies P, MacDonald ME, Gusella JF, Harper PS, Shaw DJ (1993) Relationship between trinucleotide repeat expansion and phenotypic variation in Huntington's disease. Nat Genet 4:393–397
122. Solomon AC, Stout JC, Johnson SA, Langbehn DR, Aylward EH, Brandt J, Ross CA, Beglinger L, Hayden MR, Kieburtz K et al (2007) Verbal episodic memory declines prior to diagnosis in Huntington's disease. Neuropsychologia 45:1767–1776
123. Soulet D, Cicchetti F (2011) The role of immunity in Huntington's disease. Mol Psychiatry 16:889–902
124. Stadelmann C, Kerschensteiner M, Misgeld T, Bruck W, Hohlfeld R, Lassmann H (2002) BDNF and gp145trkB in multiple sclerosis brain lesions: neuroprotective interactions between immune and neuronal cells? Brain 125:75–85
125. Tabrizi SJ, Langbehn DR, Leavitt BR, Roos RA, Durr A, Craufurd D, Kennard C, Hicks SL, Fox NC, Scahill RI et al (2009) Biological and clinical manifestations of Huntington's disease in the longitudinal TRACK-HD study: cross-sectional analysis of baseline data. Lancet Neurol 8:791–801
126. Tong X, Ao Y, Faas GC, Nwaobi SE, Xu J, Haustein MD, Anderson MA, Mody I, Olsen ML, Sofroniew MV et al (2014) Astrocyte Kir4.1 ion channel deficits contribute to neuronal dysfunction in Huntington's disease model mice. Nat Neurosci 17:694–703
127. Trushina E, Singh RD, Dyer RB, Cao S, Shah VH, Parton RG, Pagano RE, McMurray CT (2006) Mutant huntingtin inhibits clathrin-independent endocytosis and causes accumulation of cholesterol in vitro and in vivo. Hum Mol Genet 15:3578–3591
128. Turley SD, Burns DK, Rosenfeld CR, Dietschy JM (1996) Brain does not utilize low density lipoprotein-cholesterol during fetal and neonatal development in the sheep. J Lipid Res 37:1953–1961
129. Valenza M, Leoni V, Karasinska JM, Petricca L, Fan J, Carroll J, Pouladi MA, Fossale E, Nguyen HP, Riess O et al (2010) Cholesterol defect is marked across multiple rodent models of Huntington's disease and is manifest in astrocytes. J Neurosci 30:10844–10850
130. Valenza M, Leoni V, Tarditi A, Mariotti C, Bjorkhem I, Di Donato S, Cattaneo E (2007) Progressive dysfunction of the cholesterol biosynthesis pathway in the R6/2 mouse model of Huntington's disease. Neurobiol Dis 28:133–142
131. Valenza M, Marullo M, Di Paolo E, Cesana E, Zuccato C, Biella G, Cattaneo E (2015) Disruption of astrocyte-neuron cholesterol cross talk affects neuronal function in Huntington's disease. Cell Death Differ 22:690–702

132. Valenza M, Rigamonti D, Goffredo D, Zuccato C, Fenu S, Jamot L, Strand A, Tarditi A, Woodman B, Racchi M et al (2005) Dysfunction of the cholesterol biosynthetic pathway in Huntington's disease. J Neurosci 25:9932–9939
133. van den Bogaard SJ, Dumas EM, Acharya TP, Johnson H, Langbehn DR, Scahill RI, Tabrizi SJ, van Buchem MA, van der Grond J, Roos RA et al (2011) Early atrophy of pallidum and accumbens nucleus in Huntington's disease. J Neurol 258:412–420
134. Vance JE (2006) Lipid imbalance in the neurological disorder, Niemann-Pick C disease. FEBS Lett 580:5518–5524
135. Vonsattel JP, DiFiglia M (1998) Huntington disease. J Neuropathol Exp Neurol 57:369–384
136. Vonsattel JP, Keller C, Pilar Amaya MD (2008) Neuropathology of Huntington's Disease. Handb Clin Neurol 89:599–618
137. Vonsattel JP, Myers RH, Stevens TJ, Ferrante RJ, Bird ED, Richardson EP Jr (1985) Neuropathological classification of Huntington's disease. J Neuropathol Exp Neurol 44:559–577
138. Vorisek I, Syka M, Vargova L (2017) Brain diffusivity and structural changes in the R6/2 mouse model of Huntington disease. J Neurosci Res 95:1474–1484
139. Wang L, Lin F, Wang J, Wu J, Han R, Zhu L, Difiglia M, Qin Z (2012) Expression of mutant N-terminal huntingtin fragment (htt552-100Q) in astrocytes suppresses the secretion of BDNF. Brain Res 1449:69–82
140. Wang L, Lin F, Wang J, Wu J, Han R, Zhu L, Zhang G, DiFiglia M, Qin Z (2012) Truncated N-terminal huntingtin fragment with expanded-polyglutamine (htt552-100Q) suppresses brain-derived neurotrophic factor transcription in astrocytes. Acta Biochim Biophys Sin (Shanghai) 44:249–258
141. Wang N, Gray M, Lu XH, Cantle JP, Holley SM, Greiner E, Gu X, Shirasaki D, Cepeda C, Li Y et al (2014) Neuronal targets for reducing mutant huntingtin expression to ameliorate disease in a mouse model of Huntington's disease. Nat Med 20:536–541
142. Wexler NS, Lorimer J, Porter J, Gomez F, Moskowitz C, Shackell E, Marder K, Penchaszadeh G, Roberts SA, Gayan J et al (2004) Venezuelan kindreds reveal that genetic and environmental factors modulate Huntington's disease age of onset. Proc Natl Acad Sci USA 101:3498–3503
143. Wilson CJ, Xu ZC, Emson PC, Feler C (1990) Anatomical and physiological properties of the cortical and thalamic innervations of neostriatal tissue grafts. Prog Brain Res 82:417–426
144. Wood TE, Barry J, Yang Z, Cepeda C, Levine MS, Gray M (2019) Mutant huntingtin reduction in astrocytes slows disease progression in the BACHD conditional Huntington's disease mouse model. Hum Mol Genet 28:487–500
145. Yang SH, Cheng PH, Banta H, Piotrowska-Nitsche K, Yang JJ, Cheng EC, Snyder B, Larkin K, Liu J, Orkin J et al (2008) Towards a transgenic model of Huntington's disease in a non-human primate. Nature 453:921–924
146. Yu ZX, Li SH, Evans J, Pillarisetti A, Li H, Li XJ (2003) Mutant huntingtin causes context-dependent neurodegeneration in mice with Huntington's disease. J Neurosci 23:2193–2202
147. Zeron MM, Hansson O, Chen N, Wellington CL, Leavitt BR, Brundin P, Hayden MR, Raymond LA (2002) Increased sensitivity to N-methyl-D-aspartate receptor-mediated excitotoxicity in a mouse model of Huntington's disease. Neuron 33:849–860
148. Zhang Y, Chen K, Sloan SA, Bennett ML, Scholze AR, O'Keeffe S, Phatnani HP, Guarnieri P, Caneda C, Ruderisch N et al (2014) An RNA-sequencing transcriptome and splicing database of glia, neurons, and vascular cells of the cerebral cortex. J Neurosci 34:11929–11947
149. Zoghbi HY, Orr HT (2000) Glutamine repeats and neurodegeneration. Annu Rev Neurosci 23:217–247
150. Zuccato C, Ciammola A, Rigamonti D, Leavitt BR, Goffredo D, Conti L, MacDonald ME, Friedlander RM, Silani V, Hayden MR et al (2001) Loss of huntingtin-mediated BDNF gene transcription in Huntington's disease. Science 293:493–498
151. Zuccato C, Liber D, Ramos C, Tarditi A, Rigamonti D, Tartari M, Valenza M, Cattaneo E (2005) Progressive loss of BDNF in a mouse model of Huntington's disease and rescue by BDNF delivery. Pharmacol Res 52:133–139
152. Zuccato C, Marullo M, Vitali B, Tarditi A, Mariotti C, Valenza M, Lahiri N, Wild EJ, Sassone J, Ciammola A et al (2011) Brain-derived neurotrophic factor in patients with Huntington's disease. PLoS ONE 6:e22966

Chapter 15
Induced Pluripotent Stem Cell-Derived Astroglia: A New Tool for Research Towards the Treatment of Alzheimer's Disease

Rebecca Atkinson-Dell and Lisa Mohamet

Abstract Despite over a century of research into Alzheimer's disease (AD), progress in understanding the complex aetiology has been hindered, in part, by a lack of human, disease relevant, cellular models, reflected in an inability to translate results from animal studies to successful human therapies. Induced pluripotent stem cell (iPSC) technology, in which somatic cells are reprogrammed to pluripotent stem cells, creates an ideal physiologically relevant model as they maintain the genetic identity of the donor. These iPSCs can self-renew indefinitely in vitro and have the capacity to differentiate into any cell type, opening up new discovery and therapeutic opportunities. Despite a plethora of publications indicating the generation and utility of iPSC-derived neurones for disease modelling to date, in comparison only a limited number of studies have described generation of enriched astroglia from patients with early- or late-stage onset of AD. We recently reported that iPSC-astroglia derived from these patients are capable of mimicking a wide variety of deficits in homeostatic molecular cascades, intimately associated with AD, that are routinely observed in vivo. This review examines the opportunities and limitations of this innovative technology in the context of AD modelling and uses for preclinical discovery to improve our success for an efficacious therapeutic outcome.

Keywords Alzheimer's disease · Induced pluripotent stem cell · Astroglia · Neurodegeneration · Therapeutics

15.1 Introduction

Despite over a century of research into Alzheimer's disease (AD), progress in understanding AD complex aetiology has been hindered in part, by a lack of human, disease relevant, cellular models reflected in an inability to translate results from animal studies to successful human therapies. In the 1980s Glenner and Wong [18] identified one of the central components of the disease, namely, beta-amyloid protein

R. Atkinson-Dell · L. Mohamet (✉)
StrataStem Ltd., Suite 112, 4a Rylands Street, Warrington WA1 1EN, UK
e-mail: lisamohamet@stratastem.com

A. Verkhratsky et al. (eds.), *Neuroglia in Neurodegenerative Diseases*,
Advances in Experimental Medicine and Biology 1175,
https://doi.org/10.1007/978-981-13-9913-8_15

plaques that were considered key to triggering neuronal cell damage. Later in the same decade, researchers discovered a second key component, tau protein tangles to cause neuronal cell degeneration. In 1987, Pfizer began clinical trials of the first specific drug, Tacrine, to target the symptoms of AD, which was later approved by the FDA in 1993. Yet, to date, there is still no drug or treatment that will cure AD or any other form of dementia. Despite a number of potential treatments in the pipeline, current treatment paradigms fall into two categories: acetylcholinesterase inhibitors or N-methyl-D-aspartate (NMDA)—receptor antagonists, both of which, only help treat symptoms, have modest clinical efficacy and do not treat the underlying cause(s) of AD. Moreover, no new AD drug treatments have been approved since 2003. Data shows that between 2002 and 2012, the attrition rate for novel AD drugs was 99.6%, attributed to a lack of clinical efficacy [11, 12].

Human pathology is dominated by the late-onset or sporadic form of AD (SAD), which disease variant does not display significant Mendelian genetic bias. At the same time, absolute majority of cell- and animal-based models of AD employ mutant genes isolated from the clinically rare and dominantly inherited familial AD (FAD), characterised by early onset. Whilst the use of transgenic animal models of AD is crucial to preclinical drug development and may recapitulate some key human AD biological components, such as amyloidogenesis, neuronal loss and cognitive deficits, which are the hallmarks of AD in humans, are rather limited in animal models [23]. Subsequently, there is a compelling need for superior complex human models that offer confidence in disease relevance and improved clinical translation.

Induced pluripotent stem cells (iPSCs) offer a promising advantage over existing models for humanising drug development earlier in the pipeline and offering the potential to provide a cellular-based model for research, preclinical drug efficacy and toxicity. This chapter will focus on highlighting the progress and achievements of human iPSC-derived astroglia in AD modelling, their limitations and future challenges and opportunities both in research and translation to the clinic.

15.2 Astroglia and Their Role in Health and Disease

In order to understand the pathophysiology of any disease, one must comprehensively understand the role and contribution of any cellular component(s) in a typical physiological setting. Astroglia (astrocytes) are an abundant cell type in the central nervous system (CNS) and are heterogeneous in structure and molecular profile. A single astrocyte creates a distinct non-overlapping territory that encompasses thousands of synapses. Their extensive branches and fine processes allow direct communication over long distances, as well as indirect communication through secretion of chemokines and cytokines. To this end, astrocytes are the major homeostatic cells of the CNS, executing their diverse functions at molecular, cellular, tissue and organ levels; furthermore, astrocytes contribute to several systemic functions such as regulation of Ca^{+} balance, energy nd food intake [20, 75, 81]. Multiple homeostatic pathways expressed in astroglial cells represented, for example, by membrane transporters for neurotransmitters [89], a complex system of secretion of neuromodulators, neurotransmitter precursors and hormones [72, 77, 76] or aerobic glycolysis producing energy substrates [53] maintain CNS functional activity and provide essential neuroprotection. Astrocytes are also a significant component of the neurovascular unit as their endfeet processes terminate directly onto cerebral vessels, regulating cerebral blood flow according to metabolic demand.

Human astroglia have distinct features over those of rodents, in both their morphological and molecular heterogeneity. For example, the protoplasmic and fibrous astrocytes of the human brain are substantially (10–20 times) larger and protoplasmic astrocytes are markedly more complex than the astrocytes of rodents. As a result, a human protoplasmic astroglial cell covers ~20 times more synapses than the same cell in the rodent brain [49, 50]. In addition, human brains contain several unique types of astrocytes, which are absent in non-primate CNS. An abundant type of human astroglia is represented by interlaminar astrocytes [3, 10]. The second type of astroglia, found only in humans and higher primates, are polarised astrocytes. Further heterogeneity is revealed by the relative protein expression profiles human astroglia. Whilst all of the human astroglial subtypes outlined above are positive for canonical astrocyte marker, glial fibrillary acidic protein (GFAP), their expression of other astrocyte-associated proteins such as the calcium-binding protein, s100B, excitatory amino acid transporters (EAA) 1 and 2 and glutamine synthetase (GS) vary markedly [50, 63]. Indeed, interlaminar astrocytes, are unique in their reactivity to antibodies against the extracellular matrix receptor, CD44 [2, 63]. The significant differences that exist between rodent and human astrocytes and regional specificity represent another obstacle in understanding translational pathophysiology of human astroglia.

15.3 Current Human Cellular Models

Largely due to technical limitations, we still know relatively little about a major cell type of the brain, how it develops and their functional properties in both health and disease. Traditional cell-based approaches utilise assays based on primary cells, which have restricted use due to limited supply and/or transformed and immortalised cell lines, such as human brain microvascular endothelial cells or SH-SY5Y neuroblastoma cell line, which fail to offer a physiologically relevant in vitro model that captures the specific genomic information of the patient, and certainly are unable to capture the biology for complex diseases of ageing with environmental and genetic risk factors, such as AD.

Human astrocyte development comprises two distinct types: first, a foetal, proliferative astrocyte progenitor and an adult non-proliferative mature astroglia [86]. The current state-of-the-art involves isolation of the progenitor cell, which can be subsequently matured in vitro. However, these cells can quickly become quiescent in long-term cultures, and when cultured in the presence of serum show a morphological phenotype markedly different from those in vivo. For example, an eloquent comparative study on human microglia demonstrated significant transcriptional and enhancer remodelling of microglia when transitioned from the brain to in vitro culture [19]. A more recently developed technique known as 'immunopanning' provides improved isolation and purification of mature human astrocytes. The immunopanned cells were used to assess gene expression signatures in human and mouse astrocytes, showing that only a third of the genes most enriched in human astrocytes were expressed in mouse [86]. Of further significance is that the authors also demonstrated functional hominid distinctions to rodents highlighted by differences in astroglial responses to exogenous glutamate. Therefore, current studies on models employing rodent cells should be considered with caution when their application is to human health and disease. Nonetheless, these types of invasive methods for human adult astrocyte sample collection remain challenging [86].

Advances in RNA-sequencing and single-cell biology have not only allowed us to gain insights into the role of astroglia, but also demonstrated marked gene expression profiles dependent on their origin, development and environmental niche [47, 85, 86]. This further highlights the limited biological relevance of culturing isolated cell types, in monolayers. The 3D structure of the CNS comprises a cellular component (neurones, astrocytes, microglia, oligodendrocytes) and an extracellular matrix component with an integral role in facilitating cell–cell interactions, cell viability, cell morphology, cell differentiation and ultimately influencing disease advancement. Consequently, bioengineering of applicable 3D in vitro culture matrices, scaffolds and co-culture systems, to recapitulate complex in vivo modelling are a current focus within the field; however, challenges outlined above still remain.

Together, these limitations underscore a pressing need for new technologies that can replicate human (patho)physiology, at scale, to provide means for improved disease modelling, and a better understanding of human astrocyte development and their role in disease. Thus, efforts to study astrocyte physiology should be directed as much as possible to the most physiologically relevant system: the intact human brain.

15.4 Induced Pluripotent Stem Cell Technology—A Humanised Platform to Study Health and Disease in a Dish

A ground-breaking study by Evans and Kaufman [16] demonstrated generation and in vitro propagation of mouse embryonic stem (ES) cells. It took a further decade, for the field to understand the key factors critical to support ex vivo culture of human ES cells and in 1998 Thomson et al. [68], defined the necessary culture conditions and transcription factors critical for the maintenance of pluripotency. A pluripotent stem cell is defined as a stem cell that has the capacity, given the appropriate cues, to form any of the >200 cell types in the human body. These cells also possess the ability to self-renew in vitro and therefore, potentially permitting an infinite supply of human cells that can be matured to any cell type(s) of interest. Pluripotent stem cells comprise an expanding number of different cell types, but largely speaking comprise two main cell types: ES cells, active during early development, and induced pluripotent stem cells (iPSCs) that are artificially generated in vitro, using nuclear reprogramming. Nuclear reprogramming, by which nuclei of differentiated somatic cells are reprogrammed by injection into an undifferentiated cell type to induce a pluripotent or embryonic-like cell state, was first described by Gurdon et al. [22], in which an enucleated oocyte was successfully injected with the nucleus from a somatic cell to create a cloned Xenopus. This scientific breakthrough led to mammalian cloning of the infamous Dolly, the sheep [80]. These seminal studies led Takashi and Yamanaka to identify the key transcriptional regulators, namely, octamer-binding transcription factor-3/4 (Oct3/4), sex-determining region Y-box 2 (Sox-2), kruppel-like factor-4 (KLF-4) and c-Myc (now referred to as 'Yamanaka factors') to successfully reprogram adult human dermal fibroblasts cells to the pluripotent stem cell state and the first to report an 'induced' pluripotent stem cell (iPSCs) [64]. A number of publications thereafter have reported various cocktails of transcription factors able to reprogram a variety of adult cell types [68, 79], including keratinocytes [46], peripheral blood mononuclear cells [44, 64] to human urine-derived cells [78, 88]. This innovation revolutionised the stem cell field as it circumvented ethical apprehension associated with the use of human ESCs, which ultimately required destruction of human embryos. Importantly, it has opened new therapeutic opportunities, creating an ideal physiologically relevant model; whereby cells can be derived from any individual; are genetically identical to the donor; can self-renew indefinitely and have the capacity to differentiate into any cell type. Therefore, providing a platform that offers significant advantages over existing models, by delivering the only source of clinically relevant, healthy and diseased human-cell types amenable to various regenerative medicine applications and the ability to study 'disease in a dish' (Fig. 15.1).

Over the last decade, researchers have focused on improving the efficiency of reprogramming by introduction of alternative delivery of episomal, non-integrating and more recently, the use of small molecules for chemical reprogramming and reducing the likelihood of tumorigenicity, making them potentially safer for patient-specific cellular therapy [1]. Moreover, transdifferentiation (or direct conversion), which permits reprogramming of somatic cells to mature committed cell types or multipotent progenitors was first demonstrated by direct conversion of astrocytes into neurones by exogenous expression of four neural transcription factors [6, 26]. The advantages of transdifferentiation over reprogramming are two folds: first, it permits reduced timelines for generation as negates the necessity for generation of a pluripotent cell stage, and second maintains the 'age' of the cells from the donor source. Cells undergoing reprogramming to an ES–cell-like state could effectively wipe the age of the cells and this is particularly relevant for diseases of ageing. The disadvantage of transdifferentiation however, is that it results in progeny that are non-proliferative or possess limited multipotency. For the purposes of this review, we will focus on the use of iPSC-derived cell types and therefore refer the reader to an excellent review of alternative reprogramming strategies [21]. iPSC-derived astroglia for modelling human development.

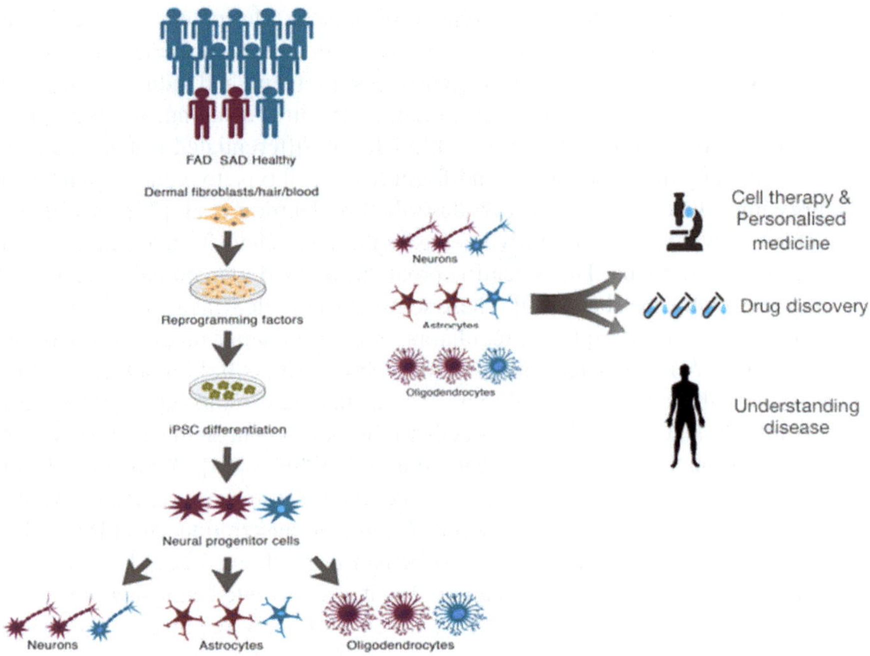

Fig. 15.1 Principles of stem cell technology. Human iPSCs, non-invasively generated from somatic cells (e.g., skin fibroblasts), have the capacity to self-renew indefinitely, can generate all cells of the body and retain the genetic information of the donor. Robust generation of specific mature neural cells (such as astroglia) from iPSCs created from patients with brain pathologies offers an unrivalled platform for the study of human brain disorders in vitro, including the screening of potential novel therapeutics, more accurate disease modelling and in cellular and personalised therapy. FAD: Familial AD; iPSC: Induced pluripotent stem cell; SAD: Sporadic form of AD. Reproduced with permission from [48]

In order for us to learn about pathophysiology of any disease, one must understand the physiology of that cellular system in health. Despite a significant number of publications on iPSC-derived neurones in the last decade, by comparison, only a limited number of studies to date have described homogenous generation of enriched astroglia from healthy patients (Fig. 15.2). More recently, the use of iPSC-derived models has permitted a snapshot of the development of healthy astroglia. A number of groups have developed modified protocols and methods to generate human astroglia from various sources of pluripotent stem cells [14, 34, 37, 43, 57, 61, 66]. The majority of these publications describe modifications of media composition; extracellular matrix, seeding density and timings of morphogen(s) (detailed later), that result in significant improvements in differentiation efficiency, functionality and maturity of iPSC-derived astroglia. Current methods exploit the gliogenic switch observed from in vivo developmental cues, whereby iPSC-astroglia can be derived from either patterned neural progenitor cells or from committed glial lineage [55, 58].

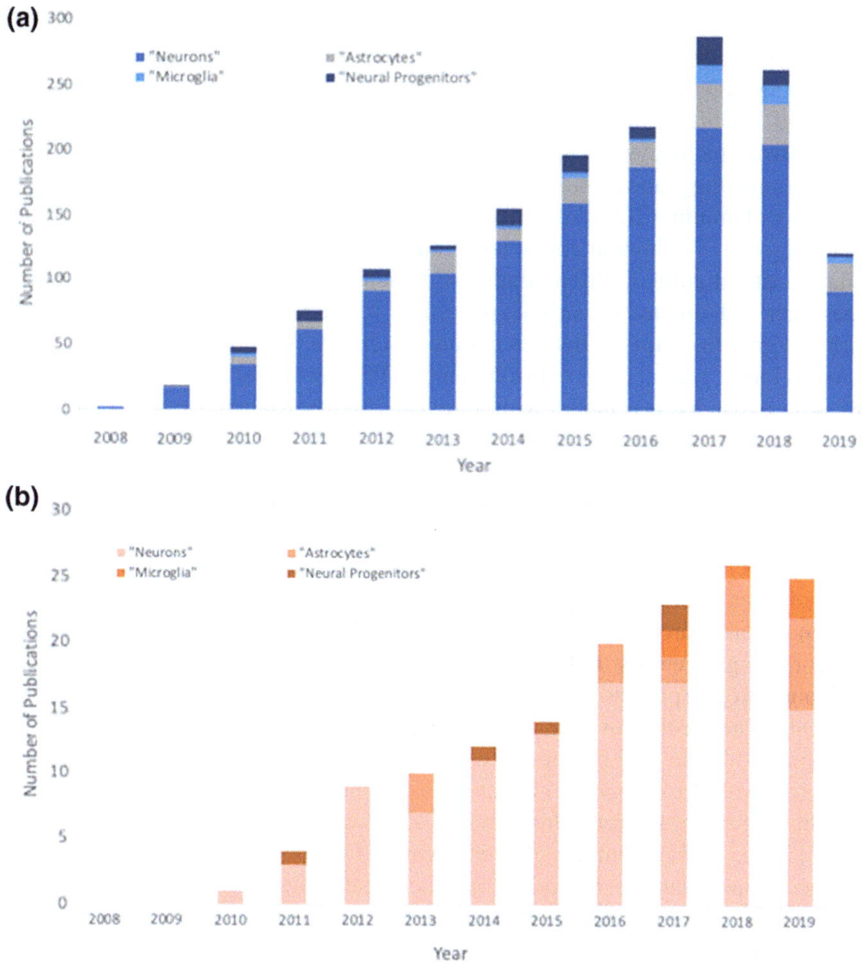

Fig. 15.2 Publication data on original research papers using induced pluripotent stem cells between 2008 and 2019 (May). Analysis of the search terms, title and abstract only publication search terms: ('induced pluripotent stem cells' and "neurons/astrocytes/microglia/neural progenitors") (**a**—blue bars) and ('induced pluripotent stem cells' and 'Alzheimer's Disease' and "neurons/astrocytes/microglia/neural progenitors") (**b**—red bars) for research papers published on NCBI database (PubMed) between 2008 and 2019

Typically, the first step to establish an iPSC-astroglia platform is to generate an intermediate neural progenitor cell (NPC) population that can be expanded and cryopreserved. These cells retain a multipotent, lineage-restricted and differentiative capacity. This recapitulates known cues from in vivo development and patterning that is achieved along the neural tube in primitive neuroepithelial cells, which give rise to different classes of glia. Broadly speaking, there are two predominant methods for the derivation of NPCs from pluripotent stem cells: first, via an embryoid body intermediate (a 3D cluster of stem cells that mimics embryo development) either in the presence or absence of SMAD inhibition, or via a monolayer-based method with dual SMAD inhibition [62]. Specifically, undifferentiated iPSCs are dissociated and plated in culture medium comprising Noggin to inhibit bone morphogenetic protein (BMP) pathways and SB431542, an antagonist of transforming growth factor β (TGFβ) signalling, and these cells are cultured in suspension or plated onto adherent matrix (e.g., laminin) (2D culture) to promote neuroectodermal lineage commitment. The emergence of columnar epithelial neural rosette structures (polarised cells) (10-15 days post initiation of differentiation) is then selected and expanded for a number of passages and directed to astroglia progenitor formation by the addition of specific cytokines/growth factors in defined culture conditions (detailed below).

Astroglia progenitors derived from human iPSCs was first described by Hu et al. [28], who expanded precursors with mitogens thought to increase oligodendrocyte transcription factor 2 positive progenitors and reduce the development of post-mitotic neurones (e.g., cAMP, T3, insulin growth factors (IGFs), platelet-derived growth factor (PDGF)). Krencik and Zhang [37] were one of the first to demonstrate that repression of neurogenesis in neuroepithelial cells that could be accomplished by regular dissociation of neural rosettes, and this triggered differentiation of the cells into glial progenitors. Subsequently, neuroepithelial cells were expanded in suspension (known as neurospheres) and triturated weekly to reduce neuron differentiation and promote gliogenesis. Following around 90 days in culture, emergence of predominant astroglia progenitors was observed by their positive demarcation of S100B and CD44. These cells have limited self-renewal capacity and can be cryopreserved and expanded for several passages. Terminal astroglial differentiation was achieved by the removal of exogenous mitogens to prohibit mitosis and supplementation of medium with ciliary neurotrophic factor (CTNF) to promote gliogeneic gene expression. CTNF and other members of interleukin (IL)-6 type cytokines have been shown to induce Janus kinase (JAK)/signal transducer and activator of transcription (STAT) signal transduction to activate phosphorylation of STAT3. However, this protocol is labour intensive and time-consuming (~180 days). Shaltouki et al. [61] demonstrated generation of functional astrocytes within in 60–90 days in culture using defined, xeno-free conditions from both iPSC- and ESC-derived NPCs. However, this method resulted in 20–30% contamination with unwanted cell types. Recently published data has shown that transient expression of the nuclear factor 1A (NF1A) was sufficient to induce conversion of iPSC-neural stem cells to astroglia within 5 days. Chromatin remodelling and GFAP promoter demethylation permitted generation of region-specific or reactive astrocytes [66].

An important consideration for any cellular ex vivo model is how does it recapitulate in vivo identity and function. Astroglia morphological changes are closely correlated to their activated or resting status but are typically identified by their stellate profile. Lessons learned from in vivo development also help us to identify astrocyte subtypes. To date, numerous studies, including our own, have demonstrated iPSC-derived astroglia exhibit a number of characteristics to those in vivo; for example, are immunoreactive for the canonical markers GFAP and S100B. More recent efforts to find a comprehensive and definite set of standardised markers to delineate mature astrocytes have described additional markers such as aquaporin-4 GS, glutamate transporters GLT-1and GLAST1; Aldehyde Dehydrogenase 1 Family Member A1 (ALDH1A1) [30, 33, 36, 38, 66]. We recently demonstrated highly efficient generation of enriched populations of mature human cortical astroglia (with less than 5% of neuron contamination) from a seemingly healthy donor iPSC within 30 days of induction from cortical neural stem cells (NSCs). Based on the method described in Shaltouki et al. [61], NSCs were exposed to CTNF, to activate downstream JAK/STAT signalling that transcriptionally activates astrocyte-specific loci, GFAP and S100B; BMP2; epidermal growth factor (EGF) and insulin, all of which have been shown to drive astroglial commitment and proliferation. We assessed a panel of known functional astrocyte markers namely, GFAP, GS, S100B and EAAT1 comparable to those reported in the literature for adult astrocytes in vivo [30].

A known limitation of using pluripotent stem cells models is that often the terminally differentiated cell types resemble more foetal-like phenotypes and the above protocols, as well as others, report expression profiles consistent with an immature astrocyte phenotype [14, 31, 60, 61]. More recently, a comprehensive study of different culture conditions for the derivation of human iPSC-derived astrocytes from 30 independent donors generated as three independent cohorts (both healthy and diseased samples) showed derivation of astrocytes expressing classical markers, S100B, GFAP, GLAST, Vimentin, ALDH1L1 and APOE using a commercially available media within 30 days. It must be noted that the authors however, found significant differences in robustness of generation and heterogeneity of expression. Transcriptional profiling of these iPSC-derived astrocytes and primary human foetal astrocytes from two brain regions (cerebral cortex and midbrain) and comparisons to in vivo human astrocyte transcriptomic metadata revealed that iPSC-astrocytes are highly analogous to primary foetal astrocytes. Differential expression of genes associated with regulation of neuronal maturation, such as synapse or ion channel formation was underpresented in iPSC-astrocytes, whereas signals promoting extracellular cell adhesion and interaction were upregulated compared to human in vivo astrocytes [67].

Of further interest, is the examination of astroglia regional subtypes throughout the CNS. During neuroepithelial differentiation the cells can undergo directed specification to regional progenitor cells for the generation of different neuronal subtypes. This is also thought to be true for astroglial subtypes, and therefore addition of exogenous cues will determine patterning of astroglia, for example, sonic hedgehog (SHH) directs the formation of ventral astroglia, whilst retinoic acid (RA) derived progenitors exhibit spinal cord phenotypes and those without additional morphogens display characteristic cortical astroglia. However, it is important to note that a significant limitation in our knowledge of markers to specifically identify each astroglial subtype remains controversial. For example, classical delineation of astrocyte identity is by GFAP positivity, particularly relevant for reactive astrocytes, yet some evidence shows that a subpopulation of non-reactive astrocytes do not express GFAP [38], this is confounded by a lack of clearly defined markers that are able to delineate their identity.

Astrocytes may adopt either a quiescent state with protoplasmic morphology, characterised by low GFAP and high GLT1, or a fibrous, reactive phenotype characterised by high GFAP and low GLT1 and traditional culture systems reflect the latter [84]. Therefore, a robust cell model of mature quiescent astrocytes would be beneficial to progress studies of human neural function. TCW et al. [67] showed that iPSC-astrocyte gene expression was closely clustered to those of a quiescent state rather than reactive astrocytes. However, some caution should be given to the interpretation as comparisons were made to datasets from murine astrocytes due to a lack of availability of human samples.

15.5 Astroglia in AD Pathogenesis

Diseases of the CNS can be generally defined as a homeostatic failure of nerve tissue, and hence they are directly associated with the functional performance of homeostatic astroglia. Since these pathologies are the result of neuronal death, a neuro-centric focus in the search for mechanisms and therapeutic approaches has prevailed for some time. The classical neurocentric view of neurodegeneration has, in recent years, been challenged with a plethora of information to support the role of astroglia in non-cell-autonomous mechanisms in neurodegenerative diseases. Astrocyte pathology contributes to all types of neurological disorders and this contribution is complex and disease specific. In AD, astroglia undergoes complex and regionally specific pathological changes. A decrease in astroglial profiles (indicative of atrophic changes) at the early stages of disease progression have been observed in studies on transgenic animals harbouring AD-related mutant human genes [4, 5, 39, 52, 82]. At the later stages of disease, emergence of senile plaques elicited astroglial reactivity, with hypertrophic cells accumulating around the plaques [27, 52]. In the human brain, astrogliosis is prominent at the early stages of the disease, but decreases in advanced stages of AD, when overall glial paralysis contributes to severe brain atrophy [56, 73]. In summary, various aspects of astrogliopathic changes contribute to the progression of AD, and conceivably, astrocytes define both early cognitive deficits due to deficient synaptic support, whilst astroglial reactivity defines resilience of nervous tissue to the pathology. Please refer to preceding chapters for in-depth reviews of the contribution of astroglia to the pathophysiology of AD.

15.6 iPSC-Derived Astroglia for AD Modelling

Despite a plethora of publications indicating the generation and utility of iPSC-derived neurones for disease modelling to date, in comparison only a limited number of studies have described consistent generation of enriched astrocytes from patients with AD [17, 29, 34, 35, 51] (Fig. 15.2). Generation of functional astroglia from healthy iPSCs has previously been reported to be time-consuming, with further limitations in purity (summarised above), which denotes a significant challenge in delineating autonomous contribution of astroglia in disease pathogenesis. Although studies of cell-autonomous pathobiology of human astrocytes derived from patient stem cells are in the nascent state, several lines of evidence show that these astroglia retain some pathological disease signatures, including AD (Fig. 15.3). Importantly,

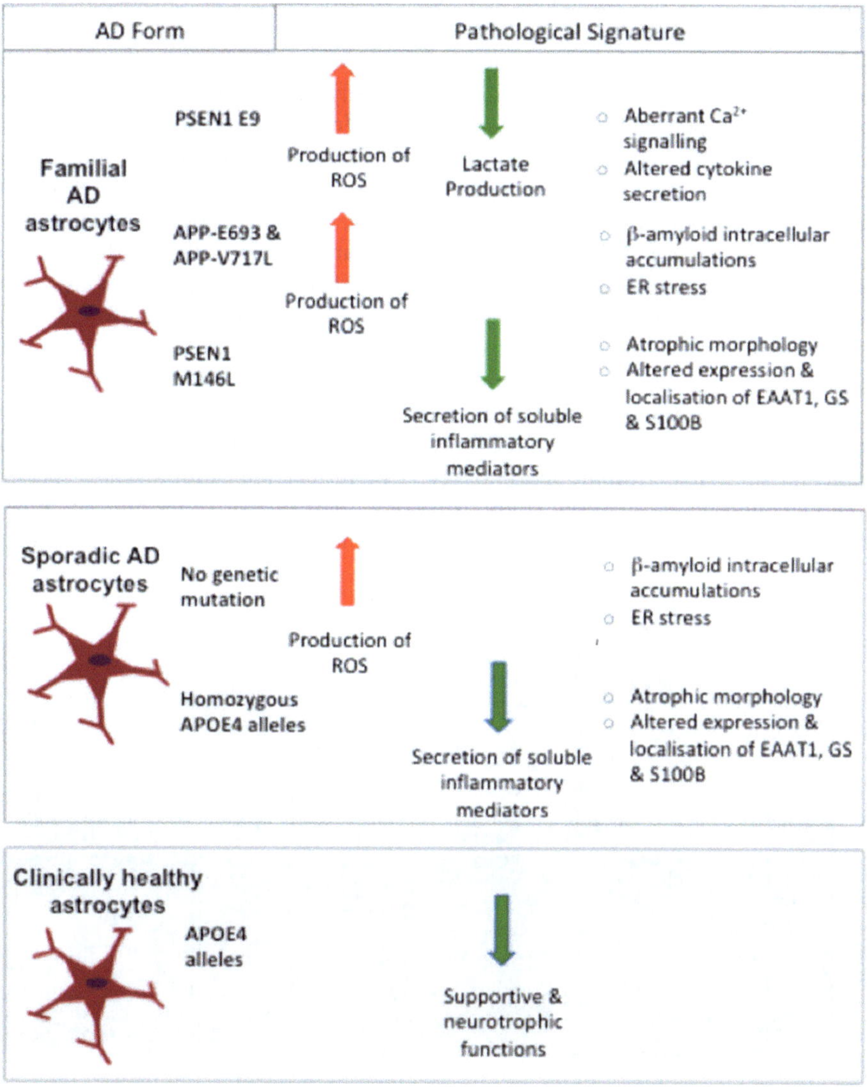

Fig. 15.3 Comparison of AD pathological signatures in iPSC-derived astroglia. AD modelling in iPSC-astroglia derived from patients with either late-stage (sporadic) or early-onset (familial) forms of the disease and a summary of key pathological changes from [30, 34, 51, 87]

advances in genome engineering technologies, such as CRISPR/Cas9, allows us to engineer parental cells to generate control or reference cells that are almost biologically identical (apart from the amino acid changes used to introduce or repair the genetic mutation). This allows researchers to minimise the clonal and donor variations exhibited in iPSC-derived cell types, permitting robust endpoints analyses. Availability of such cells is becoming more accessible as the field expands, with large consortia such as European Stem Cell Bank Initiative and academic groups providing open access.

We recently reported that human astroglia reprogrammed from dermal fibroblasts of a 53-year-old male donor with type III early-onset FAD (bearing an M146L mutation in the presenellin-1 gene, *PSEN1*) and from an 87-year-old female clinically affected with late-onset SAD (homozygous for the four allele of apolipoprotein E, *ApoE4*$^{+/+}$; the single-largest genetic factor determining SAD risk) exhibited pathological phenotypes when compared with iPSC-derived astroglia reprogrammed from a healthy control [30]. This pathological signature comprised (i) mislocalisation and abnormal expression of mature astrocyte markers, (ii) compromised astrocyte heterogeneity and (iii) astroglial atrophy. Astrocytic dysgenesis, manifested as an almost complete loss of processes and overall reduction in cell size was significant (Fig. 15.4); a finding that strongly correlates with observations of morphological astroglial atrophy in early-stage AD pathology in mouse models [52, 73]. In contrast, astrocytes derived from iPSCs isolated from patients with fronto-temporal dementia show hypertrophic morphology [24], further indicating disease-specific glial metamorphoses. A recent study using iPSC-derived astrocytes reprogrammed from a patient exhibiting a PSEN1 E9 mutation, known to cause FAD, supported our

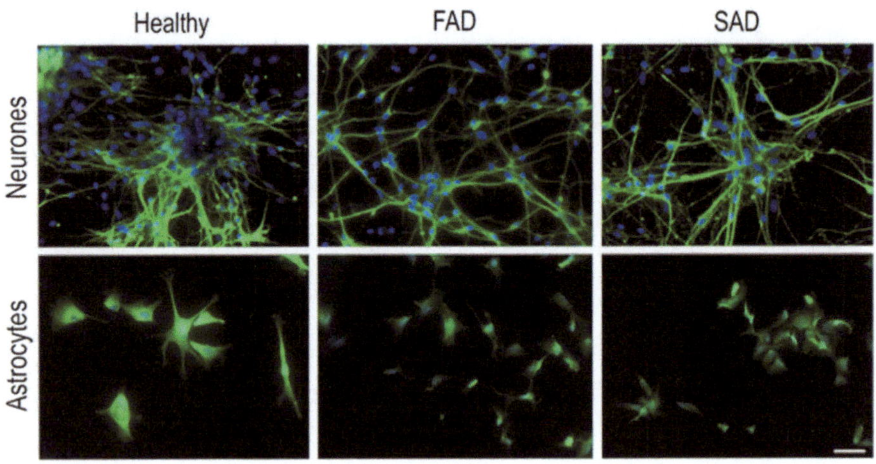

Fig. 15.4 Comparison of healthy control, FAD and SAD patient-derived βIII-tubulin immunoreactive neurones and GFAP immunoreactive astrocytes. Whilst early neuronal appearance is indistinguishable across the groups, AD astrocytes show markedly reduced heterogeneity of morphology and striking atrophy compared to healthy cells. Scale bar = 50 μm. Reproduced with permission from [30]

initial observation of a lack of difference in the differentiation potential of AD and control (healthy) iPSC-derived astrocytes. However, the author reports that stellate morphology was observed across both healthy- and AD-derived astrocytes. This discrepancy may be down to significantly different culture methods, reflecting a more mature phenotype or indeed a donor/mutation-specific phenotype [51]. It must be noted that our study was based on non-isogenic comparisons with a single clone per donor and although, the only publication to date, to provide an in-depth morphological and immunocytochemical characterisation of iPSC-astrocytes in AD, additional quantitative data must be performed to provide more conclusive evidence.

Distribution of astroglial markers was similarly aberrant in cells derived from AD patients in our study. For example, S100B in AD iPSC-derived astroglia was almost entirely confined to large nuclear inclusions, in contrast to healthy iPSC-derived astroglia, which show typical cytoplasmic localisation (Fig. 15.4). This was accompanied by significant decrease in total S100B expression in AD astroglia when compared to control astroglia. Since S100B is known to interact with various cytoskeletal components, this phenomenon may represent a novel and early mechanism underlying SAD- and FAD-induced astrocytic atrophy. Mislocalisation was also observed for glutamine synthetase in both FAD and SAD astroglia, whilst the glutamate transporter EAAT1 was misplaced only in SAD iPSC-derived astroglia. Furthermore, the expression levels of both EAAT1 and glutamine synthetase were decreased only in FAD-derived astroglia [30]. Intriguingly, other iPSC-based studies have shown that astroglia derived from iPSCs from SAD and FAD backgrounds exhibit no significant deficit in their overall abilities to sequester extracellular glutamate compared to controls [34, 87]. Whether these findings are a result of the application of glutamate assay concentrations being orders of magnitude higher than predicted physiological [59], and hence masking nuanced alterations in uptake, or whether they imply that it is the processing rather than gross uptake of glutamate that is altered in AD, remains to be established.

Three human isoforms of apolipoprotein E (ApoE) exist; with homozygous expression of the E4 alleles conferring significantly increased risk for late-onset SAD in contrast to the E2 isoform, which has been reported to confer a protective mechanism against AD. Whilst it is known that ApoE contributes to neuronal health by transporting the cholesterol required for cellular repair, synaptic plasticity and dendritic spine integrity [8], the precise mechanisms by which the ApoE isoforms contribute to AD pathogenesis or protection are not fully understood. Astroglia generated from iPSCs derived from APOE4 donors secrete active ApoE, but exhibited profound hypolipidation in astrocytes $ApoE4^{+/+}$ SAD-prone background compared to $ApoE3^{+/+}$ controls [87], consistent with findings in mice and humans [25, 83]. Thus, human iPSC-derived astrocyte models provide an excellent platform to study relative contribution of its isoforms to AD progression.

One of the most impactful and early pathological characteristics of AD is progressive loss of synapse/synaptic function albeit through a yet unknown mechanism. Of further interest is that iPSC-derived *ApoE4* astrocytes show an impaired ability to support neuronal survival and synaptogenesis, compared to controls; indicating the loss of neurotrophic and neuroprotective functions [87]. This functional deficit appears independent of neuronal contact or glutamate scavenging, hence implicating alterations in the astrocytic secretome. An increasing area of interest in AD progression is neuroinflammation, in which the cytokine mileu is thought to drive amyloid deposition and hyperactivation/recruitment of microglia. It can be characterised by the accumulation of pro-inflammatory cytokines such as IL-6, IL-1β and tumour necrosis factor (TNFα) amongst others. Pro-inflammatory astrocyte reactivity contributes to glutamate imbalance and release of mediators that cause synapse death [15]. A number of studies have observed altered cytokine secretion from iPSC-astrocytes derived from AD patients, compared to be healthy. Interestingly, Oksanen et al. [51] showed partial rescue of this phenotype using a γ secretase inhibitor, indicating a role for amyloidosis in neuroinflammation. In our study, we revealed significant alteration of the constitutive secretion of the pro-inflammatory mediators IL-8 and membrane cofactor protein-1(MCP-1) in FAD- and SAD iPSC-derived astroglia when compared to seemingly healthy controls [30]. However, it must be noted that this was constitutive secretion and not under disease-relevant stimulatory conditions. These observations may further indicate a potential glial paralysis that has been postulated to be a fundamental factor in the evolution of AD [73]. It is thought that Ca^{2+} homeostasis could play a role in neuroinflammation, and astroglia derived from human iPSCs are capable of generating spontaneous Ca^{2+} signals [58], although proper physiological characterisation of these cells is yet to be produced. A recent study goes some way to address some of the questions around functional iPSC-derived astroglia Ca^{2+} activity [67].

At the earliest stages of AD pathology, activated reactive astrocytes are predominant in the molecular layer of the cerebral cortex and close to amyloid plaques. A recent study has also shown that in conjunction with microglia, reactive astrocytes form a 'net' over the amyloid-beta plaques and astrocytic processes invade the plaque, directly interacting with amyloid-β (Aβ) protein [7]. Accumulation of Aβ oligomers in intracellular organelles has been described in some, but not all, iPSC-derived SAD- and FAD-astrocytes, concomitant with the induction of both ER and oxidative stress [34]. This was recently corroborated in PSEN1 E9 mutant iPSC-astrocytes, but a discrepancy in Aβ1-42 secretion was seen between these studies [87]. Specifically, astrocytes derived from patients with the *APP*-E693Δ mutation demonstrated large inclusions of Aβ, which were shown to co-localise to the endoplasmic reticulum, early endosomes and lysosomes, whilst those from *APP*-V717L patients showed no accumulation at all. This finding sheds new insight into disease heterogeneity, both within and between SAD and FAD, and is supportive of our own findings of subtle variations at the cellular level between the two forms of the disease [43].

A new paradigm in the field is the link of neurodegenerative disease as a metabolic disorder. This hypothesis addresses some of the most important features of AD that include mitochondrial dysregulation, oxidative stress and diminished cerebral glucose metabolism [70, 65]. Glucose hypometabolism in the brain appears early in the genesis of AD [71], and in fact presents a common phenomenon with other neurodegenerative diseases [4]. Mitochondrial dysfunction, which is strongly associated with age-related neurodegeneration, is particularly prevalent in AD [5, 39]. Metabolic changes in the cerebrospinal fluid (CSF) of AD patients includes elevated methionine (MET), 5-hydroxyindoleacetic acid (5-HIAA), vanillylmandelic acid, xanthosine and glutathione have been linked to accumulation and abnormal tau metabolism [82]. In addition, decreased blood flow in specific brain areas and reduction of glucose transporters at the blood-brain barrier were also shown to contribute to the hypometabolic state in AD [27]. As a result, a significant correlation between diminished cerebral glucose metabolism and cognitive performance has been shown in AD patients [56, 74]. In line with these findings, AD is closely linked to dysfunction in the regulation of energy metabolism, production of radical oxygen species (ROS) and mitochondrial defects [74]. Recently, PSEN1 mutant iPSC-astroglia exhibited oxidative phosphorylation versus glycolytic metabolism seen in healthy iPSC-astroglia. However, γ-secretase inhibitor treatment was unable to reverse mitochondrial metabolism, indicating this is likely to be independent of Aβ induced pathophysiology [51].

In summary, astroglia derived from iPSCs are capable of mimicking in vitro, a wide variety of deficits in homeostatic molecular cascades intimately associated with AD, that is routinely observed in vivo and in patients. Collectively, these results show that iPSC-derived astrocytes secrete elevated levels of Aβ1-42, show altered Ca^{2+} homeostasis, reactive oxygen species/mitochondrial metabolism switching thus providing a platform for preclinical screening. To this end, Thorne et al., utilised human ES-derived astrocytes to execute a phenotypic assay for compounds that prevent oxidative stress. This study established a scalable system to support high-throughput screening of over 4000 compounds in stem cell-derived astrocytes [69].

15.7 Therapeutic Strategies Towards Neuroregeneration

The WHO reports the global burden of neurological diseases that affects up to 1 billion people and accounts for 12% of all deaths worldwide and 72% of this total burden is caused by four disorders: cerebrovascular disease, epilepsy, dementia and migraine. The global cost of mental health conditions alone was estimated at US$2.5 trillion in 2010, with a projected increase of >US$6 trillion in 2030 and in Europe alone, the total cost of diseases of the brain is estimated at €800 billion per year. A significant contributing factor to one of the largest medical burdens currently facing public health is a lack of knowledge as to the cause(s) of such diseases and pathways that could be manipulated to reduce their progression. A gap in the translation from existing animal models to success in human trials has been identified as a major factor in lack of effective therapeutic interventions in a number of diseases, including AD. The exponential increase in the expected number of patients presenting with AD in particular, not only represents a major area of unmet medical need, but it also represents a significant market opportunity. For example, to delay the onset of AD by 5 years could save $50 billion in annual healthcare costs in the US alone. There have been no new drug approvals for treatment of AD since 2003 and existing treatment paradigms fall into two categories: acetylcholinesterase inhibitors or NMDA receptor antagonists, both of which, only help treat symptoms, have modest clinical efficacy and do not treat the underlying cause(s) of AD. Data shows that between 2002 and 2012, the attrition rate for novel AD drugs was 99.6%, attributed to a lack of sufficient target engagement or adverse toxicity. The development of new AD drugs to market are confounded by our incomplete understanding of AD, but also challenges in the clinical setting as overt clinical symptoms are not evident until the latter stages of the disease; meaning that many therapeutics may be refractive. The field is moving towards treating patients earlier in AD progression, but expose associated risks with drug dosing failures and side effects, which are yet still poorly defined. As of 2018, there were 112 agents in the current pipeline for AD treatment distributed across early and late stages. The mechanism of action of the majority of these therapeutics is disease modifying [11]. However, recent trials such as Lanabecestat that failed at phase II/III, pharma giants Eli Lilly and Astra Zeneca announcing the end of two phase III trials for similar β-secretase inhibitors and more strikingly, Pfizer ending all drug discovery programmes into neurological diseases clearly shows that a new strategy towards neurological disease discovery and patient treatment is critical.

Previously, astrocytes were not considered as targets for neurodegenerative drug discovery, as such disorders were previously viewed as neuronal in their pathology. However, progress in astroglia research has revealed key functional roles of astrocytes in the CNS and their neuroprotective or neurotoxic attributes in disease states. The advent of novel technologies available (including improved human cellular models such as iPSCs) enables phenotypic high-throughput compound and genetic screening campaigns for drug discovery. Development of therapeutics towards astrocytes, or perhaps more importantly, targeting both astrocytes and neuron degeneration may provide new opportunities to generate efficacious and ultimately disease-modifying medicines for an ever-growing unmet medical need. The shift from a neuron-centric view to one that incorporates, not only astrocytes, but also other key cells of the CNS (e.g., microglia), is a crucial step into rejuvenating neurodegenerative drug discovery to treat disease.

15.8 Limitations

The promise of iPSCs to revolutionise regenerative medicine has become clear in the preceding decade by enabling generation of patient-specific cells for cell- and disease-specific pathogenesis modelling and for cell-based therapeutic advances. However, careful consideration of the limitations such as, chromosomal instability, genetic differences between donors and epigenetic memory from the parental cell means that thorough characterisation is necessary prior to use as a cellular model. Furthermore, significant challenges to their use in the clinic exist such as the use of oncogenes (e.g., c-Myc) for reprogramming that could lead to tumours, viral transduction reactivation and incomplete differentiation of iPSCs. The limitations of current models are further highlighted when specifically applied to neuroglia. Very little is known about pathological neuroglial phenotypes in the context of human AD, again largely due to limitations of animal models and lack of access to human samples. This is even more pronounced as remarkable differences between rodent and human astrocytes, hinder in-depth characterization of translational pathophysiology of human astroglia. However, as a note to the reader, it should be emphasised that cultured astrocytes discussed during this chapter represent a simplified model relative to that of astrocytes in the CNS, whereby additional interactions with other cell types and matrix components are likely to influence the astrocytic phenotype. Though, these highly purified human astrocyte culture models and combined precision genome editing represent a unique system to delineate the autonomic responses of astrocytes to be defined stimuli/matrix/co-cultures in both healthy and AD-affected cells in an unparalleled manner.

15.9 Future Perspectives

'Humanisation' of neuropathological research is the main challenge which theoretical medicine is facing. A large majority of neurological diseases do not have an effective cure and with only symptomatic therapies available at best. There are classes of neurological disorders, which do not develop in animals (rare occurrences of neurodegenerative phenotype in lemurs or in some canines remain exceptions), and hence require development of artificial animal models. These are commonly produced in model organisms such as mice, zebrafish and *Drosophila* due to their amenability for genetic manipulation. However, the brains of these animals are not even remotely close to the brain of humans, their lifespan is significantly shorter and their social interactions are much inferior and fundamentally distinct to humans. These limitations underlie the slow progress of neurological therapies.

Another salient revolution developing over last decade concerns a fundamental shift in the understanding of cellular pathophysiology of the brain. The classical paradigm that regards neurones as the cell-autonomous substrate of neuropathology has shifted towards neuroglial mechanisms, that, by virtue of homeostatic and defensive capabilities, seems to determine the resistivity of nervous tissue to pathological insults and chronic neuropathologies. Evolution of astroglia from lesser mammals to humans is remarkable and human astroglia are unique in their complexity compared to rodents. Poor translation from animal models to clinical outcomes has severely limited the development of effective therapeutics for neurological disorders to date, not least AD which remains incurable. The idiosyncratic astroglia of the human brain might be the key to better understanding of uniquely human neurological diseases; hence the development of such human-based models as described here to more accurately study such diseases is essential. Emerging evidence using iPSC-astroglial models is beginning to uncover subtle variations in individual molecular and cellular phenotypes not only between but also within FAD and SAD classifications. Invariably these models pave the way towards the stratification of patient treatment regimes and personalised medicine. Arguably, the most influential development in stem cell culture is the ability to derive in vitro tissues termed organoids that capture more of the complexity of 3D tissues such as multicellular components and functional characteristics of organs. This is particularly relevant for the study of neurodegeneration as it possible to generate cerebral organoids [9, 45, 13, 32, 40–42].

Generation of brain organoids from human pluripotent stem cells exploits EB formation, which permits differentiation into cells of the three lineages, closely mimicking the developing brain. The outer layer forms ectodermal properties, latterly forming neural progenitor cells. Culture of these EBs in matrigel (an extracellular matrix preparation) promotes self-organisation (the defining feature of organoids versus a spheroid/3D culture) and polarisation of neuroepithelium. Subsequent propagation in bioreactor culture induces luminal structures resulting in a fluid-filled cavity. A principal study demonstrated the use of such cerebral organoids for AD modelling; generating iPSC-derived organoids from patients with FAD, recapitulated some of the key hallmarks of AD pathology (including hyperphosphorylated tau and amyloid aggregation). Of further relevance, following treatment β- and γ-secretase inhibitors, FAD organoids showed significantly reduced Aβ and tau pathology [54]. This demonstrates an exciting opportunity for organoids to increase the translational and clinical predictivity of neurological disease discovery and personalised medicine.

References

1. Hou P, Li Y, Zhang X, Liu C, Guan J, Li H, Zhao T, Ye J, Yang W, Liu K, Ge J, Xu J, Zhang Q, Zhao Y, Deng H (2013) Pluripotent stem cells induced from mouse somatic cells by small-molecule compounds. Sci 341(6146):651–654. https://doi.org/10.1126/science.1239278. 9 Aug
2. Akiyama H, Tooyama I, Kawamata T, Ikeda K, McGeer PL (1993) Morphological diversities of CD44 positive astrocytes in the cerebral cortex of normal subjects and patients with Alzheimer's disease. Brain Res 632(1–2):249–259
3. Andriezen WL (1893) The neuroglia elements in the human brain. Br Med J 2(1700):227–230
4. Beauquis J, Pavía P, Pomilio C, Vinuesa A, Podlutskaya N, Galvan V, Saravia F (2013) Environmental enrichment prevents astroglial pathological changes in the hippocampus of APP transgenic mice, model of Alzheimer's disease. Exp Neurol 239:28–37
5. Beauquis J, Vinuesa A, Pomilio C, Pavía P, Galván V, Saravia F (2014) Neuronal and glial alterations, increased anxiety, and cognitive impairment before hippocampal amyloid deposition in PDAPP mice, model of Alzheimer's disease. Hippocampus 24(3):257–269
6. Berninger B, Costa MR, Koch U, Schroeder T, Sutor B, Grothe B, Götz M (2007) Functional properties of neurons derived from in vitro reprogrammed postnatal astroglia. J Neurosci 27(32):8654–8664
7. Bouvier DS, Jones EV, Quesseveur G, Davoli MA, Ferreira T, Quirion R, Mechawar N, Murai KK (2016) High resolution dissection of reactive glial nets in Alzheimer's disease. Sci Rep 6:24544
8. Bu G (2009) Apolipoprotein E and its receptors in Alzheimer's disease: pathways, pathogenesis and therapy. Nat Rev Neurosci 10(5):333–344
9. Clevers H (2016) Modeling development and disease with organoids. Cell 165(7):1586–1597
10. Colombo JA (2017) The interlaminar glia: from serendipity to hypothesis. Brain Struct Funct 222(3):1109–1129
11. Cummings J, Lee G, Ritter A, Zhong K (2018) Alzheimer's disease drug development pipeline: 2018. Alzheimers Dement (NY) 4:195–214
12. Cummings JL, Morstorf T, Zhong K (2014) Alzheimer's disease drug-development pipeline: few candidates, frequent failures. Alzheimers Res Ther 6(4):37

13. Dyer MA (2016) Stem cells expand insights into human brain evolution. Cell Stem Cell 18(4):425–426
14. Emdad L, D'Souza SL, Kothari HP, Qadeer ZA, Germano IM (2012) Efficient differentiation of human embryonic and induced pluripotent stem cells into functional astrocytes. Stem Cells Dev 21(3):404–410
15. Esler WP, Wolfe MS (2001) A portrait of Alzheimer secretases–new features and familiar faces. Science 293(5534):1449–1454
16. Evans MJ, Kaufman MH (1981) Establishment in culture of pluripotential cells from mouse embryos. Nature 292(5819):154–156
17. Fong LK, Yang MM, Dos Santos Chaves R, Reyna SM, Langness VF, Woodruff G, Roberts EA, Young JE, Goldstein LSB (2018) Full-length amyloid precursor protein regulates lipoprotein metabolism and amyloid-β clearance in human astrocytes. J Biol Chem 293(29):11341–11357
18. Glenner GG, Wong CW (1984) Alzheimer's disease: initial report of the purification and characterization of a novel cerebrovascular amyloid protein. Biochem Biophys Res Commun 120(3):885–890
19. Gosselin D, Skola D, Coufal NG, Holtman IR, Schlachetzki JCM, Sajti E, Jaeger BN, O'Connor C, Fitzpatrick C, Pasillas MP, Pena M, Adair A, Gonda DD, Levy ML, Ransohoff RM, Gage FH, Glass CK (2017) An environment-dependent transcriptional network specifies human microglia identity. Science 356(6344)
20. Gourine AV, Kasymov V, Marina N, Tang F, Figueiredo MF, Lane S, Teschemacher AG, Spyer KM, Deisseroth K, Kasparov S (2010) Astrocytes control breathing through pH-dependent release of ATP. Science 329(5991):571–575
21. Graf T (2011) Historical origins of transdifferentiation and reprogramming. Cell Stem Cell 9(6):504–516
22. Gurdon JB, Elsdale TR, Fischberg M (1958) Sexually mature individuals of Xenopus laevis from the transplantation of single somatic nuclei. Nature 182(4627):64–65
23. Götz J, Ittner LM (2008) Animal models of Alzheimer's disease and frontotemporal dementia. Nat Rev Neurosci 9(7):532–544
24. Hallmann AL, Araúzo-Bravo MJ, Mavrommatis L, Ehrlich M, Röpke A, Brockhaus J, Missler M, Sterneckert J, Schöler HR, Kuhlmann T, Zaehres H, Hargus G (2017) Astrocyte pathology in a human neural stem cell model of frontotemporal dementia caused by mutant TAU protein. Sci Rep 7:42991
25. Hanson AJ, Bayer-Carter JL, Green PS, Montine TJ, Wilkinson CW, Baker LD, Watson GS, Bonner LM, Callaghan M, Leverenz JB, Tsai E, Postupna N, Zhang J, Lampe J, Craft S (2013) Effect of apolipoprotein E genotype and diet on apolipoprotein E lipidation and amyloid peptides: randomized clinical trial. JAMA Neurol 70(8):972–980
26. Heinrich C, Blum R, Gascón S, Masserdotti G, Tripathi P, Sánchez R, Tiedt S, Schroeder T, Götz M, Berninger B (2010) Directing astroglia from the cerebral cortex into subtype specific functional neurons. PLoS Biol 8(5):e1000373
27. Heneka MT, Sastre M, Dumitrescu-Ozimek L, Dewachter I, Walter J, Klockgether T, Van Leuven F (2005) Focal glial activation coincides with increased BACE1 activation and precedes amyloid plaque deposition in APP[V717I] transgenic mice. J Neuroinflammation 2:22
28. Hu BY, Weick JP, Yu J, Ma LX, Zhang XQ, Thomson JA, Zhang SC (2010) Neural differentiation of human induced pluripotent stem cells follows developmental principles but with variable potency. Proc Natl Acad Sci USA 107(9):4335–4340
29. Israel MA, Yuan SH, Bardy C, Reyna SM, Mu Y, Herrera C, Hefferan MP, Van Gorp S, Nazor KL, Boscolo FS, Carson CT, Laurent LC, Marsala M, Gage FH, Remes AM, Koo EH, Goldstein LS (2012) Probing sporadic and familial Alzheimer's disease using induced pluripotent stem cells. Nature 482(7384):216–220
30. Jones VC, Atkinson-Dell R, Verkhratsky A, Mohamet L (2017) Aberrant iPSC-derived human astrocytes in Alzheimer's disease. Cell Death Dis 8(3):e2696
31. Juopperi TA, Kim WR, Chiang CH, Yu H, Margolis RL, Ross CA, Ming GL, Song H (2012) Astrocytes generated from patient induced pluripotent stem cells recapitulate features of Huntington's disease patient cells. Mol Brain 5:17

32. Kadoshima T, Sakaguchi H, Nakano T, Soen M, Ando S, Eiraku M, Sasai Y (2013) Self-organization of axial polarity, inside-out layer pattern, and species-specific progenitor dynamics in human ES cell-derived neocortex. Proc Natl Acad Sci USA 110(50):20284–20289
33. Kleiderman S, Gutbier S, Ugur Tufekci K, Ortega F, Sá JV, Teixeira AP, Brito C, Glaab E, Berninger B, Alves PM, Leist M (2016) Conversion of nonproliferating astrocytes into neurogenic neural stem cells: control by FGF2 and interferon-γ. Stem Cells 34(12):2861–2874
34. Kondo T, Asai M, Tsukita K, Kutoku Y, Ohsawa Y, Sunada Y, Imamura K, Egawa N, Yahata N, Okita K, Takahashi K, Asaka I, Aoi T, Watanabe A, Watanabe K, Kadoya C, Nakano R, Watanabe D, Maruyama K, Hori O, Hibino S, Choshi T, Nakahata T, Hioki H, Kaneko T, Naitoh M, Yoshikawa K, Yamawaki S, Suzuki S, Hata R, Ueno S, Seki T, Kobayashi K, Toda T, Murakami K, Irie K, Klein WL, Mori H, Asada T, Takahashi R, Iwata N, Yamanaka S, Inoue H (2013) Modeling Alzheimer's disease with iPSCs reveals stress phenotypes associated with intracellular Aβ and differential drug responsiveness. Cell Stem Cell 12(4):487–496
35. Konttinen H, Gureviciene I, Oksanen M, Grubman A, Loppi S, Huuskonen MT, Korhonen P, Lampinen R, Keuters M, Belaya I, Tanila H, Kanninen KM, Goldsteins G, Landreth G, Koistinaho J, Malm T (2019) PPARβ/δ-agonist GW0742 ameliorates dysfunction in fatty acid oxidation in PSEN1ΔE9 astrocytes. Glia 67(1):146–159
36. Krencik R, Ullian EM (2013) A cellular star atlas: using astrocytes from human pluripotent stem cells for disease studies. Front Cell Neurosci 7:25
37. Krencik R, Zhang SC (2011) Directed differentiation of functional astroglial subtypes from human pluripotent stem cells. Nat Protoc 6(11):1710–1717
38. Kuegler PB, Baumann BA, Zimmer B, Keller S, Marx A, Kadereit S, Leist M (2012) GFAP-independent inflammatory competence and trophic functions of astrocytes generated from murine embryonic stem cells. Glia 60(2):218–228
39. Kulijewicz-Nawrot M, Verkhratsky A, Chvátal A, Syková E, Rodríguez JJ (2012) Astrocytic cytoskeletal atrophy in the medial prefrontal cortex of a triple transgenic mouse model of Alzheimer's disease. J Anat 221(3):252–262
40. Lancaster MA, Knoblich JA (2014) Generation of cerebral organoids from human pluripotent stem cells. Nat Protoc 9(10):2329–2340
41. Lancaster MA, Knoblich JA (2014) Organogenesis in a dish: modeling development and disease using organoid technologies. Science 345(6194):1247125
42. Lancaster MA, Renner M, Martin CA, Wenzel D, Bicknell LS, Hurles ME, Homfray T, Penninger JM, Jackson AP, Knoblich JA (2013) Cerebral organoids model human brain development and microcephaly. Nature 501(7467):373–379
43. Liao MC, Muratore CR, Gierahn TM, Sullivan SE, Srikanth P, De Jager PL, Love JC, Young-Pearse TL (2016) Single-cell detection of secreted Aβ and sAPPα from human IPSC-derived neurons and astrocytes. J Neurosci 36(5):1730–1746
44. Loh YH, Agarwal S, Park IH, Urbach A, Huo H, Heffner GC, Kim K, Miller JD, Ng K, Daley GQ (2009) Generation of induced pluripotent stem cells from human blood. Blood 113(22):5476–5479
45. Di Lullo E, Kriegstein AR (2017) The use of brain organoids to investigate neural development and disease. Nat Rev Neurosci 18(10):573–584
46. Maherali N, Ahfeldt T, Rigamonti A, Utikal J, Cowan C, Hochedlinger K (2008) A high-efficiency system for the generation and study of human induced pluripotent stem cells. Cell Stem Cell 3(3):340–345
47. Matyash V, Kettenmann H (2010) Heterogeneity in astrocyte morphology and physiology. Brain Res Rev 63(1–2):2–10
48. Mohamet L, Jones VC, Dayanithi G, Verkhratsky A (2018) Pathological human astroglia in Alzheimer's disease: opening new horizons with stem cel technology. Future Neurol 13(2). Review. https://doi.org/10.2217/fnl-2017-0029
49. Oberheim NA, Goldman SA, Nedergaard M (2012) Heterogeneity of astrocytic form and function. Methods Mol Biol 814:23–45
50. Oberheim NA, Takano T, Han X, He W, Lin JH, Wang F, Xu Q, Wyatt JD, Pilcher W, Ojemann JG, Ransom BR, Goldman SA, Nedergaard M (2009) Uniquely hominid features of adult human astrocytes. J Neurosci 29(10):3276–3287

51. Oksanen M, Petersen AJ, Naumenko N, Puttonen K, Lehtonen Š, Gubert Olivé M, Shakirzyanova A, Leskelä S, Sarajärvi T, Viitanen M, Rinne JO, Hiltunen M, Haapasalo A, Giniatullin R, Tavi P, Zhang SC, Kanninen KM, Hämäläinen RH, Koistinaho J (2017) PSEN1 mutant iPSC-derived model reveals severe astrocyte pathology in Alzheimer's disease. Stem Cell Rep 9(6):1885–1897
52. Olabarria M, Noristani HN, Verkhratsky A, Rodríguez JJ (2010) Concomitant astroglial atrophy and astrogliosis in a triple transgenic animal model of Alzheimer's disease. Glia 58(7):831–838
53. Pellerin L, Magistretti PJ (2012) Sweet sixteen for ANLS. J Cereb Blood Flow Metab 32(7):1152–1166
54. Raja WK, Mungenast AE, Lin YT, Ko T, Abdurrob F, Seo J, Tsai LH (2016) Self-organizing 3D human neural tissue derived from induced pluripotent stem cells recapitulate Alzheimer's disease phenotypes. PLoS ONE 11(9):e0161969
55. Rao MS, Noble M, Mayer-Pröschel M (1998) A tripotential glial precursor cell is present in the developing spinal cord. Proc Natl Acad Sci USA 95(7):3996–4001
56. Rodriguez-Vieitez E, Saint-Aubert L, Carter SF, Almkvist O, Farid K, Schöll M, Chiotis K, Thordardottir S, Graff C, Wall A, Långström B, Nordberg A (2016) Diverging longitudinal changes in astrocytosis and amyloid PET in autosomal dominant Alzheimer's disease. Brain 139(Pt 3):922–936
57. Roybon L, Lamas NJ, Garcia AD, Yang EJ, Sattler R, Lewis VJ, Kim YA, Kachel CA, Rothstein JD, Przedborski S, Wichterle H, Henderson CE (2013) Human stem cell-derived spinal cord astrocytes with defined mature or reactive phenotypes. Cell Rep 4(5):1035–1048
58. Santos R, Vadodaria KC, Jaeger BN, Mei A, Lefcochilos-Fogelquist S, Mendes APD, Erikson G, Shokhirev M, Randolph-Moore L, Fredlender C, Dave S, Oefner R, Fitzpatrick C, Pena M, Barron JJ, Ku M, Denli AM, Kerman BE, Charnay P, Kelsoe JR, Marchetto MC, Gage FH (2017) Differentiation of inflammation-responsive astrocytes from glial progenitors generated from human induced pluripotent stem cells. Stem Cell Rep 8(6):1757–1769
59. Scimemi A, Beato M (2009) Determining the neurotransmitter concentration profile at active synapses. Mol Neurobiol 40(3):289–306
60. Serio A, Bilican B, Barmada SJ, Ando DM, Zhao C, Siller R, Burr K, Haghi G, Story D, Nishimura AL, Carrasco MA, Phatnani HP, Shum C, Wilmut I, Maniatis T, Shaw CE, Finkbeiner S, Chandran S (2013) Astrocyte pathology and the absence of non-cell autonomy in an induced pluripotent stem cell model of TDP-43 proteinopathy. Proc Natl Acad Sci USA 110(12):4697–4702
61. Shaltouki A, Peng J, Liu Q, Rao MS, Zeng X (2013) Efficient generation of astrocytes from human pluripotent stem cells in defined conditions. Stem Cells 31(5):941–952
62. Shi Y, Kirwan P, Livesey FJ (2012) Directed differentiation of human pluripotent stem cells to cerebral cortex neurons and neural networks. Nat Protoc 7(10):1836–1846
63. Sosunov AA, Wu X, Tsankova NM, Guilfoyle E, McKhann GM, Goldman JE (2014) Phenotypic heterogeneity and plasticity of isocortical and hippocampal astrocytes in the human brain. J Neurosci 34(6):2285–2298
64. Takahashi K, Tanabe K, Ohnuki M, Narita M, Ichisaka T, Tomoda K, Yamanaka S (2007) Induction of pluripotent stem cells from adult human fibroblasts by defined factors. Cell 131(5):861–872
65. Tay TL, Savage JC, Hui CW, Bisht K, Tremblay M (2017) Microglia across the lifespan: from origin to function in brain development, plasticity and cognition. J Physiol 595(6):1929–1945
66. Tchieu J, Calder EL, Guttikonda SR, Gutzwiller EM, Aromolaran KA, Steinbeck JA, Goldstein PA, Studer L (2019) NFIA is a gliogenic switch enabling rapid derivation of functional human astrocytes from pluripotent stem cells. Nat Biotechnol 37(3):267–275
67. Tcw J, Wang M, Pimenova AA, Bowles KR, Hartley BJ, Lacin E, Machlovi SI, Abdelaal R, Karch CM, Phatnani H, Slesinger PA, Zhang B, Goate AM, Brennand KJ (2017) An efficient platform for astrocyte differentiation from human induced pluripotent stem cells. Stem Cell Rep 9(2):600–614
68. Thomson JA, Itskovitz-Eldor J, Shapiro SS, Waknitz MA, Swiergiel JJ, Marshall VS, Jones JM (1998) Embryonic stem cell lines derived from human blastocysts. Science 282(5391):1145–1147

69. Thorne N, Malik N, Shah S, Zhao J, Class B, Aguisanda F, Southall N, Xia M, McKew JC, Rao M, Zheng W (2016) High-throughput phenotypic screening of human astrocytes to identify compounds that protect against oxidative stress. Stem Cells Transl Med 5(5):613–627
70. Tian L, Hui CW, Bisht K, Tan Y, Sharma K, Chen S, Zhang X, Tremblay ME (2017) Microglia under psychosocial stressors along the aging trajectory: Consequences on neuronal circuits, behavior, and brain diseases. Prog Neuropsychopharmacol Biol Psychiatry 79(Pt A):27–39
71. Verkhratsky A, Marutle A, Rodríguez-Arellano JJ, Nordberg A (2015) Glial asthenia and functional paralysis: a new perspective on neurodegeneration and Alzheimer's disease. Neuroscientist 21(5):552–568
72. Verkhratsky A, Matteoli M, Parpura V, Mothet JP, Zorec R (2016) Astrocytes as secretory cells of the central nervous system: idiosyncrasies of vesicular secretion. EMBO J 35(3):239–257
73. Verkhratsky A, Nedergaard M (2014) Astroglial cradle in the life of the synapse. Philos Trans R Soc Lond B Biol Sci 369(1654):20130595
74. Verkhratsky A, Rodríguez JJ, Steardo L (2014) Astrogliopathology: a central element of neuropsychiatric diseases? Neuroscientist 20(6):576–588
75. Verkhratsky A, Zorec R, Parpura V (2017) Stratification of astrocytes in healthy and diseased brain. Brain Pathol 27(5):629–644
76. Verkhratsky A, Zorec R, Rodríguez JJ, Parpura V (2016) Astroglia dynamics in ageing and Alzheimer's disease. Curr Opin Pharmacol 26:74–79
77. Verkhratsky A, Nedergaard M (2016) The homeostatic astroglia emerges from evolutionary specialization of neural cells. Philos Trans R Soc Lond B Biol Sci 371(1700)
78. Wang L, Huang W, Su H, Xue Y, Su Z, Liao B, Wang H, Bao X, Qin D, He J, Wu W, So KF, Pan G, Pei D (2013) Generation of integration-free neural progenitor cells from cells in human urine. Nat Methods 10(1):84–89
79. Wernig M, Meissner A, Foreman R, Brambrink T, Ku M, Hochedlinger K, Bernstein BE, Jaenisch R (2007) In vitro reprogramming of fibroblasts into a pluripotent ES-cell-like state. Nature 448(7151):318–324
80. Wilmut I, Schnieke AE, McWhir J, Kind AJ, Campbell KH (1997) Viable offspring derived from fetal and adult mammalian cells. Nature 385(6619):810–813
81. Yang L, Qi Y, Yang Y (2015) Astrocytes control food intake by inhibiting AGRP neuron activity via adenosine A1 receptors. Cell Rep 11(5):798–807
82. Yeh CY, Vadhwana B, Verkhratsky A, Rodríguez JJ (2011) Early astrocytic atrophy in the entorhinal cortex of a triple transgenic animal model of Alzheimer's disease. ASN Neuro 3(5):271–279
83. Youmans KL, Tai LM, Nwabuisi-Heath E, Jungbauer L, Kanekiyo T, Gan M, Kim J, Eimer WA, Estus S, Rebeck GW, Weeber EJ, Bu G, Yu C, Ladu MJ (2012) APOE4-specific changes in Aβ accumulation in a new transgenic mouse model of Alzheimer disease. J Biol Chem 287(50):41774–41786
84. Zamanian JL, Xu L, Foo LC, Nouri N, Zhou L, Giffard RG, Barres BA (2012) Genomic analysis of reactive astrogliosis. J Neurosci 32(18):6391–6410
85. Zhang Y, Barres BA (2010) Astrocyte heterogeneity: an underappreciated topic in neurobiology. Curr Opin Neurobiol 20(5):588–594
86. Zhang Y, Sloan SA, Clarke LE, Caneda C, Plaza CA, Blumenthal PD, Vogel H, Steinberg GK, Edwards MS, Li G, Duncan JA, Cheshier SH, Shuer LM, Chang EF, Grant GA, Gephart MG, Barres BA (2016) Purification and characterization of progenitor and mature human astrocytes reveals transcriptional and functional differences with mouse. Neuron 89(1):37–53
87. Zhao J, Davis MD, Martens YA, Shinohara M, Graff-Radford NR, Younkin SG, Wszolek ZK, Kanekiyo T, Bu G (2017) APOE ε4/ε4 diminishes neurotrophic function of human iPSC-derived astrocytes. Hum Mol Genet 26(14):2690–2700

88. Zhou T, Benda C, Duzinger S, Huang Y, Li X, Li Y, Guo X, Cao G, Chen S, Hao L, Chan YC, Ng KM, Ho JC, Wieser M, Wu J, Redl H, Tse HF, Grillari J, Grillari-Voglauer R, Pei D, Esteban MA (2011) Generation of induced pluripotent stem cells from urine. J Am Soc Nephrol 22(7):1221–1228
89. Zhou Y, Danbolt NC (2013) GABA and glutamate transporters in brain. Front Endocrinol (Lausanne) 4:165

MIX
Papier aus verantwortungsvollen Quellen
Paper from responsible sources
FSC® C105338

If you have any concerns about our products,
you can contact us on
ProductSafety@springernature.com

In case Publisher is established outside the EU,
the EU authorized representative is:
Springer Nature Customer Service Center GmbH
Europaplatz 3, 69115 Heidelberg, Germany

Printed by Libri Plureos GmbH
in Hamburg, Germany